Lecture Notes in Statistics   128

Edited by P. Bickel, P. Diggle, S. Fienberg, K. Krickeberg,
I. Olkin, N. Wermuth, S. Zeger

Springer
New York
Berlin
Heidelberg
Barcelona
Budapest
Hong Kong
London
Milan
Paris
Santa Clara
Singapore
Tokyo

L. Accardi
C.C. Heyde (Editors)

Probability Towards 2000

 Springer

L. Accardi
Centro Vito Volterra
Universita degli Studi di Roma Tor Vergata
Via della Ricerca Scientifica
00133 Roma
Italy

C.C. Heyde
Department of Statistics
Columbia University
2990 Broadway
Mail Code 4403
New York, NY 10027
U.S.A.

Library of Congress Cataloging-in-Publication Data
Probability towards 2000 / L. Accardi, C.C. Heyde (editors).
      p.   cm. -- (Lecture notes in statistics ; 128)
    ISBN 0-387-98458-5 (softcover : alk. paper)
    1. Probabilities--Congresses.   I. Accardi, L. (Luigi), 1947-   .
  II. Heyde, C. C. III. Series: Lecture notes in statistics (Springer
  -Verlag) ; v. 128.
  QA273.A1P784   1998
  519.2--dc21   97-48856

Printed on acid-free paper.

Camera ready copy provided by the editors.
Printed and bound by Braun-Brumfield, Ann Arbor, MI.
Printed in the United States of America.

9 8 7 6 5 4 3 2 1

ISBN 0-387-98458-5 Springer-Verlag New York Berlin Heidelberg   SPIN 10663680

PREFACE

This volume is the endproduct of a Symposium titled *Probability Towards 2000* held at Columbia University, New York from October 2-6, 1995. The Symposium was generously sponsored by the Istituto dell'Enciclopedia Italiano in Rome and organized in New York through the cooperation of the Centro Vito Volterra, University of Rome, Tor Vergata, the Italian Academy for Advanced Study in America at Columbia and the Center for Applied Probability at Columbia.

A key objective of the Symposium was to obtain a broad view of probability and where the subject is heading. This matter was addressed both through 34 talks and at a round table discussion on the last afternoon, at which it was decided to produce this volume of perspectives. The intention is to chart a course ahead for probability and versions of selected conference talks plus some additional commissioned material are included.

To elucidate the perspective of the Symposium and this volume, the Manifesto of the Symposium, written by L. Accardi, and an Opening Address to the Symposium by J.L. Teugels, then President of the Bernoulli Society for Mathematical Statistics and Probability are included in the Preface.

GOAL OF THE SYMPOSIUM: MANIFESTO

Luigi ACCARDI

The fact that nowdays there exists no scientific field, from biology to economics, from physics to social sciences, from medicine to complexity theory, from meteorology to decision theory,... in which probability theory does not play a major role, should not let one forget that only the period between the two world wars marks the definitive entrance of probability theory among the fundamental mathematical disciplines such as geometry, analysis, algebra, ... In these years P. Lévy, A.N. Kolmogorov and N. Wiener opened the way to the establishment of strong connections of probability theory with several branches of classical mathematics: combinatorial theory, classical analysis, in particular measure theory, elliptic and parabolic equations, potential theory, harmonic analysis, dynamical systems - through the von Neumann - Birkhoff ergodic theorem - and what is now called infinite dimensional analysis (mainly based on functional integration)....

The birth of information theory, with C. Shannon in the fifties, and the subsequent results of Kolmogorov, both in this theory and in classical mechanics, were the crucial steps in the establishment of the now flourishing *chaos theory*: the study of the stochastic propoerties of deterministic dynamical systems.

The transition from random processes to the systematic study of random fields benefited from fruitful interactions with classical statistical mechanics and quantum field theory which have left their trace in several deep results and techniques (hyperconductivity, logarithmic Sobolev inequalities,..., as well as in some new problems such as percolation, phase transition, Dobrushin's theory, hydrodynamical limit,..., which have now enriched the language, tools and scope of probability.

The traditional interaction with the biological and social sciences, limited in the early days mainly to statistics, is now ramified into a multiplicity of mathematical models, involving not only the scientific disciplines, but also important industrial and economic activities.

Starting from the seventies of the present century, a new impetus has come to probability theory from its interaction with quantum physics, which has in its turn mediated several connections with disciplines such as pure physics, algebraic and differential geometry, functional analysis,... and obliged the students of this discipline to undertake a deep rethinking of the very axioms of the theory, which for several aspects is similar to the one undergone by geometry between 1830 and the early decades of the present century.

The reason why the interaction of probability with quantum physics is different from the abovementioned ones is that the problem here is not only to apply classical techniques or to extend them to situations which, being even more general, still remain within the same qualitative type of intuition, language and techniques. The probabilistic apparatus needed in quantum theory, although strictly related to the classical one, has some qualitatively new features, requires a new language, a new type of probabilistic intuition, and is modeled on a new class of basic examples. Furthermore, the formalism of quantum theory, with its complex wave functions and Hilbert spaces, operators instead of random variables,..., creates a distance between the mathematical model and the physical phenomena which is certainly greater than that of classical physics. For these reasons these new languages and techniques might be perceived as extraneous by many classical probabilists and researchers in mathematical statistics. However, since the developments motivated by quantum theory provide not only powerful new theoretical tools to probability, but also some conceptual challenges which can enter into the common education of all mathematicians in the same way as happened for the basic qualitative ideas of the non-euclidean geometries, it is natural to expect that in the first decades of the XXI–st century shall see, in probability, developments analogous to those seen in geometry in the period between the two world wars.

Among the new connections between probability and other branches of mathematics, one might mention the intersections between probability and geometry, the probabilistic applications of quantum groups, some probabilistically flavoured papers in operator theory,....

The purpose of the present Symposium is to stimulate the reflections of probabilists on the future trends of probability theory through a dialogue involving the most advanced areas of classical probability as well as the more recent developments.

The quantum probability section of the Symposium is dedicated, on the occasion of his sixtieth birthday, to K.R. Parthasarathy who, for his contributions to both classical and quantum probability, symbolizes the continuity between innovation and tradition better than anyone else.

PROBABILITY TOWARDS 2000

Opening Ceremony
Jozef L. TEUGELS

To me, the *Symposium Towards 2000* comes at the right time. Over the past two years I have come across a number of situations where issues were raised similar to those hinted at in the *Manifesto of the Symposium*. Let me give three explicit illustrations.

About two years ago I participated in a meeting in Bologna - again the Italian spirit is apparent - on environmetrics. During the opening ceremony Jean-Pierre Contzen from Eurostat delivered a splendid keynote lecture that ended with the lapidary question: *What are the real issues that should be taken up by probabilists and statisticians in connection with the protection of the environment?* By the end of the meeting some tentative answers have been given: the greenhouse effect, the ozone layer, the possibility of a second Tchernobyl, river and air pollution. The discussion on the relevance of each of these topics continues till today.

Not even a year ago some 20 stochasticians from Western Europe gathered in Amsterdam. Under the driving impulse of a number of Dutch colleagues we are currently investigating the possibility to create *EURANDOM*. This is an acronym for a truly international institute where researchers work and interact together on a daily basis for an extended period at a single location. Research would be directed in fundamental as well as in applied aspects of probability theory, statistics and operations research. At the end of the meeting in Amsterdam the panel was asked to formulate concrete research topics within the realm of Eurandom. A potential list of over a dozen topics was ready in no time. But it is obvious that a careful selection of just the right topics will be vital for the success of the institute.

My third example does not come from the scientific environment of established meetings of scholars. Nevertheless, to me this third example is the most significant. The European regional committee of the Bernoulli Society organizes every second year a meeting for *Young Statisticians*. These gatherings are meant for some 40 students, graduates or fresh doctorates. It is by now a tradition that the opening lecture for these fresh and unspoiled stochasticians is given by a not-so-fresh scholar. This year the organisers of the European Young Statisticeans Meeting asked me to give a lecture on *"What are the important developments in stochastics needed for the next century?"*

In all of these examples the crucial question has been essentially the same: are we ready to endorse the purpose of the present symposium, namely *"to stimulate the reflections of probabilists - I would rather use stochasticians - on the future trends through a dialogue involving the most advanced areas of classical theory as well as the more recent developments."*

It is fortunate that the initiative for the symposium comes from the direction of physics. Physics has had the good fortune to quickly absorb an elaborate mathematical apparatus which has been operational for more than a century. With the development of quantum physics also a genuine stochastic component was added. Nowadays, many problems originating from physics are challenging the probabilists and a equal footing collaboration has emerged from that. Prof. Parthasarathy indeed will remain an inspiring example of what is possible in a collaborative atmosphere.

Apart from physics, probably no other science has seen such a fruitful development. Or shall I say has *not yet* seen? Things are changing rapidly in biology, chemistry, industrial sciences etc. On the one hand it is fortunate that the relevance of a stochastic component in all of these sciences has been recognized; on the other hand the same relevance requires from the stochasticians a honest attempt for intensive collaboration with their colleagues from other sciences.

This symposium *almost surely* has a positive probability to become a milestone. We owe it to our profession to use this welcome occasion for reflection. We should convince ourselves of the need for internal cohesion of our discipline. But we also need to inform our colleagues in other fields if and when we are ready for collaboration. Even more importantly we should tell the next generations what we feel are some of the crucial issues that have to be taken up in the near future.

To enhance its importance to the broader forum of scholars and students we have suggested to the organisers of the symposium to allow some time at the end of the meeting for a round table discussion. The residue of that discussion could then appear in one of the forthcoming issues of *BERNOULLI NEWS* and reach a wide readership all over the world.

We all are looking forward to a challenging meeting.

L.ACCARDI　　　　　C.C.HEYDE

Rome　　　　　　　　New York and Canberra

October 1997

CONTENTS

NON-LINEAR EXTENSIONS OF CLASSICAL AND QUANTUM STOCHASTIC CALCULUS AND ESSENTIALLY INFINITE DIMENSIONAL ANALYSIS

LUIGI ACCARDI,* *Nagoya University*

YUN–GANG LU,** *Università di Bari*

IGOR VOLOVICH,*** *Russian Academy of Sciences*

Dedicated to K.R. Parthasarathy on the occasion of his 60th birthday

1. Introduction

It is likely (at least for its proponent) that quantum probability, or more generally *algebraic probability* shall play for probability a role analogous to that played by algebraic geometry for geometry: many will complain against a loss of *immediate intuition*, but this is compensated for by an increase in power, the latter being measured by the capacity of solving old problems, not only inside probability theory, or at least of bringing non-trivial contributions to their advancement. The present, reasonably satisfactory, balance between developement of new techniques and problems effectively solved by these new tools should be preserved in order to prevent implosion into a self-substaining circle of problems and the main route to achieve this goal is the same as for classical probability, namely to keep a strong contact with advanced mathematical developement on one side and with real statistical data, wherever they come from, on the other.

The spectrum of quantum probability is very broad ranging from a new approach to the axioms of probability and to the interpretational problems of quantum theory (cf. [Ac97a] and references therein), to the classical theorems of probability related to the notion of *independence* (laws of large numbers, central limit theorems, invariance principles, De Finetti theorem, ...), Markov chains, conditioning and statistical dependence [Ac90], stochastic calculus [HuPa84], [Par92], entropy and information [OhyPet93]. In all these fields the quantum probabilistic approach has brought some nontrivial feedback to classical probability, of both technical and conceptual nature.

For example the notion of *free independence* [Voic91], emerged from a circle of ideas born in harmonic analysis and the theory of random walks on graphs gave rise to a new generation of central limit theorems in which the semi-circle law plays the

---

* Postal address: Graduate School of Polymathematics, Nagoya University, Nagoya, 464-01, Japan and Centro Matematico Vito Volterra, Università di Roma, Roma, 00133, Italia.

** Postal address: Dipartimento di Matematica, Università di Bari, Bari, 70125, Italia.

*** Postal address: Steklov Mathematical Institute, Russian Academy of Sciences, Vavilov St. 42, 117966, Moscow, Russia.

1

role which, in the usual (classical or quantum) central limit theorems, is played by the Gaussian (classical, boson or fermion) distribution. In less than three years it turned out that this new discovery, originated as a purely mathematical developement, not only stimulated probabilists to a radical rethinking of apparently established concepts, such as *statistical independence* or *Gaussianity*, but also provided a natural technical and conceptual tool which allowed solution of some long standing open problems in different branches of physics such as quantum electrodynamics, quantum chromodynamics, solid state physics. In their turn these applications gave a feed-back to pure mathematics by showing that the notion of free independence and the related central limit theorems were only the first step of an infinite hierarchy of notions of *statistical independence* and of associated central limit theorems [Lu96], [LuDeG95] from which it emerged, with the notion of *interacting Fock space* a third class of explicitly tractable probabilistic models beyond the familiar classes of Gaussian and Markovian ones. In Section 12 below it is explained in what sense these models can be considered as nonlinear deformations of the usual Gaussian models.

More generally, the applications to physics, emerging from these new developements are now so numerous that a whole monograph [AcLuVo97a] is devoted only to a particular class of them, the so-called *stochastic limits of quantum theory.*

The present paper is not a survey of these developements but rather, in the spirit of the New York Symposium, a speculation about possible non-trivial developements of probability theory. The claim that such a speculation is not totally unweary shall be based on the following three facts: (i) the expectation for the realization of these developements is not lost in the fog of an indeterminate future, but concerns the next two or three decades; (ii) these developements were born from the solutions of several problems posed to mathematics by various branches of applied sciences mainly physics and quantum communication (we underline here that these achievements are not restricted to the rigorous proof of things somehow understood by physicists, but in some cases have lead to the discovery of new phenomena in fields, such as quantum electrodynamics, which were considered completely understood by physicists); (iii) the very first steps towards the realization of these developements have already been taken and they seem to suggest the feasibility of the program that we shall try to outline in what follows.

We shall concentrate on only three topics:

(i) the merging of white noise analysis with quantum probability;

(ii) the role of essentially infinite dimensional analysis in this picture;

(iii) the notion of interacting Fock space as a nonlinear deformation of the usual Gaussianity and the related central limit theorems.

Each of these topics would require a separate overview (cf. [AcLuVo97b], [Ac95], [AcLuVo97c]), but our main goal is just to outline the main ideas and possible developements and to show with these three examples how the solution of specific problems posed by applications (from physics in our case) may lead to new ways of looking at familiar objects as well as to new interesting mathematical structures.

The importance of the nonlinear functionals of white noise was first recognized in the physical literature, especially in connection with quantum field theory. Almost fifty years of attempts to construct a manageable nonlinear quantum field theory have taught us that nonlinear fields, whatever they are, shall be very singular objects. Since, as shown below, a quantum field can be looked at as a non-commuting pair of classical random fields, it follows that the problem of dealing with singular objects involves classical probability directly and is not restricted to quantum theory. The fact that this problem, in the more specific form of dealing with nonlinear functionals of white noise, arose also in different fields, such as electrical engineering and mathematical biology, is further support in this direction.

The usual stochastic calculi (both classical and quantum) do not give a satisfactory answer to this problem: they can be thought to give a rigorous meaning to nonlinear functionals only of the *first power of the white noise* and the reason why this is not the same thing as giving a meaning to nonlinear functionals of, say, the square of white noise is explained in Section 10 below.

The very meaning of an expression such as *the square of white noise* is questionable because the white noise is not a bona fide random variable, but a distribution valued random variable, and there are well known problems with the definition of nonlinear functions (such as products or powers) of distributions. A large literature now exists concerning the attempts to give a meaning to various classes of functionals of this type. The main techniques developed up to now to achieve this goal are either based on *discretization* or on *regularization* with subsequent removal of the regularizing factor (i.e. either removal of the cut-off or continuum limit). In order to perform this step one has to subtract some infinite (or tending to infinity) quantities with a procedure called *renormalization* in the physical literature.

From the probabilistic point of view one could say that the regularization approach corresponds to the *Stratonovich approach* to stochastic integration.

The program of formulating a direct Itô approach to the nonlinear functionals of the white noise has been initiated by Hida [Hi].

On the other hand the so-called *stochastic limit of quantum theory* (cf. Section 4 below) suggested the construction of a *white noise approach to (classical and quantum) stochastic differential equations*. Such an approach is now well established from a mathematical point of view and the next natural step after that is the question: *is it possible to introduce, within the context of white noise calculus, what has been the fundamental tool in (both classical and quantum) stochastic calculus, namely the Itô formula?* This too can be done in a rather elegant way (cf. Section 9 below) which unifies *in a single formula* all the known Itô tables: classical and quantum, for Brownian motion and for point processes. Another *bonus* of this formula is that it naturally suggests a *nonlinear extension of the (classical and quantum) Itô formulae*. This leads to the problem of specifying what we mean by a *nonlinear Itô table*. In Section 11 we give an answer to this problem in the case of the powers of white noise and we produce the first examples of nonlinear Itô tables. Not surprisingly the solution involves a probabilistic analogue of the physicists *renormalization* procedure.

In the present approach the *renormalization problem* arises in a purely mathematical context (in our case as the divergence of a mutual quadratic variation) independent of any physical constant such as mass, charge, ..., which usually accompany the renormalization problem in the physical literature. Experience from this literature suggests that probably the construction of higher order Itô tables shall require non-trivial modifications of the procedure used for the second order one. However we feel that, in order to attack this problem, a preliminary step is to acquire a full mathematical understanding of the second order stochastic calculus: an objective still not achieved.

It is a remarkable fact that, using Kuo formulation of the Levy Laplacian [Kuo96] and the equivalence of the Yang–Mills equations with the Levy–Laplace equation, established in [AcGiVo94a], we obtain a formulation of the Yang–Mills equations in terms of the *square of white noise*. In Section 13 below we describe this connection, for the moment only at an intuitive, heuristic level.

Finally, the notion of interacting Fock space, discussed in Section 14 below describes another kind of nonlinearity, not necessarily related to singular random variables, but emerging already at the level of usual, classical and quantum, Brownian motion.

2. Classical and quantum probability

Classical complex valued random variables can be added, multiplied and multiplied by complex numbers, all these operations being meant pointwise, $(XY(\omega) = X(\omega)Y(\omega)$, etc. ...). This means that they form a (complex) algebra. In this algebra multiplication is (associative and) commutative and the identity is the constant random variable equal to 1. Complex conjugation $(X^*(\omega) = \overline{X(\omega)})$ is defined on this algebra and satisfies $(X^*)^* = X$, $(\lambda X)^* = \bar{\lambda} X^*$ (λ a complex number). Such an operation is called an *involution*. An associative (but not necessarily commutative) algebra with an *involution* and an identity is called a $*$-algebra. The $n \times n$ matrices are an example.

Statistics enters this picture through the expectation value: if $(\Omega, \mathcal{F}, P)$ is the probability space of a stochastic process, then to every functional F of the process one can associate its expetation (whenever it exists)

$$E(F) = \langle F \rangle = \int_\Omega F dP$$

(the notation $\langle F \rangle$ for an expectation is common in the physical and engineering literature and we shall often use it in the following). The expectation functional $E(\cdot)$ is linear, positive and normalized:

$$(2.1) \qquad\qquad F \geq 0 \;\Rightarrow\; E(F) \geq 0 \quad ; \qquad E(1) = 1$$

The two conditions (2.1) make sense for any linear functional $E(\cdot)$ on a $*$-algebra if 1 is meant as the identity of the algebra and $F \geq 0$ is meant in the sense that F is a sum of elements of the form x^*x or a limit thereof. In quantum theory such a linear functional is called a *state* and this terminology has been adopted by quantum probability. An example is given by the normalized trace $(E(A) = \sum_{j=1}^n a_{jj}/n)$ on the $n \times n$ matrices.

Classical real valued random variables are particular cases of complex valued ones and vector valued random variables are reduced to scalar valued ones by taking components. In fact, given any stochastic process with values in any state space, its scalar valued functions determine its finite dimensional distributions and hence, up to stochastic equivalence, the process itself. Thus to any stochastic process we can associate the $*$-algebra of all its complex valued functions (always measurable in what follows). If $\mathcal{A}$ is any $*$-sub-algebra of this algebra with the property that the restriction of the expectation functional $E(\cdot)$ on $\mathcal{A}$ uniquely determines the distribution of the proces (e.g. the algebra of step functions, the linear combinations of the trigonometric exponentials, the polynomials in the random variables — if the moments uniquely determine the distibution —, ...) then the pair $\{\mathcal{A}, E(\cdot)\}$ contains all the statistical information on the process. Such a pair (a $*$-algebra plus a state on it) is the prototype of what is now called an *algebraic probability space* (*quantum* if $\mathcal{A}$ is non-commutative, *classical* if it is commutative).

In all the applications we shall consider the elements of $\mathcal{A}$ are realized concretely as operators acting on a Hilbert space $\mathcal{H}$.

3. White noise and Brownian motion

In this Section we recall some standard notions of white noise analysis, as pioneered by Hida [Hi75] in mathematics and as usually employed in the physical literature.

In the following, when speaking of operator valued distributions, we shall refer them, unless explicitly mentioned otherwise, to the Schwartz test function space $\mathcal{S}(R^d)$.

<u>*DEFINITION*</u> (3.1) Let G be a positive distribution on $\mathbb{R}^d$ (e.g. a positive function). A *scalar Boson Fock field* on $\mathbb{R}^d$ with *covariance* G is defined by:

– a Hilbert space $\mathcal{H}$, called the *Fock space*;

– a pair of operator valued distributions $a(k), a^+(k)$ ($k \in \hat{\mathbb{R}}^d$) called respectively *creation* and *annihilation* densities;

– a unit vector Φ in $\mathcal{H}$, called *the vacuum vector* and satisfying

$$a(k)\Phi = 0, \qquad \Phi \in Dom\,(a^+(k)), \qquad \forall k$$

Moreover, the *n-particle* (or *number*) vectors

$$\{(a^+(k_1)\ldots a^+(k_n))\Phi \;:\; n \in \mathbb{N}\}$$

are total in $\mathcal{H}$ and in the domain of $a(k)$ for any $k_1, \ldots, k_n \in \mathbb{R}^d$ and on the linear span of these vectors they satisfy

$$< \xi, a(k)\eta >=< a^+(k)\xi, \eta >$$

and the so-called *canonical commutation relations (CCR)*

$$[a(k), a^+(k')] = G(k)\delta(k - k')$$

$$[a(k), a(k')] = 0$$

Definition 1 In the notations of Definition (3.1), a Boson Fock white noise on $\mathbb{R}^d$ is a Boson Fock field on $\mathbb{R}^{d+1}$ with commutator of the form:

$$(3.1) \qquad [b(t,k), b^+(t',k')] = \delta(t-t')G(k)\delta(k'-k)$$

$$[b(t,k), b(t',k')] = 0 \ .$$

Definition 2 A classical white noise on $\mathbb{R}^{d+1}$, with covariance G, is a classical self-adjoint stochastic process $w(t,k)$ indexed by $\mathbb{R}^{d+1}$ with a cyclic vector Φ such that

(i) The family $w(t,k)$ is Φ-Gaussian

(ii) The $w(t,k)$ are δ-correlated in time, i.e.

$$< \Phi, w(t,k)w(t',k')\Phi >= \delta(t-t')G(k)\delta(k'-k) \ .$$

Definition (3.2) is justified by the following remark: one can prove that the two stochastic processes

$$q(t,k) := b(t,k) + b^+(t,k), \qquad p(t,k) := \frac{1}{i}\big(b(t,k) - b^+(t,k)\big)$$

are isomorphic to classical white noises on $\mathbb{R}^{d+1}$. Moreover, if b satisfies (3.1), then

$$(3.2) \qquad [q(t,k), p(t',k')] = 2i\delta(t-t')G(k)\delta(k'-k)$$

In other terms a quantum white noise is a pair of classical white noises with commutator given by (3.2). The *standard white noise* on $\mathbb{R}$ is obtained when $d = 1$ and

$G = 1$. In this case $b(t,k)$ does not depend on k and we simply write $b(t)$ or b_t. The corresponding classical white noise shall be denoted w_t:

$$w_t = b_t + b_t^+$$

4. White noises and their integrals

A scalar Boson Fock white noise over $\mathbf{R}$ can be concretely realized as follows (symmetric representation). Denote

$$\mathcal{F} = \bigoplus_{n=0}^{\infty} L^2_{\mathrm{Sym}}(\mathbf{R}^n)$$

where $L^2_{\mathrm{Sym}}(\mathbf{R}^n)$ are the square integrable functions on $\mathbf{R}^n$ symmetric under permutation of their arguments. An element ψ on $\mathcal{F}$ is given by a sequence of functions $\psi = \{\psi^{(n)}\}_{n=0}^{\infty}$ where $\psi^{(0)} \in \mathbf{C}$, $\psi^{(n)} \in L^2_{\mathrm{Sym}}(\mathbf{R^n})$ and

$$\|\psi\|^2 = \sum_{n=0}^{\infty} \|\psi^{(n)}\|^2_{L^2(\mathbf{R^n})} < \infty \ .$$

More explicitly

$$\|\psi\|^2 = |\psi^{(0)}|^2 + \sum_{n=1}^{\infty} \int_{\mathbf{R^n}} |\psi^{(n)}(s_1, \ldots, s_n)|^2 ds_1 \ldots ds_n \ .$$

The elements of $L^2_{\mathrm{sym}}(\mathbf{R}^n) = L^2_{\mathrm{sym}}(\mathbf{R})^{\otimes_s}$ are called n-particle vectors.

Denote by $\mathcal{S} \subset L^2(R)$ the Schwarz space of smooth functions decreasing at infinity faster than any polynomial and define

$$\mathcal{D}_{\mathcal{S}} := \{\psi \in \mathcal{F} | \psi^{(n)} \in \mathcal{S}(R^n)\}$$

$$\mathcal{D}_{\mathcal{S}}^o := \{\psi \in \mathcal{D}_{\mathcal{S}} | \psi^{(n)} = 0 \ \text{ for almost all } n \in \mathbb{N} \ \}$$

$$\mathcal{D}(b) := \left\{ \psi \in \mathcal{D}_{\mathcal{S}} \,\middle|\, \sum_{h=1}^{\infty} n\|\psi^{(n)}\|^2 < \infty \right\}$$

On the subspace $\mathcal{D}(b)$ of $\mathcal{D}_{\mathcal{S}}$ define, for each $s \in \mathbf{R}$, the linear operator $b(s)$, from $\mathcal{D}(b)$ to $\mathcal{F}$ by

$$(4.1) \qquad (b(s)\psi)^{(n)}(s_1, \ldots, s_n) = \sqrt{n+1}\psi^{(n+1)}(s, s_1, \ldots, s_n) \ .$$

The annihilator $b(t)$ is a densely defined operator for any $t \in \mathbb{R}$ and, for any square integrable function g, the integral

$$B(g) = \int_{\mathbf{R}} ds \ \ \overline{g}(s)b(s)$$

is well defined as a Bochner integral on the exponential or number vectors.

On the contrary we shall see that the creation density $b^+(t)$ is not an operator but a sesquilinear form defined as follows. For $\eta, \psi \in \mathcal{D}(b)$ define the sesquilinear forms

$$b_t^+(\eta, \psi) = (b(t)\eta, \psi) \qquad , \qquad b_t(\eta, \psi) = (\eta, b(t)\psi)$$

and denote

$$b_t^+(\eta, \psi) \equiv (\eta, b^+(t)\psi)$$

With these notations one has, on $\mathcal{D}(b)$:

$$(\eta, b^+(t)\psi) = (b(t)\eta, \psi)$$

The sesquilinear form $b_t^+(\eta, \psi)$ is generated by the operator valued distribution

$$(b^+(t)\psi)^{(n)}(s_1 \ldots s_n) = \frac{1}{\sqrt{n}} \sum_{i=1}^{n} \delta(t - s_i)\psi^{(n-1)}(s_1 \ldots \hat{s}_i \ldots s_n) \ .$$

One proves that:

Proposition 1 For any square integrable function g there exists an operator $B^+(g)$, defined on the domain $\mathcal{D}(b)$ by the relation

$$(B^+(g)\psi)^{(n)}(s_1,\ldots,s_n) = \frac{1}{\sqrt{n}} \sum_{i=1}^{n} g(s_i)\psi^{(n-1)}(s_1,\ldots,\hat{s}_i,\ldots,s_n) .$$

Moreover, on this domain $(B^+(g))$ satisfies the relation

$$\langle B^+(g)\psi, \psi'\rangle = \langle \psi, B(g)\psi'\rangle .$$

So $B^+(g)$ is a well defined operator on the domain $\mathcal{D}(b)$ and one denotes it with the symbolic notation

$$B^+(g) := \int_{\mathbf{R}} ds \quad g(s)b^+(s) .$$

By integrating a white noise (classical or quantum) over an interval $[0, t]$ one gets the

corresponding increment process of a Brownian motion:

$$(4.2) \qquad B_t := B_{(0,t]} = \int_0^t b(s)ds$$

$$(4.3) \qquad B_t^+ := B_{(0,t]}^+ := \int_0^t b^+(s)ds .$$

Similarly one shows that the sesquilinear form, weakly defined on $\mathcal{D}(b)$ by the formal identity

$$(4.4) \qquad N_t := \int_0^t b^+(s)b(s)ds$$

defines, in fact, a bona fide operator for each t. The family of operators (N_t) is called the *number process*.

For any classical (real or complex valued) stochastic process (X_t), indexed by (a subset of) $\mathbb{R}$, we denote by dX or dX_t the associated *increment process*, i.e. the random variable valued measures on the real line which assigns the (random) measure $X_T - X_S$ to the interval $[S, T] \subseteq \mathbb{R}$. The measure dX_t is also called a *stochastic differential* and this notation makes a connection with stochastic calculus (cf. Section 6 below).

The connection between the quantum Brownian motion B, B^+ and number process N with the classical Wiener and Poisson processes is given by the identities

$$(4.5) \qquad dW = dB + dB^+$$

$$(4.6) \qquad dP = dN + dB + dB^+ .$$

5. A new approach to stochastic calculus: motivation

In the algebraic framework described in Section 2 the notion of *random variable* is translated as follows: let (X_t) be the real valued solution of the classical stochastic differential equation

$$(5.1) \qquad d\,X_t \;=\; l\,d\,t \;+\; a\,d\,w \;:\; X(0) = X_o$$

driven by a classical one-dimensional Brownian motion (W_t) and with adapted coefficients l, a which guarantee the existence and uniqueness of the solution of (5.1) for all initial data X_o in $L^2(\mathbf{R})$.

If $f : \mathbf{R} \to \mathbf{R}$ is a smooth (say $\mathcal{C}^2$) bounded function, in the spirit of quantum probability, we identify $f(X_o)$ as a multiplication operator on $L^2(\mathbf{R})$. Therefore $f(X_t)$ is realized as a multiplication operator on $L^2(\mathbf{R}) \otimes L^2(\Omega, \mathcal{F}, P)$, where $(\Omega, \mathcal{F}, P)$ is the probability space of the increment process of the Brownian motion.

Defining

$$(5.2) \qquad j_t(f) := f(X_t) \,,$$

then, for each $t \geq 0$, j_t is a $*-$homomorphism

$$j_t : \mathcal{C}^2(\mathbf{R}) \subseteq \mathcal{B}(L^2(\mathbf{R})) \to \mathcal{B}(L^2(\mathbf{R}) \otimes L^2(\Omega, \mathcal{F}, P))$$

where for any Hilbert space H we denote $\mathcal{B}(H)$ the algebra of all bounded operators on H (another example of $*$-algebra).

Since the random variable X_t is uniquely determined, up to stochastic equivalence, by the homomorphism j_t, we can identify the two objects. Since any operator acting on $L^2(\mathbf{R})$ can be identified to the operator $T \otimes 1$, acting on $L^2(\mathbf{R}) \otimes L^2(\Omega, \mathcal{F}, P)$, in particular the multiplication operators $f(X_t)$ can be realized as operators acting on the product space $L^2(\mathbf{R}) \otimes L^2(\Omega, \mathcal{F}, P)$. If there exists a unitary operator U_t, acting on the space $L^2(\mathbf{R}) \otimes L^2(\Omega, \mathcal{F}, P)$ such that

$$(5.3) \qquad j_t(f) = f(X_t) = U_t \; f(X_o) \; U_t^*$$

then we say that the homomorphism j_t is *implementable*. If this is the case then, simply by replacing $f(X_o)$ in (5.3) by an arbitrary operator $a \in \mathcal{B}(L^2(\mathbf{R}))$, we extend the random variable X_t (identified to the homomorphism j_t) to a homomorphism from the whole algebra $\mathcal{B}(L^2(\mathbf{R}))$, into $\mathcal{B}(L^2(\mathbf{R}) \otimes L^2(\Omega, \mathcal{F}, P))$. Such a homomorphism is the prototype of an *algebraic (or quantum) random variable*.

Since $f(X_t)$ satisfies a stochastic differential equation, easily deduced from (5.1) and the Itô formula and since U_t and $f(X_t)$ are related by (5.3), it is natural to conjecture that U_t should also satisfy a stochastic differential equation. This conjecture turns out to be correct and in fact one can show that the equation satisfied by U_t has the form

$$(5.4) \qquad dU_t = \{i\left(D + D^+\right) dW_t + \left(-\frac{\gamma}{2}\, D^+ D + iH\right) dt\}U_t$$

where γ is the covariance of the Brownian motion W_t and D^+, D, H are operators that, with some algebra, one can explicitly determine in terms of the coefficients l and a of the stochastic equation (5.4).

The fundamental result of the Hudson and Parthasarathy [HuPa84] quantum stochastic calculus is that it gives a meaning to equations of the form

$$(5.5) \qquad dU_t = \{i\left(DdB_t^+ + D^+dB_t\right) + \left(-\frac{\gamma}{2}D^+D + iH\right)dt\}U_t$$

as integral equations

$$(5.6) \qquad U_t = 1 + i\int_0^t \left(DdB_s^+ + D^+dB_s\right)U_s + \int_0^t \left(-\frac{\gamma}{2}D^+D + iH\right)ds)U_s$$

where B_t^+, B_t is the Fock Brownian motion with variance $\gamma > 0$ acting on the Boson Fock space $\Gamma(L^2(\mathbb{R})) \otimes \mathcal{K}$, $\mathcal{K}$ is a Hilbert space, $D, H = H^*$ are operators on a Hilbert space $\mathcal{H}_S$ and the first integral in (5.6) is a quantum stochastic integral with respect to the pair B_t^+, B_t. Moreover the theory shows that, whenever the operators D, H satisfy some regularity conditions (e.g. they are bounded), equation (5.5), with the initial condition $U(0) = 1$, admits a unique solution, which is a unitary operator on the space $\mathcal{H}_S \otimes \Gamma(L^2(\mathbb{R}))$.

Since we know (cf. Section 4) that $dB_t^+ + dB_t = dW_t$, we can understand in which sense a Hudson–Parthasarathy equation of the form (5.5) generalizes the usual stochastic differential equation (5.1).

The relevance of this result for quantum physics has been shown in a long series of papers starting from [AcFriLu87] whose main achievement consists in the proof of the fact that equations of the form (5.5) or generalizations thereof can be obtained as appropriate limits of *the usual Hamiltonian equations of quantum physics*. This is called *the stochastic limit of quantum theory.*

Recently it has been realized that the stochastic limit of quantum theory not only gives a physical meaning to the classical and quantum stochastic calculus but also naturally suggests a new approach to this calculus according to the following ideas.

Equation (5.5), with $H = \kappa D^+D$ where κ is a real number whose explicit form will be given later, is the stochastic limit of the Hamiltonian equation

$$(5.7) \qquad \partial_t U_{t/\lambda^2}^{(\lambda)} = -i(Da_t^{(\lambda)+} + D^+a_t^{(\lambda)})U_{t/\lambda^2}^{(\lambda)}$$

where $a_t^{(\lambda)+}, a_t^{(\lambda)}$ are usual Boson Fock creation and annihilation operators on the Fock space $\Gamma(L^2(\mathbb{R}^d))$ (notice that the Fock space $\Gamma(L^2(\mathbb{R})) \otimes \mathcal{K}$, is also obtained from the stochastic limit and not put in expressly from the beginning). Equation (5.7) is widely studied in physics in connection with a multiplicity of different models; in particular there is a rich literature on it in quantum optics, where it enters as the basic model equation in laser theory (a description of how this equation is deduced from the standard quantum electrodynamics Hamiltonian is in [Haak].

The explicit form of the operators $a_t^{(\lambda)+}, a_t^{(\lambda)}$ will not be relevant here. What is important is that, as $\lambda \to 0$, they converge in the sense of mixed momenta (correlators) to the annihilation and creation operators b_t^+, b_t of a Boson Fock white noise, $a_t^{(\lambda)} \to b_t$.

Because of this convergence it is quite natural to conjecture that the solution $U_{t/\lambda^2}^{(\lambda)}$ will converge in an appropriate sense to the solution U_t of the equation

$$(5.8) \qquad \partial_t U_t = -i(Db_t^+ + D^+b_t)U_t \qquad ; \qquad U_0 = 1\,.$$

But we know from the theory of stochastic limits that $U^{(\lambda)}_{t/\lambda^2}$ converges to the solution of the quantum stochastic equation of the form (5.9), i.e.

$$(5.9) \qquad dU_t = \{i\left(DdB_t^+ + D^+dB_t\right) + \left(-\frac{\gamma}{2}\,D^+D + i\kappa D^+D\right)dt\}U_t\ .$$

Therefore it is natural to conjecture that (5.8) and (5.9) are not two different equations, but only different ways of writing the same equation.

The first step to answer this conjecture is to give an independent meaning to the equation (5.8). This is not a trivial point even in the simplest case in which $D = 1$ because in this case equation (5.8) becomes

$$(5.10) \qquad \partial_t U_t = -i(b_t^+ + b_t)U_t$$

whose meaning is not clear since the (classical) white noise $w_t = b_t^+ + b_t$ is not an operator but only an operator valued distribution.

In this particular case (i.e. in which only the first powers of the white noise appear in the formal Hamiltonian equation) it is effectively possible to give a meaning to the integral equation associated to equation (5.8) as a weak equation on the domain of number vectors (i.e. one considers the equation not for U itself but for its matrix elements with respect to these vectors).

This method however cannot be generalized to higher powers of the white noise and, even in the linear case, it does not give much insight on the connection between the white noise and the stochastic equation. In order to get such an insight one has to work directly on the iterated series solution of equation (5.8) and this requires the developement of two new tools:

   i) the causal commutator rule, and

   ii) the theory of distributions on the standard simplex.

Item (i) has to do with the quantum manifestation of the difference between the Itô and the Stratonovich stochastic integration. It has important physical implications (generalization of the notion of Lamb shift) but we shall not discuss it here (cf. [Ac-Nag96]). Item (ii) is the basic technical tool both in the solution of singular ordinary differential equations of the form (5.4) and in the proof of their equivalence with stochastic differential equations. We shall discuss it in the following section.

6. The theory of distributions on the standard simplex

The theory of distributions on the standard simplex is based on a notion of δ-functions on the semi-axis which differs from the standard one (cf. e.g. [Cho–Br]) because of the space of test functions considered here is more irrugular.

Such a δ-function is called *causal* because, as will be clear from the following developments, its emergence is the expression of the *causality condition* which is coded into the decreasing order $(t \geq t_1 \geq \cdots \geq t_n)$ of the time variables in the iterated series.

Our definition of δ_+ is motivated by the following lemma.

Lemma 1 Let $F \in L^1(\mathbf{R}_+)$ be an integrable function. Then for all $t, c \in \mathbb{R}_+$ and any function $\varphi : \mathbb{R}_+ \to \mathbf{C}$, continuous at zero and left-continuous at any $t > 0$ one has

$$\lim_{\lambda \to 0} \int_0^t d\tau \frac{1}{\lambda^2} F\left(\frac{t-\tau}{\lambda^2}\right) \varphi(\tau) \chi_{[0,\tau]}(c) = \lim_{\lambda \to 0} \int_0^t d\tau \frac{1}{\lambda^2} F\left(\frac{t-\tau}{\lambda^2}\right) \varphi(\tau) \chi_{[c,\infty)}(\tau)$$

$$(6.1) \qquad\qquad\qquad\qquad = \chi_{(c,+\infty)}(t)\varphi(t) \int_0^{+\infty} F(\sigma)d\sigma .$$

Definition 3 Denote

(6.2)
$$\mathcal{C} := \{\varphi : \mathbf{R}_+ \to \mathbf{C} \ s.t. \ \varphi \ is \ continuous \ at \ zero \ and \ left\text{-}continuous \ at \ any \ t > 0\}$$

$$(6.3) \qquad\qquad \tilde{\mathcal{C}} := \ Linear \ span \ of \ \{\varphi\chi_{[0,\cdot]}(c) : c \in \mathbf{R}_+, \varphi \in \mathcal{C}\}$$

(Choosing $c = 0$ in (6.3), we see that $\mathcal{C} \subseteq \tilde{\mathcal{C}}$).

For $t \geq 0$ define $\delta_+(\cdot - \cdot)$ as the unique linear extension of the map:

$$(6.4) \qquad \delta_+(t - \cdot) : \ \varphi\chi_{[0,\cdot]}(c) = \varphi\chi_{[c,+\infty)}(\cdot) \in \tilde{\mathcal{C}} \to \chi_{(c,+\infty)}(t)\varphi(t) = \chi_{[0,t)}(c)\varphi(t)$$

In particular, for any $t > 0$

$$(6.5) \qquad\qquad\qquad \delta_+(t - \cdot) : \varphi \in \mathcal{C} \to \varphi(t) .$$

Remark 1 In the following we shall use exchangeably the notations

$$(6.6) \quad \delta_+(t - \cdot)[\varphi\chi_{[0,\cdot]}(c)] =: \langle \delta_+(t - \cdot), \varphi\chi_{[0,\cdot]}(c)\rangle =: \int_0^{+\infty} \delta_+(t - \tau)\varphi(\tau)\chi_{[0,\tau]}(c)d\tau$$

and also the notation

$$(6.7) \qquad\qquad \int_0^{+\infty} \delta_+(t - \tau)\varphi(\tau)\chi_{[c,d]}(\tau)d\tau = \int_c^d \delta_+(t - \tau)\varphi(\tau)d\tau .$$

Remark 2 From (6.6) and the identities

$$(6.8) \qquad\qquad \chi_{[c,d)} = \chi_{[c} - \chi_{[d} \qquad ; \qquad \chi_{(c,d]} = \chi_{(c} - \chi_{(d} ,$$

one deduces that

$$(6.9) \qquad\qquad\qquad \langle \delta_+(t - \cdot), \varphi\chi_{[c,d)}\rangle = \varphi(t)\chi_{(c,d]}(t) .$$

So the practical rule to work with $\delta_+(t - \cdot)$ can be formulated as follows: use the same rule as for the usual δ-function with the exception that, whenever there is an expression of the form $\chi_{[c,d)}(\tau)$ (τ is the variable of integration) it has to be replaced not by $\chi_{[c,d)}(t)$ but by $\chi_{(c,d]}(t)$.

The main result of the theory of distributions on the standard simplex given in the following lemma.

Lemma 2 Let $m \geq 2$ be a natural integer and let $\{p_1, q_1, \ldots, p_m, q_m\}$ be a permutation of the set $\{1, 2, \ldots, 2m\}$. Then, for any $\varphi_1, \ldots, \varphi_m \in \mathcal{C}$, $t > 0$ $m \geq 2$, one has

(6.10)

$$\int_0^t dt_1 \int^{t_1} dt_2 \cdots \int_0^{t_{2m-1}} dt_{2m} \prod_{k=1}^m \delta_+(t_{q_k} - t_{p_k}) \varphi_k(t_{p_k})$$

$$= \begin{cases} \int_0^t ds_1 \int_0^{s_1} ds_2 \cdots \int_0^{s_{m-1}} ds_m \prod_{k=1}^m \varphi_k(s_k), & \text{if } (q_k, p_k) \text{ is the identity permutation} \\ 0, & \text{otherwise.} \end{cases}$$

In the stochastic limits of quantum theory, the integrals corresponding to the identity permutation are called *time-consecutive* or *type I* and those corresponding to any other permutation are called *non-time-consecutive* or *type II*. The identity shows the drastic simplification of the white noise approach with respect to the standard quantum field theoretical context, where all the permutations give a non-zero contribution. On the other hand, the experience with the stochastic limit shows that, in this limit, the type II terms tend to zero, therefore the white noise approach effectively captures the leading contribution to the theory.

Using Lemma (6.5) above it becomes easy to compute matrix elements of products of creation and annihilation operators of a quantum white noise and to prove directly, i.e. without appealing to stochastic calculus, the convergence of the iterated series solution of the singular equation (5.6) as well as the unitarity of the solution.

7. Stochastic integrals with respect to white noises

In the notation of Section 4 and

(7.1)
$$dB_f(t) := f(t) b_t dt ,$$

the left stochastic integrals are defined by

$$\left(\int dB_f(t) F(t) \psi \right)^{(n)} (s_1, \ldots, s_n,) = \sqrt{n+1} \int dt\, f(t) (F(t)\psi)^{(n+1)} (t, s_1, \ldots, s_n,)$$

$$\left(\int dB_f^+(t) F(t) \psi \right)^{(n)} (s_1, \ldots, s_n,)$$

(7.2)
$$= \frac{1}{\sqrt{n}} \sum_{i=1}^n f(s_i) (F(s_i)\psi)^{(n-1)} (s_1, \ldots, \hat{s}_i, \ldots, s_n,)$$

where $F(t)$ is a stochastic process (not necessarily adapted) and for any vector φ in the Fock space we denote by $\varphi^{(n)}$ its component on the n-particle space. The white

noise approach to stochastic integrals extends the known (classical and quantum) approaches (with simple variants to include the Fermi and free cases) including the non-adapted Itsuda–Skorokhod integrals. Even in the simplest case of the Brownian motion itself, it allows the obtaining of some results technically stronger than those obtained by standard methods. As an example of these results we mention the *uniform convergence of the mutual quadratic variation.*

Lemma 3 *Assume that $\psi^{(n)}$ is bounded, fix a bounded interval (S, T) and consider a partition of (S, T) into intervals of equal width Δt. Then*

(7.3)
$$|\sum_t (\Delta B_t \Delta B_t^+ \psi)^{(n)}(s_1, \ldots, s_n) - (T - S) \cdot \psi^{(n)}(s_1, \ldots, s_n)| \leq \Delta t \cdot \| \psi^{(n)} \|_\infty \ .$$

In particular the limit

(7.4)
$$\lim_{\Delta t \to 0} \sum_t (\Delta B_t \Delta B_t^+ \psi)^{(n)}(s_1, \ldots, s_n) = (T - S) \cdot \psi^{(n)}(s_1, \ldots, s_n)$$

holds uniformly in $s_1, \ldots, s_n$.

Also the symbolic relation between Brownian motion and white noise can be given a rigorous meaning in this context, as shown by the following theorem.

Theorem 1 *Let $\psi \in \mathcal{D}_S^o$ be such that, for each n, $\psi^{(n)}$ is continuous with compact support. Then one has the following:*

(i)

(7.5)
$$\lim_{\Delta t \to 0} \left\| \left(\frac{\Delta B_t}{\Delta t} - b(t) \right) \psi \right\| = 0$$

where the operator $b(t)$ is defined in (4.1).

(ii) The strong limit, as $\Delta t \to 0$, of $\Delta B_t^+/\Delta t - b^+(t)$ does not exist on the number vectors. However the weak limit of this expression on $\mathcal{D}_S^o$ does exist, i.e.
$\forall \psi_1, \psi_2 \in \mathcal{D}_S^o$

(7.6)
$$\lim_{\Delta t \to 0} \left(\psi_1, \frac{\Delta B_t^+}{\Delta t} \psi_2 \right) = (\psi_1, b^+(t)\psi_2)$$

$$\equiv \frac{1}{\sqrt{n}} \sum_{i=1}^n \int_{\mathbf{R}^{n-1}} ds_1 \ldots d\hat{s}_i \ldots ds_n \psi_1^{(n)}(s_1, \ldots, t, s_n) \psi_2^{(n-1)}(s_1, \ldots, \hat{s}_i, \ldots, s_n) \ .$$

(iii)

(7.7)
$$\lim_{\Delta t \to 0} \left\| \left(\frac{\Delta B_t \cdot \Delta B_t^+}{\Delta t} - 1 \right) \psi \right\| = 0 \ .$$

(iv)

$$(7.8) \qquad \lim_{\Delta t \to 0} \left\| \frac{\Delta B_t^+}{\Delta t} \, \Delta B_t \psi \right\| = 0 \ .$$

For the proofs of these results we refer to [AcLuVo-Na].

8. Stochastic integrals with respect to nonlinear white noise

First note that b_t^k for $k = 1, 2, \ldots$ is an operator in the Fock space which acts on continuous functions $\psi^{(n)}(s_1, \ldots, s_n)$ as follows:

$$(b_t^k \psi)^{(n)}(s_1, \ldots, s_n) = c_{k,n} \psi^{(n+k)}(t, t, \ldots, t, s_1, \ldots, s_n) \ .$$

Here $c_{k,n} = ((n+1)\ldots(n+k))^{1/2}$. We define the right b_t^k-stochastic integral on the continuous functions as follows:

$$\left(\int dt F_t b_t^k \psi \right)^{(n)}(s_1, \ldots, s_n) = \int dt (F_t \varphi_t)^{(}n)(s_1, \ldots, s_1)$$

where $\varphi_t^{(}n)(s_1, \ldots, s_1) = c_{k,n} \psi^{(n+k)}(t, t, \ldots, t, s_1, \ldots, s_n)$ and where the integral on the right hand side is a usual Bochner integral.

The *creation operator* b_t^{+k} is not a *bona fide* operator but a sesquilinear form defined by analogy with the first order case (i.e. $k = 1$, see Section 4) that is $b_t^{+k}(\eta, \psi) = (b_t^k \eta, \psi)$. Accordingly we define the left stochastic integral with respect to b_t^{+k} as the sesquilinear form

$$(\eta, \int dt b_t^{+k} F_t \psi) = \left(\int dt F_t^+ b_t^k \eta, \psi \right) \ .$$

The higher singularity of the case $k > 1$ is reflected by the fact that now, even after integration with respect to a test function, the result is not an operator but remains a sesquilinear form.

We have defined here the right stochastic integral over annihilation operators and the left stochastic integral over creations operators. Analogously one can define the left stochastic integrals over annihilation operators and right stochastic integrals over creation operators (but in this case some domain conditions have to be introduced).

9. Classsical and quantum Itô tables

A *classical Itô table* is a rule for multiplying real or complex random variable valued to obtain a measures on the real line a measure of the same type. Replacing classical random variable valued measures on $\mathbb{R}$ by operator valued measures one obtains a *quantum Itô table.*

The simplest examples are the rules giving the square (with respect to the Itô multiplication) of the Lebesgue measure (*Newton table*)

$$(9.1) \qquad dt^2 = 0$$

or for the increment Wiener process (*Itô table*)

$$(9.2) \qquad dW^2 = dt \quad ; \quad dW\,dt = 0$$

or for the increment compensated Poisson process

$$(9.3) \qquad dP^2 = dP + dt$$

However it should be noted that the precise definition of the Itô multiplication is given in terms of *mutual quadratic variation* and is independent of any notion of stochastic calculus (cf. [DeMe] for the classical case and [AcQu88] for the quantum one).

Remark 3 A more intuitive way to interpret the identities (9.1), (9.2), (9.3) is the following: consider the stochastic differentials as denoting the increments of the corresponding processes in a small but finite interval $[t, t + dt]$ and interpret these identities as valid up to terms of order $o(dt)$ where $o(dt) = o([t, t + dt])$ denotes a quantity which, when summed over all the intervals $[t_j, t_j + dt_j]$ of a partition of a given interval $[S, T]$ tends to zero in some topology. This interpretation remains exactly the same in the quantum case, the only difference between the classical and the quantum case being the topologies considered.

The classical Itô table has been generalized by Hudson and Parthasarathy [HuPa] to the quantum, more precisely the *one-dimensional Boson Fock Itô table* given by:

$$(9.4)$$

	dB^+	dN	dB	dt
dB	dt	dB	0	0
dN	dB^+	dN	0	0
dB^+	0	0	0	0
dt	0	0	0	0

where dB, dB^+, dN denote respectively the annihilation, creation and number process on the Fock space over $L^2(\mathbb{R})$ (more frequently one uses $L^2(\mathbb{R}_+)$). The connection between the quantum table (9.4) and the classical rules (9.2), (9.3) is given by the identities (4.2), (4.3) of Section 4 which allow one to recover the classical Itô table from the quantum one just by using the distributivity of addition with respect to the Itô multiplication.

10. White noise approach to the classical and quantum Itô table

In terms of stochastic differentials the identities (4.2), (4.3), (4.4) become respectively

$$(10.1) \qquad dB_t = b(t)dt \quad ; \quad dB_t^+ = b^+(t)dt \quad ; \quad dN_t := b^+(t)b(t)dt \ .$$

It is therefore natural to ask oneself if also the various Itô tables can be expressed in terms of white noise. The following theorem, formulated for the standard quantum white noise of covariance γ, answers this question.

Theorem 2 The Itô tables for the standard classsical and quantum Wiener and Poisson processes become unified by the following rule:

i) write the standard stochastic differentials in white noise terms;

ii) multiply them following the usual operator rules and considering dt as a scalar;

iii) put the result in normal order by applying the commutation rules;

iv) replace the expression $\delta(0)dt$ by 1;

v) replace the product of any normally ordered expression times dt^2 by 0.

Remark 4 The above rules can be symbolically expressed by the following table

$$\delta(0)dt = 1 \tag{10.2}$$

$$(normally\ ordered\ expression)\cdot dt^2 = 0 \tag{10.3}$$

where the identities are to be interpreted as explained in the Remark 9.1.

Proof. Applying the above listed rules to the product $dB_t dB_t^+$ one finds

$$dB_t dB_t^+ = b_t dt \cdot b_t^+ dt = [b_t, b_t^+]dt^2 + b_t^+ b_t dt^2 = \gamma[\delta(0)dt]dt + o(dt)$$
$$= \gamma dt + o(dt)\ , \tag{10.4}$$

which gives $dB_t dB_t^+ = \gamma dt$. Similarly,

$$dB_t^+ dB_t = b_t^+ dt\, b_t dt = b_t^+ b_t dt^2 = o(dt) \tag{10.5}$$

corresponding to $dB_t^+ dB_t = 0$. With the same rules we find, for $dN_t dN_t$,

$$dN_t dN_t = b_t^+ b_t dt \cdot b_t^+ b_t dt = b_t^+[b_t, b_t^+]b_t dt^2 + b_t^{+2} b_t^2 dt^2$$
$$= \gamma[\delta(0)dt]b_t^+ b_t dt + o(dt) = \gamma b_t^+ b_t dt + o(dt) = \gamma dN_t \tag{10.6}$$

and for $dB_t dN_t$,

$$dB_t dN_t = b_t dt \cdot b_t^+ b_t dt = [b_t, b_t^+]b_t dt^2 + b_t^+ b_t^2 dt^2$$
$$= \gamma[\delta(0)dt]b_t dt + o(dt) = \gamma b_t dt + o(dt) \tag{10.7}$$

corresponding to

$$dB_t dN_t = \gamma dB_t \tag{10.8}$$

and

$$dN_t dB_t^+ = b_t^+ b_t dt\, b_t^+ dt = b_t^+[b_t, b_t^+]dt^2 + b_t^{+2} b_t dt^2 = b_t^+ dt + o(dt)$$

corresponding to

$$dN_t dB_t^+ = dB_t^+\ . \tag{10.9}$$

11. Nonlinear Itô tables and renormalization theory

The white noise expressions of the stochastic differentials of the quantum Wiener and number processes are respectively

$$(11.1) \qquad b_t dt \quad ; \quad b_t^+ dt \quad ; \quad b_t^+ b_t dt$$

while for the classical Wiener and Poisson processes these expressions are

$$(11.2) \qquad (b_t + b_t^+) dt \quad ; \quad (b_t + b_t^+ + b_t^+ b_t) dt \ .$$

Notice that these expressions are either *first order* or *normally ordered second order*. This means that the functionals one can construct by means of the usual (classical or quantum) stochastic calculus are nonlinear functionals of the integrals

$$(11.3) \qquad \int b_t dt \quad ; \quad \int b_t^+ dt \quad ; \quad \int b_t^+ b_t dt$$

(cf. Section 6 on stochastic integrals for a precise meaning of this statement). These types of functionals are only the first floor of an infinite hierarchy of functionals depending on higher powers of the white noise, e.g.

$$(11.4) \qquad \int b_t^p dt \quad ; \quad \int b_t^{p+} b_t^q dt$$

which are examples of *nonlinear functionals of the white noise*. Notice the difference between the two integrals

$$(11.5) \qquad \left(\int_0^T (b_t + b_t^+) dt \right)^2 \quad ; \quad \int_0^T (b_t + b_t^+)^2 dt \ .$$

The former is simply the square of the classical Wiener process W_T, the latter depends on the square of classical white noise w_t and cannot be dealt with the usual stochastic calculus.

In the following we shall describe the simplest example of such a formula.

The advantage of the unified Itô table (10.2), (10.3) versus the usual classical and quantum ones is that, at least from a formal point of view nothing prevents the possibility it to the higher powers of of applying white noise such as $b_t^2, b_t^3, (b_t^3)^+ b_t^2, \ldots$. We shall refer to such tables as *nonlinear Itô tables*. Let us begin to investigate these tables in the simplest non-classical case, corresponding to the square of the quantum white noise, i.e. $b_t^2, (b_t^2)^+$. By applying the rules of Theorem (10.1) and using the notation

$$dB_2(t) := b_t^2 dt \quad ; \quad dB_2^+(t) := (b_t^2)^+ dt \ ,$$

we find

$$(11.6) \qquad dB_2(t) dB_t = b_t^2 dt b_t dt = b_t^3 dt^2 = 0 \ ,$$

$$(11.7) \qquad dB_2^+(t) dB_t = b_t^{2+} dt b_t dt = b_t^{2+} b_t dt^2 = 0 \ ,$$

$$dB_t dB_2^+(t) = b_t dt b_t^{+2} dt = [b_t, b_t^{+2}] dt^2 + b_t^{+2} b_t dt^2$$

$$= 2\gamma \delta(0) b_t^+ dt^2 + o(dt)$$

$$= 2\gamma b_t^+ dB_t^+ ,$$

$$(11.8) \qquad dB_t^+ dB_t^+ = 0 \ .$$

So there is a non-trivial Itô table between first and second order white noise. More interesting is to apply our rules within the second order white noise:

$$dB_2(t)dN_t = b_t^2 dt \cdot b_t^+ b_t dt = -[b_t^+, b_t^2]b_t dt^2 + b_t^+ b_t^3 dt^2$$
$$= -[b_t^+, b_t]b_t dt^2 - b_t[b_t^+, b_t]dt^2 + o(dt)$$
(11.9)
$$= \gamma b_t dt + \gamma b_t dt + o(dt) = 2\gamma dB_2(t) ,$$

$$dN_t dB_2^+(t) = b_t^+ b_t dt \cdot b_t^{+2} dt = b_t^+[b_t, b_t^{+2}]dt^2 + b_t^{+3} b_t dt^2 =$$
(11.10)
$$= 2\gamma dB_2^+(t)$$

(11.11)
$$dN_t dB_2(t) = b_t^+ b_t dt \cdot b_t^2 dt = b_t^+ b_t^3 dt^2 = 0 .$$

However the mutual quadratic variation of $B_2(t)$ and $B_2^+(t)$ leads to an infinity, more precisely:

(11.12)
$$dB_2(t)dB_2^+(t) = 2\gamma^2 + 4\gamma dN_t ;$$

in fact:

$$b_t^2 dt b_t^{+2} dt = b_t^2 b_t^{+2} dt^2 = b_t[b_t, b_t^{+2}]dt^2 + b_t b_t^{+2} b_t dt^2 =$$
$$= 2\gamma[b_t, b_t^+]dt + 2\gamma b_t^+ b_t dt + \gamma b_t^+ b_t dt + \gamma b_t^+ b_t dt + o(dt) =$$
$$= 2\gamma^2 + 2\gamma b_t^+ b_t dt + 2\gamma b_t^+ b_t dt + o(dt) =$$
$$= 2\gamma^2 + 4\gamma b_t^+ b_t dt + o(dt) .$$

Formula (11.12) shows that if, instead of the usual mutual quadratic variation, we introduce the *renormalized mutual quadratic variation* defined by

(11.13)
$$: [[B_2, B_2^+]] : (t, t+dt) \; := \; dB_2(t)dB_2^+(t) - 2\gamma^2 = 4\gamma dN_t$$

then the limit

(11.14)
$$\lim_{max|t_{j+1}-t_j|\to 0} \sum_j : [[B_2, B_2^+]] : (t_j, t_{j+1})$$

taken over the partitions $t_1 < \ldots < t_{j+1} < \ldots$ of a given interval $[S, T]$ exists in the topology of weak convergence on exponential vectors (the same topology used by Hudson and Parthasarathy to define their quantum Itô table) and is equal to the classical random variable valued measure on $\mathbb{R}$:

$$4\gamma(N_T - N_S) .$$

It is clear, in view of (11.13), that the limit (11.14) corresponds to the usual mutual quadratic variation to which one has formally subtracted the *infinite constant* $\infty \cdot 2\gamma\alpha$. Since in physics subtraction of infinite constants is called *renormalization*, it is natural, by analogy, to call the expression (11.13) *renormalized mutual quadratic variation*.

We can sum up the above considerations in the following:

Theorem 3　The Itô table for the square of the standard quantum white noise process B_2, B_2^+ becomes closed after addition of the usual number process N and has the form

(11.15)

	$dB_2^+(t)$	dN	dB_2		
dB_2	$4\gamma dN$	$2\gamma dB_2$	0		
$dB_2^+(t)$	0	0	0		
dN	$2\gamma dB_2^+$	γdN	0	0

where the product $dB_2 dB_2^+$ should be interpreted in the sense of the renormalized mutual quadratic variation and all the other products in the sense of the usual mutual quadratic variation.

Remark 5　Let us note that our Itô table generates a non-associative algebra because one has $(dNdN)dB_2^+ = 2dB_2^+$ but $dN(dNdB_2^+) = 4dB_2^+$.

From now on we set $\gamma = 1$. For higher powers one can prove that the following general renormalized Ito table holds for the stochastic differentials

(11.16)
$$dB_{(m,n)} = b_t^{+m} b_t^n dt$$

defined as sesquilinear forms on the domain $\mathcal{D}(b)$ in the Fock space.

Theorem 4　The following multiplication table holds for the renormalized products of stochastic differentials

(11.17)
$$dB_{(m,n)} dB_{(k,l)} = nk dB_{(m+k-1, n+l-1)}$$

for any natural integers m, n, k and l such that $m + k - 1 \geq 0$ and $n + l - 1 \geq 0$.

Proof. We have to compute

$$dB_{(m,n)} dB_{(k,l)} = (b_t^{+m} b_t^n b_t^{+k} b_t^l dt^2)_{ren} \ .$$

Let us bring this expression to the normal order by using the Wick theorem. We have

$$b_t^n b_t^{+k} = b_t^{+k} b_t^n + nk\delta(0) b_t^{+(k-1)} b_t^{n-1} + \ldots$$

The terms denoted $\ldots$ include some formal power of the type $\delta(0)^p$, $p > 1$ and these are renormalized simply by subtracting them from the original expression. Then we get

$$(b_t^{+m} b_t^n b_t^{+k} b_t^l dt^2)_{ren} = b_t^{+(m+k)} b_t^{n+l} dt^2 + nk\delta(0) dt b_t^{+(m+k-1)} b_t^{n+l-1} dt$$
$$= nk dB_{(m+k-1, n+l-1)}$$

Remark 6 We have obtained an algebra with involution with generators $[m, n]$ satisfying to relations

$$(11.18) \qquad [m, n][k, l] = nk[m + k - 1, n + l - 1] \ .$$

This algebra is non-associative because of the factor nk. In principle we can redefine the renormalization also by dividing by these factors, then we can get an associative multiplication table.

Remark 7 The renormalization procedure used in Theorem (11.4) above is equivalent to the following generalization of the basic identity (10.2):

$$(11.19) \qquad \delta(0)^m dt^n = \delta_{m,1} \delta_{n,1}$$

in the sense that the identity (10.2) plus the subtraction procedure gives the same result as the evaluation of the normally ordered form using the rule (11.19).

12. A quantum approach to the square of classical white noise

In this Section we discuss a non-trivial consequence of the second order Itô table (11.15), namely the fact that, in order to close the nonlinear Itô table for some highly singular classical process, it is necessary to introduce some other classical process not commuting with the initial one. In other words: if we start from a classical (singular) process X and we look for the smallest family of processes which contains X and have a closed Itô table, then this family must be a quantum stochastic process.

The conclusion is that if a stochastic calculus for the higher powers of white noise exists at all (and the indications emerging from the present paper suggest it does) then it *must be* a quantum stochastic calculus.

The simplest example where the above mentioned phenomenon occurs can be constructed by considering the process

$$(12.1) \qquad w_2(t) := b_t^{+2} + b_t^2 \ .$$

A calculation of the formal commutator $[b_t^2, b_t^{+2}]$ leads to the result

$$[w_2(s), w_2(t)] = 2\delta(t - s)\{b_s b_t^+ + b_t^+ b_s - b_t b_s^+ - b_s^+ b_t\}$$

which is zero, being the product of two distributions with disjoint support. Thus the distribution random variables of the process (12.1) commute, i.e. $w_2(t)$ is a classical singular process. However, computing the mutual quadratic variation of $w_2(t)$ with the nonlinear Itô table (11.15) leads to

$$dw_2(t)^2 = 4\gamma dN_t$$

and dN_t does not commute with $dw_2(t)$ as easily verified using the Itô table (11.15).

We should expect such a result because the pair $b_t^{+2} dt, b_t^2 dt$ has not a closed Itô table. The situation is different if we consider the renormalized square of the classical white noise.

Because of the parity of the δ-function, the white noise $w_t = b_t^+ + b_t$ is a classical process in the sense that, for every s, t one has $[w_t, w_s] = 0$. Therefore its square is also a classical process and its normally ordered form is

$$(b_t^+ + b_t)^2 = b_t^{+2} + b_t^2 + 2b_t^+ b_t + \gamma\delta(0) \ .$$

Therefore its renormalized form

$$w_2(t) =: (b_t^+ + b_t)^2 := b_t^{+2} + b_t^2 + 2b_t^+ b_t \ ,$$

is also a classical process and, in the notation of Section 11, one has

$$dw_2(t) = dB_2^+(t) + dB_2(t) + 2dN(t) \ .$$

Therefore, using the nonlinear table (11.15) or (11.18) we find

$$\begin{aligned}
dw_2 dw_2 &= ([2,0] + [0,2] + 2[1,1])([2,0] + [0,2] + 2[1,1]) \\
&= 4[1,1] + 4[0,2] + 4[2,0] + 4[1,1] = 4([2,0] + [0,2] + 2[1,1]) = 4dw_2 \ .
\end{aligned}$$

We have also

$$dw_2 dw = ([2,0] + [0,2] + 2[1,1])([1,0] + [0,1]) = 2[0,1] + 2[1,0] = 2dw$$

Therefore the renormalized Ito table of the classical quadratic white noise is closed in itself and is also closed if we also include the classical (first order) white noise. In fact one has:

$$(12.2) \qquad\qquad dw_2 dw_2 = 4dw_2 \qquad , \qquad dw_2 dw = 2dw \ .$$

<u>Final remark</u>. The Itô table is interesting in itself, but of course its main utility is the possibility to use it to solve equations. In the nonlinear quadratic case, a first example of flow equation, generalizing the usual quantum flows has been studied in [AcLuOb96]. At the moment the existence of the solution is proved only in the space of Hida distributions, but we conjecture that the solution is a *bona fide* operator also in this case.

13. White noise approach to the Yang -Mills gauge theory

In this section as an application of the nonlinear *quantum* stochastic calculus discussed in Section 11, we will show how the *classical* (i.e. not quantized) Yang-Mills equations can be written in terms of this calculus.

The Yang-Mills field (gauge potential or connection) is a Lie-algebra valued differential form. It is given as a smooth map $A_\mu : \mathbf{R}^n \to \mathcal{A}$ where $\mathcal{A}$ is a Lie algebra (for instance, the algebra of $N \times N$ matrices) and $\mu = 1, 2, ..., n$. The Yang-Mills equations have the form

$$(13.1) \qquad\qquad \partial_\mu F_{\mu\nu} + [A_\mu, F_{\mu\nu}] = 0$$

where we assume here summation over repeating indices, $A_\mu = A_\mu(x)$, $\partial_\mu = \partial/\partial x^\mu$ and

$$F_{\mu\nu} = \partial_\mu A_\nu - \partial_\nu A_\mu + [A_\mu, A_\nu]$$

Let $\gamma : [0,1] \to \mathbf{R}^n$ be a piecewise smooth path in $\mathbf{R}^n$, with components, $\gamma = (\gamma_1(\tau), ..., \gamma_n(\tau))$ and let us consider the following differential equation in the space of $N \times N$ matrices (or, more generally, on a Lie algebra):

$$(13.2) \qquad \frac{dg_t}{dt} = V_t g_t \quad , \qquad g_0 = I$$

where

$$V_t = \int_0^t d\tau \, A_\mu(\gamma(\tau)) \dot{\gamma}_\mu(\tau)$$

and $\dot{\gamma}_\mu(\tau) = d\gamma_\mu(\tau)/d\tau$. The solution g_t of (13.2) for $t = 1$ is called a parallel transport along the path γ and it will be denoted $g(\gamma) = g_1$. One also uses the notation

$$(13.3) \qquad g(\gamma) = P exp \left\{ \int_0^1 d\tau \, A_\mu(\gamma(\tau)) \dot{\gamma}_\mu(\tau) \right\}$$

where P means an ordering along the path. In the Abelian case ($N = 1$) we have just the ordinary exponential function. In [AcGiVo94a] it was proved that the Yang-Mills equations are equivalent to the Lévy-Laplace equation for the parrallel transport. More precisely:

Theorem 5 The gauge potantial A_μ satisfies the Yang–Mills equations (13.1) if and only if the parallel transport $g(\gamma)$ satisfies the equation

$$(13.4) \qquad \Delta_L g(\gamma) = 0$$

where Δ_L denotes the Lévy–Laplacian.
We call (13.4) **the Lévy–Yang–Mills equation**.

The Lévy–Laplacian was introduced by P. Lévy (cf. [Lev51] for early history) and it has been much studied by T. Hida, H.H. Kuo, N. Obata and other mathematicians of the white noise group, for a review see [Kuo96] or [Hi91], [HiObSa92] for more informations. It can be defined as follows (we only outline the basic points; for an exact definition in terms of Fréchet derivatives cf. [AcGiVo94a]).
Let $F = F(\gamma)$ be a matrix valued functional on the Wiener path space with values in $\mathbf{R}^n$ and define the operators

$$(13.5) \qquad b_\mu(\tau) F(\gamma) = \frac{\delta}{\delta \gamma_\mu(\tau)} F(\gamma), \quad b_\mu^+(\tau) F(\gamma) = \gamma_\mu(\tau) F(\gamma)$$

where $\delta/\delta\gamma_\mu(\tau)$ is the functional derivative Then one has the canonical commutation relations

$$(13.6) \qquad [b_\mu(\tau), b_{nu}^+(\sigma)] = \delta_{\mu\nu} \delta(\tau - \sigma) \ .$$

Following [Kuo96] the Lévy–Laplacian can be defined as the limit

$$(13.7) \qquad \Delta_L F = \lim_{\epsilon \to 0} \int_{-\epsilon}^{\epsilon} b_\mu(t + \tau) b_\mu(t - \tau) d\tau F \ .$$

Moreover Kuo writes (13.6) in the following suggestive form

$$(13.8) \qquad \Delta_L = b(t)^2 dt^2 \ .$$

Using our notations from Section 11 one can rewrite (13.8) as

$$(13.9) \qquad \Delta_L = dB_2(t)dt \ .$$

If one takes a Fock representation of the canonical commutation relations (13.6) then one can consider the expression (13.3) for $g(\gamma)$ as *a simbol* of an operator in the Fock space and one can write the following equation

$$(13.10) \qquad \Delta_L g(b^+)\Phi = 0$$

where

$$(13.11) \qquad g(b^+) = Pexp\{ \int_0^1 d\tau A_\mu(b_\mu^+(\tau))\dot{b}_\mu^+(\tau)\}$$

Here Φ is the Fock vacuum and Δ_L is defined in (13.7). By using (13.9) one can write the Levy–Yang–Mills equations (13.10) as the following quantum stochastic equation

$$(13.12) \qquad dB_2 dt g(b^+)\Psi = 0 \ .$$

The equivalence between the Yang–Mills equations and (13.12) is checked by combining the form (13.10), obtained from the Kuo prescription rewritten in our notations (13.9) and the rule $\delta(0)dt = 1)$ from Section 11.

Let us consider the Abelian case and show that the Levy–Maxwell equation $\Delta_L g = 0$, i.e. the quantum stochastic equation

$$(13.13) \qquad dt b_\mu(t) b_\mu(t) \int_0^1 A_\lambda(b^+(\sigma))\dot{b}_\lambda^+(\sigma)d\sigma \Phi = 0$$

is equivalent to the Maxwell equations. We have to bring (13.13) to the normal order. To this end we will use the following formula for the commutator

$$[b_\mu(t), f(b^+(\sigma))] = \partial_\mu f(b^+(\sigma))\delta(t - \sigma)$$

Let us take $0 < t < 1$. We have

$$b_\mu(t) \int_0^1 A_\lambda(b^+(\sigma))\dot{b}_\lambda^+(\sigma)d\sigma \Phi =$$

$$= \int_0^1 \{\partial_\mu A_\lambda(b^+(\sigma))\delta(t - \sigma)\dot{b}_\lambda^+(\sigma) + A_\mu(b^+(\sigma))\dot{\delta}(t - \sigma)d\sigma\}\Phi = F_{\mu\lambda}(b^+(t)\dot{b}_\lambda^+(t)\Phi$$

Now one has

$$dt b_\mu(t) b_\mu(t) \int_0^1 A_\lambda(b^+(\sigma))\dot{b}_\lambda^+(\sigma)d\sigma \Phi = dt b_\mu(t) F_{\mu\lambda}(b^+(t)\dot{b}_\lambda^+(t)\Phi =$$

$$= dt\delta(0)\partial_\mu F_{\mu\nu}(b^+(t)\dot{b}^+_\lambda(t)\Phi + dt F_{\mu\mu}(b^+(t))\dot{\delta}(0)\Phi = \partial_\mu F_{\mu\lambda}(b^+(t)\dot{b}^+_\lambda(t)\Phi$$

Here we have used the rule $\delta(0)dt = 1$ and also the fact that $F_{\mu\mu} = 0$ because $F_{\mu\nu}$ is antisymmetric. Therefore the Levy–Maxwell equation (13.13) is equivalent to

$$\partial_\mu F_{\mu\lambda}(b^+(t)\dot{b}^+_\lambda(t)\Phi = 0$$

Now one shows that the last equation is equivalent to the Maxwell equations

$$\partial_\mu F_{\mu\lambda} = 0$$

Since we have reformulated the nonlinear Yang–Mills equations as an infinite-dimensional linear Lévy–Laplace equation we have to develop tools in infinite-dimensional analysis and especially in nonlinear stochastic calculus, which will permit us to study the properties of the Yang–Mills equations such as the existence and regularity of a general solution with various boundary condition, both in Euclidean and Lorentzian cases, as well as special solutions (for instance instantons and their moduli) and also Yang–Mills fields on manifolds, guantization,... in terms of the Lévy–Laplacian equation. For a discussion of applications of the Lévy–Laplace–Yang–Mills equation (13.4) in quantum chromodynamics see [ArVo96].

14. Interacting Fock spaces

Roughly speaking, an interacting Boltzmannian (we use *Boltzmannian* for what is usually called *free* because the pairing interacting-free sounds strange) Fock space over a given pre-Hilbert space $\mathcal{K}$ is a space which, as a vector space, coincides with the usual Free Fock space but in which, for any n, the n-particle space has its own scalar product $\langle\cdot,\cdot\rangle_n$. The scalar products on different n-particle spaces being related by the only three conditions that:

(i) linear combinations of the n-particle vectors are dense;

(ii) it is possible to (densely) define in the usual way the (free) creation operator associated to a given test function f;

(iii) the adjoint of the creation operator (annihilation) exists on a dense subset of the n-particle vectors.

For any vector space $\mathcal{H}$ denote $\mathcal{L}(\mathcal{H})$ the family of densely defined linear operators on $\mathcal{H}$ and we call an *n-particle vector* any element of the algebraic tensor product $\otimes^n\mathcal{H}$ of the form $f_1 \otimes \cdots \otimes f_n$. We adopt the usual convention that

$$(14.1) \qquad\qquad \otimes^0\mathcal{H} = \mathbf{C}\cdot\Phi$$

where Φ is a fixed unit vector called *the vacuum*. When $\mathcal{H}$ is a pre-Hilbert space and confusion might arise between the the algebraic and the Hilbert space tensor product, we shall denote the former by $\odot^n\mathcal{H}$ and the latter by $\otimes^n\mathcal{H}$. A linear operator A on a pre–Hilbert space $\mathcal{H}$ is said to have an adjoint if it has one on the Hilbert space completion of $\mathcal{H}$.

Definition 4 Let $\mathcal{K}$ be a vector space. An interacting Fock space over $\mathcal{K}$ is defined by the assignment, for each $n \in \mathbf{N}$, of a pre-scalar product $(\,\cdot\,|\,\cdot\,)_n$ on $\otimes^n \mathcal{K}$ with the following properties:

(i) For each n the n-particle vectors are dense in each n-particle space for the topology induced by the scala product $(\,\cdot\,|\,\cdot\,)_n$. Moreover

$$(14.2) \qquad\qquad (z\Phi|z'\Phi)_0 = \overline{z}z' \,.$$

(ii) The creation operator

$$(14.3) \quad A^+(f)\Phi = A^+(f)1 = f$$

$$(14.4)$$
$$A^+(f) : f_1 \otimes ... \otimes f_n \in \{\otimes^n \mathcal{K}, (\cdot|\cdot)_n\} f \otimes f_1 \otimes ... \otimes f_n \in \{\otimes^{n+1}\mathcal{K}, (\cdot|\cdot)_{n+1}\}$$

is well defined on the n-particle vectors. This means that, if F_n is such a vector and $f \in \mathcal{K}$, then if $(F_n \mid F_n)_n = 0$, it follows that also $(f \otimes F_n \mid f \otimes F_n)_{n+1} = 0$.

(iii) On a dense sub-domain of the n-particle vectors each creation operator $A^+(f)$ has an adjoint operator, denoted $A(f)$ satisfying the condition

$$(14.5) \qquad\qquad A(f)\Phi = 0$$

This means that, if F_n and F_{n+1} are respectively n- and $(n+1)$-particle vectors then there exists a constant $C(f, F_{n+1})$ such that

$$| (f \otimes F_n \mid F_{n+1})_{n+1} | \leq C(f, F_{n+1})(F_n \mid F_n)_n^{1/2} \,.$$

Remark 8 When for each n, the nth scalar product $\langle\cdot,\cdot\rangle_n$ coincides with the usual one

$$(14.6) \qquad\qquad \langle f_1 \otimes \cdots \otimes f_n, g_1 \otimes \cdots \otimes g_n\rangle_n = \prod_{h=1}^{n} \langle f_h, g_h \rangle \,,$$

one recovers the free Fock space. The symmetrized and antisymmetrized scalar products give the Boson and Fermion Fock spaces respectively. LEMMA 14.1. *The sequence*

$$(14.7) \qquad\qquad \{\otimes^n \mathcal{K}, (\cdot|\cdot)_n\}_{n\in\mathbb{N}}$$

defines an interacting Fock space over $\mathcal{K}$ if and only if, for any $n \in \mathbb{N}$, there exists a sesquilinear map

$$(14.8) \qquad\qquad T_n : \mathcal{K} \times \mathcal{K} \to \mathcal{L}(\{\otimes^n \mathcal{K}, (\cdot|\cdot)_n\})$$

which is defined on the n-particle vectors has the property that for all $n \in \mathbb{N}$ and for all $f, f_1, \ldots, f_{n+1},\, g_1, \ldots, g_n \in \mathcal{K}$ one has

$$(14.9)$$
$$(f \otimes g_1 \otimes \cdots \otimes g_n | f_1 \otimes \cdots \otimes f_{n+1})_{n+1} = (g_1 \otimes \cdots \otimes g_n | T_n(f, f_1)f_2 \otimes \cdots \otimes f_{n+1})_n \,.$$

Lemma 14.1 shows that the creation and annihilation operators on an interacting Fock

space satisfy the relation

$$(14.10) \qquad A(f) : \otimes^{n+1} \mathcal{K} \to \otimes^n \mathcal{K}$$

characterized by

(14.11)
$$A(f)A^+(f_1)f_2 \otimes \cdots \otimes f_{n+1}) = A(f)(f_1 \otimes \cdots \otimes f_{n+1}) = T_n(f, f_1)f_2 \otimes \cdots \otimes f_n$$

for each $n \in \mathbb{N}$. Introducing the number operator

$$(14.12) \qquad N(f_1 \otimes f_2 \otimes \cdots \otimes f_n) = n(f_1 \otimes f_2 \otimes \cdots \otimes f_n) \qquad ; \qquad n \in \mathbb{N} \ .$$

one can write (14.11) as the purely algebraic relation

$$(14.13) \qquad A(f)A^+(g) = T_N(f, g)$$

which is an operator generalization of the zero-deformation of the usual quantum mechanical commutation relations.

Now we give some examples to show how the conditions of Definition (14.1) can be satisfied.

Example 1 Let $\mathcal{H} := L^2(M, d\mu)$, where, M is a measurable space and μ is a σ-finite measure on M and let be given a sequence of positive functions $\{\lambda_n(x_1, x_2, \cdots, x_n)\}_{n=1}^{\infty}$

$$(14.14) \qquad \lambda_n \ : \ M^n \longrightarrow \ \mathbf{R}_+$$

with the property that for μ-almost all $x_o, x_1, \ldots, x_n \in M$:

$$(14.15) \qquad \lambda_n(x_1, \ldots, x_n) = 0 \Rightarrow \lambda_{n+1}(x_o, \ldots, x_n) = 0 \ .$$

Define for each n the scalar product

(14.16) $\langle f_o \otimes \cdots \otimes f_n, \ g_o \otimes \cdots \otimes g_n \rangle_n$

$$:= \int \mu(dx_o) \ \ldots \mu(dx_n)\lambda_{n+1}(x_o, \ldots, x_n)\bar{f}_o(x_o) \ldots \bar{f}_n(x_n)g_o(x_o) \ldots g_n(x_n) \ .$$

Condition (14.15) implies that the creator is well defined and one has

(14.17)
$\langle f_o \otimes \cdots \otimes f_n, A^+_{g_o} g_1 \otimes \cdots \otimes g_n \rangle_n$

$$= \int \lambda_{n+1}(x_o, \ \ , x_n)\bar{f}_o(x_o) \ldots \bar{f}_n(x_n)g_o(x_o) \ldots g_n(x_n)\mu(dx_o) \ldots \mu(dx_n) \ ,$$

but, again by condition (14.15), we can divide λ_{n+1} by λ_n almost everywhere, so the expression (14.17) is equal to

$$(14.18) \qquad \int \mu(dx_1)\dots\mu(dx_n)\lambda_n(x_1,\dots,x_n)\bar{f}_1(x_1)\dots\bar{f}_n(x_n)g_1(x_1)\dots g_n(x_n) \cdot$$

$$\cdot \int \mu(dx_o)\frac{\lambda_{n+1}(x_o,\dots,x_n)}{\lambda_n(x_1,\dots,x_n)}\,\bar{f}_o(x_o)g_o(x_o) \ .$$

Thus ( ) is equal to

$$(14.19) \qquad\qquad \langle A_{g_o}f_o \otimes \cdots \otimes f_n, g_1 \otimes \cdots \otimes g_n\rangle_n$$

with A_{g_o} given by

(14.20)

$$A_{g_o}f_o \otimes \cdots \otimes f_n(x_1,\dots,x_n)$$
$$= \left[\int \mu(dx_o)\frac{\lambda_{n+1}(x_o,\dots,x_n)}{\lambda_n(x_1,\dots,x_n)}\,\bar{f}_o(x_o)g_o(x_o)\right] f_1(x_1)\dots\dots f_n(x_n) \ .$$

Example 2 Let $T \in (0,+\infty]$ and in Example (14.1) choose $M = [0,T]$ with the Lebesgue measure, with the convenction that, if $T = \infty$, then for any $a \in \mathbf{R}$ the interval $(a,T]$ (resp. $[a,T]$) is understood as (a,∞) (resp. $[a,\infty)$). With these choices $\mathcal{H} := L^2([0,1],dx)$. Endow the n-fold algebraic tensor product $\mathcal{H}_n := \mathcal{H} \otimes \cdots \otimes \mathcal{H}$ with the scalar product

$$(14.21) \quad \langle f_1 \otimes f_n, g_1 \otimes g_n\rangle_n := \int_{[0,1]^n} \prod_{h=1}^{n}(\bar{f}g)(x_h) \cdot \chi_{\nabla_n}(x_1,\cdots,x_n)dx_1\cdots dx_n\rangle \ .$$

This means that in our case

$$(14.22) \qquad\qquad \lambda_n = \chi_{\nabla_n}$$

where for any $n \in \mathbf{n}$, we use ∇_n to denote the set

$$(14.23) \qquad\qquad \{(x_1,\cdots,x_n) \in [0,T]^n \ : \ x_1 > \cdots > x_n\}$$

and, as usual, χ_A is the indicator function of the set A. Note that the scalar product defined above is not the usual one due to the presence of the interaction term $\chi_{\Delta_n(1)}$. Clearly condition (14.15) of Example (14.1) is satisfied by the choice (14.22).

Example 3 Let $\mathcal{H}$ be as in Example (14.2) but, on the n-fold algebraic tensor product we choose the scalar product

$$(14.24)$$
$$< f_1 \otimes \cdots \otimes f_n, g_1 \otimes \cdots \otimes g_n > := \int_{[0,1]^n} \prod_{k=1}^{n}(\bar{f}_k g_k)(x_k)\chi_{\Delta_n}(x_1,\cdots,x_n)dx_1\dots dx_n$$

where, $\chi_{\Delta_n(1)}$ *is the indicator function of the standard simplex, i.e. the set*

$$(14.25) \qquad \Delta_n(1) := \left\{ (x_1, \cdots, x_n) : \ 0 \leq x_1 \leq x_2 \leq \cdots \leq x_n \leq 1 \right\}$$

i.e. in this case

$$(14.26) \qquad\qquad\qquad \lambda_n = \chi_{\Delta_n(1)} \ .$$

Clearly condition (14.15) of Example (14.2) is satisfied by this choice.

Remark 9 The above example can also be used to illustrate how, starting from an interacting Boltzmanian Fock space, one can define an interacting Boson (or Fermion) Fock space by the usual symmetrization (or anti-symmetrization) procedure. The situation however is more delicate in the Hilbert module case and it is for this reason that Skeide introduced the notion of centered Hilbert module to which the symmetrization (and presumably also the anti-symmetrization) procedure can be applied) (cf. [Ske96]).

15. Central limit theorems for interacting Fock Spaces

From the central limit of the classical Bernoulli process one obtains a Gaussian random variable. The pioneering work of W. von Waldenfels on quantum central limit theorems [voWa78] allowed extension of this result to the quantum case (cf. [AcBa87a], [AcBa87b], [Lu89]) showing that the Boson and Fermi Fock spaces and the associated creation–annihilation processes can be obtained as central limits of natural quantum (resp. Boson and Fermion) generalizations of the Bernoulli process.

As explained in the previous section, the stochastic limit of quantum electrodynamics lead to the notion of interacting Fock space. Since the stochastic limit involves a generalization of the techniques used in the quantum central limit theorems, it is natural to ask the question whether also the interacting Fock spaces (or at least some of them) can be obtained from some kind of central limit theorem. An affirmative answer to this question was given in series of papers (cf [AcLuVo97c] for a survey and references) which produced a wealth of examples of interacting Fock spaces arising from central limit theorems and, in several cases, gave the explicit form of the limit law (the analogue of the Gaussian).

In this Section we shall describe the first of these central limit theorems, based on the notion of *chronological independence* [LuDeG95], [Lu96] (different from Voiculescu's free independence) and in which it turned out later that the analogue of the Gaussian of the Gaussian law is just the familiar *arcsine law*. The latter result has been independently obtained by N. Muraki [Mur96] who also provided an interesting interpretation of the Lu–De Giosa central limit theorem in terms of random walks on graphs.

The notion of *quantum Bernoulli process* appropriate to the central limit theorem we are going to describe is the following.

Fix the usual basis of $\mathbf{R}^2$:

$$(15.1) \qquad\qquad\qquad e_1 := \begin{pmatrix} 1 \\ 0 \end{pmatrix}, \ e_2 := \begin{pmatrix} 0 \\ 1 \end{pmatrix}$$

and, for any $N \geq 1$, $1 \leq n \leq N$, $1 \leq k_1 < k_2 < \cdots < k_n \leq N$ define

$$(15.2) \qquad \delta_{\emptyset}^{(N)} := e_1^{\otimes N} := e_1 \otimes e_1 \otimes \ldots \otimes e_1 \qquad (N - times)$$

$$(15.3) \qquad \delta_{(k_1,\ldots,k_n)}^{(N)} := e_1^{\otimes(k_1-1)} \otimes e_2 \otimes e_1^{\otimes(k_2-k_1-1)} \otimes e_2 \otimes \cdots \otimes e_2 \otimes e_1^{\otimes(N-k_n)} .$$

From now on, when $n = 0$, $(k_1, \ldots, k_n)$ is understood as $\emptyset$. The family

$$(15.4) \qquad \{\delta_{(k_1,\ldots,k_n)}^{(N)} \otimes \delta_{(h_1,\ldots,h_m)}^{(N)} \\ : 0 \leq n, m \leq N, 1 \leq k_1 < \cdots < k_n \leq N, 1 \leq h_1 < \cdots < h_m \leq N\}$$

can be identified to a basis of the space $\mathcal{M}_2^{\otimes N}$, where $\mathcal{M}_2$ is the space of the 2×2-matrices.

It is convenient to generalize the previous definitions by putting, for any $f_1, \ldots, f_n \in L_{Riem}^1([0,1])$ — the set of all complex Riemann integrable functions on the interval $[0,1]$ —

$$(15.5) \qquad \delta_{(k_1,\cdots,k_n)}^{(N)}(f_1,\ldots,f_n) := \Big(\prod_{h=1}^{n} f_h\Big(\frac{k_h - 1}{N}\Big)\Big) \cdot \delta_{(k_1,\ldots,k_n)}^{(N)} .$$

In this notation, for any $k = 1, \ldots, N$ and $f \in L_{Riem}^1([0,1])$, we define an operator $\mathbf{T}_N^{+}(f,k)$ (discrete chronological creator) from $(\mathbf{R}^2)^{\otimes N}$ onto itself as follows:

$$(15.6) \qquad \mathbf{T}_N^{+}(f,k)\left[\delta_{(k_1,\ldots,k_n)}^{(N)}\right] = \begin{cases} f(\frac{k-1}{N}) \cdot \delta_{(k)}^{(N)}, & \text{if } n = 0 \\ 0, & \text{if } n \geq 1 \text{ and } k_1 \leq k \\ f(\frac{k-1}{N}) \cdot \delta_{(k,k_1,\ldots,k_n)}^{(N)} & \text{if } n \geq 1 \text{ and } k < k_1 . \end{cases}$$

Denoting by $\mathbf{T}_N(f,k)$ the adjoint of $\mathbf{T}_N^{+}(f,k)$ (discrete chronological annihilator), we see that

$$(15.7) \qquad \mathbf{T}_N(f,k)\left[\delta_{(k_1,k_2,\ldots,k_n)}^{(N)}\right] = \begin{cases} 0 & \text{if } n = 0, \\ \bar{f}(\frac{k-1}{N}) \cdot \delta_{\emptyset}^{(N)} & \text{if } n = 1 \text{ and } k = k_1, \\ 0 & \text{if } n \geq 1 \text{ and } k < k_1, \\ \bar{f}(\frac{k-1}{N}) \cdot \delta_{(k_2,\ldots,k_n)}^{(N)} & \text{if } n \geq 2 \text{ and } k = k_1 . \end{cases}$$

With the convention that, for any operator X, $X^0 = X^+$ and $X^1 = X$, we define the operators

$$(15.8) \qquad \mathbf{S}_N^{\varepsilon}(f) := \frac{1}{\sqrt{N}} \sum_{1 \leq k \leq N} \mathbf{T}_N^{\varepsilon}(f,k)$$

With the previous definitions and notations, the following result was proved in [DegLu95]:

Theorem 6 For any $m \in \mathbf{N}$, $g_1, \cdots, g_m \in L^1_{Riem}([0,1])$ and $\varepsilon = (\varepsilon(1), ..., \varepsilon(m)) \in \{0,1\}^m$, the mixed moments

$$(15.9) \qquad \langle \delta_\emptyset^{(N)}, \mathbf{S}_N^{\varepsilon(1)}(g_1) \ldots \mathbf{S}_N^{\varepsilon(m)}(g_m) \delta_\emptyset^{(N)} \rangle$$

converge, as $N \to \infty$, to the corresponding momenta

$$(15.10) \qquad \langle \Phi, A^{\varepsilon(1)}(g_1) \ldots A^{\varepsilon(m)}(g_m) \Phi \rangle$$

where, A^+, A, Φ are the creation, annihilation operators and the vacuum vector of the interacting Fock space space described in Example (14.3) of Section (14.).

Theorem (15.1) shows that the sequences $\mathbf{S}_N(f)$, $\mathbf{S}_N^+(f)$, $\delta_\emptyset^{(N)}$ converge respectively, in the sense of mixed momenta, to the annihilation operator $A(f)$, the creation operator $\mathbf{A}^+(f)$ and the vacuum vector Φ defined in Example (14.3).

References

[1] ACCARDI, L., BACH, A. (1987) *The harmonic oscillator as quantum central limit of Bernoulli processes.* Accept by: Prob. Th. and Rel. Fields, Volterra Preprint.

[2] ACCARDI, L., BACH, A. (1987) *Central limits of squeezing operators.* in: Quantum Probability and Applications IV Springer LNM N. 1396 7–19.

[3] ACCARDI, L., QUAEGEBEUR, J. (1988) *Ito algebras of Gaussian quantum fields.* Journ. Funct. Anal. 85 213–263.

[4] ACCARDI L. (1990) *An outline of quantum probability.* Unpublished manuscript (a Russian version was accepted for Publication in Uspehi Matem. Nauk. The second half of the paper has appeared in Trudi Moscov. Matem. Obsh. 56 (1995) english translation in: Trans. Moscow Math. Soc. 56 (1995) 235–270.)

[5] ACCARDI, L., GIBILISCO, P., VOLOVICH I.V. (1994) *Yang–Mills gauge fields as harmonic functions for the Lévy Laplacian* Russian Journal of Mathematical Physics 2 235–250.

[6] ACCARDI, L., LU, Y.G., VOLOVICH I. *Non–Commutative (Quantum) Probability, Master Fields and Stochastic Bosonization* Volterra preprint, CVV-198-94, hep-th/9412246.

[7] ACCARDI, L. (1995) *Yang–Mills equations and Levy laplacians.* in: Dirichlet forms and stochastic processes, Eds. Ma Z.M., Röckner M., Yan J.A., Walter de Gruyter 1–24.

[8] ACCARDI, L. (1997) *On the axioms of probability theory.* Plenary talk given at the Annual Meeting of the Deutsche Mathematiker–Vereiningung. To appear in: Jaheresberichte der Deutsche Mathematiker–Vereiningung.

[9] ACCARDI, L. (1996) *Applications of Quantum Probability to Quantum Theory.* Lectures delivered at Nagoya University (based on [AcLuVo97a]).

[10] ACCARDI, L., LU, Y.G., OBATA, N. *Towards a nonlinear extension of stochastic calculus.*

[11] ACCARDI, L., LU, Y.G., VOLOVICH, I. *Quantum theory and its stochastic limit.* Monograph in preparation.

[12] ACCARDI, L., LU, Y.G., VOLOVICH, I. *White noise approach to stochastic calculus and nonlinear Ito tables* Submitted to: Nagoya Journal of Mathematics.

[13] ACCARDI, L., LU, Y.G., VOLOVICH, I. (1997) *Interacting Fock spaces and Hilbert module extensions of the Heisenberg commutation relations.* to appear in: Preprint IIAS.

[14] AREFEVA, I., VOLOVICH, I. (1996) *The master field in QCD and q–deformed qauntum field theory.* Nucl. Phys. B 462 600–613.

[15] CHOQUET–BRUHAT, Y. (1973) *Distributions, Theorie et Problemes*, Masson et C., Editeurs.

[16] DELLACHERIE, C., MEYER, P.A. (1975) *Probabilites et potentiel.* Hermann.

[17] HIDA. T. (1975) *Analysis of Brownian Functionals.* Carleton Mathematical Lecture notes 13.

[18] HIDA, T. (1991) *A Role of the Levy Laplacian on the Causal Calculus of Generalized White Noise Functionals.* Preprint.

[19] HIDA, T., OBATA, N., SAITO, K. (1992) *Infinite dimensional rotations and Laplacians in terms of white noise calculus.* Nagoya Math. J. **128** 65–93.

[20] HUDSON R.L., PARTHASARATHY K.R. (1994) *Quantum Ito's formula and stochastic evolutions*, Commun. Math. Phys. **93** 301–323.

[21] KUO, H.-H. (1996) *White Noise Distribution Theory*, CRC Press.

[22] LEVY, P. (1951) *Problemes concrets d'analyse fonctionnelle* Gauthier Villars, Paris.

[23] LU, Y.G. (1992) *The Boson and Fermion Brownian Motion as Quantum Central limits of the Quantum Bernoulli Processes.* Bollettino UMI, (7) 6–A, 245–273. Volterra preprint (1989).

[24] LU, Y.G., DE GIOSA, M. (1995) *The free creation and annihilation operators as the central limit of quantum Bernoulli process.* Preprint Dipartimento di Matematica Università di Bari 2.

[25] LU Y.G. *The interacting Free Fock Space and the Deformed Wigner Law.*

[26] MEYER P.A. (1993) *Quantum Probability for Probabilists*, Lect. Notes in Math. Vol. 1538, Springer–Verlag.

[27] MURAKI, N. *Noncommutative Brownian motion in monotone Fock space,* to appear in Commun. Math. Phys.

[28] OBATA N. (1994) *White Noise Calculus and Fock Space*, Lect. Notes in Math. Vol. 1577, Springer–Verlag.

[29] OBATA N. (1995) *Generalized quantum stochastic processes on Fock space*, Publ. RIMS **31** 667–702.

[30] OHYA, M., PETZ, D. (1993) *Quantum entropy and its use*, Springer, Texts and Monographs in Physics.

[31] PARTHASARATHY K.R. (1992) *An Introduction to Quantum Stochastic Calculus*, Birkhäuser.

[32] SKEIDE M. (1996) *Hilbert modules in quantum electro dynamics and quantum probability.* Volterra Preprint N. 257.

[33] VOICULESCU, D. (1991) *Free noncommutative random variables, random matrices and the II_1 factors of free groups.* in: Quantum Probability and related topics, World Scientific VI.

[34] VON WALDENFELS, W., GIRI, N. (1978) *An Algebraic Version of the Central Limit Theorem.* Z. Wahrscheinlichkeitstheorie verw. Gebiete 42, 129–134.

TRENDS AND OPEN PROBLEMS IN THE THEORY OF RANDOM DYNAMICAL SYSTEMS

LUDWIG ARNOLD,* *Universität Bremen*

1. Introduction

The area of *random dynamical systems* (henceforth abbreviated as 'RDS') can be superficially described as the 'intersection' of stochastic processes with dynamical systems. It is an example for the fact that a symbiosis of two mathematical disciplines at the right moment amounts to opening a scientific gold mine, both conceptually and as far as significant applications are concerned.

Roughly speaking, an RDS is a 'combination' of a measure-preserving dynamical system $(\Omega, \mathcal{F}, \mathbb{P}, \theta)$ in the sense of ergodic theory (modeling random noise), and a smooth (or topological) dynamical system, typically generated by a differential or difference equation $\dot{x} = f(x)$ or $x_{n+1} = \varphi(x_n)$ (the system being perturbed by noise).

This concept is a simultaneous generalization of ergodic theory and smooth (or topological) dynamics. It opens new and unveiling views of both subjects. It also creates a circle of problems non–existent in the nonrandom case: the interplay of measurability and dynamics.

One can say that the new subject was born around 1980 when several people (see below for references) realized that a stochastic differential equation generates not just a family of Markov processes, each solving the equation for a given fixed initial value, but a *flow* of random diffeomorphisms. This crucial fact, together with a random substitute of linear algebra (called *multiplicative ergodic theorem*, proved by Oseledets [37] in 1968, and made popular in the West by Ruelle [40] in 1979), makes it possible to develop a *dynamical* as well as a *geometric* theory of these equations (for the latter see the article of Elworthy in this volume).

This article is based on and assumes knowledge of my *Six Lectures on Random Dynamical Systems* [4]. Since contained in a recent Springer Lecture Notes in Mathematics volume, it can be probably assumed to be universally available.

The books by the following authors treat aspects of RDS: Kifer [28] (products of iid random mappings), Bougerol and Lacroix [17] (products of iid random matrices), Carmona and Lacroix [19] (with a chapter on products of random matrices), and Liu and Qian [31] (smooth ergodic theory, Pesin theory). For a systematic account the reader probably has to wait for my forthcoming monograph [1].

The aim of this article is basically described by its title: to present a program which should (and most probably will) occupy researchers for several years to come, and which will undoubtedly be beneficial for users.

---

*Postal address: Institut für Dynamische Systeme, Universität Bremen, D-28334 Bremen, Germany.

Due to lack of space, I will concentrate on *smooth dynamics* of RDS. I will almost completely exclude *topological dynamics* of RDS, and refer to the forthcoming books by Nguyen Dinh Cong [33] and Gundlach [26].

2. Basic facts

2.1 Dynamical systems We will need a set $\mathbb{T}$ called *time*, which in this paper is exclusively $\mathbb{Z}$ (*two-sided discrete time*), or $\mathbb{R}$ (*two-sided continuous time*).

A *dynamical system* on the set $X \neq \emptyset$ is a mapping $\varphi : \mathbb{T} \times X \to X$, $(t, x) \mapsto \varphi(t)x$, for which the family of mappings $\varphi(t) : X \to X$, $x \mapsto \varphi(t)x$, forms a *flow*, i.e., it satisfies (i) $\varphi(0) = \mathrm{id}_X$, id_X the identity mapping on X, (ii) $\varphi(t + s) = \varphi(t) \circ \varphi(s)$. Here $\circ$ denotes composition.

According to the choice of the category of the space X and its self-mappings, we have three giant sub-disciplines: metric dynamics or ergodic theory, topological dynamics, and differentiable or smooth dynamics.

A *metric* dynamical system $(\Omega, \mathcal{F}, \mathbb{P}, (\theta(t))_{t \in \mathbb{T}})$ is a flow of mappings of a probability space $(\Omega, \mathcal{F}, \mathbb{P})$, such that

(i) $(t, \omega) \mapsto \theta(t)\omega$ is measurable,

(ii) $\theta(t) : \Omega \to \Omega$ is measure preserving, i.e., $\theta(t)\mathbb{P} = \mathbb{P}$.

Clearly every stationary stochastic process $(\xi_t)_{t \in \mathbb{T}}$ with state space $(E, \mathcal{E})$ generates a probability measure $\mathbb{P}$ on the product space $(\Omega, \mathcal{F}) = (E^{\mathbb{T}}, \mathcal{E}^{\mathbb{T}})$ which is invariant with respect to the shift $\theta(t)\omega := \omega(t + \cdot)$, hence a metric dynamical system (*real noise case*).

Example 1 (White noise case) Brownian motion induces a metric dynamical system as follows: Let $(W_t)_{t \in \mathbb{R}}$ be standard Brownian motion in $\mathbb{R}^m$. Put $\Omega = \{\omega \in C(\mathbb{R}, \mathbb{R}^m) : \omega(0) = 0\}$, $\mathcal{F}$ the Borel σ-algebra of Ω, $\mathbb{P}$ the (Wiener) measure on $\mathcal{F}$ generated by W. The shift $\theta(t)\omega(\cdot) := \omega(t + \cdot) - \omega(t)$ is measure preserving and ergodic, and $W_t(\omega) = \omega(t)$ is Brownian motion. We will always work with this canonical model.

2.2 Definition of an RDS

Definition 1 Let $(\Omega, \mathcal{F}, \mathbb{P}, (\theta(t))_{t \in \mathbb{T}})$ be a metric DS, and $(X, \mathcal{B})$ a measurable space. Let

$$\varphi : \mathbb{T} \times \Omega \times X \to X, \quad (t, \omega, x) \mapsto \varphi(t, \omega)x,$$

be a mapping with the following properties:

(i) $\varphi(0, \omega) = \mathrm{id}_X$,

(ii) Cocycle property: For all $s, t \in \mathbb{T}$ and all $\omega \in \Omega$

$$\varphi(t + s, \omega) = \varphi(t, \theta(s)\omega) \circ \varphi(s, \omega).$$

(1) If φ is measurable, then it is called a measurable RDS over θ.

(2) If, in addition to (1), X is a topological space, and φ satisfies*

$$(t, x) \mapsto \varphi(t, \omega)x \quad continuous \ for \ all \ \omega \in \Omega,$$

then φ is called a continuous *or topological RDS over θ.*

(3) If, in addition to (2), X is a smooth manifold, and φ satisfies for some k, $1 \leq k \leq \infty$,

$$x \mapsto \varphi(t, \omega)x \quad is \ C^k \quad for \ all \ (t, \omega) \in \mathbb{T} \times \Omega$$

(i.e., it is k times differentiable with respect to x, and the derivatives are continuous with respect to (t, x)), then φ is called smooth, *more precisely a C^k RDS over θ.*

(4) If, in addition to (3), $X = \mathbb{R}^d$ and $\varphi(t, \omega) \in Gl(d, \mathbb{R})$, the RDS is called linear.

We remark that

$$(\omega, x) \mapsto \Theta(t)(\omega, x) := (\theta(t)\omega, \varphi(t, \omega)x)$$

is a (measurable) flow on $\Omega \times X$, the so–called *skew–product* of θ and φ. We use the terms 'RDS', 'cocycle' and 'skew–product flow' synonymously.

2.3 Examples of RDS As in the deterministic case, all reasonably regular RDS have (infinitesimal) generators.

(i) For discrete time $\mathbb{T} = \mathbb{Z}$, all RDS are *products of random mappings*

$$\varphi(n, \omega) = \begin{cases} \varphi(\theta^{n-1}\omega) \circ \cdots \circ \varphi(\omega), & n \geq 1, \\ \mathrm{id}_X, & n = 0, \\ \varphi(\theta^n\omega)^{-1} \circ \cdots \circ \varphi(\theta^{-1}\omega)^{-1}, & n \leq -1. \end{cases}$$

where $\varphi(\omega) = \varphi(1, \omega)$ denotes the time one mapping.

(ii) In the continuous time case $\mathbb{T} = \mathbb{R}$, we ask for stochastic equivalents of the symbolic relation 'flow = exp(vector field)'. There are two basically different answers:

(ii)(a) RDS φ for which $t \mapsto \varphi(t, \omega)x$ is absolutely continuous are solution flows of pathwise *random* differential equations of the form $\dot{x} = f(\theta(t)\omega, x)$, and conversely.

(ii)(b) The theory of RDS would be much less interesting without the following fact: There is a huge class of important RDS which can principally *not* be obtained from a *pathwise* differential equation, but have *stochastic differential equations* as their generators.

The classical case is a (Stratonovich) stochastic differential equation in $\mathbb{R}^d$ of the form

$$(1) \qquad\qquad dx = f_0(x)dt + \sum_{j=1}^{m} f_j(x) \circ dW_t^j,$$

where $f_0, \ldots, f_m$ are smooth vector fields, and W is standard Brownian motion on $\mathbb{R}^m$. The door from stochastic analysis to dynamical systems was opened by the

---

* If X is a topological space, then $\mathcal{B}$ is always taken to be its Borel σ-algebra generated by the open sets.

discovery that equation (1) generates a two-parameter flow $\varphi_{s,t}(\omega)$ of diffeomorphisms (see Baxendale [15], Elworthy [24], Kunita [29]).

However, only very recently it was shown that there exists a solution of (1) for which $\varphi(t,\omega) := \varphi_{0,t}(\omega)$ is a cocycle over Brownian motion θ (Arnold and Scheutzow [9]), i.e., (1) indeed generates an RDS.

The inverse problem of which RDS φ has a generator goes beyond the classical model (1) and is described in section 3.

2.4 The multiplicative ergodic theorem The field of RDS would virtually not exist without the celebrated *multiplicative ergodic theorem* (henceforth abbreviated as 'MET') of Oseledets [37].

Theorem 1 (Multiplicative ergodic theorem) Let $\Phi(t,\omega)$ be a linear cocycle in $\mathbb{R}^d$ over the ergodic metric dynamical system $(\Omega, \mathcal{F}, \mathbb{P}, (\theta(t))_{t\in\mathbb{T}})$. Assume the integrability conditions $\alpha^\pm \in L^1(\mathbb{P})$, where

$$\alpha^\pm(\omega) := \begin{cases} \log^+ \|\Phi(1,\omega)^{\pm 1}\|, & \mathbb{T} = \mathbb{Z}, \\ \sup_{0 \le t \le 1} \log^+ \|\Phi(t,\omega)^{\pm 1}\|, & \mathbb{T} = \mathbb{R}. \end{cases}$$

Then there exist an invariant set $\tilde{\Omega} \in \mathcal{F}$ of full measure, constants $\lambda_1 > \ldots > \lambda_p$ (the Lyapunov exponents of the cocycle*) and $d_i \in \mathbb{N}$ with $\sum_{i=1}^{p} d_i = d$ (their* multiplicities*) such that for each $\omega \in \tilde{\Omega}$ the following holds: There exists a splitting*

$$\mathbb{R}^d = E_1(\omega) \oplus \cdots \oplus E_p(\omega)$$

of $\mathbb{R}^d$ into random measurable subspaces $E_i(\omega)$ (so-called Oseledets splitting*) with $\dim E_i(\omega) = d_i$ which is invariant, i.e.,*

$$\Phi(t,\omega)E_i(\omega) = E_i(\theta(t)\omega).$$

The splitting is dynamically characterized as follows:

$$\lim_{t \to \pm\infty} \frac{1}{t} \log \|\Phi(t,\omega)x\| = \lambda_i \iff x \in E_i(\omega) \setminus \{0\}.$$

For a detailed proof and historical remarks, see Arnold [1], Chapter 3.

The set $\mathcal{S}(\theta, \Phi) = \{(\lambda_i, d_i) : i = 1, \ldots, p\}$ is called the *Lyapunov spectrum* of the linear cocycle Φ.

2.5 Invariant measures for RDS For all further steps of the theory of RDS we need *invariant measures*.

Definition 2 Let φ be an RDS on $(X, \mathcal{B})$ over θ. A probability measure μ on $(\Omega \times X, \mathcal{F} \otimes \mathcal{B})$ is said to be an invariant measure *of φ, or φ–invariant, if it satisfies*
1. *$\Theta(t)\mu = \mu$ for all $t \in \mathbb{T}$,*
2. *the marginal of μ on Ω is $\mathbb{P}$.*

Invariance of μ, rewritten in terms of its factorization μ_ω, where $\mu(d\omega, dx) = \mu_\omega(dx)\mathbb{P}(d\omega)$ (such a factorization exists and is $\mathbb{P}$-a.s. unique if $(X, \mathcal{B})$ is a standard measurable space), is

$$\varphi(t, \omega)\mu_\omega = \mu_{\theta(t)\omega} \quad \mathbb{P}\text{-a.s. for all } t \in \mathbb{T}.$$

2.6 The MET for RDS on manifolds Let M be a smooth Riemannian manifold, and φ be a $\mathcal{C}^1$ RDS on M over θ. We can differentiate (linearize) $\varphi(t, \omega)$ at $x \in M$ to obtain a linear mapping

$$T\varphi(t, \omega, x) : T_x M \to T_{\varphi(t,\omega)x} M.$$

$T\varphi$ is a continuous linear cocycle on the tangent bundle TM over the skew-product flow Θ (*not θ!*).

Theorem 2 (MET for RDS on manifolds) Let φ be a $\mathcal{C}^1$ RDS on a Riemannian manifold M of dimension d. Let μ be an ergodic invariant measure of φ, such that $\alpha^\pm \in L^1(\mu)$, where

$$\alpha^\pm(\omega, x) := \begin{cases} \log^+ \|T\varphi(1, \omega, x)^{\pm 1}\|, & \mathbb{T} = \mathbb{Z}, \\ \sup_{0 \le t \le 1} \log^+ \|T\varphi(t, \omega, x)^{\pm 1}\|, & \mathbb{T} = \mathbb{R}. \end{cases}$$

Then there exist a Θ-invariant set $\Delta \subset \Omega \times M$ of full μ measure, and a finite list of fixed numbers $\lambda_1 > \lambda_2 > \ldots > \lambda_p$ (the Lyapunov exponents of φ under μ) with multiplicities $d_1 + d_2 + \cdots + d_p = d$ (the λ_i and d_i form the Lyapunov spectrum of φ under μ), such that for each fixed $(\omega, x) \in \Delta$ the following holds: There is a measurable splitting

$$T_x M = E_1(\omega, x) \oplus E_2(\omega, x) \oplus \cdots \oplus E_p(\omega, x)$$

with $\dim E_i(\omega, x) = d_i$ which is invariant, i.e.,

$$T\varphi(t, \omega, x)E_i(\omega, x) = E_i(\Theta(t)(\omega, x)).$$

The splitting is dynamically characterized by

$$\lim_{t \to \pm\infty} \frac{1}{t} \log \|T\varphi(t, \omega, x)v\| = \lambda_i \iff v \in E_i(\omega, x) \setminus \{0\}.$$

The moral of the MET is that for a $\mathcal{C}^1$ RDS φ with invariant measure μ, things are as nice and as simple as for a fixed point in the deterministic autonomous case: The λ_i are the stochastic analogues of the (real parts of) eigenvalues, and the E_i are the analogues of the generalized eigenspaces.

3. Present state of the theory

We now give a brief description of important accomplishements in the theory of RDS, and of problems which can be considered settled.

3.1 Generation of RDS The problem of which RDS with continuous time $\mathbb{T} = \mathbb{R}$ can be produced by an infinitesimal recipe has a satisfactory answer. The result, in intuitive terms, is an 'engine' which turns an additive vector field–valued cocycle into a multiplicative mapping–valued cocycle. We believe we have described the most general reasonable 'engine' [9], based on Kunita's [29] general theory of stochastic differential equations.

The crucial additional statistical property which an RDS φ has to have to be generated by a general stochastic differential equation is that for each x, the stochastic process $t \mapsto \varphi(t, \omega)x$ is a semimartingale. To obtain such a φ from a stochastic differential equation, Brownian motion has to be replaced by a vector field–valued semimartingale with stationary increments (a *helix*). The result is succinctly written as 'semimartingale cocycle $=$ exp(semimartingale helix)'.

For a survey see Arnold [3].

3.2 Perfection of crude cocycles Uniqueness of solutions of stochastic differential equations (1) only gives a *crude* cocycle in the sense that

$$\varphi(t + s, \omega) = \varphi(t, \theta(s)\omega) \circ \varphi(s, \omega)$$

is only valid for fixed $s, t \in \mathbb{R}$ and $\omega \notin N_{s,t}$, where $N_{s,t}$ is a $\mathbb{P}$ null set which might depend on s and t. The question is whether φ can be *perfected*, i. e., whether there is a cocycle ψ which is indistinguishable from φ and for which the cocycle property holds identically. This is a subtle question of considerable technical importance. For example, only for perfect cocycles the MET holds on an *invariant* set of full measure.

While for (1) the dependence of $N_{s,t}$ on t can be easily removed by continuity, the removal of s is highly non–trivial, but possible in all relevant cases, see [9].

3.3 Invariant manifolds Let φ be a $\mathcal{C}^1$ RDS on a Riemannian manifold M with (ergodic) invariant measure μ satisfying the integrability conditions of the MET Theorem 2. The construction of (global and local) random submanifolds of M which are invariant for φ and tangent to the splitting of $T\varphi$ is one of the basic tasks of the local theory of RDS. This task can be considered completed, thanks to the work of Ruelle [40], Carverhill [20], Boxler [18], Dahlke [23], and Wanner [45].

3.4 Pesin's formula One of the deepest and most exciting results in the theory of smooth RDS is the celebrated *Pesin's formula*: If M is a compact Riemannian manifold, φ a $\mathcal{C}^2$ RDS and μ a φ-invariant measure (such that $T\varphi$ and $T^2\varphi$ satisfy certain integrability conditions with respect to μ) then

$$h_\mu(\varphi) = \int_{\Omega \times M} \sum d_i(\omega, x) \lambda_i^+(\omega, x) d\mu(\omega, x)$$

if and only if μ is a *Sinai–Bowen–Ruelle measure* (i. e., is absolutely continuous with respect to the Riemannian volume on the unstable invariant manifolds). Here $x^+ :=$ $\max(x, 0)$, and $h_\mu(\varphi) = h_\mu(\Theta | \mathcal{F} \otimes M)$ denotes the *fiber* (or relativized) *entropy* (see Bogenschütz [16]).

This theorem was proved by Liu and Qian [31] for products of iid random diffeomorphisms, generalizing the deterministic result by Ledrappier and Young [30], and for the general case of C^2 RDS by Bahnmüller and Liu [14].

We warmly recommend the book by Liu and Qian [31] which, in my opinion, is the first really complete presentation including RDS of what is called *Pesin theory*.

3.5 Random Hartman–Grobman theorem   Let φ be a C^1 RDS in $\mathbb{R}^d$ with discrete time $\mathbb{T} = \mathbb{Z}$ (say), time-one map $\varphi(\omega) := \varphi(1, \omega)$ and $\varphi(\omega, 0) = 0$. The point 0 is called *hyperbolic* if the cocycle generated by $A(\omega) := D\varphi(\omega, 0)$ is hyperbolic, i.e., has only non-vanishing Lyapunov exponents. We seek a random homeomorphism $h : \Omega \to \mathrm{Homeo}(\mathbb{R}^d)$ with $h(\omega, 0) = 0$ such that

$$A(\omega) = h(\theta\omega)^{-1} \circ \varphi(\omega) \circ h(\omega).$$

The crucial new phenomenon in the random case is the appeerence of θ in the above equation which makes the problem infinite-dimensional.

A global random Hartman–Grobman theorem was proved by Wanner [45] for C^1 RDS, and a local version for $\mathbb{T} = \mathbb{Z}$, and for $\mathbb{T} = \mathbb{R}$ if φ is generated by a random differential equation. Hence in a neighborhood of a hyperbolic fixed point, a C^1 RDS is topologically conjugate to its own linearization.

The local Hartman–Grobman theorem for stochastic differential equations is still open, see the subsection on anticipative problems.

3.6 Random normal forms   The task of smooth normal form theory is to seek $h(\omega) \in \mathrm{Diff}^\infty(\mathbb{R}^d)$ (which also is near identity, i.e., satisfies $Dh(\omega, 0) = \mathrm{id}$) such that

$$\psi(\omega) = h(\theta\omega)^{-1} \circ \varphi(\omega) \circ h(\omega)$$

is 'as simple as possible', the best–possible case being $\psi(\omega) = D\varphi(\omega, 0)$. Which terms of the Taylor expansion of φ cannot be transformed away (so–called *resonant* terms) depends on the interplay of the linear cocycle generated by $D\varphi(\omega, 0)$ and the metric dynamical system θ. Normal form theory can be considered complete (except for stochastic differential equations, see section 4), see Arnold and Xu ([11],[12],[13]).

4. Research program

We now present a list of problems which are still far from being settled, but where substantial progress should be expected in the next few years.

4.1 Theory of stochastic bifurcation   Bifurcation theory of RDS studies qualitative changes in parametrized families φ_α of RDS, e.g. those generated by a family of stochastic differential equations

$$dx = f_0^\alpha(x) + \sum_{j=1}^{m} f_j^\alpha(x) \circ dW_t^j.$$

In [4] we have described the problems in more detail. Our philosophy still is to search for bifurcations on the level of invariant measures, and to detect critical parameter values by changes of the sign of Lyapunov exponents of the reference measure.

However, very little progress has been made recently (see e.g. Arnold and Schmalfuß [10]), so that stochastic bifurcation theory remains to be one of the great challenges to future research.

4.2 Anticipative problems For RDS generated by stochastic differential equations many problems have been blocked by the following 'collision' of stochastic analysis and multiplicative ergodic theory: The random Oseledets spaces $E_i(\omega, x)$ of the MET, used in virtually all constructions for smooth RDS, are typically not measurable with respect to the past, hence with respect to the information available in stochastic analysis at time $t = 0$. This calls for utilizing anticipative calculus. Another method would be enlarging the canonical Wiener filtration by the Oseledets spaces, proving that semimartingales remain semimartingales, and calculating the new compensator.

For a more detailed description and a catalogue of problems see Arnold [2]. Some problems have been settled: Arnold and Imkeller [5] calculated the Malliavin derivative of the Oseledets spaces, and in [6] gave an anticipative Fubini theorem by which a linear stochastic differential equation can be diagonalized.

A lot remains to be done, in particular a local Hartman–Grobman theorem, and, most urgently, smooth normal form theory for stochastic differential equations (the latter was formally 'done' by engineers who just ignored the fact that certain terms are not adapted).

4.3 Random attractors The investigation of random sets which attract many other (deterministic or random) sets is an important new line of research on (finite– and) infinite–dimensional systems under random perturbations.

Let φ be a topological RDS on a Polish space. A mapping $\omega \mapsto K(\omega)$ into the non–void compact subsets of X for which $\omega \mapsto d(x, K(\omega))$ is measurable for any $x \in X$ is called a *random compact set*. Let $\mathcal{D}$ be a family of those sets which is closed with respect to set inclusion.

A *random attractor* of φ in the 'universe' $\mathcal{D}$ is a random compact set $A \in \mathcal{D}$ with the following properties:

(i) A is φ–invariant, i. e., $\varphi(t, \omega)A(\omega) = A(\theta(t)\omega)$,
(ii) A is attracting all sets in $\mathcal{D}$, i. e.,

$$\lim_{t \to \infty} d(\varphi(t, \theta(-t)\omega)D(\theta(-t)\omega)|A(\omega)) = 0 \quad \text{for any } D \in \mathcal{D},$$

where $d(A|B) := \sup_{x \in A} d(x, B)$ is the semi–Hausdorff metric. The union of all random compact sets being attracted by A is the *domain of attraction $D(A)$* of A.

Random attractors have been studied by Crauel and Flandoli [22], Crauel [21], Schmalfuß [41],[43], and others. The questions investigated so far were: existence and uniqueness, estimates of the (finite!) Hausdorff dimension, invariant measures on the attractor.

What we need are more case studies. Also, the generator of the RDS restricted to the attractor should be determined and studied.

4.4 Global and topological methods for RDS Research efforts for RDS have concentrated so far on taking profit from the MET, i. e., on local theory. We believe that global methods are underdeveloped and deserve much more attention.

For example, *Lyapunov function* techniques should be developed as a simultaneous generalization of the deterministic concept (where a compact set is asymptotically stable for a continuous dynamical system if and only if it has a Lyapunov function, see Mischaikow [32], chapter 3), and the concept for Markov and diffusion processes based on the infinitesimal generator (and heavily used in the qualitative theory of stochastic differential equations, see e. g. Khasminskii [27]).

Another promising project would be to develop a *Conley index theory* for RDS, e. g. following the presentation by Mischaikow [32]. The outcome will certainly *not* be an ω–wise version of the deterministic theory, as the following shows: Ochs and Oseledets [36] construct a continuous RDS φ on a compact interval without a random fixed point, i. e., there is no random variable x for which

$$\varphi(t,\omega)x(\omega) = x(\theta(t)\omega).$$

On the basis of this result, one can construct continuous RDS on the closed unit ball of any dimension without random fixed points. In other words, it is impossible to generalize the Brouwer fixed point theorem, hence the Lefschetz fixed point theorem, hence McCord's theorem (Mischaikow [32], theorem 2.2.11) to the random case.*

4.5 Representation of Markov chains as RDS It is well–known that a measurable cocycle φ with time $\mathbb{T} = \mathbb{Z}$ which is a product of iid random mappings $\psi_n = \psi(\theta^n\cdot)$ generates a homogeneous Markov chain (x_n), $x_{n+1} = \psi_n x_n$, in the state space X with transition probability

$$P(x, B) = \mathbb{P}\{\omega : \psi(\omega)x \in B\},$$

see Kifer [28].

The inverse problem of constructing a measurable/continuous/C^k cocycle of iid mappings (necessarily invertible due to two–sided time) with a prescribed transition probability is largely unsolved. Note that it follows from the theory of non–negative matrices that if X is countable then a transition matrix (p_{ij}) can be represented by random permutations of X if and only if it is bistochastic.

However, constructions available for one–sided time (Kifer [28] for measurable mappings, Quas [39] for smooth mappings) do not yield invertible mappings.

4.6 Classification of RDS One of the basic tasks of the theory of RDS is *classification*, for which we need a concept of *isomorphy*. Assuming that our RDS are defined over the same θ, this isomorphism has the general form

$$\psi(t,\omega) = h(\theta(t)\omega)^{-1} \circ \varphi(t,\omega) \circ h(\omega),$$

where $h(\cdot)$ is a random (measure–preserving) bimeasurable bijection in the metric/measurable case, a random homeomorphism in the continuous case, a random diffeomorphism in the smooth case, and a random linear isomorphism in the linear case.

---

* The Banach fixed point theorem, however, has a random version, see Schmalfuß [42].

There has been considerable progress on the various levels of the classification problem in recent years.

For the metric isomorphism problem see Gundlach [26] who proved that if an RDS φ is 'weak–Bernoulli' then the fiber entropy $h_\mu(\varphi)$ is a complete isomorphism invariant, hence the RDS φ is isomorphic to a standard Bernoulli shift, and the isomorphism decouples the system from the noise.

On the topological level, the random Hartman-Grobman theorem of Wanner [45] allows one to linearize hyperbolic cocycles by a random homeomorphism.

The random normal form theory can also be considered as a contribution to the smooth classification problem.

The topological classification of hyperbolic linear cocycles in $\mathbb{R}^d$ was solved by Nguyen Dinh Cong [35],[34],[33]. For example, if θ is an irrational rotation on (S^1, Leb) there are infinitely many classes of hyperbolic cocycles which are *not* pairwise topologically equivalent.

There remains the largely unsolved problem of *linear* classification of linear cocycles, in other words, the problem of a 'random Jordan canonical form'. Clearly the Lyapunov spectrum is an invariant. For the two–dimensional case, Oseledets [38] proved (recovering statements by Thieullen) that every linear cocycle is linearly conjugate to one of the following four types: orthogonal, triangular, diagonal with one–point Lyapunov spectrum, and diagonal with simple spectrum.

By choosing an appropriate random basis, the MET permits one to block–diagonalize a cocycle, where the blocks just have one Lyapunov exponent. The task is now to classify those blocks. This amounts to improving the MET by describing what is going on *inside* an Oseledets space, which is a challenging, but probably very difficult problem.

4.7 The MET under discretization and perturbation It is an important problem of numerical analysis of dynamical systems to study the effect of discretization on invariant objects. For example, only if at a hyperbolic point the discretized versions of the stable and unstable manifolds are close to their originals one has a chance of long time shadowing of respective orbits. Reasonable numerical procedures will have this property, which basically follows from the continuous dependence of the eigenvalues and eigenspaces of a matrix on its entries.

This is not at all the case for RDS (see Arnold and Nguyen Dinh Cong [8] for genericity results, and Arnold and Kloeden [7] for numerical questions): Neither do Lyapunov exponents depend continuously on the data of the cocycle, nor are the Oseledets spaces of two cocycles close provided their Lyapunov exponents are. As a result, it is the MET which causes trouble in the numerical analysis of RDS.

In spite of certain preliminary results by Talay [44], Wihstutz [46] and Grorud and Talay [25] on the numerical approximation of the top Lyapunov exponent of a stochastic differential equation, the main question of a general perturbation theory (in particular, a numerical analysis) for the MET is untouched.

Results are urgently needed, in particular to support the credibility of computer simulations for RDS.

References

[1] L. Arnold. *Random dynamical systems.* Springer, Berlin Heidelberg New York, 1997 (to appear).

[2] L. Arnold. Anticipative problems in the theory of random dynamical systems. In M. C. Cranston and M. A. Pinsky, editors, *Stochastic Analysis (Proceedings of Symposia in Pure Mathematics, Volume 57)*, pages 529–541. American Mathematical Society, Providence, Rhode Island, 1995.

[3] L. Arnold. Generation of random dynamical systems. In W. Kliemann and N. Sri Namachchivaya, editors, *=Nonlinear Dynamics and Stochastic Mechanics*, CRC Mathematical Modelling Series, pages 203–230, Boca Raton, Florida, USA, 1995. CRC Press.

[4] L. Arnold. Six lectures on random dynamical systems. In R. Johnson, editor, *Dynamical Systems (CIME Summer School 1994)*, volume 1609 of *Lecture Notes in Mathematics*, pages 1–43. Springer, Berlin Heidelberg New York, 1995.

[5] L. Arnold and P. Imkeller. Furstenberg-Khasminskii formulas for Lyapunov exponents via anticipative calculus. *Stochastics and Stochastics Reports*, 54:127–168, 1995.

[6] L. Arnold and P. Imkeller. Stratonovich calculus with spatial parameters and anticipative problems in multiplicative ergodic theory. *Stochastic Processes and their Applications*, 62:19–54, 1996.

[7] L. Arnold and P. Kloeden. Discretization of a random dynamical system near a hyperbolic point. *Mathematische Nachrichten*, 181:43–72, 1996.

[8] L. Arnold and Nguyen Dinh Cong. On the simplicity of the Lyaponov spectrum of products of random matrices. *Ergodic Theory and Dynamical Systems*, 17:1–21, 1997.

[9] L. Arnold and M. Scheutzow. Perfect cocycles through stochastic differential equations. *Probab. Theory Relat. Fields*, 101:65–88, 1995.

[10] L. Arnold and B. Schmalfuß. Fixed points and attractors for random dynamical systems. In A. Naess and S. Krenk, editors, *IUTAM Symposium on Advances in Nonlinear Stochastic Mechanics*, pages 19–28. Kluwer, Dordrecht, 1996.

[11] L. Arnold and Xu Kedai. Normal forms for random diffeomorphisms. *J. Dynamics and Differential Equations*, 4:445–483, 1992.

[12] L. Arnold and Xu Kedai. Simultaneous normal form and center manifold reduction for random differential equations. In C. Perelló, C. Simó, and J. Solá-Morales, editors, *EQUADIFF-91*, volume 1, pages 68–80. World Scientific, Singapore, 1993.

[13] L. Arnold and Xu Kedai. Normal forms for random differential equations. *Journal of Differential Equations*, 116:484–503, 1995.

[14] J. Bahnmüller and Liu Pei–Dong. Characterization of measures satisfying Pesin's entropy formula for random dynamical systems. Submitted, 1995.

[15] P. Baxendale. Brownian motion in the diffeomorphism group I. *Compositio Mathematica*, 53:19–50, 1984.

[16] T. Bogenschütz. Entropy, pressure, and a variational principle for random dynamical systems. *Random & Computational Dynamics*, 1:99–116, 1992.

[17] P. Bougerol and J. Lacroix. *Products of random matrices with applications to Schrödinger operators*. Birkhäuser, Boston Basel Stuttgart, 1985.

[18] P. Boxler. A stochastic version of center manifold theory. *Probab. Th. Rel. Fields*, 83:509–545, 1989.

[19] R. Carmona and J. Lacroix. *Spectral theory of random Schrödinger operators*. Birkhäuser, Boston Basel Stuttgart, 1990.

[20] A. Carverhill. Flows of stochastic dynamical systems: ergodic theory. *Stochastics*, 14:273–317, 1985.

[21] H. Crauel. Global random attractors are uniquely determined by attracting deterministic compact sets. Submitted, 1995.

[22] H. Crauel and F. Flandoli. Attractors for random dynamical systems. *Probab. Theory Relat. Fields*, 100:365–393, 1994.

[23] S. Dahlke. Invariant manifolds for products of random diffeomorphisms. *J. Dynamics and Differential Equations*, 1996.

[24] K. D. Elworthy. *Stochastic differential equations on manifolds*. Cambridge University Press, Cambridge, 1982.

[25] A. Grorud and D. Talay. Approximation of Lyapunov exponents of nonlinear stochastic differential systems. *SIAM J. Appl. Math.*, 1996.

[26] V. M. Gundlach. Random homoclinic orbits. *Random & Computational Dynamics*, 3:1–33, 1995.

[27] R. Z. Khasminskii. *Stochastic stability of differential equations*. Sijthoff and Noordhoff, Alphen, 1980. (Translation of the Russian edition, Nauka, Moscow 1969).

[28] Y. Kifer. *Ergodic theory of random transformations*. Birkhäuser, Boston Basel Stuttgart, 1986.

[29] H. Kunita. *Stochastic flows and stochastic differential equations*. Cambridge University Press, Cambridge, 1990.

[30] F. Ledrappier and L.-S. Young. Entropy formula for random transformations. *Probab. Theory Relat. Fields*, 80:217–240, 1988.

[31] P.-D. Liu and M. Qian. *Smooth ergodic theory of random dynamical systems*, volume 1606 of *Lecture Notes in Mathematics*. Springer, Berlin Heidelberg New York, 1995.

[32] K. Mischaikow. Conley index theory. In R. Johnson, editor, *Dynamical Systems (CIME Summer School 1994)*, volume 1609 of *Lecture Notes in Mathematics*, pages 119–207. Springer, Berlin Heidelberg New York, 1995.

[33] Nguyen Dinh Cong. *Topological dynamics of random dynamical systems*. Oxford University Press, 1997.

[34] Nguyen Dinh Cong. Structural stability of linear random dynamical systems. Report 312, Institut für Dynamische Systeme, Universität Bremen, June 1994.

[35] Nguyen Dinh Cong. Topological classification of linear hyperbolic cocycles. Report 305, Institut für Dynamische Systeme, Universität Bremen, May 1994.

[36] G. Ochs and V. Oseledets. On recurrent cocycles and the non–existence of random fixed points. Report 382, Institut für Dynamische Systeme, Universität Bremen, 1996.

[37] V. I. Oseledets. A multiplicative ergodic theorem. Lyapunov characteristic numbers for dynamical systems. *Trans. Moscow Math. Soc.*, 19:197–231, 1968.

[38] V. I. Oseledets. Classification of Gl(2,R)–valued cocycles of dynamical systems. Report 360, Institut für Dynamische Systeme, Universität Bremen, 1995.

[39] A. Quas. On representations of Markov chains by random smooth maps. *Bull. London Math. Soc.*, 23:487–492, 1991.

[40] D. Ruelle. Ergodic theory of differentiable dynamical systems. *Publ. Math., Inst. Hautes Etud. Sci.*, 50:275–306, 1979.

[41] B. Schmalfuß. Measure attractors and stochastic attractors. Report 332, Institut für Dynamische Systeme, Universität Bremen, 1995.

[42] B. Schmalfuß. A random fixed point theorem based on Lyapunov exponents. Report 349, Institut für Dynamische Systeme, Universität Bremen, 1995.

[43] B. Schmalfuß. The random attractor of the stochastic Lorenz system. Report 367, Institut für Dynamische Systeme, Universität Bremen, 1996.

[44] D. Talay. Approximation of upper Lyapunov exponents of bilinear stochastic differential equations. *SIAM J. Numer. Anal.*, 28:1141–1164, 1991.

[45] T. Wanner. Linearization of random dynamical systems. In C. Jones, U. Kirchgraber, and H. O. Walther, editors, *Dynamics Reported*, volume 4, pages 203–269. Springer, Berlin Heidelberg New York, 1995.

[46] V. Wihstutz. Numerics for Lyapunov exponents of degenerate linear stochastic systems. In W. Kliemann and N. Sri Namachchivaya, editors, =Nonlinear *Dynamics and Stochastic Mechanics*, CRC Mathematical Modelling Series, Boca Raton, Florida, USA, 1995. CRC Press.

PROBABILITY AND STATISTICS:
SELF-DECOMPOSABILITY, FINANCE AND TURBULENCE

OLE E. BARNDORFF-NIELSEN,* *Aarhus University*

Abstract

After some general remarks about the relationship between probability and statistics, a discussion is given of closely similar, key features of empirical data from finance and from turbulence, and this is followed by an account of recent work on stochastic modelling incorporating those features.

INFINITE DIVISIBILITY; INVERSE GAUSSIAN DISTRIBUTIONS; NORMAL INVERSE GAUSSIAN DISTRIBUTIONS; ORNSTEIN–UHLENBECK TYPE PROCESSES; SUBORDINATION; STYLIZED FEATURES

MAITLAND: What rings true then?
MADRIGAL: The probability. The likelihood.
    They work to make things fit together.
Enid Bagnold: The Chalk Garden (1956)

1. Introduction

Statistical Science builds, of course, to a significant extent on probability theory. However, even a quarter of a century ago the part of probability theory that was of essential importance and use in statistics was quite limited and relatively elementary. This picture has changed dramatically in recent years concurrently with a general development of mathematical statistics towards employing advanced techniques and results from many different fields of mathematics. In fact, some areas of mathematics, including probability, that were considered by many as highly abstract and far removed from any statistical relevance have turned out to provide just the right concepts and tools for statistical analysis of data of widely differing character and origin. To some extent the situation is, as noted elsewhere (Barndorff-Nielsen, 1993) reminiscent of that which characterised Physics at the beginning of this century, where a split developed between those physicists that followed the new mathematically dominated trends and the more traditionally oriented physicists.

Two striking examples, of recent origin, of the impact of advanced probabilistic developments on statistical theory and practice are the use of stochastic analysis in survival analysis (see Andersen, Borgan, Gill and Keiding, 1993), and the role of point processes and interacting particle systems in spatial statistics and image analysis (cf., for instance, Ripley, 1988, and Winkler, 1995). In both fields the probability and the likelihood do indeed "work to make things fit together".

---

* Postal address: Department of Mathematical Sciences, Aarhus University, Denmark.

Other examples, though on a much more modest scale, relate to infinite divisibility and Lévy processes. These concepts have recently turned out to be suitable tools for nonparametric inference in Bayesian statistics, see Damien, Laud and Smith (1995). And in section 3 of the present paper an outline will be given of other uses of Lévy processes and infinite divisibility, in particular self-decomposability, now in statistical model building of stochastic processes for the analysis of extensive data sets from finance and turbulence. Observed time series from these two fields exhibit a number of very similar key features, as will be discussed in section 2.

2. Stylized Features of Finance and Turbulence

A number of characteristic features of observational series from finance and from turbulence are summarised in table 1. The features are widely recognized as being esssential for understanding and modelling within these two, quite different, subject areas. In finance the observational series concerned consist of values of assets such as stocks or stocks returns or exchange rates, while in turbulence the series typically show the velocities or velocity derivatives (or differences), in the mean wind direction of a large Reynolds number wind field.

	Finance	Turbulence
semiheavy tails	+	+
asymmetry	(+)	+
varying activity	volatility	intermittency
aggregational Gaussianity	+	+
quasi long range dependence	+	(+)

TABLE 1. Stylised features.

Before commenting on table 1 we show in figures 1 and 2 some typical examples of empirical probability densities of velocity differences in large Reynolds number wind fields. Note that the ordinate axes are logarithmic.

Figure 1 is based on Wyngaard and Tennekes (1970; Fig. 1).

The data in figure 2 were collected over the Pacific Ocean by an airplane and were discussed in van Atta and Park (1972). Each of the eight distributions exhibited is based on about a quarter of a million observations. The figure, which is similar to Barndorff-Nielsen (1979; Figure 10), also shows the result of fitting the hyperbolic distribution to the data. The hyperbolic distribution is characterised by the fact that the graph of the log density function is a hyperbola. In particular, then, the tails of the hyperbolic distribution are log linear. The fitting is rather succesful except for the cases (a) and (b) that correspond to small distances between the measurement points and in which the tails are heavier than log linear. Further examples of a similar nature can be found in Barndorff-Nielsen (1979) and Barndorff-Nielsen, Jensen and Sørensen (1989).

For similar figures in the field of finance, see for instance Eberlein and Keller (1995) and Shephard (1995).

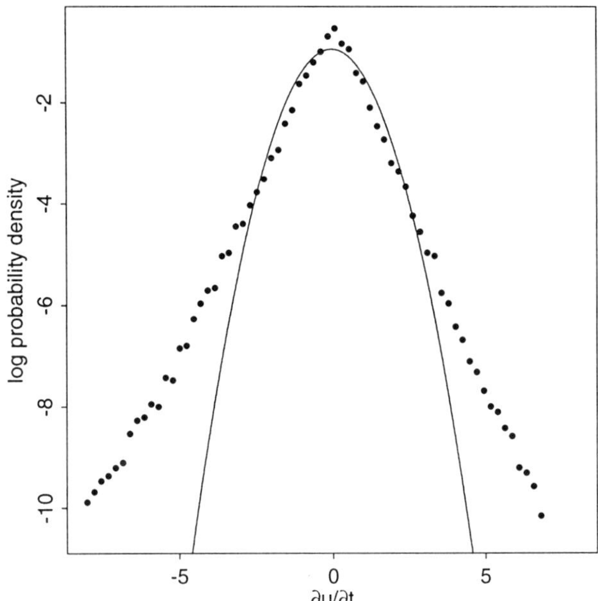

FIGURE 1. Probability density of first derivative of principal velocity component in a turbulent mixing layer.

The term 'semiheavy tails', in table 1, is intended to indicate that the data suggest modelling by probability distributions whose densities behave as

$$\text{const.}\,|x|^{\rho\pm}\exp\{-\sigma_\pm\,|x|\}$$

for some $\rho_+,\rho_- \geq 0$ and $\sigma_+,\sigma_- > 0$ and for $x \to \pm\infty$.

Distributions of financial asset returns are generally close to symmetric around 0, but velocity differences in turbulence show an inherent asymmetry consistent with Kolmogorov's modified theory of homogeneous high Reynolds number turbulence. The shape parameter estimates from the fitting of hyperbolic distributions to the data from figure 2 exhibit this clearly, see Barndorff-Nielsen (1986).

A very characteristic trait of time series from turbulence as well as finance is that there seems to be a kind of switching regime between periods of relatively small random fluctuations and periods of high 'activity'. In turbulence this phenomenon is known as intermittency, see e.g. Frisch (1995; chapter 8) for a thorough discussion, whereas in finance one speaks of stochastic volatility or conditional heteroscestadisticity.

By aggregational Gaussianity is meant the fact that, for instance, long term aggregation of financial asset returns, in the sense of summing the returns over longer periods, will lead to approximately normally distributed variates, i.e. a normal central limit effect rules. For an illustration of this, see for instance Eberlein and Keller (1995). In the turbulence context the effect is demonstrated by figure 2.

Finally in table 2 we refer, by 'quasi long range dependence', to the patterns of

  OLE E. BARNDORFF-NIELSEN

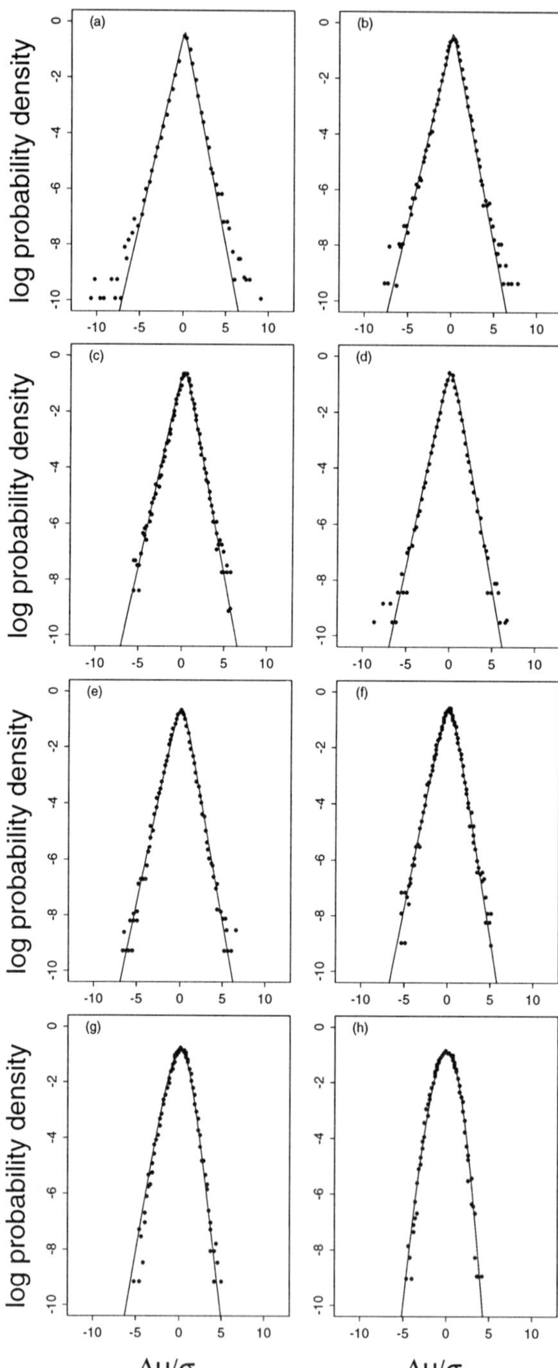

FIGURE 2. Probability densities of the velocity differences for various values of distance r, and fitted hyperbolic distributions. (a) $r = 1.38\ cm$ (b) $r = 4.14\ cm$ (c) $r = 9.67\ cm$ (d) $r = 20.7\ cm$ (e) $r = 42.8\ cm$ (f) $r = 87\ cm$ (g) $r = 7.06\ m$ (h) $r = 28.3\ m$.

autocorrelations so typically observed in time series from both of the two fields of investigation and which indicate dependence structures near to long range dependence. There is no single mathematical formulation of the latter concept but an often used definition, when the concern is with stationary series or processes, is to require that the autocorrelation function $r(u)$, say, behaves as $r(u) \sim L(u)u^{-2\bar{H}}$ where $\bar{H} \in (0, \frac{1}{2})$ is a constant and $L(u)$ is a slowly varying function. A behaviour close to this can be obtained by adding independent processes of the autoregressive type, and the resulting sum process provides an instance of what the term 'quasi long range dependence' is meant to indicate.

3. Parametric modelling

Some recent approaches to the modelling of turbulent wind fields and, separately, of the fluctuations of financial assets have build on self-decomposability in combination with normal variance-mean mixtures, autoregressions, and state space structures, see Barndorff-Nielsen, Jensen and Sørensen (1990,1993,1995) and Barndorff-Nielsen (1995,1996a,b). These approaches have aimed at incorporating the stylized features discussed in section 2, as far as possible.

Considering the discrete time case, suppose the data are from a time series x_t, $t \in \mathbb{Z}$. We wish the theoretical one-dimensional marginal distributions to have the characteristics discussed in section 2, in particular their tail behaviour should be log linear or somewhat heavier than that. Among the further desiderata, indicated in section 2, is tractable modelling of possible quasi long range dependence.

As a step towards these desiderata, let us consider the possibility of constructing a first order stationary autoregressive model

$$(3.1) \qquad x_t = \rho x_{t-1} + u_t$$

with i.i.d. innovations u_t. Expressing the characteristic function $\phi(s)$ of x_t by this relation shows that

$$(3.2) \qquad \phi(s) = \phi(\rho s)\phi_\rho(s)$$

where $\phi_\rho(s)$ denotes the characteristic function of u_t. If a decomposition of ϕ as in (3.2) is to be possible for all $\rho \in (0,1)$ then and only then is the distribution of x_t self-decomposable in the sense of Lévy. We are thus led to search for statistically and probabilistically tractable distributions that are self-decomposable and exhibit semiheavy tail behaviour.

Two such types are the hyperbolic distributions and the generalised logistic distributions and for applications of these to turbulence or finance we refer to Barndorff-Nielsen (1979,1986), Barndorff-Nielsen, Jensen and Sørensen (1989, 1990, 1993), Eberlein and Keller (1995) and Küchler, Neumann, Sørensen and Streller (1994).

Both the hyperbolic and the generalised logistic distributions have log linear tails and are therefore not capable of describing the the kind of tail behaviour seen in figure 1 and figures 2(a) and 2(b). Instead we may turn to the normal inverse Gaussian distribution (Barndorff-Nielsen, 1978, 1995, 1996a,b) which decreases, for $x \to \pm\infty$ as

$$const. \, |x|^{-3/2} \exp(-\sigma_\pm |x|) \, .$$

Probabilistically, this distribution may be defined via subordination of Brownian motion by the homogeneous inverse Gaussian Lévy process z_t. The transition probability density of the latter process for unit time differences is of the form

$$(3.3) \qquad f(z; \delta, \gamma) = (2\pi)^{-1/2} \delta e^{\delta\gamma} z^{-3/2} \exp\left\{ -\frac{1}{2}\left(\delta^2 z^{-1} + \gamma^2 z \right) \right\}$$

where $\delta > 0$ and $\gamma > 0$ are parameters. In other words, (3.3) is the density of z_1. The distribution determined by (3.3) is the *inverse Gaussian distribution* and is denoted $IG(\delta, \gamma)$. It equals the distribution of the first passage time to level δ of a Brownian motion with drift γ and diffusion coefficient 1. The Brownian motion b_t to be subordinated has drift β and diffusion coefficient 1, and we define the *normal inverse Gaussian Lévy process* x_t by

$$(3.4) \qquad x_t = b_{z_t} + \mu t$$

for arbitrary drift parameters β and μ. Let $\alpha = \sqrt{\beta^2 + \gamma^2}$. The *normal inverse Gaussian distribution* with parameters $\mu, \delta, \alpha, \beta$ is denoted by $NIG(\alpha, \beta, \mu, \delta)$ and is given as the distribution of x_1. The corresponding density is expressible as

$$(3.5) \qquad g(x; \alpha, \beta, \mu, \delta) = a(\alpha, \beta, \mu, \delta) q\left(\frac{x-\mu}{\delta} \right)^{-1} K_1\left\{ \delta\alpha q\left(\frac{x-\mu}{\delta} \right) \right\} \exp(\beta x)$$

where

$$(3.6) \qquad a(\alpha, \beta, \mu, \delta) = \pi^{-1} \alpha \exp\left(\delta\sqrt{\alpha^2 - \beta^2} - \beta\mu \right)$$

and

$$(3.7) \qquad q(x) = \sqrt{1 + x^2}$$

and where K_1 is the modified Bessel function of third order and index 1. The distribution is symmetric around μ provided $\beta = 0$. The normal distribution $N(\mu, \sigma^2)$ appears as a limiting case for $\beta = 0$, $\alpha \to \infty$ and $\delta/\alpha = \sigma^2$, and the Cauchy distribution is the limit case of $NIG(\alpha, 0, 0, 1)$ for $\alpha \to 0$. Further, by suitable choice of the parameters, the normal inverse Gaussian can be made to look essentially as any hyperbolic distribution (Barndorff-Nielsen, 1995) so that previous statistical results of applying the hyperbolic distributions can be incorporated in reanalysis based on the normal inverse Gaussian. Figure 3 shows the result of fitting the normal inverse Gaussian distribution to the data from figure 1. The estimated values of the parameters, as determined by maximum likelihood estimation, are $\hat{\alpha} = 0.90$, $\hat{\beta} = -0.09$, $\hat{\mu} = 0.08$, $\hat{\delta} = 0.94$. (The estimates and the figure were produced by means of a computer program developed by Preben Blæsild and Michael K. Sørensen; this program provides estimates and associated graphs for one-, two- and three-dimensional hyperbolic and normal inverse Gaussian distributions, as well as for some closely related distribution types).

In the financial context the use of process subordination was proposed by Clark (1973). He, however, did not consider the inverse Gaussian as subordinator but focused instead on the log normal distribution, which leads to less tractable results.

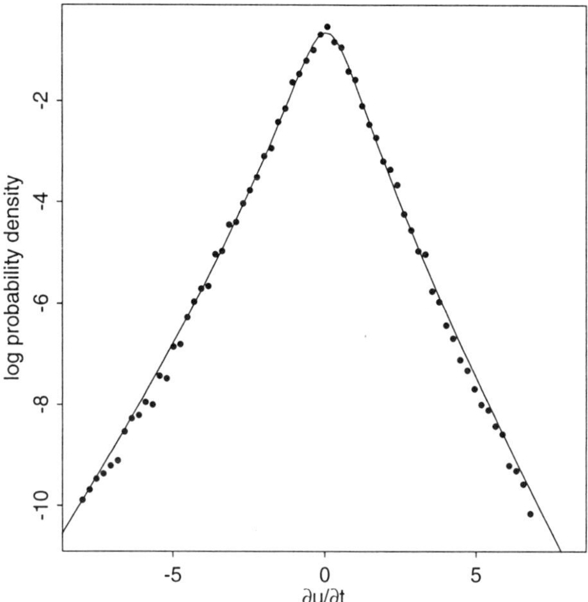

FIGURE 3. The normal inverse Gaussian distribution fitted to the turbulence data from figure 1.

Earlier Mandelbrot and Taylor (1967) had argued that stable laws should constitute natural models for logarithmic stock returns. This has plausibility both because of the very definition of stability and because the stable distributions can be represented as mixtures of normal distributions. Indeed, while initially a normal model would seem reasonable, due to a central limit effect, the fact that the intensity of trading - or, in other words, the number of trading events - varies randomly over time indicates that the variance of the normal should be considered as a random variable. Later work has shown that the assumption of stability is empirically untenable (cf., for instance, Eberlein and Keller, 1995) but other specifications in the form of mixing of normals are used commonly in stochastic volatility modelling, see in particular Shephard (1995). For an approach to pricing of assets under models of such types see Rachev and Ruschendorf (1994).

From this perspective it is relevant to note that the normal inverse Gaussian, the hyperbolic and the generalised logistic distributions are all normal variance-mean mixtures, i.e. they can be represented as the distribution of a random variable x for which there is an associated random variable z such that x given z follows a normal law with variance z and mean of the form $\mu + \beta z$ (cf. Barndorff-Nielsen, 1977; Barndorff-Nielsen, Kent and Sørensen, 1982).

Returning to the subordination setting, we have (Barndorff-Nielsen, 1996b) that the joint distribution of (z_1, x_1), which is infinitely divisible, has Lévy-Khintchine representation

$$(3.8) \qquad \log \phi(u,s) = i\mu s + \int_{-\infty}^{\infty} \int_{0}^{\infty} \left(e^{i(uz+sx)} - 1 \right) g(z,x;\alpha,\beta,\delta) dz dx ,$$

$\phi(u,s)$ denoting the characteristic function of (z_1, x_1) and where the Lévy density $g(z,x;\alpha,\beta,\delta)$ is given by

$$(3.9) \qquad g(z,x;\alpha,\beta,\delta) = (2\pi)^{-1} \delta z^{-2} \exp\left\{ -\frac{1}{2} \left(z^{-1} x^2 + \alpha^2 z - 2\beta x \right) \right\} .$$

Incidentally, the distribution of (z_1, x_1) and multivariate versions of that has been studied from a viewpoint very different from that taken here in Hassaïri (1992,1993) and Letac and Seshadri (1995).

It follows in particular from (3.8) together with a standard integral expression for the Bessel function K_1 that the $NIG(\alpha,\beta,\mu,\delta)$ distribution has Lévy-Khintchine representation

$$(3.10) \qquad \log \varphi(s) = \int_{|x| \geq 1} (e^{isx} - 1) f(x;\alpha,\beta,\delta) dx$$

$$+ \int_{|x| < 1} (e^{isx} - 1 - isx) f(x;\alpha,\beta,\delta) dx$$

$$+ is \left\{ \mu + 2\pi^{-1} \delta\alpha \int_{0}^{1} sinh(\beta x) K_1(\alpha x) dx \right\}$$

where

$$(3.11) \qquad f(x;\alpha,\beta,\delta) = \pi^{-1} \delta\alpha |x|^{-1} K_1(\alpha|x|) \exp(-\beta x) .$$

In addition to being normal variance-mean mixtures, the normal inverse Gaussian, the hyperbolic and the generalised logistic distributions are self-decomposable, as shown in Halgreen (1979) and Barndorff-Nielsen, Kent and Sørensen (1982). The self-decomposability of the normal inverse Gaussian distribution can also be seen directly from (3.11), using a wellknown property of the Bessel function K_1. Thus all three distribution types allow autoregressive type modelling in discrete time, as in (3.1).

In relation to the fact, mentioned earlier, that in turbulence the distributions of velocities and of first and second order derivatives of velocities tend to have very similar distributional shape the following curious property (Barndorff-Nielsen, Jensen and Sørensen, 1990) is noteworthy. With x_t as in (3.1) and k a positive integer, let $a^{[k]}(t)$ be the lag k difference process

$$a^{[k]}(t) = x_t - x_{t-k} .$$

Then, provided the distribution of x_t is self-decomposable and symmetric around 0, one has that the distribution of $a^{[k]}(t)$ is identical to that of x_t if the regression coefficient ρ happens to equal $2^{-1/k}$.

In continuous time the analogue of autoregressions of order 1 are the Ornstein–Uhlenbeck type processes. These have the form

$$(3.12) \qquad x_t = e^{-\lambda t} x_0 + \int_0^t e^{-\lambda(t-\tau)} dw_\tau$$

where $\lambda > 0$ and where w_t is a homogeneous Lévy process.

Now, let $c(\zeta)$ be the characteristic function of a self-decomposable distribution, denote the corresponding cumulant generating function by $\kappa(\zeta)$, i.e. $\kappa(\zeta) = \log c(\zeta)$, and suppose that $c(\zeta)$ is differentiable and $\zeta \kappa'(\zeta)$ is continuous at 0. Then, as shown in Barndorff-Nielsen, Jensen and Sørensen (1995), $\exp\{\zeta \kappa'(\zeta)\}$ is an infinitely divisible characteristic function and if we let x_0 have characteristic function $c(\zeta)$ and take w_t to be the homogeneous Lévy process for which w_1 has characteristic function $\exp\{\lambda \zeta \kappa'(\zeta)\}$, the process x_t determined by (3.12) is stationary and x_t has characteristic function $c(\zeta)$.

The conditions are satisfied, in particular, if $c(\zeta)$ is the characteristic function of the normal inverse Gaussian distribution. To understand the nature of the associated Ornstein–Uhlenbeck process, as well as for the purpose of further constructions, it is of interest to know the Lévy density of the background driving Lévy process w_t. Writing $q(y; \alpha, \beta, \delta)$ for this density it can be shown that

$$(3.13) \qquad q(y; \alpha, \beta, \delta) = (1 + \beta x) f(x; \alpha, \beta, \lambda \delta) + \pi^{-1} \lambda \delta \alpha^2 K_0(\alpha |x|) e^{-\beta x}$$

where K_0 is a Bessel function. For the derivation and some consequences of this formula see Barndorff-Nielsen (1996b).

Processses with normal inverse Gaussian marginals that exhibit (quasi) long range dependence can be constructed by summing of normal inverse Gaussian processes of Ornstein–Uhlenbeck type or by a more direct self-similarity approach, see Barndorff-Nielsen (1996b).

Some further, closely related, work concerning state space modelling, to capture dependence features of the conditional heteroscedasticity type, multivariate extensions, and continuous processes is discussed in Barndorff-Nielsen (1996a,b), Barndorff-Nielsen, Jensen and Sørensen (1995) and Rydberg (1996a,b,c).

References

[1] Andersen, P.K., Borgan, Ø., Gill, R.D. and Keiding, N. (1993): *Statistical Models based on Counting Processes.* Heidelberg: Springer-Verlag.

[2] Barndorff-Nielsen, O.E. (1977): Exponentially decreasing distributions for the logarithm of particle size. *Proc. R. Soc. London* **A 353**, 401-419.

[3] Barndorff-Nielsen, O.E. (1978): Hyperbolic distributions and distributions on hyperbolae. *Scand. J. Statist.* **5**, 151-157.

[4] Barndorff-Nielsen, O.E. (1979): Models for non-Gaussian variation; with applications to turbulence. *Proc. R. Soc. London* **A 368**, 501-520.

[5] Barndorff-Nielsen, O.E. (1986): Sand, wind and statistics. *Acta Mechanica* **64**, 1-18.

[6] Barndorff-Nielsen, O.E. (1993): Important areas for future statistical research. *Statist. Comp.* **3**, 181.

[7] Barndorff-Nielsen, O.E. (1995): Normal inverse Gaussian processes and the modelling of stock returns. Research Report 300. Dept. Theor. Statistics, Aarhus University.

[8] Barndorff-Nielsen, O.E. (1996a): Normal inverse Gaussian distributions and stochastic volatility modelling. *Scand. J. Statist.* (To appear.)

[9] Barndorff-Nielsen, O.E.(1996b): Processes of normal inverse Gaussian type. *Finance and Stochastics.* (To appear.)

[10] Barndorff-Nielsen, O.E., Jensen, J.L. and Sørensen, M. (1989): Wind shear and hyperbolic distributions. *Boundary-Layer Meteorology* **46**, 417-431.

[11] Barndorff-Nielsen, O.E., Jensen, J.L. and Sørensen, M. (1990): Parametric modelling of turbulence. *Phil. Trans. R. Soc. London* **A 332**, 439-455.

[12] Barndorff-Nielsen, O.E., Jensen, J.L. and Sørensen, M. (1993): A statistical model for the streamwise component of a turbulent wind field. *Ann. Geophysicae* **11**, 99-103.

[13] Barndorff-Nielsen, O.E., Jensen, J.L. and Sørensen, M. (1995): Some stationary processes in discrete and continuous time. Research Report 241. Dept. Theor. Statist., Aarhus University.

[14] Barndorff-Nielsen, O.E., Kent, J. and Sørensen, M. (1982): Normal variance-mean mixtures and the z-distributions. *Int. Statist. Review* **50**, 145-159.

[15] Clark, P.K. (1973): A subordinated stochastic process model with finite variance for speculative prices. *Econometrica* **41**, 135-159.

[16] Damien, P., Laud, P.W. and Smith, A.F.M. (1995): Approximate random variable generation from infinitely distributions with applications to Bayesian inference. *J. R. Statist. Soc. B* **57**, 547-563.

[17] Eberlein, E. and Keller, U. (1995): Hyperbolic distributions in finance. *Bernoulli* **1**, 281-299.

[18] Frisch, U. (1995): *Turbulence. The legacy of A.N. Kolmogorov.* Cambridge University Press.

[19] Halgreen, C. (1979): Self-decomposability of the generalized inverse Gaussian and hyperbolic distributions. *Z. Wahrscheinlichkeitstheorie verw. Gebiete* **47**, 13-17.

[20] Hassaïri, A. (1992): La classification des familles exponentielles naturelles sur R^d par l'action du groupe linéaire de R^{d+1}. *C. R. Acad. Sc. Paris* **315**, 207-210.

[21] Hassaïri, A. (1993): Les $(d+3)G$-orbites de la classe de Morris-Mora des familles exponentielles de R^d. *C. R. Acad. Sc. Paris* **317**, 887-890.

[22] Küchler, U., Neumann, K, Sørensen, M. and Streller, A. (1994): Stock returns and hyperbolic distributions. Discussion Paper No. 23, Sonderforschungsbereich 373, Humboldt Universität zu Berlin.

[23] Letac, G. and Seshadri, V. (1995): A random continued fraction in R^{d+1} with an inverse Gaussian distribution. *Bernoulli* **1**, 381-393.

[24] Mandelbrot, B. and Taylor, H.M. (1967): On the distribution of stock price differences. *Operations Research* **15**, 1057-1062.

[25] Rachev, S.T. and Ruschendorf, L. (1994): Models for option prices. *Theory Probab. Appl.* **39**, 120-152.

[26] Ripley, B.D. (1988): *Statistical Inference for Spatial Processes.* Cambridge University Press.

[27] Rydberg, T.H. (1996a): The normal inverse Gaussian Lévy process: approximation and simulation. Revised version to appear in *Stochastic Models*.

[28] Rydberg, T.H. (1996b): Generalized hyperbolic diffusions with applications towards finance. Submitted to *Math. Finance.*

[29] Rydberg, T.H. (1996c): Existence of unique equivalent martingale measures in a Markovian setting. *Finance and Stochastics.* (To appear.)

[30] Shephard, N. (1995): Statistical aspects of ARCH and stochastic volatility. In D.R. Cox, D.V. Hinkley and O.E. Barndorff-Nielsen (Eds.): *Time Series Models. In econometrics, finance and other fields.* London: Chapman and Hall. Pp. 1-67.

[31] Van Atta, C.W. and Park, J. (1972): Statistical self-similarity and inertial subrange turbulence. In M. Rosenblatt and C. W. van Atta (Eds.): *Statistical Models and Turbulence.* Lecture Notes in Physics 12. Springer-Verlag: New York. Pp. 402-426.

[32] Winkler, G. (1995): *Image Analysis, Random Fields and Dynamic Monte Carlo Methods. A mathematical introduction.* Berlin: Springer-Verlag.

[33] Wyngaard, J.C. and Tennekes, H. (1970): Measurements of the small scale structure of turbulence at moderate Reynolds numbers. *Phys. Fluids* **13**, 1962-1969.

SOME THOUGHTS ON EXTREME VALUES

J. BEIRLANT,[*] *Katholieke Universiteit Leuven*

J.L. TEUGELS,[*] *Katholieke Universiteit Leuven*

P. VYNCKIER,[*] *Katholieke Universiteit Leuven*

Abstract

Extreme value theory is becoming a vital ingredient of current day probability and statistics. After dealing quickly with some of the crucial aspects of extreme value analysis, we scan the literature for open problems; they abound both in the theory and in the applications.

EXTREME VALUE THEORY;TAIL INDEX;WIND SPEED DATA

AMS 1991 SUBJECT CLASSIFICATION: PRIMARY 62G30

SECONDARY 62P

1. Introduction

In statistics curricula the major emphasis is righteously concentrated on what one could call *central limit thinking*. The overwhelming role of the normal distribution can be appreciated in all basic and applied courses in probability and statistics.

There are however cases where the information contained in a set of data can not be restricted to the information contained in the central terms. The extreme values in the sample often provide additional if not substantial insight in the data. If this happens to be the case, then it can hardly be denied that the extremes deserve special attention and that central limit thinking should be accompanied by *extreme value thinking*. For an earlier account on some of the issues involved, see [22]. A recent volume which reflects the diversity of applications involved is [21].

This discussion is not meant to cover the entire field of extreme value theory which

---

[*] Postal address: Celestijnenlaan 200B, B-3001 Heverlee, Belgium.

by now has been recognized as an important part of probability and statistics. We rather consider it as a tutorial: we start out with a number of relevant examples. We give an indication of the probabilistic modeling and state the most important statistical issues. We then illustrate these findings on a specific example. At the end we outline a series of problems and issues that seem important possibilities for concrete work in the future.

We will stick to the i.i.d. case. This is basically due to space limitations. In the last section we will hint at generalizations that have been partly tackled in the literature or that will prove to be of vital importance in the future. As a slightly overstated paradigm one could say that whenever central limit theory makes sense, extreme value theory adds something, perhaps even something new, to the insight in the data.

2. Examples

We give several illustrations where extreme value thinking is crucial.

Example 1: *Meteorological data* generally have no alarming aspect as long as they are situated in a narrow band around the average. The situation however changes dramatically when a river level overshoots a certain safety threshold and causes floods; also when the level drops down too much, resulting in river drought.

Another illustration is provided by *wind speed data* where mainly high speeds are to be considered. It is somewhat surprising that building codes in neighboring countries can be based on a different mathematical model for maximal wind speeds: in the US one uses the Fréchet distribution while in Canada one advocates the Gumbel distribution. Tail behaviour and quantiles differ considerably between these two models.

In the same vein Dekkers et al. in [16] have investigated height requirements for *sea dikes*. It is clear that central limit thinking hardly contributes essential features to this kind of problem.

Example 2: *Exposure data* have an important central limit aspect. Accumulation of dangerous radiation can cause fatal side effects. Nevertheless the effect of radiation changes dramatically once a certain threshold is reached. Similar aspects apply to ozone measurements in large cities where too high an ozone level causes a multitude of inconveniences to the inhabitants. An application to trend detection in ground-level ozone can be found in Smith [46].

Environmetrics is a scientific area where the role of extremes can hardly be underestimated. Pollution levels for rivers do not give rise to specific actions as long as they are considered to be about average. Extreme value theory comes into play as soon as the pollution level hits a too high value. Many similar examples can be found when one is interested in the values of environmental indicators.

Example 3: One of the prime applications of extreme value thinking can be found in *non-life insurance*. Some portfolios seem to have a tendency to occasionaly include a large claim that jeopardizes the solvency of a portfolio if not of a substantial part of the company. Apart from major accidents as earthquakes, hurricanes, airplane crashes, etc. there are a vast number of occasions where large claim occur. See for example [3].

Once in awhile automobile insurance leads to excessive claims. More often fire portfolios encounter large claims. Industrial fires, especially, cause a lot of side effects in loss of property, temporary unemployment and lost contracts.

A special mention should be given to lawsuits resulting from medical malpractice that often lead to excessive claims from the insured.

3. Notation

As we restrict ourselves to the case of i.i.d. random variables the notational complexity can be kept to a minimum.

Let $X_1, X_2, \ldots, X_n$ be a sequence of independent random variables with a common distribution $F(x) = P\{X \le x\}$ which is concentrated on an interval $-\infty \le x_- \le x_+ \le \infty$ where x_- (x_+) is the left (right) endpoint of the support of F.

With the above sequence we associate the sequence of *order statistics*

$$X_1^* \le X_2^* \le \ldots \le X_n^*$$

that will be crucial in what follows. For simplicity, we will focus our attention on the right tail behaviour of F and the corresponding maximum X_n^*.

It turns out the majority of properties of the extremes in the sample is controlled by the tail behaviour of F or equivalently by that of its *quantile function* which is defined by $Q(p) = F^{\leftarrow}(p) = \inf\{y : F(y) \geq p\}$ for $p \in (0,1)$. We will make constant use of the *tail (quantile) function* defined by $U(x) = Q(1 - \frac{1}{x})$ for $x > 0$.

4. Maxima

In [4] we gave a quick procedure on how to arrive at the solution of the *domain of attraction problem* for the maximum, which is the *central* result underlying extreme value methods. By this we mean:

(i) the derivation of all possible limit laws for $a_n^{-1}(X_n^* - b_n)$ where $\{a_n\}$ are positive norming constants while $\{b_n\}$ are real centering constants,

and

(ii) the determination of the conditions on F that lead to one of the above limits.

The solution to the first problem is that the possible limiting distributions are given by the one-parameter family of *extreme value distributions* or *extremal laws*

$$G_\gamma(y) = P\{Y \leq y\} = \exp{-(1 + \gamma y)^{-\frac{1}{\gamma}}}, \quad \gamma \in \mathbf{R}$$

where γ is called the *extreme value index*. The solution to the second problem depends on the value of this index: a distribution F belongs to the *domain of attraction* of the limit distribution G_γ (often written in the form $F \in \mathcal{D}(G_\gamma)$) if and only if

$$(1) \qquad \{U(xu) - U(x)\}/g(x) \to h_\gamma(u) = \int_1^u v^{\gamma-1}\, dv$$

for all $u \geq 1$ as $x \to x_+$, for some *auxiliary function* $g > 0$. In particular one can take $a_n = g(n)$. The explicit form on the right of the above equation is a consequence of the existence of the limit. See [9], p.139.

- If $\gamma > 0$, then we find the *Fréchet case* which is typically a very heavy tailed distribution. The prime example that actually characterizes $\mathcal{D}(G_\gamma)$ with $\gamma > 0$

is offered by the *Pareto-type distributions* where $1 - F(x) \sim x^{-\frac{1}{\gamma}} \ell(x)$ and ℓ is a slowly-varying function.

- When $\gamma < 0$ then we obtain the *Weibull (extremal) case* which appears typically with distributions for which $x_+ < \infty$ and hence can be interpreted as light-tailed. Beta distributions are among the most popular members of the corresponding domain of attraction.

- The intermediate case $\gamma = 0$ is the most popular and is called the *Gumbel case*. The domain $\mathcal{D}(G_0)$ contains inter alia the lognormal distribution, the gamma distribution and the Weibull distribution $1 - F(x) = \exp -cx^\tau$ with $\tau > 0$.

5. Statistical Aspects

The determination of the extreme value index is of course of vital importance. Basically two approaches are available in literature for this problem.

First, a threshold approach has been proposed in which all observations which exceed a high threshold u are modelled parametrically. As shown by Pickands in [38], consistency is achieved within the model (1) by specifying the conditional distribution of threshold exceedances to follow the Generalized Pareto Distribution. Smith [44] has discussed maximum likelihood estimation in this setting. See also [45].

Second, several authors have advocated a nonparametric approach in which the corresponding estimators of γ are based on nonparametric estimations of quantities involved in (1). We outline this setup in somewhat more detail.

If there is prior knowledge about the positivity of γ then a variety of estimation procedures may be invoked. Under the Pareto-type assumption it is easy to show that

$$\lim_{x \uparrow \infty} \frac{\int_x^\infty \frac{1-F(y)}{y}\, dy}{1 - F(x)} = \gamma \, .$$

Hence we can get an estimator for γ by replacing F by its empirical version and x by a large order statistic. This leads to

$$H_{k,n} := \frac{1}{k} \sum_{i=1}^{k} \log X_{n-i+1}^* - \log X_{n-k}^*,$$

which was proposed by Hill [28]. This estimator together with generalizations like the kernel versions treated by Csörgő et al. [14] are well covered in [7].

Turning to the general case, the choice for appropriate estimators gets much slimmer. The current most popular are an estimator by Pickands [39]

$$\frac{1}{\log 2} \log \frac{X^*_{n-k+1} - X^*_{n-2k+1}}{X^*_{n-2k+1} - X^*_{n-4k+1}}$$

and the *moment estimator* introduced by Dekkers et al. [16]

$$H_{k,n} + 1 - \frac{1}{2}\left(1 - \frac{(H_{k,n})^2}{H^{(2)}_{k,n}}\right)^{-1}$$

where in turn $H^{(m)}_{k,n} := \frac{1}{k}\sum_{i=1}^{k}\{\log X^*_{n-i+1} - \log X^*_{n-k}\}^m, m > 1$. Drees [17] proposed a refined Pickands estimator.

In [5] we constructed an alternative that has been defined by

$$(2)\qquad \frac{1}{k}\sum_{i=1}^{k}\log UH_{i,n} - \log UH_{k+1,n} = \tilde{H}_{k,n} + \frac{1}{k}\sum_{j=1}^{k}\log H_{j,n} - \log H_{k+1,n}$$

where $UH_{i,n} = X^*_{n-i}H_{i,n}$, and $\tilde{H}_{k,n}$ is the Hill estimator after deletion of the largest observation.

One crucial matter in the above estimators is the choice of the free quantity k. In general k should be small enough to avoid large bias in the estimation. On the other hand k should be large enough to avoid large variances. In [5] and [6] we have settled for a choice of k that minimizes a nonparametric estimator of the mean squared error. The resulting algorithms and illustrations of the method are amply explained in [7], sections 2.4 and 3.2. Other solutions to this problem are presented in [15] and [18].

Another statistical issue is the determination of sharp confidence intervals. Correspondingly of vital importance is of course whether we can safely test an hypothesis of the form $\gamma = \gamma_0$ for a fixed γ_0. This problem is not yet fully solved but there are conservative alternatives available. See again [7], section 2.5, for a survey of the existing material.

The ultimate goal of extreme value analysis often is to estimate high quantiles, small exceedance probabilities and endpoints of distributions. Here we refer to [16], [26], and [7] sections 2.6, 3.3 and 4.3.

6. An Example

To illustrate the above procedures we have looked into *wind speed data* that have been provided to us by *NIST* in Gaithersburg, USA. In [42] Simiu et al. have reported on extreme wind speeds in the contiguous United States. Out of the 129 airport stations, 49 provided daily fastest-mile speeds over a period somewhere between 1965 and 1992. The measurements have been made (or calibrated if necessary) by anemometers 10m above ground. Wind speeds from hurricanes and tornadoes have not been incorporated.

We have calculated the extreme value indices for the 49 stations using the estimator (2). If the estimated γ is positive at a certain location then very high wind speeds can be expected there. In our illustration we have drawn an open circle with center at the location and with radius proportional to the estimated value of γ. If the estimated value is negative we similarly have drawn a shaded circle around the location. On some sites we could not really make a decision on whether γ was positive or negative. We have chosen to put small stars at the latter locations.

The figure illustrates the results of our analysis. The geographical layout of the US shows clearly. Unfortunately we do not have measurements in the wide region of the Gulf of Mexico nor at the westcoast. Also the region around the Great Lakes deserved rather more measurements. We should point out that the illustration is really static as no time component has been brought into the analysis.

A Bayesian approach to model extreme wind speeds can be found in [11].

7. Further Theoretical Issues

As we have indicated above many problems remain, even for the case of i.i.d. random variables. Here is an incomplete list of some of these theoretical issues.

- If we drop the assumption of independence or/and of identical distribution then of course a wide variety of new problems emerges. It has been shown in [33] that the extremal laws are rather robust in the sense that under mild regularity conditions, the same laws appear for the case of weakly dependent

EXTREMAL INDEX

random variables. Nevertheless the statistical determination of the analogue of the extreme value index has only been started recently. See [29].

- The incorporation of a further *time* or/and *spatial component* would of course make the above more applicable to data dealing with exposures to health hazards, rain fall (as treated for example in [48]), hurricanes, floods, volcanic eruptions among others. Recent work in this area can be found in [13] and [47]. For a theoretical account on results involving measures of threshold exceedance by high values of stationary stochastic sequences see [32].

- Let us mention that in the above examples we should be aware that *extremal behaviour* may not be synonymous with *catastrophical behaviour*. For the latter the number of measurements is (fortunately) very small which poses the statistician new and hard problems. Perhaps in some cases one might be able to expand the meager data set by proper inclusion and calibration of data referring to other time scales and/or geographical areas.

- The number of scientific papers on *multivariate extremes* is currently restricted to a few dozen. Here we refer to [24], [49], [50], [12], [31], [25], and [19]. The problem of estimating multivariate exceedance probabilities was initiated in [23]. It is quite clear that - given an acceptable interpretation to the meaning of *multivariate extreme* - problems to be tackled abound.

8. Further Practical Issues

Let us mention that there is a vast set of other practical problems where *extremal thinking* should be applied, jointly with central limit thinking. Here are a few of them.

- There are areas where observed bivariate values show an apparant simple relationship after appropriate *scaling*. As a result there often appears a single quantity, called the *scaling constant* for simplicity, that seems to capture the most important feature of the data.

In many applications the determination of this scaling constant is crucial. In particular the appearance of *long-range dependency* seems to be a recurrent theme in all applications. Here is an incomplete set of examples.

- The prime example is probably the *fractal dimension*. There exists a wealth of different definitions such as Hausdorff dimension, correlation dimension, box counting dimension, capacity dimension, etc. Depending on its definition the value of the fractal dimension will differ. However in the gaussian case all definitions give probabilistically the same value. See [41], [20]. A Hill-type estimator for the correlation exponent of a stationary ergodic sequence is proposed in [37].

- Very much related is the concept of the *self-similarity* or *Hurst index* that appears in stable, not necessarily gaussian, processes. This example is not only significant for the theoretical aspects but also for its increasing list of statistical studies. See [8].

- This item picks up an idea from previous sections. In insurance the *Pareto-index* plays the role of indicator on how dangerous a certain risk is to the insurer. More generally the *extremal index* from extreme value theory has proved to be a decisive quantity in assessing the global behaviour of the underlying stochastic model.

- A related but seemingly different quantity is *Ling's scaling index* that tries to pin down a universal relationship between a line transect approximation of a planar curve and the measuring unit. This empirical law has found applications in the study of rough surfaces. See [34].

- The *elasticity modulus* is used in economic models to forecast the length of business-cycles and medium-term growth. This area provides another example of models where long-range dependency is an essential feature of the process.

- Among the recent applications of these scaling laws we mention flood analysis, earthquake prediction, materials science and all applications of time series with long-range dependencies. See [41], [51].

- A theory for extremes in chaotic dynamics was started in [2].

- As remarked by Riley in [40], one of the key Black-Scholes assumptions is that price movements in financial markets follow the same kind of probability distribution that applies to natural phenomena, i.e. the bell-shaped lognormal distribution. In practice however this is not so. In particular, empirical distributions exhibit far more frequent extreme outcomes of *fat tails*. This means that in financial mathematics it is time to introduce concepts like heavy-tailed distributions, long-range dependency and extreme value behaviour. This whole area of course opens up a wide variety of unexpected challenges to the probabilistic model builder. Some available references are [27], [30], [35] and [36].

- In the area of genetic algorithms one only seems to use random variables with normal-like behaviour. Nevertheless the performance of these algorithms often proves to be controlled by extremal rather than central values. See [1].

- Digging for precious stones is basically searching for extremes. In recent work we have been successful in applying extreme value thinking to come up with results that are pleasing to the geologists. See [10]. More concrete problems have been posed and await professional tackling.

Acknowledgement

The authors take great pleasure to thank E. Simiu for providing the data used in our main example.

References

[1] ADLER, R., FELDMAN, R. & TAQQU, M. (EDS.), *A Practical Guide to Heavy Tailed Data*, Birkhäuser, Boston, MA.

[2] BALAKRISHNAN, V., NICOLIS, C. & NICOLIS, G. (1995) Extreme value distributions in chaotic dynamics, *J. Statist. Physics*, **80**, 307–336.

[3] BEIRLANT, J., TEUGELS, J.L. & VYNCKIER, P. (1993) Extremes in non-life insurance. In *Extremes Value Theory and Applications*, J. Galambos, J. Lechner, and E. Simiu Eds., 489–510, Kluwer, Dordrecht.

[4] BEIRLANT, J. & TEUGELS, J.L. (1996) A simple approach to classical extreme value theory. In *Exploring Stochastic Laws, Festschrift for Academician V.S. Korolyuk*, 457–468, VSP, Zeist.

[5] BEIRLANT, J., VYNCKIER, P. & TEUGELS, J.L. (1996) Excess functions and estimation of the extreme-value index, *Bernoulli*, **2**, 293–318.

[6] BEIRLANT, J., VYNCKIER, P. & TEUGELS, J.L. (1996) Tail index estimation, Pareto quantile plots and regression diagnostics, *J. Amer. Statist. Assoc.*, **91**, 1659–1667.

[7] BEIRLANT, J., TEUGELS, J.L. & VYNCKIER, P. (1996) *Practical Analysis of Extreme Values*, Leuven University Press, Leuven.

[8] BERAN, J. (1992) Statistical methods for data with long-range dependency, *Statist. Science*, **7**, 404–427.

[9] BINGHAM, N.H., GOLDIE C.M. & TEUGELS, J.L. (1987) *Regular Variation*, Cambridge University Press, Cambridge.

[10] CAERS, J., VYNCKIER, P., BEIRLANT, J. & ROMBOUTS, L. (1996) Extreme value analysis of diamond size distributions, *Math. Geology*, **28**, 25–43.

[11] COLES, S.G. AND POWELL, E.A. (1996) Bayesian methods in extreme value modelling: a review and new developments, *Int. Statist. Rev.*, **64**, 119–136.

[12] COLES, S.G. AND TAWN, J.A. (1991) Modelling extreme multivariate events, *J. R. Statist. Soc. Ser. B.*, **53**, 377–392.

[13] COLES, S.G. AND TAWN, J.A. (1996) Modelling extremes of the areal rainfall process, *J. R. Statist. Soc. Ser. B*, **58**.

[14] CSÖRGŐ, S., DEHEUVELS, P. & MASON, D. (1985) Kernel estimates of the tail index of a distribution, *Ann. Statist.*, **13**, 1050–1077.

[15] DANIELSSON, J., DE HAAN, L., PENG, L. & DE VRIES, C.G. (1997) Using a bootstrap method to choose the sample fraction in tail index estimation. Preprint, Erasmus University Rotterdam.

[16] DEKKERS, A.L.M., EINMAHL, J.H.J. & DE HAAN, L. (1989) A moment estimator for the index of an extreme-value distribution, *Ann. Statist.*, **17**, 1833–1855.

[17] DREES, H. (1995) Refined Pickands estimators of the extreme value index, *Ann. Statist.*, **23**, 2059–2080.

[18] DREES, H. & KAUFMANN, E. (1997) Selecting the optimal sample fraction in univariate extreme value estimation. Preprint, University of Cologne.

[19] EINMAHL, J.H., DE HAAN, L. & HUANG, X. (1993). Estimating a multidimensional extreme value distribution, *J. Multivar. Anal.*, **47**, 35–47.

[20] FEUERVERGER, A., HALL, P. & WOOD, A.T.A. (1994) Estimation of fractal index and fractal dimension of a Gaussian process by counting the number of level crossings, *J. Time Series Anal.*, **15**, 587–606.

[21] GALAMBOS, J., LECHNER, J. & SIMIU, E. (1994). *Extreme Value Theory and Applications*, Kluwer, Dordrecht.

[22] GUMBEL, E.J. (1958) *Statistics of Extremes*, Columbia University Press, New York.

[23] DE HAAN, L. (1994) Estimating exceedence probabilities in higher-dimensional space, *Commun. Statist. - Stochastic Models*, **10**, 765–780.

[24] DE HAAN, L. & RESNICK, I.S. (1977) Limit theory for multivariate sample extremes, *Z. Wahrscheinlichkeitsth.*, **40**, 317–337.

[25] DE HAAN, L. & RESNICK, I.S. (1993) Estimating the limit distribution of multivariate extremes, *Commun. Statist. - Stochastic Models*, **9**, 275–309.

[26] DE HAAN, L. & ROOTZEN, H. (1992) On the estimation of high quantiles, *J. Statist. Plann. Inf.*, **35**, 1–13.

[27] DE HAAN, L., RESNICK, I.S., ROOTZEN, H. & DE VRIES, C.G. (1989) Extremal behaviour of solutions to a stochastic difference equation with application to ARCH process, *Stoch. Proc. Appl.*, **32**, 213–224.

[28] HILL, B.M. (1975) A simple general approach to inference about the tail of a distribution, *Ann. Statist.*, **3**, 1163–1174.

[29] HSING, T.L. (1991) Tail index estimation using dependent data, *Ann. Statist.*, **19**, 1547–1569.

[30] JANSEN, D.W. & DE VRIES, C.G. (1991) On the frequency of large stock returns: Putting booms and busts into perspective, *Review of Economics and Statistics*, **73**, 18–24.

[31] JOE, B., SMITH, R.L. & WEISSMAN, I. (1992) Bivariate threshold methods for extremes, *J. R. Statist. Soc. Ser. B*, **54**, 171–183.

[32] LEADBETTER, M.R. (1995) On high level exceedance modelling and tail inference, *J. Statist. Plann. Inf.*, **45**, 247–260.

[33] LEADBETTER, M.R., LINDGREN, G. & ROOTZÉN, H. (1982) *Extremal and Related Properties of Stationary Processes*, Springer-Verlag, New York.

[34] LING, F.F. (1990) Fractals, engineering surfaces and tribology, *Wear*, **136**, 141–156.

[35] LONGIN, F.M. (1996) The asymptotic distribution of extreme stock market returns, *Journal of Business*, **69**, 383–408.

[36] McCULLOCH, J.H. (1996) Financial applications of stable distributions, *Statistical Methods in Finance, Handbook of Statistics*, **14**, Maddala, G.S. and Rao, C.R (eds.) North-Holland, NY.

[37] MIKOSCH, T. & WANG, Q. (1995) A Monte Carlo method for estimating the correlation exponent, *J. Statist. Physics*, **78**, 799–813.

[38] PICKANDS III, J. (1971) The two-dimensional Poisson process and extremal processes, *J. Appl. Prob.*, **8**, 745–756.

[39] PICKANDS III, J. (1975) Statistical inference using extreme order statistics, *Ann. Statist.*, **3**, 119–131.

[40] RILEY, B. (1996) When rocket scientists crash out of orbit, *Financial Times*, April 17, p.17.

[41] SAMORODNITSKY, G. & TAQQU, M.S. (1994) *Stable non-Gaussian Random Processes: Stochastic Models with Infinite Variance*, Chapman & Hall, New York.

[42] SIMIU, E., CHANGERY, M.J. & FILLIBEN, J.J. (1979) *Extreme Wind Speeds at 129 Stations in the Contiguous United States*, NBS Building Science Ser. 118, National Bureau of Standards, Washington, DC.

[43] SMITH, R.L. (1984) Threshold methods for sample extremes. In *Statistical Extremes and Applications*, (ed. Tiago de Oliveira), 621–638, Reidel, Dordrecht.

[44] SMITH, R.L. (1985) Maximum likelihood estimation in a class of non-regular cases, *Biometrika*, **72**, 67–90.

[45] SMITH, R.L. (1987) Estimating tails of probability distributions, *Ann. Statist.*, **15**, 1174–1207.

[46] SMITH, R.L. (1989) Extreme value analysis of environmental time series: an application to trend detection in ground-level ozone, *Statist. Science*, **4**, 367–393.

[47] SMITH, R.L. (1995) Regional estimation from spatially dependent data. Preprint, University of North Carolina.

[48] SMITH, R.L., TAWN, J. & YUEN, H.K. (1990) Statistics of multivariate extremes, *Int. Statist. Rev.*, **58**, 47–58.

[49] TAWN, J.A. (1988) Bivariate extreme value theory: models and estimation, *Biometrika*, **75**, 397–415.

[50] TAWN, J.A. (1990) Modelling multivariate extreme value distributions, *Biometrika*, **77**, 245–253.

[51] TURCOTTE, D.L. (1994) Fractal theory and the estimation of extreme floods, *J. Res. Nat. Inst. Stand. Technol.*, **99**, 377–389.

STOCHASTIC POSITIVE FLOWS AND QUANTUM FILTERING EQUATIONS

V. P. BELAVKIN,* *University of Nottingham*

Abstract

Quantum stochastic completely positive flows and Radon-Nikodym derivatives in Fock scale are defined. A characterization of the unbounded stochastic generators of completely positive flows is given and the minimal solutions to the quantum stochastic evolution equations are constructed. This suggests the general form of classical as well as quantum stochastic filtering evolutions with respect to the Wiener (diffusion), Poisson (jumps), or general quantum noise.

COMPLETELY POSITIVE FLOWS, QUANTUM STOCHASTIC EQUATIONS, QUANTUM FILTERING DYNAMICS

Introduction.

The quantum stochastic filtering theory, announced in [1, 2] and developed then in [3], provides the derivations for new types of irreversible stochastic quantum dynamics, some particular types of which have been discovered recently in the phenomenological theories of quantum permanent reduction [4, 5], continuous measurement collapse [6, 7], spontaneous jumps [8, 9], diffusions and localizations [10, 11]. The main feature of such dynamics is their Markovianity such that the irreversible evolution can be described in terms of a linear stochastic differential operator equation, the solution to which is normalized only in the mean square sense.

In quantum theory of open systems there is a well known Lindblad's form [12] of quantum Markovian master equation, satisfied by the one-parameter semigroup of completely positive (CP) maps. This is non-stochastic equation obtained by averaging stochastic Langevin equation for quantum diffusion [13] over the driving quantum noises. On the other hand the Langevin equation is satisfied by a quantum stochastic process of dynamical representations, which are obviously completely positive due to *-multiplicativity of the representations. The representations give the examples of pure CP maps, but among pure CP maps there are not only the representations. This means a possibility to construct a pure irreversible quantum stochastic CP dynamics, which can not be driven by a Langevin equation.

This paper is devoted to the mathematical derivation of the general structure for the quantum stochastic CP flows and their generators. Here we would like to outline this structure on the formal differential level in terms of the stochastic differentials,

---

* Postal address: Mathematics Department, University of Nottingham, NG7 2RD, UK.

generating an Itô $*$–algebra

(0.1)
$$\mathrm{d}\Lambda\left(a\right)^{*}\mathrm{d}\Lambda\left(a\right)=\mathrm{d}\Lambda\left(a^{\star}a\right),\quad \sum\lambda_{i}\mathrm{d}\Lambda\left(a_{i}\right)=\mathrm{d}\Lambda\left(\sum\lambda_{i}a_{i}\right),\quad \mathrm{d}\Lambda\left(a\right)^{*}=\mathrm{d}\Lambda\left(a^{\star}\right)$$

with given mean values $\langle\mathrm{d}\Lambda\left(t,a\right)\rangle=l\left(a\right)\mathrm{d}t$, $a\in\mathfrak{a}$. Here $\mathfrak{a}$ is in general a noncommutative $*$-algebra with a self-adjoint annihilator (death) $d=d^{\star}$, $\mathfrak{a}d=0$, corresponding to $\mathrm{d}t=\mathrm{d}\Lambda\left(t,d\right)$, and $l:\mathfrak{a}\to\mathbb{C}$ is a positive $l\left(a^{\star}a\right)\geq 0$ linear functional, normalized as $l\left(d\right)=1$, corresponding to the determinism $\langle\mathrm{d}t\rangle=\mathrm{d}t$. The functional l defines the GNS representation $a\mapsto\boldsymbol{a}=(a_{\nu}^{\mu})_{\nu=+,\bullet}^{\mu=-,\bullet}$ of $\mathfrak{a}$ in terms of the quadruples

(0.2)
$$a_{\bullet}^{\bullet}=j\left(a\right),\quad a_{+}^{\bullet}=k\left(a\right),\quad a_{\bullet}^{-}=k^{*}\left(a\right),\quad a_{+}^{-}=l\left(a\right),$$

where $j\left(a^{\star}a\right)=j\left(a\right)^{*}j\left(a\right)$ is the operator representation $j\left(a\right)^{*}k\left(a\right)=k\left(a^{\star}a\right)$ on the pre-Hilbert space $\mathcal{K}$ of the Kolmogorov decomposition $l\left(a^{\star}a\right)=k\left(a\right)^{*}k\left(a\right)$, and $k^{*}\left(a\right)=k\left(a^{\star}\right)^{*}$.

As was proved in [14], the quantum stochastic processes $t\in\mathbb{R}_{+}\mapsto\Lambda\left(t,a\right),a\in\mathfrak{a}$ with independent increments $\mathrm{d}\Lambda\left(t,a\right)=\Lambda\left(t+\mathrm{d}t,a\right)-\Lambda\left(t,a\right)$, forming an Itô $\star$-algebra, can be represented in the Fock space $\mathfrak{F}$ over the space of $\mathcal{K}$-valued square-integrable functions on $\mathbb{R}_{+}$ as $\Lambda_{\nu}^{\mu}\left(t,a_{\nu}^{\mu}\right)=a_{\nu}^{\mu}\Lambda_{\mu}^{\nu}\left(t\right)$. Here

(0.3)
$$a_{\nu}^{\mu}\Lambda_{\mu}^{\nu}\left(t\right)=a_{\bullet}^{\bullet}\Lambda_{\bullet}^{\bullet}\left(t\right)+a_{+}^{\bullet}\Lambda_{\bullet}^{+}\left(t\right)+a_{\bullet}^{-}\Lambda_{-}^{\bullet}\left(t\right)+a_{+}^{-}\Lambda_{-}^{+}\left(t\right),$$

is the canonical decomposition of Λ into the exchange $\Lambda_{\bullet}^{\bullet}$, creation $\Lambda_{\bullet}^{+}$, annihilation $\Lambda_{-}^{\bullet}$ and preservation (time) $\Lambda_{-}^{+}=tI$ processes of quantum stochastic calculus [15], [16] having the mean values $\langle\Lambda_{\mu}^{\nu}\left(t\right)\rangle=t\delta_{+}^{\nu}\delta_{\mu}^{-}$ with respect to the vacuum state in $\mathfrak{F}$. Thus the parametrising algebra $\mathfrak{a}$ can be always identified as in the finite-dimensional case [14] with a $\star$-subalgebra of the algebra $\mathcal{Q}\left(\mathcal{K}\right)$ of all quadruples $\boldsymbol{a}=(a_{\nu}^{\mu})_{\nu=+,\bullet}^{\mu=-,\bullet}$, where $a_{\nu}^{\mu}:\mathcal{K}_{\nu}\to\mathcal{K}_{\mu}$ are the linear operators on $\mathcal{K}_{\bullet}=\mathcal{K},\mathcal{K}_{+}=\mathbb{C}=\mathcal{K}_{-}$, having the adjoints $a_{\nu}^{\mu*}\mathcal{K}_{\mu}\subseteq\mathcal{K}_{\nu}$, with the Hudson–Parthasarathy (HP) multiplication table [17]

(0.4)
$$a\bullet b=(a_{\bullet}^{\mu}b_{\nu}^{\bullet})_{\nu=\bullet,+}^{\mu=\bullet,-},$$

the death $\boldsymbol{d}=\left(\delta_{-}^{\mu}\delta_{\nu}^{+}\right)_{\nu=\bullet,+}^{\mu=-,\bullet}$, and the involution $a_{-\nu}^{\star\mu}=a_{-\mu}^{\nu*}$, where $-(-)=+$, $-\bullet=\bullet$, $-(+)=-$.

The main result of this paper is the construction of the CP flows for the linear unbounded generators $\lambda_{\nu}^{\mu}:\mathcal{B}\to\mathcal{B}$ of the linear quantum stochastic CP evolutions ϕ_{t} over $\mathcal{B}=\mathcal{L}\left(\mathcal{H}\right)$ in terms of the quantum stochastic differentials $\mathrm{d}\phi=\phi\circ\lambda_{\nu}^{\mu}\mathrm{d}\Lambda_{\mu}^{\nu}$ with $\phi_{0}=\imath$ at $t=0$, where $\imath\left(B\right)=B$ is the identical representation of $\mathcal{B}$. As in the bounded case and finite dimensional Itô algebra [19] this can be written in the "Lindblad" form $\boldsymbol{\lambda}\left(B\right)=\boldsymbol{L}^{\star}\jmath(B)\boldsymbol{L}-\boldsymbol{K}^{\star}B-B\boldsymbol{K}$, defining the quantum stochastic differential equation as

$$\mathrm{d}\phi_{t}\left(B\right)+\phi_{t}\left(K^{*}B+BK-L^{*}\jmath\left(B\right)L\right)\mathrm{d}t=\phi_{t}\left(L_{\bullet}^{\bullet}\jmath\left(B\right)L_{\bullet}-B\otimes\delta_{\bullet}^{\bullet}\right)\mathrm{d}\Lambda_{\bullet}^{\bullet}$$

(0.5)
$$+\phi_{t}\left(L_{\bullet}^{\bullet}\jmath\left(B\right)L-K^{\bullet}B\right)\mathrm{d}\Lambda_{\bullet}^{+}+\phi_{t}\left(L^{*}\jmath\left(B\right)L_{\bullet}-BK_{\bullet}\right)\mathrm{d}\Lambda_{-}^{\bullet},$$

where $\jmath$ is an operator representation of $\mathcal{B}$, and $\delta_\bullet^\bullet$ is the identity operator in $\mathcal{K}$. Such an extension of Lindblad's form for quantum stochastic generators was discovered recently in [18] even for a nonlinear case. We shall prove that this structure is necessary at least in the case of the w*-continuous generators, which are extendable to the covariant ones over the algebra of all bounded operators $\mathcal{L}(\mathcal{H})$. The existence of minimal CP solution which is constructed under certain continuity conditions proves that this structure is also sufficient for the CP property of any solution to this stochastic equation.

The Lindblad case $\Lambda(t,a) = \alpha t \mathrm{I}$ is described by the simplest one-dimensional Itô algebra $\mathfrak{a} = \mathbb{C}d$ with $l(a) = \alpha \in \mathbb{C}$ and the nilpotent multiplication $\alpha^\star \alpha = 0$ corresponding to the non-stochastic (Newton) calculus $(\mathrm{d}t)^2 = 0$ in $\mathcal{K} = 0$. The standard Wiener process $\mathrm{Q} = \Lambda_-^\bullet + \Lambda_\bullet^+$ in Fock space is described by the second order nilpotent algebra $\mathfrak{a}$ of pairs $a = (\alpha, \xi)$ with $d = (1,0)$, $\xi \in \mathbb{C}$, represented by the quadruples $a_+^- = \alpha$, $\quad a_\bullet^- = \xi = a_+^\bullet$, $\quad a_\bullet^\bullet = 0$ in $\mathcal{K} = \mathbb{C}$, corresponding to $\Lambda(t,a) = \alpha t \mathrm{I} + \xi \mathrm{Q}(t)$. The unital $\star$-algebra $\mathbb{C}$ with the usual multiplication $\zeta^\star \zeta = |\zeta|^2$ can be embedded into the two-dimensional Itô algebra $\mathfrak{a}$ of $a = (\alpha, \zeta)$, $\alpha = l(a)$, $\zeta \in \mathbb{C}$ as $a_\bullet^\bullet = \zeta$, $a_+^\bullet = +i\zeta$, $a_\bullet^- = -i\zeta$, $a_+^- = \zeta$. It corresponds to $\Lambda(t,a) = \alpha t \mathrm{I} + \zeta \mathrm{P}(t)$, where $\mathrm{P} = \Lambda_\bullet^\bullet + i\left(\Lambda_\bullet^+ - \Lambda_-^\bullet\right)$ is the representation of the standard Poisson process, compensated by its mean value t. Thus our results are applicable also to the classical stochastic differentials of completely positive processes, corresponding to the commutative Itô algebras, which are decomposable into the Wiener, Poisson and Newton orthogonal components.

1. Quantum completely positive flows

Throughout the complex pre-Hilbert space $\mathcal{D} \subseteq \mathcal{H}$ is a reflexive Fréchet space, $\mathcal{K} \otimes \mathcal{D}$ denotes the projective tensor product (π-product) with another such space $\mathcal{K}$, $\mathcal{D}' \supseteq \mathcal{H}$ denotes the dual space of continuous antilinear functionals $\eta' : \eta \in \mathcal{D} \mapsto \langle \eta | \eta' \rangle$, with respect to the canonical pairing $\langle \eta | \eta' \rangle$ given by $\|\eta\|^2$ if $\eta' = \eta \in \mathcal{D}$, $\mathcal{B}(\mathcal{D})$ denotes the linear space of all continuous sesquilinear forms $\langle \eta | B\eta \rangle$ on $\mathcal{D}$, identified with the continuous linear operators $B : \mathcal{D} \to \mathcal{D}'$, $B^\dagger \in \mathcal{B}(\mathcal{D})$ is the Hermit conjugated form $\langle \eta | B^\dagger \eta \rangle = \langle \eta | B\eta \rangle^*$, and $\mathcal{L}(\mathcal{D}) \subseteq \mathcal{B}(\mathcal{D})$ denotes the algebra of all strongly continuous operators $B : \mathcal{D} \to \mathcal{D}$. For the definitions and properties of this standard topological spaces see for example [20]. The space $\mathcal{D}$ will be equipped with weak topology induced by its predual (= dual) $\mathcal{D}$, and $\mathcal{B}(\mathcal{D})$ will be equipped with w*-topology (induced by the predual $\mathcal{B}_*(\mathcal{D}) = \mathcal{D} \otimes \mathcal{D}$), coinciding with the weak topology on each bounded subset. Any operator $A \in \mathcal{L}(\mathcal{D})$ with $A^\dagger \in \mathcal{L}(\mathcal{D})$ can be uniquely extended to a weakly continuous operator onto $\mathcal{D}'$ as $A^{\dagger *}$, denoted again as A, where A^* is the dual operator $\mathcal{D}' \to \mathcal{D}'$, $\langle \eta | A^* \eta' \rangle = \langle A\eta | \eta' \rangle$, defining the involution $A \mapsto A^*$ for any such continuation $A : \mathcal{D}' \to \mathcal{D}'$. We say that the operator A commutes with a sesquilinear form, $BA = AB$ if $\langle \eta | BA\eta \rangle = \langle A^\dagger \eta | B\eta \rangle$ for all $\eta \in \mathcal{D}$. The commutant $\mathcal{A}^c = \{B \in \mathcal{B}(\mathcal{D}) : [A, B] = 0, \forall A \in \mathcal{A}\}$ of an operator $*$-algebra $\mathcal{A} \subseteq \mathcal{L}(\mathcal{D})$ is weakly closed in $\mathcal{B}(\mathcal{D})$, so that the weak closure $\overline{\mathcal{B}} \subseteq \mathcal{B}(\mathcal{D})$ of any $\mathcal{B} \subseteq \mathcal{A}^c$ also commutes with $\mathcal{A}$.

Let $\mathcal{B} \subseteq \mathcal{L}(\mathcal{H})$ be a unital $*$-algebra of bounded operators $B : \mathcal{H} \to \mathcal{H}$, $\|B\| < \infty$, and $(\Omega, \mathfrak{A}, P)$ be a probability space with a filtration $(\mathfrak{A}_t)_{t>0}$, $\mathfrak{A}_t \subseteq \mathfrak{A}$ of σ-

algebras on Ω. One can assume that the filtration $\mathfrak{A}_t \subseteq \mathfrak{A}_s, \forall t < s$ is generated by $x_t = \{r \mapsto x(r) : r < t\}$ of a stochastic process $x(t, \omega)$ with independent increments $\mathrm{d}x(t) = x(t + \Delta) - x(t)$, and the probability measure P is invariant under the measurable representations $\omega \mapsto \omega_s \in \Omega$, $A_s^{-1} = \{\omega : \omega_s \in A\} \in \mathfrak{A}$, $\forall A \in \mathfrak{A}$ on $\Omega \ni \omega$ of the time shifts $t \mapsto t + s, s > 0$, corresponding to the shifts of the random increments

$$\mathrm{d}x(t, \omega_s) = \mathrm{d}x(t + s, \omega), \quad \forall \omega \in \Omega, t \in \mathbb{R}_+.$$

The *filtering dynamics* over $\mathcal{B}$ with respect to the process $x(t)$ is described by a cocycle flow $\phi = (\phi_t)_{t>0}$ of linear completely positive [21] w*-continuous stochastic adapted maps $\phi_t(\omega) : \mathcal{B} \to \overline{\mathcal{B}}$, $\omega \in \Omega$ such that the stochastic process $y_t(\omega) = \langle \eta | \phi_t(\omega, B) \eta \rangle$ is causally measurable for each $\eta \in \mathcal{D}$, $B \in \mathcal{B}$ in the sense that $y_t^{-1}(B) \in \mathfrak{A}_t$, $\forall t > 0$ and any Borel $B \subseteq \mathbb{C}$. The maps ϕ_t can be extended on the $\mathfrak{A}$-measurable functions $Y : \omega \mapsto Y(\omega)$ with values $Y(\omega) \in \overline{\mathcal{B}}$ as the normal maps $\phi_t[Y](\omega) = \phi_t(\omega, Y(\omega_t))$ for each $\omega \in \Omega$, and the cocycle condition $\phi_r(\omega) \circ \phi_s(\omega_r) = \phi_{r+s}(\omega)$, $\forall r, s > 0$ reads as the semigroup condition $\phi_r[\phi_s[Y]] = \phi_{r+s}[Y]$ of the extended maps. As it was noted in the previous section, the maps $\phi_t(\omega)$ are not considered to be normalized to the identity, and can be even unbounded, but they are supposed to be normalized, $\phi_t(\omega, I) = M_t(\omega)$, to an operator-valued martingale $M_t = \epsilon_t[M_s] \geq 0$ with $M_0(\omega) = I$, or to a positive submartingale, $R_t \geq \epsilon_t[R_s], \forall s > t$ in the subfiltering case, where ϵ_t is the conditional expectation over ω with respect to $\mathfrak{A}_t$.

Now we give a noncommutative generalization of the filtering (subfiltering) CP flows for an arbitrary Itô algebra, which was suggested in [22] for a Gaussian Itô algebra of finite dimensional quantum thermal noise, and in [18] for the simple quantum Itô algebra $\mathcal{Q}(\mathbb{C}^d)$ even in the nonlinear case.

The role of the classical process $x(t)$ will play the quantum stochastic process

$$X(t) = A \otimes I + I \otimes \Lambda(t, a), \quad A \in \mathcal{A}, a \in \mathfrak{a}$$

indexed by an operator algebra $\mathcal{A} \subset \mathcal{L}(\mathcal{D})$ and a noncommutative Itô algebra $\mathfrak{a}$. Here $\Lambda(t, a)$ is the process with independent increment on a pre-Fock space $\mathfrak{F} \subset \Gamma(\mathfrak{K})$ over the space $\mathfrak{K} = L^2_{\mathcal{K}}(\mathbb{R}_+)$ of all square-norm integrable $\mathcal{K}$-valued functions on $\mathbb{R}_+$, where $\mathcal{K}$ is a pre-Hilbert space of the representation $a \in \mathfrak{a} \mapsto (a_\nu^\mu)_{\nu=+,\bullet}^{\mu=-,\bullet}$ for the Itô $\star$-algebra $\mathfrak{a}$. We define $\mathfrak{F}$ as the Fréchet space, generated by coherent vectors $f^\otimes$, with respect to the scale [14]

$$(1.1) \qquad \|f\|_p^{\otimes 2} = \int_\Gamma \|f^\otimes(\tau)\|_p^2 \, \mathrm{d}\tau := \sum_{n=0}^\infty \frac{1}{n!} \left(\int_0^\infty \|f^\bullet(t)\|_p^2 \, \mathrm{d}t \right)^n = \exp\left[\|f^\bullet\|_p^2 \right],$$

where $f^\otimes(\tau) = \bigotimes_{t \in \tau} f^\bullet(t)$ for each $f^\bullet \in \mathfrak{K}$ is represented by tensor-functions on the space Γ of all finite subsets $\tau = \{t_1, ..., t_n\} \subseteq \mathbb{R}_+$, and $\|\cdot\|_p$ is an increasing sequence of Hilbertian norms $\|k^\bullet\|_p > \|k^\bullet\|$ on $\mathcal{K}$. We shall assume that all operators $a_\bullet^\bullet$ are strongly continuous, representing the $\star$-algebra $\mathfrak{a}$ on the Fréchet space $\mathcal{K}$, and recall that $a_\bullet^{-\dagger} = a_+^{\star\bullet} \in \mathcal{K}$ for all $a \in \mathfrak{a}$.

Proposition 1 The exponential operators $W(t, a) =: \exp[\Lambda(t, a)]$: defined as the *solutions to the quantum Itô equation*

$$\text{(1.2)} \qquad dW_t(g) = W_t(g)\, d\Lambda(t, g(t)), \quad W_0(g) = I, g(t) \in \mathfrak{a}$$

with $g(t) = a$, are all in $\mathcal{L}(\mathfrak{F})$. They give an analytic representation

$$\text{(1.3)} \qquad W(t, a \star a) = W(t, a)^* W(t, a), \quad W(t, 0) = I, \quad W(t, d) = e^t I$$

of the unital $\star$-semigroup $1 + \mathfrak{a}$ for the Itô $\star$-algebra $\mathfrak{a}$ with respect to the $\star$-product $a \star a = a + a^\star a + a^\star$.

Proof. The solutions $W(t, a)$ are uniquely defined on the coherent vectors as analytic functions

$$\text{(1.4)}$$
$$W(t, a) f^\otimes(\tau) = \otimes_{r \in \tau}^{r \leq t} \left(a_\bullet^\bullet f^\bullet(r) + a_+^\bullet \right) \exp\left[\int_0^t \left(a_\bullet^- f^\bullet(r) + a_+^- \right) dr \right] \otimes_{r \in \tau}^{r \geq t} f^\bullet(r),$$

which obey the properties (1.3), see for example [14]. Thus the span of coherent vectors is invariant, and it is also invariant under $W(t, a)^* = W(t, a^\star)$. They can be extended on $\mathfrak{F}$ by continuity which follows from the continuity of $a_\bullet^\bullet$ on $\mathcal{K}$, $a_+^\bullet \in \mathcal{K}$ and boundedness of $a_\bullet^- \in \mathcal{K}'$.

Let $\mathfrak{D}$ denote the Fréchet space $\mathcal{D} \otimes \mathfrak{F}$, generated by $\psi = \eta \otimes f^\otimes$, $\eta \in \mathcal{D}$, $f^\bullet \in \mathfrak{K}$. Assuming the separability of the Itô algebra in the sense $\mathcal{K} \subseteq \ell^2$ such that $f^\bullet = (f^m)^{m \in \mathbb{N}}$, one can identify each $\psi' \in \mathfrak{D}'$ with a sequence of $\mathcal{D}'$-valued symmetric tensor-functions $\psi'_{m_1, \ldots, m_n}(t_1, \ldots t_n)$, $n = 0, 1, 2, \ldots$. Let $(\mathfrak{D}_t)_{t>0}$ be the natural filtration and $(\mathfrak{D}_{[t})_{t>0}$ be the backward filtration of the subspaces $\mathfrak{D}_t = \mathcal{D} \otimes \mathfrak{F}_t$, $\mathfrak{D}_{[t} = \mathcal{D} \otimes \mathfrak{F}_{[t}$ generated by $\eta \otimes f^\otimes$ with $f^\bullet \in \mathfrak{K}_t$ and $f^\bullet \in \mathfrak{K}_{[t}$ respectively, where $\mathfrak{K}_t = L_\mathcal{K}^2[0, t)$, $\mathfrak{K}_{[t} = L_\mathcal{K}^2[t, \infty)$ are embedded into $\mathfrak{K}$. The spaces $\mathfrak{D}_t$, $\mathfrak{D}_{[t}$ of the restrictions $E_t \psi = \psi | \Gamma_t$, $E_{[t} \psi = \psi | \Gamma_{[t}$ onto $\Gamma_t = \{\tau_t = \tau \cap [0, t)\}$, $\Gamma_{[t} = \{\tau_{[t} = \tau \cap [t, \infty)\}$ are embedded into $\mathfrak{D}$ by the isometries $E_t^\dagger : \psi \mapsto \psi_t$, $E_{[t}^\dagger : \psi \mapsto \psi_{[t}$ as $\psi_t(\tau) = \psi(\tau_t) \delta_\emptyset(\tau_{[t})$, $\psi_{[t}(\tau) = \delta_\emptyset(\tau_t) \psi(\tau_{[t})$, where $\delta_\emptyset(\tau) = 1$ if $\tau = \emptyset$, otherwise $\delta_\emptyset(\tau) = 0$. The projectors E_t, $E_{[t}$ onto $\mathfrak{D}_t$, $\mathfrak{D}^t$ are extended onto $\mathfrak{D}'$ as the adjoints to $E_t^\dagger$, $E_{[t}^\dagger$. The time shift on $\mathfrak{D}'$ is defined by the semigroup $(T^t)_{t>0}$ of adjoint operators $T^t = T_t^*$ to $T_t \psi(\tau) = \psi(\tau + t)$, where $\tau + t = \{t_1 + t, \ldots, t_n + t\}$, $\emptyset + t = \emptyset$, such that $T^t \psi(\tau) = \delta_\emptyset(\tau_t) \psi(\tau_{[t} - t)$ are isometries for $\psi \in \mathfrak{D}$ onto $\mathfrak{D}_{[t}$. A family $(Z_t)_{t>0}$ of sesquilinear forms $\langle \psi | Z_t \psi \rangle$ given by linear operators $Z_t : \mathfrak{D} \to \mathfrak{D}'$ is called *adapted* (and $(Z^t)_{t>0}$ is called *backward adapted*) if

$$\text{(1.5)} \quad Z_t(\eta \otimes f^\otimes) = \psi' \otimes E_{[t} f^\otimes \quad (Z^t(\eta \otimes f^\otimes) = \psi' \otimes E_t f^\otimes), \quad \forall \eta \in \mathcal{D}, f^\bullet \in \mathfrak{K},$$

where $\psi' = \mathfrak{D}_t'\ (= \mathfrak{D}_{[t}')$ and $E_{[t}\ (E_t)$ are the projectors onto $\mathfrak{F}_{[t}\ (\mathfrak{F}_t)$ correspondingly.

The *conditional expectation* on $\mathcal{B}(\mathfrak{D})$ with respect to the past up to a time $t \in \mathbb{R}_+$ is a positive projector, $\epsilon_t = \epsilon_t \circ \epsilon_s, \forall t < s$, which defines an adapted family of the sesquilinear forms $Z_t = \epsilon_t(Z)$, given in (1.5) for each $Z \in \mathcal{B}(\mathfrak{D})$ by $\psi' = E_t Z E_t^\dagger \psi$ with

$\psi \in \eta \otimes E_t f^{\otimes}$. The time shift $(\theta^t)_{t>0}$ on $\mathcal{B}(\mathfrak{D})$ is uniquely defined by the covariance condition $\theta^t(Z) T^t = T^t Z$ as a backward adapted family $Z^t = \theta^t(Z), t > 0$ for each $Z \in \mathcal{B}(\mathfrak{D})$. As in the bounded case [16] between the maps ϵ_t and θ^t we have the relation $\theta^r \circ \epsilon_s = \epsilon_{r+s} \circ \theta^r$ which follows from the operator relation $T^r E_s = E_{r+s} T^r$. An adapted family $(M_t)_{t>0}$ of positive $\langle \psi | M_t \psi \rangle \geq 0, \forall \psi \in \mathfrak{D}$ Hermitian $M_t^{\dagger} = M_t$ forms $M_t \in \mathcal{B}(\mathfrak{D})$ is called *martingale (submartingale)* if $\epsilon_t(M_s) = M_t \, (\epsilon_t(M_s) \leq M_t)$ for all $s \geq t \geq 0$. The bounded operator-valued martingales M_t were introduced in the case of simple HP-algebra in [23].

Let $\mathfrak{B}$ denote the space of all $Y \in \mathcal{B}(\mathfrak{D})$, commuting with all $X = \{X(t)\}$ in the sense

$$AY = YA, \quad \forall A \in \mathcal{A}, \quad YW(t,a) = W(t,a)Y, \quad \forall t > 0, a \in \mathfrak{a},$$

where $A(\eta \otimes \varphi) = A\eta \otimes \varphi$, $W(\eta \otimes \varphi) = \eta \otimes W\varphi$, and the unital $*$-algebra $\mathcal{B} \subseteq \mathcal{L}(\mathcal{H})$ be weakly dense in the commutant $\mathcal{A}^c$. The quantum filtration $(\mathfrak{B}_t)_{t>0}$ is defined as the increasing family of subspaces $\mathfrak{B}_t \subseteq \mathfrak{B}_s, t \leq s$ of the adapted sesquilinear forms $Y_t \in \mathfrak{B}$. The covariant shifts $\theta^t : Y \mapsto Y^t$ leave the space $\mathfrak{B}$ invariant, mapping it onto the subspaces of backward adapted sesquilinear forms $Y^t = \theta^t(Y)$.

The *quantum stochastic positive flow* over $\mathcal{B}$ is described by a one parameter family $\phi = (\phi_t)_{t>0}$ of linear w*-continuous maps $\phi_t : \mathcal{B} \to \mathfrak{B}$ satisfying

1. the causality condition $\phi_t(B) \subseteq \mathfrak{B}_t, \quad \forall B \in \mathcal{B}, t \in \mathbb{R}_+,$

2. the complete positivity condition $[\phi_t(B_{kl})] \geq 0$ for each $t > 0$ and for any positive definite matrix $[B_{kl}] \geq 0$ with $B_{kl} \in \mathcal{B}$,

3. the cocycle condition $\phi_r \circ \phi_s^r = \phi_{r+s}, \forall t, s > 0$ with respect to the covariant shift $\phi_s^r = \theta^r \circ \phi_s$.

Here the composition $\circ$ is understood as $\phi_r[\phi_s(B)] = \phi_{r+s}(B)$ in terms of the linear normal extensions of $\phi_t[B \otimes Z] = \phi_t(B) Z^t$ to the CP maps $\mathfrak{B} \to \mathfrak{B}$, forming a one-parameter semigroup, where $B \in \mathcal{B}$, $Z^t = \theta^t(Z)$, $Z \in \mathcal{B}(\mathfrak{F})$. These can be defined like in classical case as $\phi_t[Y](\bar{f}^{\bullet}, f^{\bullet}) = \phi_t(\bar{f}^{\bullet}, Y(\bar{f}_t^{\bullet}, f_t^{\bullet}), f^{\bullet})$ with $f_t^{\bullet}(r) = f^{\bullet}(t+r)$ by the coherent matrix elements $Y(\bar{f}^{\bullet}, f^{\bullet}) = F^* Y F$ for $Y \in \mathfrak{B}$ given by the continuous operators $F : \eta \mapsto \psi_f = \eta \otimes f^{\otimes}$, $\eta \in \mathcal{D}$ for each $f^{\bullet} \in \mathfrak{K}_t$ with the adjoints $F^* \psi' = \int_{\tau < t} f^{\otimes}(\tau)^* \psi'(\tau) \, d\tau$ for $\psi' \in \mathfrak{D}'$.

The flow is called *(sub)-filtering*, if $R_t = \phi_t(I)$ is a (sub)-martingale with $R_0 = I$, and is called contractive, if $I \geq R_t \geq R_s$ for all $0 \leq t \leq s \in \mathbb{R}_+$.

Proposition 2 The complete positivity for adapted linear maps $\phi_t : \mathcal{B} \to \mathcal{B}(\mathfrak{D})$ can be written as

(1.6)
$$\sum_{f,h \in \mathfrak{K}_t} \sum_{B,C \in \mathcal{B}} \langle \xi_B^f | \phi_t(\bar{f}^{\bullet}, B^* C, h^{\bullet}) \xi_C^h \rangle := \langle \eta^k | \phi_t(\bar{f}_k^{\bullet}, B_k^* B_l, h_l^{\bullet}) \eta^l \rangle \geq 0, \quad \forall t > 0$$

(the usual summation rule over repeated cross-level indices is understood), where $\xi_B^f = \eta^k$ if $f^{\bullet} = f_k^{\bullet}$ and $B = B_k$ with $f_k^{\bullet} \in \mathfrak{K}_t$, $B_k \in \mathcal{B}$, $k = 1, 2, ...$, otherwise $\xi_B^f = 0$, and $\phi_t(B, f^{\bullet}) = \phi_t(B) F$, $\phi_t(\bar{f}^{\bullet}, B) - F^* \psi_t(B)$.

Proof. By definition the map ϕ into the sesquilinear forms is completely positive on $\mathcal{B}$ if $\langle \psi^k | \phi(B_{kl}) \psi^l \rangle \geq 0$ whenever $\langle \eta^k | B_{kl} \eta^l \rangle \geq 0$, where η^k, ψ^k are arbitrary finite sequences. Approximating from below the latter positive forms by sums of the forms $\sum_{kl} \langle \eta^k | B_{ik}^* B_{il} \eta^l \rangle \geq 0$, the complete positivity can be tested only for the forms $\sum_{kl} \langle \eta^k | B_k^* B_l \eta^l \rangle \geq 0$ due to the additivity $\phi(\sum_i B_{ik}^* B_{il}) = \sum_i \phi(B_{ik}^* B_{il})$. If ϕ_t is adapted, this can be written as

$$\sum_{B, C \in \mathcal{B}} \langle \chi_B | \phi(B^* C) \chi_C \rangle = \langle \psi^k | \phi(B_k^* B_l) \psi^l \rangle := \sum_{k,l} \langle \psi^k | \phi(B_k^* B_l) \psi^l \rangle \geq 0,$$

where $\chi_B = \psi^k \in \mathfrak{D}_t$ if $B = B_k \in \mathcal{B}$, otherwise $\chi_B = 0$. Because any $\psi \in \mathfrak{D}_t$ can be approximated by a $\mathcal{D}$-span $\sum_f \eta^f \otimes f^\otimes$ of coherent vectors over $f_k^\bullet \in \mathfrak{K}_t$, it is sufficient to define the CP property only for such spans as

$$0 \leq \sum_{f,h} \sum_{B,C} \left\langle \xi_B^f \otimes f^\otimes | \phi(B^* C)\, (\xi_C^h \otimes h^\otimes) \right\rangle = \sum_{f,h} \sum_{B,C} \left\langle \xi_B^f | \phi(\bar{f}^\bullet, B^* C, h^\bullet)\, \xi_C^h \right\rangle.$$

Note that the subfiltering (filtering) flows can be considered as a quantum stochastic CP dilations of the quantum sub-Markov (Markov) semigroups $\theta = (\theta_t)_{t>0}$, $\theta_r \circ \theta_s = \theta_{r+s}$ in the sense $\theta_t = \epsilon \circ \phi_t$, where $\epsilon(Y)\eta = EY\psi_0$, $E\psi' = \psi'(\emptyset)$, $\forall \psi' \in \mathfrak{D}'$, with $\theta_s(I) \leq \theta_t(I) \leq I\, (\theta_t(I) = I)$, $\forall t \leq s$. The contraction $C_t = \theta_\tau(I)$ with $R_0 = I$ defines the probability $\langle \eta | C_t \eta \rangle \leq 1$, $\forall \eta \in \mathcal{H}, \|\eta\| = 1$ for an unstable system not to be demolished by a time $t \in \mathbb{R}_+$, and the conditional expectations $\langle \eta | A C_t \eta \rangle / \langle \eta | C_t \eta \rangle$ of the initial nondemolition observables $A \in \mathcal{A}$ in any state $\eta \in \mathcal{D}$, and thus in any initial state $\psi_0 \in \eta \otimes \delta_\emptyset$. The following theorem shows that the submartingale (or the contraction) $R_t = \phi_t(I)$ is the density operator with respect to $\psi_0 = \eta \otimes \delta_\emptyset$, $\eta \in \mathcal{H}$ (or with respect to any $\psi \in \mathcal{H} \otimes \mathfrak{F}$) also for the conditional state of the restricted nondemolition process $X_t = \{r \mapsto X(r) : r < t\}$.

Theorem 1 Let $t \mapsto R_t \in \mathfrak{B}_t$ be a positive (sub)-martingale and $(\mathfrak{g}_t)_{t>0}$ be the increasing family of $\star$-semigroups $\mathfrak{g}_t$ of step functions $g : \mathbb{R}_+ \to \mathfrak{a}$, $g(s) = 0$, $\forall s \geq t$ under the $\star$-product

$$(1.7) \qquad\qquad (g_k \star g_l)(t) = g_l(t) + g_k(t)^\star g_l(t) + g_k(t)^\star$$

of $g_k^\star = g_k \star 0$ and $g_l = 0 \star g_l$. The generating function $\vartheta_t(g) = \epsilon[R_t W_t(g)]$ of the output state for the process $\Lambda(t)$, defined for any $g \in \mathfrak{g}_t$ and each $t > 0$ as

$$(1.8) \qquad\qquad \langle \eta | \vartheta_t(g) \eta \rangle = \langle \psi_0 | R_t W_t(g) \psi_0 \rangle, \quad \psi_0 = \eta \otimes \delta_\emptyset,$$

is $\mathcal{B}^c$-valued, positive, $\vartheta_t \geq 0$ in the sense of positive definiteness of the kernel

$$(1.9) \qquad\qquad \langle \eta^k | \vartheta_t(g_k \star g_l) \eta^l \rangle \geq 0, \quad \forall g_k \in \mathfrak{g}_t; \eta^k \in \mathcal{D},$$

and $\vartheta_t \geq \vartheta_s | \mathfrak{g}_t$ in this sense for any $s \geq t$. If $R_0 = I$, then $\vartheta_0(0) = I \geq \vartheta_t(0)$, and if R_t is a martingale, then $\vartheta_t = \vartheta_s | \mathfrak{g}_t$ for any $s \geq t$, and $\vartheta_t(0) = I$ for all $t \in \mathbb{R}_+$. Any *family $\vartheta = (\vartheta_t)_{t \geq 0}$ of positive-definite functions $\vartheta_t : \mathfrak{g}_t \to \mathcal{B}^c$, satisfying the above consistency and normalization properties, is the state generating function of the form (1.8) iff it is absolutely continuous in the following sense*

$$(1.10) \qquad \lim_{n \to \infty} \sum_{g \in \mathfrak{g}_t} \eta_n^g \otimes g_+^{\otimes} = 0 \Rightarrow \lim_{n \to \infty} \sum_{g,h \in \mathfrak{g}_t} \langle \eta_n^g | \vartheta_t (g \star h) \eta_n^h \rangle = 0,$$

where $g_+^{\otimes}(\tau) = \otimes_{t \in \tau} g_+^{\bullet}(t)$ and $\eta_n^g = 0$ for almost all $g = 0$ (except for a finite number of $g \in \mathfrak{g}_t$).

Proof. Because the solutions $W_t(g)$ to the quantum stochastic equation (1.2) for a step function g are given by finite products of commuting exponential operators $W(t, b)$, they are multiplicative, $W_t(g_k)^* W_t(g_l) = W_t(g_k \star g_l)$, as the operators in (1.4) are. Then the positive definiteness of ϑ_t follows from their commutativity (1.7) with positive R_t:

$$\langle \eta^k | \vartheta_t (g_k \star g_l) \eta^l \rangle = \langle W_t(g_k) (\eta^k \otimes \delta_{\emptyset}) | R_t W_t(g_l) (\eta^l \otimes \delta_{\emptyset}) \rangle = \langle \psi_t | R_t \psi_t \rangle \geq 0.$$

It is $\mathcal{B}^c$-valued as

$$\langle \eta | \vartheta_t (g) B\eta \rangle = \langle \psi_0 | R_t W_t(g) B\psi_0 \rangle = \langle B\psi_0 | R_t W_t(g) \psi_0 \rangle = \langle B\eta | \vartheta_t (g) \eta \rangle, \forall B \in \mathcal{B}.$$

¿From $W_t(g_r) = W_r(g)$, $r < t$ and $W_t(0) = I$ as the case $g_0 = 0$ it follows that

$$\vartheta_t(g_r) = \langle \psi_0 | R_t W_t(g_r) \psi_0 \rangle = \langle \psi_0 | \epsilon_r (R_t) W_r(g) \psi_0 \rangle \leq \langle \psi_0 | R_r W_r(g) \psi_0 \rangle = \vartheta_r(g)$$

for any finite matrix $g = [g_k \star g_l]$, and $\vartheta_t(0) \leq 1 = \vartheta_0(0)$ if R_t is a submartingale with $R_0 = I$. This implies the normalization and compatibility conditions if R_t is martingale. The continuity condition follows from the continuity of the forms $R_t \in \mathcal{B}(\mathfrak{D})$: if $\sum_g (\eta_n^g \otimes g_+^{\otimes}) \to 0$, then

$$\sum_{g,h} \langle \eta_n^g | \vartheta_t (g \star h) \eta_n^h \rangle = \sum_{g,h} \langle W_t(g) (\eta_n^g \otimes \delta_{\emptyset}) | R_t W_t(h) (\eta_n^h \otimes \delta_{\emptyset}) \rangle$$

$$= \left\langle \sum_g \eta_n^g \otimes g_+^{\otimes} | R_t \sum_g \eta_n^g \otimes g_+^{\otimes} \right\rangle \to 0.$$

Conversely, let $(\mathcal{E}, V_t, L)$ be the GNS triple, describing the decomposition $\vartheta_t(g) = L^* V_t(g) L$ for a positive-definite kernel-function ϑ_t. It is defined by the multiplicative $*$-representation $V_t(g \star h) = V_t(g)^* V_t(h)$ of $\mathfrak{g}_t$ on a pre-Hilbert space $\mathcal{E} \subseteq \mathcal{H}$ and by a linear operator $L : \mathcal{D} \to \mathcal{E}$. The correspondence $\pi_t(B) : V_t(g) L\eta \mapsto V_t(g) LB\eta$ for $B \in \mathcal{B}$ is extended to a $*$-representation $\pi_t : \mathcal{B} \to V_t(\mathfrak{g}_t)'$ on the linear combinations $\mathcal{E}_t^{\circ} = \{\sum_k V_t(g_k) L_t \eta_k : g_k \in \mathfrak{g}_t, \eta_k \in \mathcal{D}\}$ by virtue of the commutativity of $\vartheta_t(g)$ with $\mathcal{B}$:

$$\langle B^* \eta^g | \vartheta_t (g \star h) \eta^h \rangle = \langle \pi_t(B^*) V_t(g) L_t \eta^g | V_t(h) L_t \eta^h \rangle$$

$$= \langle V_t(g) L_t \eta^g | \pi_t(B) V_t(h) L_t \eta^h \rangle = \langle \eta^g | \vartheta_t (g \star h) B\eta^h \rangle.$$

The linear correspondence $F_t : \sum_g \eta^g \otimes g_+^\otimes \mapsto \sum_g V_t(g) L\eta^g$ obviously intertwines this representation with $B \mapsto B \otimes I$ as well as the representation V_t with W_t on $\mathfrak{D}_t^\circ = \{\sum_k \eta_k \otimes W_t(g_k)\delta_\emptyset : g_k \in \mathfrak{g}_t, \eta_k \in \mathcal{D}\} \subseteq \mathfrak{D}_t$:

$$F_t\left(\eta \otimes W_t(f^\star)g_+^\otimes\right) = V_t(f \star g)L\eta = V_t(f^\star)V_t(g)L\eta = V_t(f^\star)F_t\left(\eta \otimes g_+^\otimes\right),$$

where $g_+^\otimes = W(g)\delta_\emptyset$. It is continuous operator with respect to the Hilbert space norm in $\mathcal{H}_t$ because if $\sum_g \eta_n^g \otimes g_+^\otimes \to 0$, then

$$\left\|F_t \sum_g \eta_n^g \otimes g_+^\otimes\right\|^2 = \left\|\sum_g V_t(g)L\eta_n^g\right\|^2 = \left\langle \eta_n^f | \sum_{f,h} \vartheta_t(f \star h)\eta_n^h \right\rangle \to 0$$

due to the strong absolute continuity of ϑ_t. Hence F_t can be continued to an intertwining operator $\mathfrak{D}_t^\circ \to \mathcal{H}$, and there exists the adjoint intertwining operator $F_t^* : \mathcal{E} \to \mathfrak{D}_t'$, $F_t^* V_t(g) = W_t(g)F_t^*$ such that

$$\langle \psi_0 | F_t^* F_t W_t(g)\psi_0\rangle = \langle F_t\psi_0 | V_t(g)F_t\psi_0\rangle = \langle L_t\eta | V_t(g)L_t\eta\rangle = \langle \eta | L_t^* V_t(g)L_t\eta\rangle.$$

The positive operators $F_t^* F_t \in \mathcal{B}(\mathfrak{D}_t)$ uniquely extended to the adapted ones $R_t : \mathfrak{D} \to \mathfrak{D}'$, commute with all $W_t(g), g \in \mathfrak{g}_t$. They define a submartingale (martingale) $R_t \in \mathfrak{B}_t$ due to the property $W_t = W_s|\mathfrak{g}_t$ for all $s \geq t$ and

$$\langle W_t(g_k)\psi_0^k | \epsilon_t(R_s)W_t(g_l)\psi_0^l\rangle = \langle \psi_0^k | R_s W_s(g_k \star g_l)\psi_0^l\rangle = \langle \eta^k | \vartheta_s(g_k \star g_l)\eta^l\rangle$$

$$\leq (=)\langle \eta^k | \vartheta_t(g_k \star g_l)\eta^l\rangle = \langle \psi_0^k | R_t W_t(g_k \star g_l)\psi_0^l\rangle = \langle W_t(g_k)\psi_0^k | R_t W_t(g_l)\psi_0^l\rangle$$

if $\vartheta_s|\mathfrak{g}_t \leq (=)\vartheta_t$. It is normalized, $R_0 = F_0^* F_0 = I$, as $F_0 = I$ if $\vartheta_0(0) = I$.

2. Generators of quantum CP dynamics

The quantum stochastically differentiable positive flow ϕ is defined as a weakly continuous function $t \mapsto \phi_t$ with CP values $\phi_t : \mathcal{B} \to \mathfrak{B}_t$, $\phi_0(B) = B \otimes I, \forall B \in \mathcal{B}$ such that for any product-vector $\psi_f = \eta \otimes f^\otimes$ given by $\eta \in \mathcal{D}$ and $f^\bullet = f_\bullet^* \in \mathfrak{K}$

$$(2.1)\qquad \frac{\mathrm{d}}{\mathrm{d}t}\langle \psi_f | \phi_t(B)\psi_f\rangle = \langle \psi_f | \phi_t\left(\lambda\left(\bar{f}^\bullet(t), B, f^\bullet(t)\right)\right)\psi_f\rangle, \qquad B \in \mathcal{B},$$

where $\lambda\left(\bar{k}^\bullet, B, k^\bullet\right) = \lambda(B) + k_\bullet \lambda^\bullet(B) + \lambda_\bullet(B)k^\bullet + k_\bullet \lambda_\bullet^\bullet(B)k$, $k_\bullet = \bar{k}^\bullet$ is the linear form on $\mathcal{K}$ with $k_\bullet^* = k^\bullet \in \mathcal{K}$ and $\langle \psi_f | \phi_0(B)\psi_f\rangle = \langle \eta | B\eta\rangle \exp\|f^\bullet\|^2$. The generator $\lambda(B) = \lambda(0, B, 0)$ of the quantum dynamical semigroup $\theta_t = \epsilon \circ \phi_t$ is a linear w*-continuous map $B \mapsto \in \mathcal{A}^c$, $\lambda^\bullet = \lambda_\bullet^\dagger$ is a linear w*-continuous map given by the

Hermitian adjoint values $\lambda_{\bullet}(B^*) = \lambda^{\bullet}(B)^{\dagger}$ in the continuous operators $\mathcal{K} \to \mathcal{A}^c$, and $\lambda^{\bullet}_{\bullet} : \mathcal{B} \to \mathcal{B}(\mathcal{D} \otimes \mathcal{K})$is a w*-continuous map with the values $\lambda^{\bullet}_{\bullet}(B)$ given by continuous operators $\mathcal{K} \otimes \mathcal{K} \to \mathcal{A}^c$. The differential evolution equation (2.1) for the coherent vector matrix elements $\langle \psi_f | \phi_t(B) \psi_f \rangle$ corresponds to the Itô form [17] of the quantum stochastic equation

$$(2.2) \qquad \mathrm{d}\phi_t(B) = \phi_t \circ \lambda^{\mu}_{\nu}(B) \, \mathrm{d}\Lambda^{\nu}_{\mu} := \sum_{\mu,\nu} \phi_t\left(\lambda^{\mu}_{\nu}(B)\right) \mathrm{d}\Lambda^{\nu}_{\mu}, \qquad B \in \mathcal{B}$$

with the initial condition $\phi_0(B) = B$, for all $B \in \mathcal{B}$. Here λ^{μ}_{ν} are the flow generators $\lambda^{-}_{+} = \lambda$, $\lambda^{\bullet}_{+} = \lambda^{\bullet}$, $\lambda^{-}_{\bullet} = \lambda_{\bullet}$, $\lambda^{\bullet}_{\bullet}$, called the structural maps, and the summation is taken over the indices $\mu = -, \bullet$, $\nu = +, \bullet$ of the standard quantum stochastic integrators Λ^{ν}_{μ}. For simplicity we shall assume that the pre-Hilbert Fréchet space $\mathcal{K}$ is separable, $\mathcal{K} \subseteq \ell^2$. Then the index $\bullet$ can take any value in $\{1, 2, ...\}$ and $\Lambda^{\nu}_{\mu}(t)$ are indexed with $\mu \in \{-, 1, 2, ...\}$, $\nu \in \{+, 1, 2, ...\}$ as the standard time $\Lambda^{+}_{-}(t) = t\mathrm{I}$, annihilation $\Lambda^{m}_{-}(t)$, creation $\Lambda^{+}_{n}(t)$ and exchange-number $\Lambda^{m}_{n}(t)$ operator integrators with $m, n \in \mathbb{N}$. The infinitesimal increments $\mathrm{d}\Lambda^{\mu}_{\nu}(t) = \Lambda^{t\mu}_{\nu}(\mathrm{d}t)$ are formally defined by the HP multiplication table [17] and the $\star$-property [3],

$$(2.3) \qquad \mathrm{d}\Lambda^{\alpha}_{\mu}\mathrm{d}\Lambda^{\nu}_{\beta} = \delta^{\alpha}_{\beta}\mathrm{d}\Lambda^{\nu}_{\mu}, \qquad \Lambda^{\star} = \Lambda,$$

where δ^{α}_{β} is the usual Kronecker delta restricted to the indices $\alpha \in \{-, 1, 2, ...\}$, $\beta \in \{+, 1, 2, ...\}$ and $\Lambda^{\star\mu}_{-\nu} = \Lambda^{\nu*}_{-\mu}$ with respect to the reflection $-(-) = +$, $-(+) = -$ of the indices $(-, +)$ only.

The linear equation (2.2) of a particular type, (quantum Langevin equation) with bounded finite-dimensional structural maps λ^{μ}_{ν} was introduced by Evans and Hudson [13] in order to describe the *-homomorphic quantum stochastic evolutions. The constructed quantum stochastic *-homomorphic flow (HP-flow) is identity preserving and is obviously completely positive, but it is hard to prove these algebraic properties for the unbounded case. In the general content the equation (2.2) was studied in [24], and the correspondent quantum stochastic unbounded flow was constructed even for the infinitely-dimensional non-adapted case but still with bounded λ^{μ}_{ν}. However the typical quantum filtering dynamics is not homomorphic or identity preserving, but it is completely positive and in the most interesting cases is described by unbounded generators λ^{μ}_{ν}. Here we will formulate the necessary differential conditions which follow from the complete positivity, causality, and martingale properties of the filtering flows, and which are sufficient for the construction of the quantum stochastic flows obeying these properties in the case of the bounded λ^{μ}_{ν}. As we showed in [18, 19], the found properties are sufficient to define the general structure of the bounded generators, and this will help us in construction of the minimal completely positive solutions for the quantum filtering equations also with unbounded λ^{μ}_{ν}.

Obviously the linear w*-continuous generators $\lambda^{\mu}_{\nu} : \mathcal{B} \to \mathcal{A}^c$ for CP flows $\phi^*_t = \phi_t$, where $\phi^*_t(B) = \phi_t(B^*)^{\dagger}$, must satisfy the $\star$-property $\lambda^{\star} = \lambda$, where $\lambda^{\star\nu}_{-\mu} = \lambda^{\mu*}_{-\nu}$, $\lambda^{\mu*}_{\nu}(B) = \lambda^{\mu}_{\nu}(B^*)^*$ and are independent of t, corresponding to cocycle property $\phi_s \circ \phi^s_r = \phi_{s+r}$, where ϕ^s_t is the solution to (2.2) with $\Lambda^{\mu}_{\nu}(t)$ replaced by $\Lambda^{s\mu}_{\nu}(t)$, and $\lambda^{-}_{+}(I) = 0$ if ϕ is a filtering flow, $\phi_t(I) = I$, as it is in the multiplicative case [13]. We shall assume that $\lambda = (\lambda^{\mu}_{\nu})^{\mu=-,\bullet}_{\nu=+,\bullet}$ for each $B^* = B$ defines a continuous Hermitian form $b = \lambda(B)$ on the Fréchet space $\mathcal{D} \oplus \mathcal{D}_\bullet$,

$$\langle \eta | \, b \, \eta \rangle = \sum_{m,n} \langle \eta^m | b_n^m \eta^n \rangle + \sum_m \langle \eta^m | b_+^m \eta \rangle + \sum_n \langle \eta | b_n^- \eta^n \rangle + \langle \eta | b_+^- \eta \rangle \,,$$

where $\eta \in \mathcal{D}$, $\eta^\bullet = (\eta^m)^{m \in \mathbb{N}} \in \mathcal{D}_\bullet = \mathcal{D} \otimes \mathcal{K}$. We say that an Itô algebra $\mathfrak{a}$, represented on $\mathcal{K}$, commutes in HP sense with a b, given by the form-generator λ if $(I \otimes a_\bullet^\mu) \, b_\nu^\bullet = b_\bullet^\mu (I \otimes a_\nu^\bullet)$ (For simplicity the ampliation $I \otimes a_\nu^\mu$ will be written again as a_ν^μ.) Note that if we define the matrix elements a_ν^μ, b_ν^μ also for $\mu = +$ and $\nu = -$, by the extension

$$a_\nu^+ = 0 = a_-^\mu, \qquad \lambda_\nu^+(B) = 0 = \lambda_-^\mu(B), \quad \forall a \in \mathfrak{a}, B \in \mathcal{B},$$

the HP product (0.4) of a and b can be written in terms of the usual matrix product $\mathbf{ab} = [a_\lambda^\mu b_\nu^\lambda]$ of the extended quadratic matrices $\mathbf{a} = [a_\nu^\mu]_{\nu=-,\bullet,+}^{\mu=-,\bullet,+}$ and $\mathbf{b} = b\mathbf{g}$, where $\mathbf{g} = [\delta_{-\nu}^\mu]$. Then one can extend the summation in (2.2) so it is also over $\mu = +$, and $\nu = -$, such that $b_\nu^\mu \mathrm{d}\Lambda_\mu^\nu$ is written as the trace $\mathbf{b} \cdot \mathrm{d}\mathbf{\Lambda}$ over all μ, ν. By such an extension the multiplication table for $\mathrm{d}\Lambda(a) = \mathbf{a} \cdot \mathrm{d}\mathbf{\Lambda}$, $\mathrm{d}\Lambda(b) = \mathbf{b} \cdot \mathrm{d}\mathbf{\Lambda}$ can be represented as $\mathrm{d}\Lambda(a)\,\mathrm{d}\Lambda(b) = \mathbf{ab} \cdot \mathrm{d}\mathbf{\Lambda}$, and the involution $\mathbf{b} \mapsto \mathbf{b}^\star$, defining $\mathrm{d}\Lambda(b)^\dagger = \mathbf{b}^\star \cdot \mathrm{d}\mathbf{\Lambda}$, can be obtained by the pseudo-Hermitian conjugation $b_\alpha^{\star \nu} = g_{\alpha\mu} b_\beta^{\mu *} g^{\beta\nu}$ respectively to the indefinite Minkowski metric tensor $\mathbf{g} = [g_{\mu\nu}]$ and its inverse $\mathbf{g}^{-1} = [g^{\mu\nu}]$, given by $g^{\mu\nu} = \delta_{-\nu}^\mu I = g_{\mu\nu}$.

Now let us find the differential form of the normalization and causality conditions with respect to the quantum stationary process, with independent increments $\mathrm{d}X(t) = X(t + \Delta) - X(s)$ generated by an Itô algebra $\mathfrak{a}$ on the separable space $\mathcal{K}$.

Proposition 3 Let ϕ be a flow, satisfying the quantum stochastic equation (2.2), and $[W_t(g), \phi_t(B)] = 0$ for all $g \in \mathfrak{g}, B \in \mathcal{B}$. Then the coefficients $b_\nu^\mu = \lambda_\nu^\mu(B)$, $\mu = -, \bullet$, $\nu = +, \bullet$, where $\bullet = 1, 2, ...$, written in the matrix form $b = (b_\nu^\mu)_{\nu=+,\bullet}^{\mu=-,\bullet}$, commute in the sense of the HP product with $a = (a_\nu^\mu)_{\nu=+,\bullet}^{\mu=-,\bullet}$ for all $a \in \mathfrak{a}$ and $B \in \mathcal{B}$:

$$(2.4) \qquad\qquad [a, b] := (a_\bullet^\mu b_\nu^\bullet - b_\bullet^\mu a_\nu^\bullet)_{\nu=+,\bullet}^{\mu=-,\bullet} = 0.$$

Proof. Since $\epsilon_t(\phi_s(I) - \phi_t(I))$ is a negative Hermitian form,

$$\epsilon_t(\mathrm{d}\phi_t(I)) = \epsilon_t\left(\phi_t(\lambda_\nu^\mu(I))\,\mathrm{d}\Lambda_\mu^\nu\right) = \phi_t\left(\lambda_+^-(I)\right)\mathrm{d}t \le 0.$$

Since $Y_t = \phi_t(B)$ commutes with $W_t(g)$ for all B and $g(t) = a$, we have by virtue of quantum Itô's formula

$$\mathrm{d}[Y_t, W_t] = [\mathrm{d}Y_t, W_t] + [Y_t, \mathrm{d}W_t] + [\mathrm{d}Y_t, \mathrm{d}W_t] = 0.$$

The equations (1.2), (2.2) and commutativity of a_ν^μ with Y_t and W_t imply

$$([\phi_t(b_\bullet^\mu), W_t] + [Y_t, a_\nu^\mu W_t] + \phi_t(b_\bullet^\mu)\,a_\nu^\bullet W_t - a_\bullet^\mu W_t \phi_t(b_\nu^\bullet))\,\mathrm{d}\Lambda_\mu^\nu.$$

$$= \ W_t(\phi_t(b_\bullet^\mu)\,a_\nu^\bullet - a_\bullet^\mu \phi_t(b_\nu^\bullet))\,\mathrm{d}\Lambda_\nu^\mu = W_t\phi_t(b_\bullet^\mu a_\nu^\bullet - a_\bullet^\mu b_\nu^\bullet)\,\mathrm{d}\Lambda_\mu^\nu = 0.$$

Thus $a \bullet b = b \bullet a$ by the argument [15] of independence of the integrators $\mathrm{d}\Lambda_\mu^\nu$.

In order to formulate the CP differential condition we need the notion of *quantum stochastic germ* for the CP flow ϕ at $t = 0$. It was defined in [24, 25], for a quantum stochastic differential (2.2) with $\phi_0(B) = B, \forall B \in \mathcal{B}$ as $\gamma^\mu_\nu = \lambda^\mu_\nu + \imath^\mu_\nu$, where λ^μ_ν are the structural maps $B \mapsto \lambda^\mu_\nu(B)$ given by the generators of the quantum Itô equation (2.2) and $\imath^\mu_\nu : B \mapsto B\delta^\mu_\nu$ is the ampliation of $\mathcal{B}$. Let us prove that the germ-maps γ^μ_ν of a CP flow ϕ must be conditionally completely positive (CCP) in a degenerated sense as it was found for the finite-dimensional bounded case in [18, 19]. Another, equivalent, but not so explicit characterization was suggested for this particular case in [27].

Theorem 2 If ϕ is a completely positive flow satisfying the quantum stochastic equation (2.2) with $\phi_0(B) = B$, then the germ-matrix $\gamma = (\lambda^\mu_\nu + \imath^\mu_\nu)^{\mu=-,\bullet}_{\nu=+,\bullet}$ is conditionally completely positive in the sense

$$\sum_{B \in \mathcal{B}} \iota(B)\,\zeta_B = 0 \Rightarrow \sum_{B,C \in \mathcal{B}} \langle \zeta_B | \gamma(B^*C)\,\zeta_C \rangle \geq 0.$$

Here $\zeta \in \mathcal{D} \oplus \mathcal{D}_\bullet, \mathcal{D}_\bullet = \mathcal{D} \otimes \mathcal{K}$, and $\iota = (\iota^\mu_\nu)^{\mu=-,\bullet}_{\nu=+,\bullet}$ is the degenerate representation $\iota^\mu_\nu(B) = B\delta^+_\nu \delta^\mu_-$, written both with γ in the matrix form as

$$(2.5) \qquad \gamma = \begin{pmatrix} \gamma & \gamma_\bullet \\ \gamma^\bullet & \gamma^\bullet_\bullet \end{pmatrix}, \qquad \iota(B) = \begin{pmatrix} B & 0 \\ 0 & 0 \end{pmatrix},$$

where $\gamma = \lambda^-_+$, $\gamma^m = \lambda^m_+$, $\gamma_n = \lambda^-_n$, $\gamma^m_n = \imath^m_n + \lambda^m_n$ with $\imath^m_n(B) = B\delta^m_n$ such that

$$(2.6) \qquad \gamma(B^*) = \gamma(B)^*, \qquad \gamma^n(B^*) = \gamma_n(B)^*, \qquad \gamma^m_n(B^*) = \gamma^n_m(B)^*.$$

If ϕ is subfiltering, then $D = -\lambda^-_+(I)$ is a positive Hermitian form, $\langle \eta | D\eta \rangle \geq 0$, for all $\eta \in \mathcal{D}$, and if ϕ is contractive, then $\boldsymbol{D} = -\boldsymbol{\lambda}(I)$ is positive in the sense $\langle \boldsymbol{\eta} | \boldsymbol{D\eta} \rangle \geq 0$ for all $\boldsymbol{\eta} \in \mathcal{D} \oplus \mathcal{D}_\bullet$.

Proof. The CP condition in the form (1.6) for the adapted map ϕ_t can be obviously extended on all $f^\bullet \in \mathcal{K}$ if the sesquianalytical function $f^\bullet \mapsto \phi_t(\bar{f}^\bullet, B, f^\bullet)$ is defined as the $\mathcal{K}$-function

$$(2.7) \qquad \langle \eta | \phi_t(\bar{f}^\bullet, B, f^\bullet)\,\eta \rangle = \langle \eta \otimes f^\otimes | \phi_t(B)\,\eta \otimes f^\otimes \rangle \exp\left[-\int_t^\infty \|f^\bullet(s)\|^2\,\mathrm{d}s\right],$$

where $\|f^\bullet(t)\|^2 = \sum_{n=1}^\infty |f^n(t)|^2$. It coincides with the former definition on $\mathcal{K}_t$ and does not depend on $f^\bullet(s)$, $s > t$ due to the adaptiveness (1.5) of $Y_t = \phi_t(B)$. If the $\mathfrak{D}$-form $\phi_t(B)$ satisfies the stochastic equation (2.2), the $\mathcal{D}$-form $\phi_t(\bar{f}^\bullet, B, f^\bullet)$ satisfies the differential equation [17]

$$\frac{\mathrm{d}}{\mathrm{d}t}\phi_t(\bar{f}^\bullet, B, f^\bullet) = \|f^\bullet(t)\|^2\,\phi_t(\bar{f}^\bullet, B, f^\bullet) + \phi_t(\bar{f}^\bullet, \lambda^-_+(B), f^\bullet)$$

$$+ \sum_{m=1}^{\infty} \bar{f}^m (t)\, \phi_t \left(\bar{f}^\bullet, \lambda_+^m (B), f^\bullet\right) + \sum_{n=1}^{\infty} \phi_t \left(\bar{f}^\bullet, \lambda_n^- (B), f^\bullet\right) f^n (t)$$

$$+ \sum_{m,n=1}^{\infty} \bar{f}^m (t)\, \phi_t \left(\bar{f}^\bullet, \lambda_n^m (B), f^\bullet\right) f^n (t) = \phi_t \left(\bar{f}^\bullet, \gamma \left(\bar{f}^\bullet (t), B, f^\bullet (t)\right), f^\bullet\right).$$

The positive definiteness of (2.7) ensures the conditional positive definiteness $\sum_f \sum_B B \xi_B^f = 0 \Rightarrow$

$$\sum_{B,C} \sum_{f,h} \left\langle \xi_B^f \right| \gamma_t \left(\bar{f}^\bullet, B^*C, h^\bullet\right) \xi_C^h \rangle = \frac{1}{t} \sum_{B,C} \sum_{f,h} \left\langle \xi_B^f \right| \phi_t \left(\bar{f}^\bullet, B^*C, h^\bullet\right) \xi_C^h \rangle \geq 0$$

of the form, given by $\gamma_t \left(\bar{f}^\bullet, B, f^\bullet\right) = \frac{1}{t} \left(\phi_t \left(\bar{f}^\bullet, B, f^\bullet\right) - B\right)$ for each $t > 0$. This holds also at the limit $\gamma_0 \left(\bar{f}^\bullet, B, f^\bullet\right) = \gamma \left(\bar{f}^\bullet (0), B, f^\bullet (0)\right)$, given at $t \downarrow 0$ by the $\mathcal{K}$-form

$$\gamma \left(\bar{k}^\bullet, B, k^\bullet\right) = \sum_{m,n} \bar{k}^m \gamma_n^m (B)\, k^n + \sum_m \bar{k}^m \gamma^m (B) + \sum_n \gamma_n (B)\, k^n + \gamma (B),$$

where $k^\bullet = f^\bullet (0) \in \mathcal{K}$, $\bar{k}^\bullet = k_\bullet$ and the γ's are defined in (2.5). Hence the form

$$\sum_{B,C} \sum_{\mu,\nu} \langle \zeta_B^\mu | \gamma_\nu^\mu (B^*C)\, \zeta_C^\nu \rangle := \sum_{B,C} \sum_{m,n} \langle \zeta_B^m | \gamma_n^m (B^*C)\, \zeta_C^n \rangle$$

$$+ \sum_{B,C} \left(\sum_n \langle \zeta_B | \gamma_n (B^*C)\, \zeta_C^n \rangle + \sum_m \langle \zeta_B^m | \gamma^m (B^*C)\, \zeta_C \rangle + \langle \zeta_B | \gamma (B^*C) | \zeta_C \rangle \right)$$

with $\zeta = \sum_f \xi^f$, $\zeta^\bullet = \sum_f \xi^f \otimes k_f^\bullet$, where $k_f^\bullet = f^\bullet (0)$, is positive if $\sum_B B \zeta_B = 0$. The components ζ and $\zeta^\bullet$ of these vectors are independent because for any $\zeta \in \mathcal{D}$ and $\zeta^\bullet = (\zeta^1, \zeta^2, \ldots) \in \mathcal{D} \otimes \mathcal{K}$ there exists such a function $k^\bullet \mapsto \xi^k$ on $\mathcal{K}$ with a countable support, that $\sum_k \xi^k = \zeta$, $\sum_k \xi^k \otimes k^\bullet = \zeta^\bullet$, namely, $\xi^k = 0$ for all $k^\bullet \in \mathcal{K}$ except $k^\bullet = 0$ with $\xi^0 = \zeta - \sum_{n=1}^{\infty} \zeta^n$ and $k^\bullet = e_n^\bullet$, the n-th basis element in ℓ^2, for which $\xi^k = \zeta^n$. This proves the complete positivity of the matrix form γ, with respect to the matrix representation ι defined in (2.5) on the ket-vectors $\zeta = (\zeta^\mu)$.

If $\epsilon (R_t) \leq I$, then $D = -\lambda (I) = \lim \frac{1}{t} \epsilon (I - R_t) \geq 0$, and we also conclude the dissipativity $\sum_{k,l} \langle \xi^k | D \left(\bar{k}^\bullet, l^\bullet\right) \xi^l \rangle \geq 0$ from

$$0 \leq \lim \frac{1}{t} \sum_{f,h} \left\langle \xi^f | e^{\int_0^t f_\bullet h^\bullet} I - \phi_t \left(\bar{f}^\bullet, I, h^\bullet\right) \xi^h \right\rangle = - \left\langle \xi^f | \lambda \left(\bar{f}^\bullet (0), I, h^\bullet (0)\right) \xi^h \right\rangle$$

if $\phi_t (I) \leq I$, where $\lambda \left(\bar{k}^\bullet, I, k^\bullet\right) = \gamma \left(\bar{k}^\bullet, I, k^\bullet\right) - \|k^\bullet\|^2 I = D \left(\bar{k}^\bullet, k^\bullet\right)$.

Obviously the CCP property for the germ-matrix $\boldsymbol{\gamma}$ is invariant under the transformation $\boldsymbol{\gamma} \mapsto \boldsymbol{\varphi}$ given by

$$(2.8) \qquad\qquad \boldsymbol{\varphi} (B) = \boldsymbol{\gamma} (B) + \boldsymbol{\iota} (B)\, \boldsymbol{K} + \boldsymbol{K}^\star \boldsymbol{\iota} (B),$$

where $\boldsymbol{K} = (K_\nu^\mu)_{\nu=+,\bullet}^{\mu=-,\bullet}$ is an arbitrary matrix of $K_\nu^\mu \in \mathcal{L}(\mathcal{D})$ with $K_{-\nu}^{\star\mu} = K_{-\mu}^{\nu*}$. As was proven in [18, 19] for the case of finite-dimensional matrix γ of bounded γ_ν^μ, see also [27], the matrix elements K_ν^- can be chosen in such way that the matrix map $\varphi = (\varphi_\nu^\mu)_{\nu=+,\bullet}^{\mu=-,\bullet}$ becomes CP from $\mathcal{B}$ into the quadratic matrices of $\varphi_\nu^\mu(B)$. (The other elements can be chosen arbitrarily, say as $K_+^\bullet = 0$, $K_\bullet^\bullet = \frac{1}{2}I_\bullet^\bullet$, because (2.8) does not depend on $K_+^\bullet, K_\bullet^\bullet$.) Thus the generator $\lambda = \gamma - \imath$ for a quantum stochastic CP flow ϕ can be written (at least in the bounded case) as $\varphi - \imath\boldsymbol{K} - \boldsymbol{K}^\star\imath$:

$$(2.9) \qquad \lambda_\nu^\mu(B) = \varphi_\nu^\mu(B) - B\left(\tfrac{1}{2}\delta_\nu^\mu I + \delta_-^\mu K_\nu\right) - \left(\tfrac{1}{2}\delta_\nu^\mu I + K^\mu \delta_\nu^+\right)B,$$

where $\varphi_\nu^\mu : \mathcal{B} \to \mathcal{B}(\mathcal{D})$ are matrix elements of the CP map φ and $K_\nu \in \mathcal{L}(\mathcal{D})$, $K^- = K_+^*$, $K^m = K_m^*$. Now we show that the germ-matrix of this form obeys the CCP property even in the general case of unbounded K_ν^-, $\varphi_\nu^\mu(B) \in \mathcal{B}(\mathcal{D})$.

Proposition 4 The matrix map $\gamma = (\gamma_\nu^\mu)_{\nu=+,\bullet}^{\mu=-,\bullet}$ given in (2.8) by

$$(2.10) \qquad \varphi = \begin{pmatrix} \varphi & \varphi_\bullet \\ \varphi^\bullet & \varphi_\bullet^\bullet \end{pmatrix}, \quad \text{and} \quad \boldsymbol{K} = \begin{pmatrix} K & K_\bullet \\ 0 & \frac{1}{2}I_\bullet^\bullet \end{pmatrix}, \quad \boldsymbol{K}^\star = \begin{pmatrix} K^* & 0 \\ K_\bullet^* & \frac{1}{2}I_\bullet^\bullet \end{pmatrix},$$

with $\varphi = \varphi_+^-$, $\varphi^m = \varphi_+^m$, $\varphi_n = \varphi_n^-$ and $\varphi_n^m = \gamma_n^m$ is CCP with respect to the degenerate representation $\iota = \left(\delta_-^\mu \delta_\nu^+ \iota\right)_{\nu=+,\bullet}^{\mu=-,\bullet}$, where $\iota(B) = B$, if φ is a CP map.

Proof. If $\iota(B_k)\eta^k = 0$, then

$$\langle \eta^k | \iota(B_k^* B_l)\boldsymbol{K} + \boldsymbol{K}^\star \iota(B_k^* B_l)\eta^l \rangle$$

$$= 2\mathrm{Re}\langle \iota(B_k)\eta^k | \iota(B_l)\boldsymbol{K}\eta^l \rangle = 0.$$

Hence the CCP for γ is equivalent to the CCP property for (2.8) and follows from its CP property:

$$\langle \eta^k | \gamma(B_k^* B_l)\eta^l \rangle = \langle \eta^k | \varphi(B_k^* B_l)\eta^l \rangle \geq 0$$

for such sequences $\eta^k \in \mathcal{D} \oplus \mathcal{D}_\bullet$.

3. Construction of quantum CP flows

The necessary conditions for the stochastic generator $\lambda = (\lambda_\nu^\mu)_{\nu=+,\bullet}^{\mu=-,\bullet}$ of a CP flow ϕ at $t = 0$ are found in the previous section in the form of a CCP property for the corresponding germ $\gamma = (\gamma_\nu^\mu)_{\nu=+,\bullet}^{\mu=-,\bullet}$ and $\gamma(I) \leq 0$. In the next paper we shall show, these conditions are essentially equivalent to the assumption (2.9), where $\varphi = (\varphi_\nu^\mu)_{\nu=+,\bullet}^{\mu=-,\bullet}$ is a CP map, also in the cases of unbounded λ.

Here we are going to prove under the following conditions for the operators $K, K_\bullet$ and the maps φ_ν^μ that this general form is also sufficient for the existence of the CP solutions to the quantum stochastic equation (2.2). We are going to construct the minimal quantum stochastic positive flow $B \mapsto \phi_t(B)$ for a given w*-continuous unbounded germ-matrix map of the above form, satisfying the following conditions.

1. First, we suppose that the operator $K \in \mathcal{B}(\mathcal{D})$ generates the one parametric semigroup $\left(e^{-Kt}\right)_{t>0}$, $e^{-Kr}e^{-Ks} = e^{-K(r+s)}$ of continuous operators $e^{-Kt} \in \mathcal{L}(\mathcal{D})$ in the strong sense

$$\lim_{t \searrow 0} \frac{1}{t}\left(I - e^{-Kt}\right)\eta = K\eta, \quad \forall \eta \in \mathcal{D}.$$

(A contraction semigroup on the Hilbert space $\mathcal{H}$ if K defines an accretive $K + K^\dagger \geq 0$ and so maximal accretive form.)

2. Second, we suppose that the solution $S_t^n, n \in \mathbb{N}$ to the recurrence

$$S_t^{n+1} = S_t^0 - \int_0^t S_{t-r}^0 \sum_{m=1}^{\infty} K_m S_r^n \mathrm{d}\Lambda_-^m, \quad S_t^0 = e^{-Kt} \otimes T_t,$$

where $e^{-Kt} \otimes T_t \in \mathcal{L}(\mathfrak{D})$ is the contraction given by the shift co-isometries $T_t : \mathfrak{F} \to \mathfrak{F}$, strongly converges to a continuous operator $S_t \in \mathcal{L}(\mathfrak{D})$ at $n \longrightarrow \infty$ for each $t > 0$.

3. Third, we suppose that the solution $R_t^n, n \in \mathbb{N}$ to the recurrence

$$R_t^{n+1} = S_t^* S_t + \int_0^t \mathrm{d}\Lambda_\mu^\nu\left(r, S_r^* \varphi_\nu^\mu\left(R_{t-r}^n\right) S_r\right), \quad R_t^0 = S_t^* S_t,$$

where the quantum stochastic non-adapted integral is understood in the sense [24], weakly converges to a continuous form $R_t \in \mathcal{B}(\mathfrak{D})$ at $n \longrightarrow \infty$ for each $t > 0$.

The first and second assumptions define the existence of free evolution semigroup $S^0 = \left(S_t^0\right)_{t>0}$ and its perturbation $S = (S_t)_{t>0}$ on the product space $\mathfrak{D} = \mathcal{D} \otimes \mathfrak{F}$ in the form of multiple quantum stochastic integral

(3.1)
$$S_t = S_t^0 + \sum_{n=1}^{\infty} (-1)^n \int \cdots \int_{0<t_1<\ldots<t_n<t} K_{m_n}(t-t_n) \cdots K_{m_1}(t_2-t_1) S_{t_1}^0 \mathrm{d}\Lambda_-^{m_1} \cdots \mathrm{d}\Lambda_-^{m_n},$$

iterating the quantum stochastic integral equation

(3.2)
$$S_t = S_t^0 - \int_0^t \sum_{m=1}^{\infty} K_m(t-r) S_r \mathrm{d}\Lambda_-^m, \quad S_0 = I,$$

where $K_m(t) = S_t^0 (K_m \otimes I)$. The third assumption supply the weak convergence for the series

(3.3)
$$R_t = S_t^* S_t + \sum_{n=1}^{\infty} \int \cdots \int_{0<t_1<\ldots<t_n<t} \mathrm{d}\Lambda_{\mu_1\ldots\mu_n}^{\nu_1\ldots\nu_n}\left(t_1,\ldots,t_n, \varphi_{\nu_1\ldots\nu_n}^{\mu_1\ldots\mu_n}\left(t_1,\ldots,t_n, S_{t-t_n}^* S_{t-t_n}\right)\right)$$

of non-adapted n-tuple CP integrals [24] with

$$(3.4) \qquad \varphi^{\mu_1 \cdots \mu_n}_{\nu_1 \cdots \nu_n}(t_1, \ldots, t_n) = \varphi^{\mu_1 \cdots \mu_{n-1}}_{\nu_1 \cdots \nu_{n-1}}(t_1, \ldots, t_{n-1}) \circ \varphi^{\mu_n}_{\nu_n}(t_n - t_{n-1}),$$

where $\varphi^{\mu}_{\nu}(t, B) = S_t^* \varphi^{\mu}_{\nu}(B) S_t$. The following theorem gives a characterization of the evolution semigroup S in terms of cocycles with unbounded coefficients, characterized by Fagnola [28] in the isometric and unitary case.

Proposition 5 Let the family $S^{\circ} = (S_t^{\circ})_{t>0}$ be a quantum stochastic adapted cocycle, $S_r^s S_s^{\circ} = S_{r+s}^{\circ}$, $S_t^s = \theta^s(S_t^{\circ})$, satisfying the HP differential equation

$$(3.5) \qquad dS_t^{\circ} + K S_t^{\circ} dt + \sum_{m=1}^{\infty} K_m S_t^{\circ} d\Lambda_-^m = 0, \quad S_0^{\circ} = I.$$

Then $S_t = T_t S_t^{\circ}$ is a semigroup solution, $S_r S_s = S_{r+s}$ to the non-adapted integral equation (3.2) such that $S_t \psi_f = S_t(f^{\bullet}) \eta \otimes \delta_{\varnothing}, \forall \eta \in \mathcal{D}$ on $\psi_f = \eta \otimes f^{\otimes}$ with $f^{\bullet} \in \mathfrak{K}_t$. Conversely, if $S = (S_t)_{t>0}$ is the non-adapted solution (3.1) to the integral equation (3.2), then

$$(3.6) \quad S_t^{\circ} = S_t^{\circ} + \sum_{n=1}^{\infty} \int_{0<t_1<\ldots<t_n<t} \cdots \int K_{m_n}^{\circ}(t - t_n) \cdots K_{m_1}^{\circ}(t_2 - t_1) S_{t_1}^{\circ} d\Lambda_-^{m_1} \cdots d\Lambda_-^{m_n},$$

where $K_m^{\circ}(t) = e^{-Kt} K_m \otimes I$, is the adapted solution to (3.5), defined as $S_t^{\circ} \psi_f = S_t(f^{\bullet}) \eta \otimes f^{\otimes}, \forall \eta \in \mathcal{D}$, where $S_t(f^{\bullet}) = F^* S_t F$ is given by $F\eta = \eta \otimes f^{\otimes}$ with $f^{\bullet} \in \mathfrak{K}_t$.

Proof. First let us show that the equation (3.5) is equivalent to the integral one

$$S_t^{\circ} = e^{-Kt} \otimes I - \int_0^t \sum_{m=1}^{\infty} e^{-K(t-r)} K_m S_r^{\circ} d\Lambda_-^m, \quad S_0^{\circ} = I.$$

Indeed, multiplying both parts of the integral equation from the left by $e^{K(t-s)}$ and differentiating the product $e^{K(t-s)} S_t^{\circ}$ at $t = s$, we obtain (3.5). Conversely, the integral equation can be obtained from (3.5) by the integration:

$$-\int_0^t \sum_{m=1}^{\infty} e^{-K(t-r)} K_m S_r^{\circ} d\Lambda_-^m = \int_0^t e^{-K(t-r)} (dS_r^{\circ} + K S_r^{\circ} dr)$$

$$= \int_0^t d\left(e^{-K(t-r)} S_r^{\circ}\right) = S_t^{\circ} - e^{-Kt} \otimes I.$$

The non-adapted equation (3.2) is obtained by applying the operator $T_t = T_{t-r} T_r$ to both parts of this integral equation and taking into account the commutativity of $e^{K(r-t)} K_m$ with T_r. Moreover, due to the adaptiveness of S_t°,

$$S_t \psi_f = T_t \left(E_t S_t^{\circ} \psi_f \otimes E_{[t} f^{\otimes}\right) = S_t(f^{\bullet}) \eta \otimes f_t^{\otimes},$$

where $f_t^{\otimes} = T_t f^{\otimes}$, and $S_t(f^{\bullet}) = E S_t^{\circ} F$ is the solution to the equation

$$S_t(f^{\bullet}) = e^{-Kt} + \int_0^t e^{-K(t-r)} K_{\bullet} f^{\bullet}(r) S_r(f^{\bullet}) dr, \qquad S_0(f^{\bullet}) = I.$$

Hence $S_t F = E^* S_t (f^\bullet)$ if $f^\bullet \in \mathfrak{K}_t$, and $F^* S_t F = S_t (f^\bullet)$ as $EF = I$. Since this equation is equivalent to the differential one

$$(3.7) \qquad \frac{\mathrm{d}}{\mathrm{d}t} S_t (f^\bullet) \eta + (K_\bullet f^\bullet (t) + K) S_t (f^\bullet) \eta = 0, \quad S_0 (f^\bullet) \eta = \eta, \qquad \forall \eta \in \mathcal{D},$$

the function $t \mapsto S_t (f^\bullet)$, $f^\bullet \in \mathfrak{K}$ is a strongly continuous cocycle,

$$S_r (f_s^\bullet) S_s (f^\bullet) = S_{r+s} (f^\bullet), \ \forall r, s > 0, \qquad f_s^\bullet (t) = f^\bullet (t+s), \quad S_0 (f^\bullet) = I.$$

As was proved in [24], the multiple integral (3.1) gives a solution to the integral equation (3.2), and so the multiple integral for $S_t^\circ \psi_f = S_t (f^\bullet) \eta \otimes f^\otimes$,

$$S_t (f^\bullet) \eta = e^{-Kt} + \sum_{n=1}^{\infty} (-1)^n \int \cdots \int_{0 < t_1 < \ldots < t_n < t} K (t, t_n) \cdots K (t_2, t_1) e^{-Kt_1} \eta \mathrm{d}t_1 \cdots \mathrm{d}t_n,$$

where $K (t, r) = e^{-K(t-r)} K_\bullet f^\bullet (r)$, corresponding to the iteration of the integral equation for S_t° on ψ_f, satisfies the HP equation (3.5).

The following theorem reduces the problem of solving of differential evolution equations to the problem of iteration of integral equations similar to the nonstochastic case [29, 30].

Proposition 6 Let $S_t = T_t S_t^\circ$, where $S_t^\circ \in \mathcal{L}(\mathcal{D})$ are continuous operators defining the adapted cocycle solution to the equation (3.5). Then the linear stochastic evolution equation (2.2) is equivalent to the quantum non-adapted (in the sense of [24]) integral equation

$$(3.8) \qquad \phi_t (B) = S_t^* B S_t + \int_0^t \mathrm{d}\Lambda_\mu^\nu \left(r, \phi_r \left[\varphi_\nu^\mu \left(S_{t-r}^* B S_{t-r} \right) \right] \right)$$

with $\phi_0 (B) = B \in \mathcal{B}$, where φ_ν^μ are extended onto $\mathfrak{B}$ by w-continuity and linearity as $\varphi_\nu^\mu (B \otimes Z) = \varphi_\nu^\mu (B) \otimes Z$ for $B \in \mathcal{B}$, $Z \in \mathcal{B}(\mathfrak{F})$.*

Proof. The non-adapted equation (3.8) is understood in the coherent form sense as

$$\langle \psi_f | \phi_t (B) \psi_f \rangle = \langle S_t \psi_f | B S_t \psi_f \rangle + \int_0^t \langle \psi_f | \phi_r \left[\varphi \left(\bar{f}^\bullet (r), S_{t-r}^* B S_{t-r}, f^\bullet (r) \right) \right] \psi_f \rangle \, \mathrm{d}r,$$

where $\varphi \left(\bar{k}^\bullet, B, k^\bullet \right) = \sum_{m,n} \bar{k}^m \varphi_n^m (B) k^n + \sum_m \bar{k}^m \varphi^m (B) + \sum_n \varphi_n (B) k^n + \varphi (B)$. Due to the adaptiveness of ϕ_t this can be written for $\psi_f = \eta \otimes f^\otimes = F\eta$ with $f^\bullet \in \mathfrak{K}_t$ as

$$(3.9) \qquad \phi_t \left(\bar{f}^\bullet, B, f^\bullet \right) = S_t^* \left(\bar{f}^\bullet \right) B S_t (f^\bullet)$$

$$+ \int_0^t \phi_r \left(\bar{f}^\bullet, \varphi \left(r, S_{t-r}^* \left(\bar{f}_r^\bullet \right) B S_{t-r} \left(f_r^\bullet \right) \right), f^\bullet \right) \mathrm{d}r,$$

where $S_t^* (\bar{f}^\bullet) = S_t (f^\bullet)^*$, $f_r^\bullet (t) = f^\bullet (t + r)$. Here we take into account that due to adaptiveness $F^* \phi_r [Y] F = \phi_r (\bar{f}^\bullet, F_r^* Y F_r, f^\bullet)$, where $F_r = T_r F$, and therefore

$$F^* \phi_r \left[\varphi \left(r, S_{t-r}^* BS_{t-r}\right)\right] F = \phi_r \left(\bar{f}^\bullet, F_r^* \varphi \left(r, S_{t-r}^* BS_{t-r}\right) F_r, f^\bullet\right) =$$

$$\phi_r \left(\bar{f}^\bullet, \varphi \left(r, F_r^* S_{t-r}^* BS_{t-r} F_r\right), f^\bullet\right) = \phi_r \left(\bar{f}^\bullet, \varphi \left(r, S_{t-r}^* \left(\bar{f}_r^\bullet\right) BS_{t-r} \left(f_r^\bullet\right)\right), f^\bullet\right)$$

as $F_t^* \varphi (t, B) F_t = \varphi (t, F_t^* B F_t)$ for $\varphi (t, B) = \varphi \left(\bar{f}^\bullet (t), B, f^\bullet (t)\right)$ and $S_{t-r} F_r = F_t S_{t-r} (f_r^\bullet)$, where $F_t \eta = \eta \otimes \delta_\emptyset$ for any $f^\bullet \in \mathfrak{K}_t$.

Let us prove that the operator-valued function $t \mapsto S_s (t, f^\bullet) := S_{s-t} (f_t^\bullet)$ satisfies the backward evolution equation

$$\frac{\mathrm{d}}{\mathrm{d}t} S_s (t, f^\bullet) \eta = S_s (t, f^\bullet) (K_\bullet f^\bullet (t) + K) \eta, \quad S_0 (f_s^\bullet) \eta = \eta \quad \forall t \in [0, s).$$

Indeed, taking into account the forward equation (3.7), we obtain it at $r = t$ from the cocycle property $S_s (t, f^\bullet) S_t (r, f^\bullet) = S_s (r, f^\bullet)$:

$$0 = \frac{\mathrm{d}}{\mathrm{d}t} (S_s (r, f^\bullet) \eta) = \left(\frac{\mathrm{d}}{\mathrm{d}t} S_s (t, f^\bullet) - S_s (t, f^\bullet) (K_\bullet f^\bullet (t) + K)\right) S_t (r, f^\bullet) \eta.$$

Now, replacing B in (3.9) by $Y_s (\bar{f}^\bullet, t, f^\bullet) = S_s^* (t, \bar{f}^\bullet) BS_s (t, f^\bullet)$, we can write

$$\phi_t \left(\vec{f}^\bullet, S_s^* (t, \bar{f}^\bullet) BS_s (t, f^\bullet), f^\bullet\right)$$

$$= S_s^* (\bar{f}^\bullet) BS_s (f^\bullet) + \int_0^t \phi_r \left(\vec{f}^\bullet, \varphi (r, S_s^* (r, \bar{f}^\bullet) BS_s (r, f^\bullet)), f^\bullet\right) \mathrm{d}r.$$

Calculating the total derivative $\frac{\mathrm{d}}{\mathrm{d}t} \phi_t \left(\bar{f}^\bullet, S_s^* (t, \bar{f}^\bullet) BS_s (t, f^\bullet), f^\bullet\right)$ by taking into account the backward equation, we obtain the differential equation at $s = t$:

$$\frac{\mathrm{d}}{\mathrm{d}t} \phi_t \left(\bar{f}^\bullet, B, f^\bullet\right) + \phi_t \left(\bar{f}^\bullet, K (t)^* B + BK (t), f^\bullet\right) = \phi_t \left(\bar{f}^\bullet, \varphi \left(\bar{f}^\bullet (t), B, f^\bullet (t)\right), f^\bullet\right)$$

where $K (t) = K + K_\bullet f^\bullet (t)$. This equation written for $\langle \eta | \phi_t (\bar{f}^\bullet, B, f^\bullet) \eta \rangle$ coincides with the coherent matrix form (2.1) for the quantum stochastic equation (2.2) with $\psi_f = F \eta$.

The converse is easy to show by integrating the equation for $\phi_t (\bar{f}^\bullet, B, f^\bullet)$ with B replaced by $Y (t) = S_s^* (t, \bar{f}^\bullet) BS_s (t, f^\bullet)$:

$$\phi_s (\bar{f}^\bullet, B, f^\bullet) - S_s^* (\bar{f}^\bullet) BS_s (f^\bullet) = \int_0^s \frac{\mathrm{d}}{\mathrm{d}t} \phi_t (\bar{f}^\bullet, S_s^* (t, \bar{f}^\bullet) BS_s (t, f^\bullet), f^\bullet) \mathrm{d}t$$

$$= \int_0^s \left(\frac{\mathrm{d}}{\mathrm{d}t} \phi_t (\bar{f}^\bullet, Y (r), f^\bullet) + \phi_r \left(\bar{f}^\bullet, \frac{\mathrm{d}}{\mathrm{d}t} Y (t), f^\bullet\right)\right)_{r=t} \mathrm{d}t$$

$$= \int_0^s \phi_t (\bar{f}^\bullet, \varphi (t, S_s^* (t, \bar{f}^\bullet) BS_s (t, f^\bullet)), f^\bullet) \mathrm{d}t,$$

whereas $\frac{\mathrm{d}}{\mathrm{d}t} Y(t) = (K_\bullet f^\bullet(t) + K)^* Y(t) + Y(t)(K_\bullet f^\bullet(t) + K)$.

Theorem 3 *Let φ be a w^*-continuous CP-map, and $S_t = T_t S_t^\circ$ be given by the solution to the quantum stochastic equation (3.5). Then the solutions to the evolution equation (2.2) with the generators, corresponding to (3.6), have the CP property and satisfy the submartingale (contractivity) condition $\phi_t(I) \leq \epsilon_t[\phi_s(I)]$ for all $t < s$ if $\varphi(I) \leq K + K^\dagger$ ($\phi_t(I) \leq \phi_s(I)$ if $\varphi(I) \leq \boldsymbol{K} + \boldsymbol{K}^\star$). The minimal solution can be written in the form of multiple quantum stochastic integral in the sense [24] as the series*

(3.10)

$$\phi_t(B) = \sum_{n=0}^{\infty} \int \cdots \int_{0 < t_1 < \ldots < t_n < t} \mathrm{d}\Lambda_{\mu_1 \ldots \mu_n}^{\nu_1 \ldots \nu_n}(t_1, \ldots, t_n, \varphi_{\nu_1 \ldots \nu_n}^{\mu_1 \ldots \mu_n}(t_1, \ldots, t_n, S_{t-t_n}^* B S_{t-t_n}))$$

of non-adapted n-tuple CP integrals with $S_t^* B S_t$ at $n = 0$ and

$$\varphi_{\nu_1 \ldots \nu_n}^{\mu_1 \ldots \mu_n}(t_1, \ldots, t_n) = \varphi_{\nu_1}^{\mu_1}(t_1) \circ \varphi_{\nu_2}^{\mu_2}(t_2 - t_1) \circ \ldots \circ \varphi_{\nu_n}^{\mu_n}(t_n - t_{n-1}),$$

where $\varphi_\nu^\mu(t, B) = S_t^* \varphi_\nu^\mu(B) S_t$. If φ is bounded, then the solution to the equation is unique, and $\phi_t(I) = \epsilon_t[\phi_s(I)]$ for all $t < s$ if $K + K^\dagger = \varphi(I)$ ($\phi_t(I) = I$ if $\boldsymbol{K} + \boldsymbol{K}^\star = \varphi(I)$).

Proof. The existence and uniqueness of the solutions $\phi_t(B)$ to the quantum stochastic equations (2.2) with the bounded generators $\lambda_\nu^\mu(B) = \gamma_\nu^\mu(B) - B\delta_\nu^\mu$ and the initial conditions $\phi_0(B) = B$ in an operator algebra $\mathcal{B} \subseteq \mathcal{L}(\mathcal{H})$ was proved in [24]. The CP property of the solution to this equation with the generators (2.9), given by a CP φ, can be proven in the form (3.10), which is obtained by the iteration

$$\phi_t^{n+1}(B) = S_t^* B S_t + \int_0^t \mathrm{d}\Lambda_\mu^\nu\left(r, \phi_r^n\left(\varphi_\nu^\mu\left(S_{t-r}^* B S_{t-r}\right)\right)\right), \quad \phi_t^0(B) = S_t^* B S_t$$

of the equivalent non-adapted integral equation (3.8). Indeed, in order to prove the complete positivity of the solution, written in this form, one should prove the positive definiteness of the iteration

$$
\begin{aligned}
\phi_t^{n+1}\left(\bar{f}^\bullet, B, f^\bullet\right) = {} & S_t^*\left(\bar{f}^\bullet\right) B S_t\left(f^\bullet\right) \\
& + \int_0^t \phi_s^n\left(\bar{f}^\bullet, \varphi\left(\bar{f}^\bullet(r), S_{t-r}^*\left(\bar{f}_r^\bullet\right) B S_{t-r}\left(f_r^\bullet\right), f^\bullet(r)\right), f^\bullet\right) \mathrm{d}r
\end{aligned}
$$

of the integral equation (3.9) with the CP $\phi_t^0(B) = S_t^* B S_t$. Thus, we have to test the positive definiteness of the forms

$$\sum_{B,C} \sum_{f,h} \left\langle \xi_B^f \middle| \phi_t^{n+1}(f^\bullet, B^*C, h^\bullet) \xi_C^h \right\rangle = \sum_{B,C} \left\langle B S_t(f^\bullet)\eta_B \middle| C S_t(f^\bullet)\eta_C \right\rangle$$

$$+ \int_0^t \sum_{B,C} \sum_{f,h} \left\langle \eta_B^f(r) \middle| \phi_s^n\left(\bar{f}^\bullet, \varphi\left(S_{t-r}^*\left(\bar{f}_r^\bullet\right) B^*C S_{t-r}\left(h_r^\bullet\right)\right), h^\bullet\right) \eta_C^h(r) \right\rangle,$$

where $\eta_B = \sum_f \xi_B^f$, $\eta_B^f(r) = \sum_{f(r)} \xi_B^f \otimes \boldsymbol{f}(r)$, and $\boldsymbol{f}(r) = 1 \oplus f^\bullet(r)$. It is a consequence of the CP condition for φ and the CP property for $\phi_{t_n}^n, \forall t_n < t$, which obviously follows from the positive definiteness of $\phi_r^{n-1}, r < t_n$, and so on up to $\phi_r^0, r < t_1$. The direct iteration of this integral recursion with the initial CP condition $\phi_t^0(B) = S_t^* B S_t$ gives at the limit $n \to \infty$ the solution in the form of a series

$$\phi_t\left(\bar{f}^\bullet, B, f^\bullet\right) = \sum_{n=0}^{\infty} \int \cdots \int_{0 < t_1 < \ldots < t_n < t} \varphi\left(t_1, \ldots, t_n; S_{t-t_n}^*\left(\bar{f}_{t_n}^\bullet\right) B S_{t-t_n}\left(f_{t_n}^\bullet\right)\right) dt_1 \cdots dt_n,$$

of n-tuple integrals on the interval $[0, t]$ with $S_t^*\left(\bar{f}^\bullet\right) B S_t\left(f^\bullet\right)$ at $n = 0$. The positive definite kernels

$$\varphi\left(t_1, \ldots, t_n\right) = \varphi^0\left(t_1\right) \circ \varphi^{t_1}\left(t_2\right) \circ \ldots \circ \varphi^{t_{n-1}}\left(t_n\right),$$

where $\varphi^r(t, B) = S_{t-r}^*\left(\bar{f}_r^\bullet\right) \varphi\left(\bar{f}^\bullet(t), B, f^\bullet(t)\right) S_{t-r}\left(f_r^\bullet\right)$, are obtained by the recurrence

$$\varphi\left(t_1, \ldots, t_n\right) = \varphi\left(t_1, \ldots, t_{n-1}\right) \circ \varphi^{t_{n-1}}\left(t_n\right), \qquad \varphi(t) = \varphi^0(t),$$

corresponding to (3.4). This proves the CP property for the series (3.10), which converges to a $0 \leq Y_t \leq \kappa R_t$ for any positive bounded $0 \leq B \leq \kappa I$ because of the increase $Y_t^n \leq Y_t^{n+1}$ for $Y_t^n = \phi_t^n(B)$ and the boundedness $Y_t^n \leq \kappa R_t^n$, $R_t^n \leq R_t$, where R_t is the continuous sesquilinear form (3.3).

As follows from the exponential estimate [24] for the solutions to the quantum stochastic equations (2.2) with the bounded generators, $R_t = \phi_t(I)$ might be unbounded, but strongly continuous in the Fock scale $\mathfrak{F}$. In the case of unbounded generators the solution to (3.3) might not be unique, and the iterated series (3.10) gives obviously the minimal one, which is unique among such solutions. Let us prove the submartingale property for the sesquilinear form R_t, given by the weakly convergent series (3.3). R_s for a $s > t$ is defined as the iterated solution $Y_s = R_s :=$ $\lim R_s^n$ to the backward integral equation

$$Y_s = S_s^* B S_s + \int_0^s d\Lambda_\mu^\nu\left(r, S_r^* \varphi_\nu^\mu\left(Y_{s-r}\right) S_r\right)$$

for the series $Y_s = \phi_s(B)$ with $B = I$. It satisfies the integral equation

$$R_s = S_t^* R_{s-t} S_t + \int_0^t d\Lambda_\mu^\nu\left(r, S_r^* \varphi_\nu^\mu\left(R_{s-r}\right) S_r\right),$$

where we used the semigroup property $S_{s-t} S_t = S_s$ and that

$$\int_t^s d\Lambda_\mu^\nu\left(r, S_r^* \varphi_\nu^\mu\left(Y_{s-r}\right) S_r\right) = S_t^{\circ*} \int_t^s d\Lambda_\mu^\nu\left(r, T_t^* S_{r-t}^* \varphi_\nu^\mu\left(Y_{s-r}\right) S_{r-t} T_t\right) S_t^\circ =$$

$$S_t^* \int_t^s d\Lambda_\mu^\nu\left(r - t, S_{r-t}^* \varphi_\nu^\mu\left(Y_{s-r}\right) S_{r-t}\right) S_t = S_t^* \int_0^{s-t} d\Lambda_\mu^\nu\left(r, S_r^* \varphi_\nu^\mu\left(Y_{s-t-r}\right) S_r\right) S_t.$$

This can be written in terms of the coherent matrix elements $R_s\left(\bar{f}^\bullet, f^\bullet\right) = F^* R_s F$, $f^\bullet \in \mathfrak{K}_s$ as

$$
\begin{aligned}
R_s\left(\bar{f}^\bullet, f^\bullet\right) \;=\; & S_t^*\left(\bar{f}^\bullet\right) R_{s-t}\left(\bar{f}^\bullet, f_t^\bullet\right) S_t\left(f^\bullet\right) \\
& + \int_0^t S_r^*\left(\bar{f}^\bullet\right) \varphi\left(\bar{f}^\bullet\left(r\right), R_{s-r}\left(\bar{f}^\bullet, f_r^\bullet\right), f_r^\bullet\right) S_r\left(f^\bullet\right) \mathrm{d}r.
\end{aligned}
$$

The coherent matrix elements $Y_t\left(\bar{f}^\bullet, f^\bullet\right)$ of the conditional expectation $Y_s = \epsilon_t\left(R_s\right)$ coincide with $R_s\left(\bar{f}^\bullet, f^\bullet\right)$ if $f^\bullet \in \mathfrak{K}_t$. Hence, they satisfy the integral equation

$$
\begin{aligned}
Y_t\left(\bar{f}^\bullet, f^\bullet\right) \;=\; & S_t^*\left(\bar{f}^\bullet\right) P_{s-t} S_t\left(f^\bullet\right) \\
& + \int_0^t S_r^*\left(\bar{f}^\bullet\right) \varphi\left(\bar{f}^\bullet\left(r\right), Y_{t-r}\left(\bar{f}^\bullet, f_r^\bullet\right), f_r^\bullet\right) S_r\left(f^\bullet\right) \mathrm{d}r,
\end{aligned}
$$

corresponding to the non-adapted backward equation

$$
Y_t = S_t^* P_{s-t}^\circ S_t + \int_0^t \mathrm{d}\Lambda_\mu^\nu\left(r, S_r^* \varphi_\nu^\mu\left(Y_{t-r}\right) S_r\right)
$$

where $P_s^\circ = P_s \otimes I$, $P_s = R_s\left(0,0\right)$, as $f_t^\bullet\left(r\right) = f^\bullet\left(r+t\right) = 0, \forall r \in \mathbb{R}_+$ if $f^\bullet \in \mathfrak{K}_t$. The operators $P_s = \epsilon\left(R_s\right) = \theta_s\left(I\right)$ are given by the Markov semigroup $\theta_s = \epsilon \circ \phi_s$ as the decreasing solutions to the integral equation

$$
P_s = e^{-Ks*} e^{-Ks} + \int_0^s e^{-Kr*} \varphi\left(P_{s-r}\right) e^{-Kr} \mathrm{d}r,
$$

and $P_t \leq I$ if $K + K^\dagger \leq 0$. (See, for example, [29].) Thus, the difference $\widetilde{R}_t = R_t - \epsilon_t\left(R_s\right) = R_t - Y_t$ satisfies the same equation

$$
\widetilde{R}_t = S_t^* \widetilde{I}_{s-t} S_t + \int_0^t \mathrm{d}\Lambda_\mu^\nu\left(r, S_r^* \varphi_\nu^\mu\left(\widetilde{R}_{t-r}\right) S_r\right)
$$

as R_t with $\widetilde{I}_s = I - P_s^\circ$ instead of I. The iteration of this equation defines it as the weak limit $\widetilde{R}_t = \lim \widetilde{R}_t^n$ in the form of the series (3.10) with $B = I - P_{s-t} \geq 0$. Hence $\widetilde{R}_t = \phi_t\left(I - P_{s-t}\right)$ is a positive sesquilinear form on $\mathfrak{D}$ for any $s \geq t$ due to the positivity of ϕ_t. The proof of contractivity $\phi_t(I) \leq \phi_s(I)$ for $t < s$ is similar to that one, without the vacuum averaging of R_t.

References

[1] Belavkin, V. P. Nondemolition Measurements and Nonlinear Filtering of Quantum Stochastic Processes. Lecture Notes in Control and Information Sciences, **121**, pp 245-266, Springer-Verlag, 1988.

[2] Belavkin, V.P. Nondemolition Calculus and Nonlinear Filtering in Quantum Systems. In: Stochastic Methods in Mathematics and Physics, pp 310-324, World Scientific, 1989.

[3] Belavkin, V.P. Quantum Stochastic Calculus and Quantum Nonlinear Filtering. J. Multivariate Analysis, **42** (2), pp 171-201, 1992.

[4] Gisin, N. Phys. Rev.Lett., **52**, pp 1657-60, 1984.

[5] Diosi, L. Phys Rev **A 40**, pp1165-74, 1988.

[6] Barchielli, A.and Belavkin, V.P. Measurement Continuous in Time and a Posteriori States in Quantum Mechanics. J. Phys. A: Math. Gen., **24**, pp 1495-1514, 1991.

[7] Belavkin, V.P. Quantum Continual Measurements and a Posteriori Collapse on CCR. Commun.Math.Phys., **146,** pp 611-635, 1992.

[8] Carmichael, H. Open Systems in Quantum Optics. Lecture Notes in Physics, m**18**, Springer-Verlag, 1993.

[9] Milburn, G. Phys. Rev. A, **36,** p 744, 1987.

[10] Pearle, P. Phys. Rev. D, **29,** p 235, 1984.

[11] Ghirardi, G.C., Pearle, P., Rimini A. Markov Processes in Hilbert Space and Continuous Spontaneous Localization of Systems of Identical Particles. Phys. Rev. A, **42**, pp 78-89, 1990.

[12] Lindblad, G. On the Generators of Quantum Dynamical Semigroups. Comm. Math. Phys., **48**, pp119-130, 1976.

[13] Evans, M.P., and Hudson, R.S. Multidimensional Quantum Diffusions. Lect Notes Math., 1303, pp69-88, 1988.

[14] Belavkin, V.P. Chaotic States and Stochastic Integration in Quantum Systems. Russian Math. Survey, **47**, (1), pp. 47-106, 1992.

[15] Parthasarathy, K.R. An Introduction to Quantum Stochastic Calculus. Birkhäuser, Basel, 1992.

[16] Meyer, P.A. Quantum Probability for Probabilists, Lecture Notes in Mathematics, 1538, Springer-Verlag, Heidelberg, 1993.

[17] Hudson, R.S., and Parthasarathy, K.R. Quantum Itô's Formula and Stochastic Evolution. Comm. Math. Phys., **93**, pp301-323, 1984.

[18] Belavkin, V.P. On Stochastic Generators of Completely Positive Cocycles. Russ. J. Math.Phys., 3, pp523-528, 1995.

[19] Belavkin, V.P. On the General Form of Quantum Stochastic Evolution Equation. Stochastic Analysis and Applications, Proc. of Fifth Gregynog Symposium, World Scientific, Singapore 1996, pp91-106.

[20] Obata, N. White Noise Calculus and Fock Space. Lecture Notes in Mathematics, 1577, Springer-Verlag, Heidelberg, 1994.

[21] Stinespring, W.F. Positive Functions on C*-algebras, Proc.Amer.Math.Soc. **6**, pp. 242-247, 1955.

[22] Belavkin, V.P. Continuous Nondemolition Observation, Quantum Filtering and Optimal Estimation. Lecture Notes in Physics, 378, pp151-163, Springer-Verlag, Berlin, 1991.

[23] Parthasarathy, K.R., Sinha, K.B. Stochastic Integral Representations of Bounded Quantum Martingales in Fock Space. J. Funct. Anal., 67, 126-151, 1986.

[24] Belavkin, V.P. A Quantum Nonadapted Itô Formula and Stochastic Analysis in Fock Scale. J. Funct. Anal., **102,** No.2, pp414-447, 1991.

[25] Belavkin, V.P. Positive Definite Germs of Quantum Stochastic Processes. Comptes Rendus, 1, 1996.

[26] Belavkin, V.P. A Pseudo-Euclidean Representation of Conditionally Positive Maps. Math. Notes, **49**, No.6, pp135-137, 1991.

[27] Lindsay, J.M., Parthasarathy, K.R. Positivity and Contractivity of Quantum Stochastic Flows. Stochastic Analysis and Applications, Proc.of Fifth Gregynog Symposium, World Scientific, Singapore 1996, pp315-329.

[28] Fagnola, F. Characterization of Isometric and Unitary Weakly Differentiable Cocycles in Fock Space, Quantum Probability and Related Topics, 8, World Scientific, Singapore 1993, pp143-164.

[29] Belavkin, V. P. Multiquantum Systems and Point Processes I, Rep. in Math. Phys., 28, No1, pp57-90, 1989.

[30] Chebotarev, A. M. The Theory of Conservative Dynamical Semigroup and its Applications. Preprint MIEM n.1. March 1990.

STEIN'S METHOD: SOME PERSPECTIVES WITH APPLICATIONS

LOUIS H. Y. CHEN,* *National University of Singapore*

Abstract

This paper presents Stein's method from both a concrete and an abstract point of view. A proof of the Berry-Esseen theorem using the method is given. Two approaches to the construction of Stein identities are discussed: the antisymmetric function approach and an L^2 space approach. A brief history of the developments of Stein's method and some possible prospects are also mentioned.

STEIN'S METHOD; STEIN IDENTITIES; STEIN EQUATION; ANTISYMMETRIC

FUNCTION APPROACH; L^2 SPACE APPROACH; NORMAL APPROXIMATION;

POISSON APPROXIMATION; COMPOUND POISSON APPROXIMATION ON

GROUPS

AMS 1991 SUBJECT CLASSIFICATION: PRIMARY 60F05;60F99;60B15

SECONDARY 60F10;60G55;60J25;60J27

1. Introduction

We often use approximations in the calculation of probabilities. But approximations are meaningful only if we have some knowledge of the errors involved. The use of Fourier analysis provides a means of estimating such errors. It works particularly well for probabilities involving sums of independent random variables. But if the random variables are dependent, particularly if the dependence is local or of a combinatorial nature, a method due to Stein (1972) has proved to be more fruitful.

Stein's 1972 paper deals with normal approximation. But his ideas are applicable to other probability approximations. In the past two decades, Stein's method has been

* Postal address: Department of Mathematics, National University of Singapore, Lower Kent Ridge Road, Singapore 119260, Republic of Singapore.

developed and applied to many other contexts. Notable among these are Poisson (Chen (1975)), binomial (Stein (1986)), multivariate normal (Barbour (1990), Götze (1991)), multivariate Poisson and Poisson process (Barbour (1988), Barbour and Brown (1992)), compound Poisson (Barbour, Chen and Loh (1992)), and multinomial (Loh (1992)). The application of Stein's method to Poisson approximation has been the most successful. The results obtained therefrom have diverse applications in fields ranging from random graphs to molecular biology. See, for example, Arratia, Goldstein and Gordon (1990) and Barbour, Holst and Janson (1992). While the application of Stein's method has been successful in many instances, the method's potential has not been fully realized and there is further scope for development.

Stein's method may be regarded as a method of constructing certain kinds of identities, which we call Stein identities, and making comparisons between them. In applying the method to probability approximation we construct two identities, one for the approximating distribution and the other for the distribution to be approximated. The discrepancy between the two distributions is then measured by comparing the two Stein identities through the use of the solution of an equation, called Stein equation. To effect the comparison, bounds on the solution and its smoothness are used.

In his 1986 monograph Stein proposed a general approach to the construction of Stein identities. In this approach one uses an exchangeable pair of random variables and an antisymmetric function. For the Poisson and normal approximations, Barbour (1988, 1990) expressed the Stein identities in terms of the infinitesimal generators of reversible Markov processes by transforming the first order operators in the identities to second order ones. He then solved the Stein equations in terms of the Markov processes and used coupling to bound the solutions and their smoothness. This probabilistic approach of Barbour enabled him to generalize Poisson and normal approximations to higher dimensions and to process approximation (see also Barbour and Brown (1992a)). Barbour (1997) later showed that the involvement of the infinitesimal generator of a reversible Markov process in a Stein identity comes naturally from the construction of the Stein identity using the antisymmetric function approach.

This paper is not a survey paper but an attempt to present Stein's method from both a concrete and an abstract point of view. In Section 2, we present a proof of the Berry-Esseen theorem for independent but nonidentically distributed random variables as an example of an application of Stein's method. This proof which uses the concentration inequality approach is based on an unpublished work of Chen (1986). The concentration inequality approach is due to Stein whose proof of the Berry-Esseen theorem for independent and identically distributed random variables is given in Ho and Chen (1978). For the nonidentically distributed case, additional arguments are needed to circumvent the lack of symmetry.

In Section 3, we present a systematic account of the antisymmetric function approach for constructing Stein identities by consolidating the ideas already existing in the literature. In particular, we focus our attention on a certain class of antisymmetric functions. In Section 4, we present a new approach to the construction of Stein identities. We call this an L^2 space approach and show how this approach also yields Stein identities involving the infinitesimal generators of reversible Markov processes.

In Section 5, we discuss compound Poisson approximation on groups using the L^2 space approach. This approach leads to a Stein equation different from that used by Barbour, Chen and Loh (1992) and offers prospects of viewing compound Poisson approximation in a new light. In Section 6, we give a brief history of the developments of Stein's method. In Section 7, we mention some possible prospects. It is hoped that the perspectives expressed in this paper may lead to wider and more fruitful applications of Stein's method.

2. A proof of the Berry-Esseen theorem

As an example of an application of Stein's method we give a proof of the Berry-Esseen theorem for independent but nonidentically distributed random variables. This proof uses the concentration inequality approach and is based on an unpublished work of Chen (1986).

Let C_{bd} be the class of bounded, continuous and piecewise differentiable functions f defined on $\mathbb{R}$ such that f' is bounded. We first construct a Stein identity for the standard normal distribution. By integration by parts we obtain the identity as

$$(2.1) \qquad E\left\{f'(Z) - Zf(Z)\right\} = 0$$

where Z is a standard normal random variable and $f \in C_{\mathrm{bd}}$. This identity characterizes the normal distribution in the sense that if Z is a random variable such that (2.1) holds for all $f \in C_{\mathrm{bd}}$, then Z is a standard normal random variable. The characterization is proved by substituting $f = f_z$ in (2.1) where f_z is the unique bounded solution of the Stein equation

$$(2.2) \qquad f'(w) - wf(w) = I(w \le z) - \Phi(z)$$

where Φ is the standard normal distribution function. The solution f_z is given by

$$
\begin{aligned}
f_z(w) &= -e^{w^2/2} \int_w^\infty [I(x \le z) - \Phi(z)]e^{-x^2/2}\,dx \\
&= e^{w^2/2} \int_{-\infty}^w [I(x \le z) - \Phi(z)]e^{-x^2/2}\,dx.
\end{aligned}
$$

Let $X_1, \ldots, X_n$ be independent random variables with $EX_i = 0$, $\mathrm{Var}(X_i) = \sigma_i^2$ and $E|X_i|^3 = \gamma_i < \infty$. Let $W = \sum_{i=1}^n X_i$ and without loss of generality assume $\mathrm{Var}(W) = 1$. We construct a Stein identity for $\mathcal{L}(W)$ as follows.

Let $X_1', \ldots, X_n'$ be an independent copy of $X_1, \ldots, X_n$. Define

$$
\begin{aligned}
K_i(t) &= EX_i[I(X_i \ge t \ge 0) - I(X_i < t < 0)], \\
M_i(t) &= X_i[I(X_i' \ge t \ge 0) - I(X_i' < t < 0)], \\
K(t) &= \sum_{i=1}^n K_i(t), \\
M(t) &= \sum_{i=1}^n M_i(t).
\end{aligned}
$$

Then $K(t) \ge 0$,

$$
\begin{aligned}
\int K(t)\,dt &= 1, \\
\int |t|K(t)\,dt &= \frac{1}{2}\sum_{i=1}^n \gamma_i, \\
\int E^W M(t)\,dt &= 0,
\end{aligned}
$$

and

$$\int \mathrm{Var}\, M(t)dt = \sum_{i=1}^{n} EX_i^2 E|X_i'| \le \sum_{i=1}^{n} \gamma_i,$$

where E^W denotes the conditional expectation given W.

Let $W^{(i)} = W - X_i$. Observe that for each i, $\mathcal{L}(X_i, X_i', W^{(i)}) = \mathcal{L}(X_i', X_i, W^{(i)})$. Therefore for $f \in C_{\mathrm{bd}}$,

$$\sum_{i=1}^{n} EX_i f(W + X_i') = \sum_{i=1}^{n} EX_i' f(W + X_i').$$

Rewriting the left hand side as

$$\sum_{i=1}^{n} EX_i[f(W + X_i') - f(W)] + EW f(W) = E\int f'(W + t)M(t)dt + EW f(W),$$

and the right hand side as

$$\sum_{i=1}^{n} EX_i'[f(W + X_i') - f(W)] = E\int f'(W + t)K(t)dt,$$

we obtain a Stein identity for $\mathcal{L}(W)$ as

$$(2.3) \qquad E\left\{\int f'(W + t)K(t)dt - \int f'(W + t)E^W M(t)dt - W f(W)\right\} = 0$$

where $f \in C_{\mathrm{bd}}$. Rewriting (2.3) we obtain

$$(2.4) \qquad\qquad\qquad E\left\{f'(W) - W f(W)\right\} = R(f)$$

where

(2.5)

$$R(f) = E\left\{\int [f'(W) - f'(W + t)]K(t)dt - \int [f'(W) - f'(W + t)]E^W M(t)dt\right\}.$$

To approximate $\mathcal{L}(W)$ by the standard normal distribution, we substitute $f = f_z$ (the solution of (2.2)) in (2.4). This yields

$$(2.6) \qquad\qquad\qquad F(z) - \Phi(z) = R(f_z)$$

where F is the distribution function of W. What remains to be done is to study the boundedness properties of f_z and use them to bound $R(f_z)$.

It has been proved in Stein (1986, p.23) that $0 \leq f_z(w) \leq 1$ and $|f_z'(w)| \leq 1$. Using these properties and (2.2), we obtain

$$(2.7) \qquad |f_z'(W) - f_z'(W + t)| \leq (|W| + 1)|t| + I(z - t \leq W \leq z)I(t \geq 0)$$
$$+ I(z \leq W \leq z - t)I(t < 0).$$

Combining (2.7) with (2.5) and (2.6) gives

$$(2.8) \qquad |F(z) - \Phi(z)| \leq \sum_{i=1}^{6} R_i$$

where

$$R_1 = E \int (|W| + 1)|t| K(t) dt,$$

$$(2.9) \qquad R_2 = \int_{t \geq 0} P(z - t \leq W \leq z) K(t) dt,$$

$$(2.10) \qquad R_3 = \int_{t < 0} P(z \leq W \leq z - t) K(t) dt,$$

$$R_4 = E \int (|W| + 1)|t M(t)| dt,$$

$$(2.11) \qquad R_5 = \sum_{i=1}^{n} E \int_0^{X_i'} P^{X_i}(z - t \leq W \leq z)|X_i| I(X_i' \geq 0) dt,$$

$$(2.12) \qquad R_6 = \sum_{i=1}^{n} E \int_{X_i'}^0 P^{X_i}(z \leq W \leq z - t)|X_i| I(X_i' < 0) dt.$$

Let $\alpha = \sum_{i=1}^{n} \gamma_i$. We have $R_1 \leq \alpha$ and $R_4 \leq \alpha$. But for R_2, R_3, R_5 and R_6, we need a concentration inequality. We observe that for $i \neq l$, $\mathcal{L}(X_i, X_i', W^{(i)}|X_l) = \mathcal{L}(X_i', X_i, W^{(i)}|X_l)$. Denote E^{X_l} by E_l. Then for $f \in C_{\mathrm{bd}}$,

$$(2.13) \qquad \sum_{i=1, i \neq l}^{n} E_l X_i f(W + X_i') = \sum_{i=1, i \neq l}^{n} E_l X_i' f(W + X_i').$$

Let $M^{(l)}(t) = \sum_{i=1, i \neq l}^{n} M_i(t)$. By the same arguments as for proving (2.3), (2.13) yields

$$(2.14) \quad E_l \int f'(W + t) K(t) dt = E_l W^{(l)} f(W) + E_l \int f'(W + t) M^{(l)}(t) dt$$
$$+ E_l \int f'(W + t) K_l(t) dt.$$

Now choose the f in (2.14) to be such that $f'(w) = I(a - \alpha \leq w \leq b + \alpha)$ and $f(w) = 0$ at $w = \frac{a+b}{2}$, where $a < b$. Then

$$
\begin{aligned}
\frac{1}{2}P^{X_l}(a \leq W \leq b) \;\leq\;& \int_{|t| \leq \alpha} K(t) dt P^{X_l}(a \leq W \leq b) \\
\leq\;& E_l \int I(a - \alpha \leq W + t \leq b + \alpha) K(t) dt \\
=\;& E_l \int f'(W + t) K(t) dt \\
\leq\;& E_l W^{(l)} f(W) + \frac{1}{2} E_l \int f'(W + t)^2 dt + \frac{1}{2} E_l \int M^{(l)}(t)^2 dt \\
&+ \frac{1}{2} E_l \int f'(W + t)^2 dt + \frac{1}{2} E_l \int K_l(t)^2 dt \\
\leq\;& \frac{3(b - a)}{2} + \frac{7\alpha}{2} \text{ a.s.}
\end{aligned}
$$

This gives us the following concentration inequality

$$
P^{X_l}(a \leq W \leq b) \leq 3(b - a) + 7\alpha \text{ a.s.}
$$

(which of course implies $P(a \leq W \leq b) \leq 3(b - a) + 7\alpha$). Substituting this inequality in (2.9), (2.10), (2.11) and (2.12), we obtain $R_2 + R_3 \leq 8.5\alpha$ and $R_5 + R_6 \leq 8.5\alpha$. Now that we have obtained bounds on all the error terms, (2.8) yields

$$
|F(z) - \Phi(z)| \leq 19 \sum_{i=1}^{n} \gamma_i.
$$

The concentration inequality approach exploits the principle that a Stein identity contains all the information about the distribution. Thus one can extract different aspects of the information by different choices of function in the identity. There are other proofs of the Berry-Esseen theorem using Stein's method. See Barbour and Hall (1984) and Stroock (1993). These proofs are based on the inductive approach which was also used by Bolthausen (1984) in a combinatorial central limit theorem.

3. The antisymmetric function approach

Let π be a probability measure on $\mathcal{B}(S)$. We define a Stein identity for π to be an identity of the form

$$
\text{(3.1)} \qquad\qquad \int Lf d\pi = 0
$$

where L is a linear operator and f varies over a class of functions which we denote by $\mathcal{D}(L)$. The Stein identities (2.1) and (2.3) are of the form (3.1). For (2.1), $\pi = \mathcal{L}(Z), Lf(Z) = f'(Z) - Zf(Z)$ and $\mathcal{D}(L) = C_{\mathrm{bd}}$. For (2.3), $\pi = \mathcal{L}(W), Lf(W) = \int f'(W + t)K(t)dt - \int f'(W + t)E^W M(t)dt - Wf(W)$ and $\mathcal{D}(L) = C_{\mathrm{bd}}$.

Let (W, W') be an exchangeable pair of S-valued random variables and let $F : S^2 \longrightarrow \mathbb{R}$ be an antisymmetric function (that is, $F(w, w') = -F(w', w)$) such that $E|F(W, W')| < \infty$. It can easily be verified that

$$(3.2) \qquad\qquad EE^W F(W, W') = 0.$$

In his 1986 monograph Stein proposes to use (3.2) for constructing a Stein identity for $\mathcal{L}(W)$. By an application of a lemma of Stein (1986, pp.10 and 29), we can show that the Stein identity obtained from (3.2) also characterizes $\mathcal{L}(W)$ if W takes a finite number of values and satisfies a certain connectedness condition.

We now focus on a specific class of antisymmetric functions. First we observe that $F : S^2 \to \mathbb{R}$ is antisymmetric if and only if $F(w, w') = \varphi(w, w') - \varphi(w', w)$ for some $\varphi : S^2 \to \mathbb{R}$. Thus for constructing Stein identities, one possible form of F is given by

$$F(w, w') = \psi(w, w')f(w') - \psi(w', w)f(w)$$

where $\psi : S^2 \to \mathbb{R}$ is given and $f : S \to \mathbb{R}$ is to vary over a class. The antisymmetric functions used by Stein (1986) and Diaconis (1977, 1989) are of this form. We note that there are different possible choices of ψ and so a Stein equation is not unique.

If $W = \sum_{i=1}^n X_i$ is a sum of independent random variables, then W' is usually defined by $W' = W - X_I + X_I'$ where $(X_1', \ldots, X_n')$ is an independent copy of $(X_1, \ldots, X_n)$ and I is uniformly distributed over $\{1, 2, \ldots, n\}$ and independent of $X_1, \ldots, X_n, X_1', \ldots, X_n'$. This is used in various contexts in Stein (1986) and, in particular, for constructing a Stein identity for the binomial distribution. For dependent $X_1, \ldots, X_n$, examples of (W, W') can also be found in Stein (1986, 1992) and Diaconis (1977, 1989).

In the construction of the Stein identity (2.3), the antisymmetric function approach was used in spirit and (2.3) may be expressed in the form (3.2) with $F(w, w') =$

$(w - w')g(w') - (w' - w)g(w)$ where $g(W) = E^W f(W + X_I')$ and is approximately $f(W)$.

The construction of a Stein identity for a distribution $\mathcal{L}(Z)$ may involve a fixed exchangeable pair of random variables as in the case of the binomial distribution (see Stein (1986), p.p. 44-45) or involve a sequence of exchangeable pairs as follows. First couple Z with $Z_1', Z_2', \ldots$ such that $Z_n' - Z \longrightarrow 0$ in probability as $n \longrightarrow \infty$ and for each $n \geq 1$, (Z, Z_n') is exchangeable. If there exists $\epsilon_n \downarrow 0$ such that

$$\frac{E^Z \left\{ \psi(Z, Z_n') f(Z_n') - \psi(Z_n', Z) f(Z) \right\}}{\epsilon_n} \longrightarrow Lf(Z)$$

in L^1 for $f \in \mathcal{D}(L)$ as $n \to \infty$, where ψ is given and L a nonzero operator, then a Stein identity for $\mathcal{L}(Z)$ may be taken to be $ELf(Z) = 0$.

As an example, suppose for $0 < \rho < 1, (Z, Z_\rho')$ has the bivariate normal distribution with density function $\propto \exp\{-(x^2 - 2\rho xy + y^2)/(2\sqrt{1-\rho^2})\}$. Then (Z, Z_ρ') is exchangeable and $Z_\rho - Z \to 0$ in probability as $\rho \to 1$. Take $\psi(x, y) = y - x$. It can be shown that for twice differentiable functions f such that f, f' and f'' are bounded,

$$\frac{E^Z \left\{ (Z_\rho' - Z) f(Z_\rho') - (Z - Z_\rho') f(Z) \right\}}{2(1 - \rho)} \longrightarrow f'(Z) - Z f(Z)$$

in L^1 as $\rho \to 1$. This yields a Stein identity for $\mathcal{L}(Z)$ as

$$(3.3) \qquad E\left\{ f'(Z) - Z f(Z) \right\} = 0.$$

For wider applicability, we extend (3.3) to a larger class of functions f.

Now take $\psi \equiv 1$. It can be shown that for thrice differentiable functions f such that f, f', f'' and f''' are bounded,

$$\frac{E^Z \left\{ f(Z_\rho') - f(Z) \right\}}{1 - \rho} \longrightarrow f''(Z) - Z f'(Z)$$

in L^1 as $\rho \to 1$. This yields

$$(3.4) \qquad E\left\{ f''(Z) - Z f'(Z) \right\} = 0.$$

Similarly, (3.4) can be extended to a larger class of functions. Note that the operator L in (3.4) is the infinitesimal generator of the standard Ornstein-Uhlenbeck process.

Consider another example. Let Z be a Poisson random variable with mean λ. Represent Z by $Z = X_n + Y_n$ and $Z_n' = X_n + Y_n'$ where X_n, Y_n and Y_n' are independent

Poisson random variables with means $(n-1)\lambda/n, \lambda/n$ and λ/n respectively. Then (Z, Z'_n) is exchangeable and $Z'_n - Z \to 0$ in probability as $n \to \infty$. Take $\psi(x, y) = I(y - x = 1)$. It can be shown that for bounded functions f defined on $\mathbb{Z}^+ = \{0, 1, 2, \dots\}$,

$$\frac{E^Z \{I(Z'_n - Z = 1)f(Z'_n) - I(Z - Z'_n = 1)f(Z)\}}{1/n} \longrightarrow \lambda f(Z+1) - Z f(Z)$$

in L^1 as $n \to \infty$. This yields a Stein identity for $\mathcal{L}(Z)$ as

$$(3.5) \qquad E\{\lambda f(Z+1) - Z f(Z)\} = 0.$$

Now take $\psi \equiv 1$. It can be shown that for bounded functions f defined on $\mathbb{Z}^+$,

$$\frac{E^Z \{f(Z'_n) - f(Z)\}}{1/n} \longrightarrow \lambda \Delta f(Z+1) - Z \Delta f(Z)$$

in L^1 as $n \to \infty$, where $\Delta f(x) = f(x) - f(x-1)$. This yields

$$(3.6) \qquad E\{\lambda \Delta f(Z+1) - Z \Delta f(Z)\} = 0.$$

Note again that the operator L in (3.6) is the infinitesimal generator of the immigration-death process with immigration rate λ and unit per capita death rate.

We now show that the involvement of infinitesimal generators of Markov processes in Stein identities follows naturally from the antisymmetric function approach. Suppose $\mathcal{L}(Z)$ is the stationary and initial distribution π of a reversible Markov process $\{X_t : t \geq 0\}$. Then (X_0, X_t) is exchangeable for every $t > 0$ and $X_t - X_0 \to 0$ in probability as $t \to 0$. By taking $\psi \equiv 1$, we have for f in a suitable class of functions,

$$\frac{E^{X_0} f(X_t) - f(X_0)}{t} \longrightarrow A f(X_0)$$

in L^1 as $t \to 0$, where A is the infinitesimal generator of the process. This yields a Stein identity for $\mathcal{L}(Z)$ as

$$(3.7) \qquad EAf(Z) = 0,$$

which is then extended to a larger class of functions f. In this case the solution of the Stein equation

$$Af(x) = h(x) - \pi h$$

can be expressed as

$$f(x) = - \int_0^\infty [E_x h(X_t) - \pi h]dt$$

provided that the integral exists. The solution and its smoothness can then be bounded by using coupling.

The Stein identity (3.7) allows generalization to higher dimensions. This probabilistic approach to Stein's method is due to Barbour (1988, 1990). The fact that (3.7) can be constructed by using the antisymmetric function approach was also observed by Barbour (1997). Examples of reversible Markov processes which have been used in the application of Stein's method are the Ornstein-Uhlenbeck process for normal approximation (Barbour (1990)) and immigration-death processes for Poisson approximation (Barbour (1988)) and compound Poisson approximation (Barbour Chen and Loh (1992)).

In order for a Stein identity to be useful, it has to characterize or almost characterize the underlying distribution. But not every Stein identity constructed by the antisymmetric function approach has this property. For example, consider $Z = (X, Y)$ where X and Y are independent standard normal random variables. If we use a sequence of exchangeable pairs (of random vectors) where the second components of the random vectors are identical, then a Stein identity for $\mathcal{L}(Z)$ will be

$$E\left\{ \frac{\partial}{\partial x} f(X, Y) - X f(X, Y) \right\} = 0$$

which clearly does not characterize $\mathcal{L}(Z)$. No conditions are known for determining whether a given exchangeable pair or sequence of exchangeable pairs will lead to a Stein identity which characterizes its underlying distribution. It would be of interest to find such conditions.

4. An L^2 space approach

A Stein identity may be expressed in terms of an operator and its adjoint in an L^2 space. Let π be a probability measure on $\mathcal{B}(S)$. Consider $L^2(S, \pi)$ with $(f, g) = \int_S fg d\pi$. Let A be an operator with domain $\mathcal{D}(A)$ and adjoint A^* such that 1 belongs to the domain $\mathcal{D}(A^*)$ of A^* and such that $A - A^* 1$ is not the zero operator. Then for

$f \in \mathcal{D}(A)$, $(Af, 1) = (f, A^*1) = ((A^*1)f, 1)$. This implies that $((A - A^*1)f, 1) = 0$ which yields a Stein identity for π as

$$(4.1) \qquad \int_S (A - A^*1) f \, d\pi = 0$$

where $f \in \mathcal{D}(A)$. We note that there are many possible choices of A and so a Stein identity is not unique.

Now $(A - A^*1)f \in \{1\}^{\perp} = \{h - \pi h : h \in L^2(S, \pi)\}$. Therefore $\mathcal{R}(A - A^*1) = \{h - \pi h : h \in \mathcal{C}\}$ for some $\mathcal{C} \subset \mathcal{L}^{\in}(\mathcal{S}, \pi)$ and the Stein equation

$$(A - A^*1)f = h - \pi h$$

has a solution for $h \in \mathcal{C}$.

Consider an example where $S = \mathbb{R}$ and π is the standard normal distribution. Take A to be such that $Af(x) = f'(x)$. Then $A^*g(x)$ is given by $A^*g(x) = -g'(x) + xg(x)$ and $A^*1(x) = x$. This yields a Stein identity for π as

$$\int_{\mathbb{R}} [f'(x) - xf(x)] d\pi(x) = 0.$$

Note that this identity is the same as (3.3).

Consider another example where $S = \mathbb{Z}^+$ and π is the Poisson distribution with mean λ. Take A to be such that $Af(x) = \lambda f(x + 1)$. Then A^* is given by $A^*g(x) = xg(x - 1)$ and $A^*1(x) = x$. This yields a Stein identity for π as

$$(4.2) \qquad \int_{\mathbb{Z}^+} [\lambda f(x + 1) - xf(x)] d\pi(x) = 0.$$

This identity is the same as (3.5).

Suppose S is a measurable abelian group, that is, an abelian group such that the group operation is a measurable map from S^2 to S. Let π be the compound Poisson distribution $e^{\lambda(\mu - \delta_0)}$ where $\lambda > 0, \mu$ a probability measure on $\mathcal{B}(S)$ with no atom at the identity 0, and δ_0 the Dirac measure at 0. That is, $\pi = \mathcal{L}(Y_1 + \cdots + Y_N)$ where $Y_1, Y_2, \ldots$ are independent S-valued random variables with common distribution μ, N a Poisson random variable with mean λ, and N independent of $Y_1, Y_2, \ldots$. Take A to be such that $Af(x) = \lambda \int f(x + t) d\mu(t)$. Then by the Stein identity (4.2) for the

Poisson distribution,

$$
\begin{aligned}
(Af, g) &= E\left\{\lambda \int f(Y_1 + \cdots + Y_N + t)d\mu(t)g(Y_1 + \cdots + Y_N)\right\} \\
&= EE^{Y_1, Y_2, \ldots}\left\{\lambda \int f(Y_1 + \cdots + Y_N + t)d\mu(t)g(Y_1 + \cdots + Y_N)\right\} \\
&= EE^{Y_1, Y_2, \ldots}\left\{N \int f(Y_1 + \cdots + Y_{N-1} + t)d\mu(t)g(Y_1 + \cdots + Y_{N-1})\right\} \\
&= E\left\{f(Y_1 + \cdots + Y_N)E(Ng(Y_1 + \cdots + Y_{N-1})|Y_1 + \cdots + Y_N\right\} \\
&= (f, A^*g)
\end{aligned}
$$

where $A^*g(x) = E(Ng(Y_1 + \cdots + Y_{N-1})|Y_1 + \cdots + Y_N = x)$. This gives $A^*1(x) = E(N|Y_1 + \cdots + Y_N = x)$ and yields a Stein identity for π as

$$
(4.3) \qquad E\left\{\lambda \int f(Z + t)d\mu(t) - E(N|Z)f(Z)\right\} = 0
$$

where $Z = Y_1 + \cdots + Y_N$. This identity is different from the one used in Barbour, Chen and Loh (1992) and allows compound Poisson approximation to be studied on groups. We shall return to this identity in Section 5.

Suppose π is the stationary distribution of a reversible Markov process $\{X_t : t \geq 0\}$ with state space S and infinitesimal generator A. Then A is self-adjoint in $L^2(S, \pi)$ and $A1 = A^*1 = 0$. This yields a Stein identity for π as

$$
(4.4) \qquad \int_S Af d\pi = 0.
$$

This identity is the same as (3.7) and the solution of the corresponding Stein equation can be studied as has been discussed in Section 3.

Actually $A^*1 = 0$ and (4.4) still holds even if $\{X_t : t \geq 0\}$ is not reversible so long as π is the stationary distribution and $1 \in \mathcal{D}(A^*)$. This seems to suggest that the L^2 space approach allows greater flexibility whereas the antisymmetric function approach tends to produce self-adjoint operators. The following proposition seems to support this observation.

Proposition 4.1 Suppose Z takes values in S and $\pi = \mathcal{L}(Z)$. Let (Z, Z') be exchangeable (respectively (Z, Z'_n) be exchangeable for $n \geq 1$ such that $Z'_n - Z \to 0$ in probability as $n \to \infty$). Define

$$
Lf(Z) = E^Z\left[\psi(Z, Z')f(Z') - \psi(Z', Z)f(Z)\right]
$$

(respectively L^2-$\lim_{n\to\infty} E^Z \left[\psi(Z, Z'_n)f(Z'_n) - \psi(Z'_n, Z)f(Z)\right]/\epsilon_n$ where $\epsilon_n \downarrow 0$) where ψ is given such that $E\psi(Z, Z')^2 f(Z')^2 < \infty$ (respectively $E\psi(Z_n, Z'_n)^2 f(Z'_n)^2 < \infty$ for $n \geq 1$) for $f \in \mathcal{D}(L)$. If ψ is symmetric, that is $\psi(x, y) = \psi(y, x)$, then L is self-adjoint in $L^2(S, \pi)$.

Proof. It suffices to prove the first case. We have

$$
\begin{aligned}
(Lf, g) &= E\left\{E^Z[\psi(Z, Z')f(Z') - \psi(Z', Z)f(Z)]g(Z)\right\} \\
&= E\left(\psi(Z', Z)f(Z)g(Z')\right) - E\left(\psi(Z', Z)f(Z)g(Z)\right) \\
&= E\left(\psi(Z, Z')f(Z)g(Z')\right) - E\left(\psi(Z', Z)f(Z)g(Z)\right) \\
&= E\left\{f(Z)E^Z[\psi(Z, Z')g(Z') - \psi(Z', Z)g(Z)]\right\} \\
&= (f, Lg).
\end{aligned}
$$

This proves the proposition.

In the construction of the Stein identity (2.3), the L^2 space approach was used in spirit. The arguments which led to (2.3) are essentially those that one would use for finding the adjoint of an operator in the context. Indeed,

$$
\begin{aligned}
E\int f'(W + t)K(t)dt\,g(W) &= \sum_{i=1}^{n} EX'_i f(W + X'_i)g(W) \\
&= \sum_{i=1}^{n} EX_i f(W + X'_i)g(W^{(i)} + X'_i) \\
&= EWf(W)g(W) + \sum_{i=1}^{n} EX_i\left[f(W + X'_i) - f(W)\right]g(W^{(i)} + X'_i) \\
&\quad + \sum_{i=1}^{n} EX_i f(W)\left[g(W^{(i)} + X'_i) - g(W)\right] \\
&= EWf(W)g(W) + E\left(\sum_{i=1}^{n}\int f'(W^{(i)} + X'_i + t)\tilde{M}_i(t)dt\right)g(W) \\
&\quad + Ef(W)\left(\sum_{i=1}^{n} X_i[g(W^{(i)} + X'_i) - g(W)]\right)
\end{aligned}
$$

where $\tilde{M}_i(t) = X'_i\left[I(X_i \geq t \geq 0) - I(X_i < t < 0)\right]$. Consequently, if we define A to be given by

$$Af(W) = \int f'(W + t)K(t)dt - E^W \sum_{i=1}^{n} \int f'(W^{(i)} + X_i' + t)\tilde{M}_i(t)dt,$$

then A^* is given by

$$A^*g(W) = Wg(W) + E^W \sum_{i=1}^{n} X_i \left[g(W^{(i)} + X_i') - g(W)\right].$$

This gives us a Stein identity for $\mathcal{L}(W)$ as

$$E\left(A - A^*1\right)f(W) = 0$$

which is expressible as (2.3).

The Stein identity (4.1) does not always characterize the underlying distribution π. For example, consider $Z = (X, Y)$ where X and Y are independent standard normal random variables. Take $Af(x, y) = \frac{\partial}{\partial x}f(x, y)$. Then $A^*g(x, y) = -\frac{\partial}{\partial x}g(x, y) + xg(x, y)$ and a Stein identity for $\mathcal{L}(Z)$ is

$$E\left\{\frac{\partial}{\partial x}f(X, Y) - Xf(X, Y)\right\} = 0$$

which does not characterize $\mathcal{L}(Z)$. It would be of interest to find conditions for determining whether a given operator A is such that (4.1) characterizes its underlying distribution π.

5. Compound Poisson approximation on groups

Let $\mathcal{X}$ be a measurable abelian group and let π be the compound Poisson distribution $e^{\lambda(\mu - \delta_0)}$ where $\lambda > 0$, μ a probability measure on $\mathcal{B}(\mathcal{X})$ with no atom at the identity 0, and δ_0 the Dirac measure at 0. That is, $\pi = \mathcal{L}(Y_1 + \cdots + Y_N)$ where $Y_1, Y_2, \ldots$ are independent $\mathcal{X}$-valued random variables with common distribution μ, N a Poisson random variable with mean λ, and N independent of $Y_1, Y_2, \ldots$. Define $Z = Y_1 + \cdots + Y_N$.

The Stein identity (4.3) enables us to consider compound Poisson approximation on a measurable abelian group. It also leads us to the Stein equation

$$(5.1) \qquad \lambda \int f(w + t)d\mu(t) - E(N|Z = w)f(w) = h(w) - \pi h$$

which we have to solve in order to carry out the approximation. In general, $E(N|Z = w)$ is not easy to calculate. We consider a special case which is of interest in itself.

Let $\mathcal{K}_1$ be a coset of a subgroup $\mathcal{K}_0$ of $\mathcal{X}$ such that $\mathcal{K}_1$ has infinite order in the quotient group $\mathcal{X}/\mathcal{K}_0$. For convenience, assume $\mathcal{X}$ to be an additive group. Define $\mathcal{K}_r = \mathcal{K}_{r-1} + \mathcal{K}_1$ for $r = 1, 2, \ldots$. Then $\mathcal{K}_1, \mathcal{K}_2, \ldots$ are distinct cosets of $\mathcal{K}_0$.

Assume that Y_1 takes values in $\mathcal{K}_1$, that is, $\mathrm{supp}(\mu) \subset \mathcal{K}_1$ (if there is a topology on $\mathcal{X}$). Then $N = \psi(Z)$ where $\psi(Z) = r$ if and only if $Z \in \mathcal{K}_r$ for $r = 0, 1, 2, \ldots$. Consequently $E(N|Z = w) = \psi(w)$ and the Stein equation (5.1) becomes

$$(5.2) \qquad \lambda \int f(w + t)d\mu(t) - \psi(w)f(w) = h(w) - \pi h$$

which can be solved analytically. For $w \in \mathcal{K}_r$, $r = 1, 2, \ldots$, the solution of (5.2) is given by

$$(5.3) \quad f_h(w) \;=\; -\sum_{k=0}^{\infty} \frac{\lambda^k}{\psi(w)(\psi(w) + 1)\cdots(\psi(w) + k)} \int [h(w + t) - Eh(Z)]d\mu^k(t)$$

$$=\; \frac{-E\int h(w + t)d\mu^{N - \psi(w)}(t)I(N \geq \psi(w)) + Eh(Z)P(N \geq \psi(w))}{\lambda P(N = \psi(w) - 1)}$$

where μ^k is the k-fold convolution of μ. For $w \in \mathcal{K}_0$, $f_h(w)$ is arbitrary. Since the solution f_h is bounded whenever h is bounded, the Stein identity (4.3) characterizes the compound Poisson distribution in this special setting.

We show how compound Poisson approximation on a group in this setting can be applied to multivariate Poisson approximation and compound Poisson approximation for dependent indicators. For multivariate Poisson approximation, the approximating distribution is $\mathcal{L}(Z)$ where $Z = (Z_1, \ldots, Z_d)$ and $Z_1, \ldots, Z_d$ are independent Poisson random variables with means $\lambda_1, \ldots, \lambda_d$ respectively. It has been observed in Chen and Roos (1995) that $\mathcal{L}(Z)$ is also the compound Poisson distribution $e^{\lambda(\mu - \delta_0)}$ on the additive group $\mathbb{Z}^d$ with $\lambda = \lambda_1 + \cdots + \lambda_d$ and μ given by $\mu(\{e_j\}) = \lambda_j/\lambda$, where e_j is the basis vector in $\mathbb{Z}^d$ with 1 in the jth position, $j = 1, \ldots, d$, $d = 1, 2, \ldots$.

Now define $\mathcal{K}_r = \{(x_1, \ldots, x_d) \in \mathbb{Z}^d : x_1 + \cdots + x_d = r\}$ for $r = 0, 1, 2, \ldots$. Then $\mathcal{K}_0$ is a subgroup of $\mathbb{Z}^d$ with cosets $\mathcal{K}_1, \mathcal{K}_2, \ldots$ satisfying $\mathcal{K}_r = \mathcal{K}_{r-1} + \mathcal{K}_1$ for $r = 1, 2, \ldots$. Since μ is concentrated on $\{e_1, \ldots, e_d\}$, we have $\mathrm{supp}(\mu) \subset \mathcal{K}_1$ and this shows that multivariate Poisson approximation is a special case of compound Poisson approximation on groups in the special setting.

Multivariate Poisson approximation for independent random vectors taking values $0, e_1, \ldots, e_d$ was first considered by Barbour (1988) using the probabilistic approach. The error bound he obtained was essentially the same as for Poisson approximation except for a logarithmic factor $\log \lambda$. Whether this logarithmic factor can be removed is an open question. Compound Poisson approximation on groups offers an analytic approach to addressing this question.

For compound Poisson approximation for dependent indicators, we use an example of Arratia, Goldstein and Gordon (1990, Section 4.2.1) for illustration. Let $C_1, C_2, \ldots$ be indicators for heads in a sequence of coin tosses with common success probability. Define $U = \sum_{\alpha \in \Lambda} C_\alpha C_{\alpha+1} \cdots C_{\alpha+t-1}$ where $\Lambda = \{1, 2, \ldots, n\}$ and t a positive integer. The random variable U counts the number of locations among the first n at which a head run of length at least t begins. The locations tend to occur in clumps with the number of clumps approximately Poisson distributed. Thus an appropriate approximating distribution for $\mathcal{L}(U)$ is compound Poisson.

For $\alpha \in \Lambda$, let $X_\alpha = (1 - I(\alpha > 1)C_{\alpha-1})C_\alpha C_{\alpha+1} \cdots C_{\alpha+t-1}$. Then $\sum_{\alpha \in \Lambda} X_\alpha$ counts the number of clumps. Represent each clump by a random vector in $\mathbb{Z}^\infty$ according to the clump size as follows. For $j = 1, 2, \ldots$ let

$$X_{\alpha,j} = (1 - I(\alpha > 1)C_{\alpha-1})C_\alpha C_{\alpha+1} \cdots C_{\alpha+t+j-2}(1 - C_{\alpha+t+j-1}),$$

Here the interpretation is that for $j = 1, 2, \ldots$, the clump size at α is j if and only if $X_{\alpha,j} = 1$ and that it is 0 if and only if for all $j = 1, 2, \ldots$, $X_{\alpha,j} = 0$. Define $U_\alpha = (X_{\alpha,1}, X_{\alpha,2}, \ldots)$ which represents the clump at α.

Approximate $\mathcal{L}(\sum_{\alpha \in \Lambda} U_\alpha)$ by an appropriate compound Poisson distribution $\mathcal{L}(Z)$ on the additive group $\mathbb{Z}^\infty$ and obtain an explicit bound on $d_{TV}(\mathcal{L}(\sum_{\alpha \in \Lambda} U_\alpha), \mathcal{L}(Z))$ where d_{TV} denotes total variation distance. Here $\mathcal{L}(Z) = e^{\lambda(\mu - \delta_0)}$ with

$$\text{supp}(\mu) = \{e_1, e_2, \ldots\} \subset \mathcal{K}_1 = \{(x_1, x_2, \ldots) \in \mathbb{Z}^\infty : \sum_{i=1}^\infty x_i = 1\},$$

where e_j is the jth basis vector and $\mathcal{K}_1$ a coset in $\mathbb{Z}^\infty$. Thus the approximation is a special case of compound Poisson approximation on groups in the special setting. Since $\mathcal{L}(Z)$ is a compound Poisson distribution on $\mathbb{Z}^\infty$ with $\text{supp}(\mu) = \{e_1, e_2, \ldots\}$, Z may be represented as $Z = (Z_1, Z_2, \ldots)$ where $Z_1, Z_2, \ldots$ are independent Poisson random variables.

Now

$$U = \sum_{\alpha \in \Lambda} \sum_{j=1}^{\infty} j X_{\alpha,j} = (\sum_{\alpha \in \Lambda} U_\alpha) \begin{pmatrix} 1 \\ 2 \\ \cdot \\ \cdot \\ \cdot \end{pmatrix}.$$

Therefore

$$(5.4) \qquad d_{TV} \left(\mathcal{L}(U), \mathcal{L}(\sum_{j=1}^{\infty} j Z_j) \right) \leq d_{TV} \left(\mathcal{L}(\sum_{\alpha \in \Lambda} U_\alpha), \mathcal{L}(Z) \right)$$

and we have an explicit bound for $d_{TV}\left(\mathcal{L}(U), \mathcal{L}(\sum_{j=1}^{\infty} j Z_j) \right)$ where $\mathcal{L}(\sum_{j=1}^{\infty} j Z_j)$ is a compound Poisson distribution on $\mathbb{Z}$.

Since the total variation distance on the right hand side of (5.4) is weaker than that used by Arratia, Goldstein and Gordon (1990), a better upper bound is expected. Compound Poisson approximation on groups provides an analytic approach to finding such a bound. It would be of interest to apply this approach to the various problems in, for example, Aldous (1989).

6. A brief history

Stein's method has been actively applied to normal, Poisson, Poisson process and compound Poisson approximations. Normal approximation was first considered by Stein (1972) who also introduced the method. Some recent works on dependent random variables are by Bolthausen (1984), Chen (1986), Baldi and Rinott (1989), Baldi, Rinott and Stein (1989), Rinott (1995), Dembo and Rinott (1996) and Goldstein and Reinert (1996). Works on multivariate extensions are by Barbour (1990), Götze (1991), Bolthausen and Götze (1993), Rinott and Rotar (1996) and Goldstein and Rinott (1996).

Poisson approximation was first considered by Chen (1975). There are two approaches: the local and the coupling approach. Excellent accounts of the local approach are given in Arratia, Goldstein and Gordon (1989, 1990) while the coupling approach is systematically developed in Barbour, Holst and Janson (1992). A wide range of applications from random graphs to molecular biology are also given therein. More applications to molecular biology can be found in Arratia, Gordon and Waterman (1990), Dembo and Karlin (1992), Neuhauser (1994), Dembo, Karlin and Zeitouni (1994) and Waterman (1995).

Recent papers on Poisson approximation for unbounded functions and large deviations are by Chen and Choi (1992) and Barbour, Chen and Choi (1995). Unbounded function approximation is a refinement of total variation approximation and can be applied to obtain asymptotic and large deviation results.

Poisson process approximation was first investigated by Barbour (1988) who also introduced the probabilistic approach. Many of the subsequent papers address the question of finding analogues of the multiplying constant $(1 \wedge \lambda^{-1})$ by choosing suitable metrics on the space of distributions of point processes. The papers of Barbour and Brown (1992a, 1992b), Barbour and Greenwood (1993), Brown and Xia (1995a, 1995b), Brown and Greig (1996) and Xia (1997) are works in this direction. While Brown and Greig (1996) also generalized the coupling approach to process setting, Barbour and Brown (1992b) combined Stein's method with the compensator method.

Compound Poisson approximation using Stein's method directly was first investigated by Barbour, Chen and Loh (1992) who considered the approximation for nonnegative random variables and addressed the question of generalizing the multiplying factor $(1 \wedge \lambda^{-1})$ in a correct way. Barbour and Utev (1996) addressed this question further by studying the solution of the Stein equation. More recent works on compound Poisson approximation are by Roos (1994), Chen and Roos (1995), Geske et al (1995), Schbath (1995) and Roos and Stark (1996). Although Chen and Roos (1995) considered compound Poisson approximation on groups, Stein's method was not used directly.

Stein's method has also been applied to contexts other than those discussed above. See, for example, Stein (1986), Ehm (1991) and Soon (1996) for binomial approximation, Diaconis (1989) for uniform approximation, Loh (1992) for multinomial approximation, Luk (1994) for gamma approximation, and Barbour and Grübel

(1995) and Peköz (1995) for geometric approximation.

7. Some possible prospects

For future developments, one could perhaps investigate Poisson process approximation and compound approximation for unbounded functions and apply the results to large deviations.

The Stein identity (4.3) and Stein equation (5.1) provide a framework for studying compound Poisson approximation on groups. It might be of interest to explore various possibilities apart from studying the problems discussed in Section 5.

One such possibility might be to treat Poisson process approximation as a special case of compound Poisson approximation on groups using the analytic approach. This can be done by representing a Poisson point process on a space $\mathcal{Y}$ by $\sum_{i=1}^{N} \delta_{\xi_i}$ where N is a Poisson random variable, $\xi_1, \xi_2, \ldots$ independent random elements taking values in $\mathcal{Y}$ with a common distribution, and independent of $\xi_1, \xi_2, \ldots$. Then regard $\sum_{i=1}^{N} \delta_{\xi_i}$ as a compound Poisson random element taking values in the additive group $\mathcal{X}$ generated by Dirac measures on $\mathcal{Y}$.

Although unbounded function approximation produces large deviation results, the range of the large deviation is somewhat restrictive. One should perhaps explore the possibility of developing Stein's method to treat large deviations for different ranges.

For normal approximation, one possible line of generalization might be to apply Stein's method to obtain non-uniform bounds. Another possible line might be to consider normal approximation in a Hilbert space. One could also consider extension of normal approximation to approximation by stable distributions.

Poisson process approximation has been successfully dealt with. It should be of interest to consider other process approximations.

There are also questions on the theoretical aspects of Stein's method. It has been noted that a Stein identity for a distribution is not unique. Are there canonical ones? If so, which ones?

Not every Stein identity constructed by the approaches described in this paper characterizes its underlying distribution. It would be of interest to find conditions for determining whether a construction will lead to a Stein identity which characterizes its underlying distribution.

The potential of Stein's method has not been fully realized. The theoretical aspects of the method also need to be further developed. It is hoped that the ideas and perspectives presented in this paper will help in the understanding of the method and lead to wider and more fruitful applications.

Acknowledgements

Part of this paper was written when I was visiting the Institute of Mathematics, Academia Sinica, Taipei in June 1996. I would like to thank Chii-Ruey Hwang for his invitation and hospitality. I am also thankful to the following persons for their helpful comments on an earlier draft and on my talks before this paper was completed: Andrew Barbour, Kwok-Pui Choi, Chii-Ruey Hwang, Peter Jagers, Yu-Kiang Leong, Qiman Shao, Shuenn-Jyi Sheu and Charles Stein.

References

[1] ALDOUS, D. (1989) Probability approximations via the Poisson clumping heuristic. *Appl. Math. Sci.* **77**, Springer, New York.

[2] ARRATIA, R., GOLDSTEIN, L AND GORDON, L. (1989) Two moments suffice for Poisson approximations: The Chen-Stein method. *Ann. Prob.* **17**, 9–25.

[3] ARRATIA, R., GOLDSTEIN, L AND GORDON, L. (1990) Poisson approximation and the Chen-Stein method. *Statist. Sci.* **5**, 403–434.

[4] ARRATIA, R., GORDON, L. AND WATERMAN, M. S. (1990) The Erdös-Rényi law in distribution, for coin tossing and sequence matching. *Ann. Statist.* **18**, 539–570.

[5] BALDI, P. AND RINOTT, Y. (1989) On normal approximations of distributions in terms of dependency graphs. *Ann. Prob.* **17**, 1646–1650.

[6] BALDI, P., RINOTT, Y. AND STEIN, C. (1989) A normal approximation for the number of local maxima of a random function on a graph. In *Probability, Statistics and Mathematics: Papers in Honor of Samuel Karlin* (T. W. Andersen, K. B. Athreya and D. L. Iglehart, eds.) 59–81. Academic, New York.

[7] BARBOUR, A. D. (1988) Stein's method and Poisson process convergence. *J. Appl. Prob.* **25 (A)**, 175–184.

[8] BARBOUR, A. D. (1990) Stein's method for diffusion approximations. *Prob. Theory Rel. Fields* **84**, 297–322.

[9] BARBOUR, A. D. (1997) Stein's method. In *Encyclopaedia of Statistical Science*, 2nd ed. Wiley, New York.

[10] BARBOUR, A. D. AND BROWN, T. C. (1992a) Stein's method and point process approximation. *Stoch. Proc. Appl.* **43**, 9–31.

[11] BARBOUR, A. D. AND BROWN, T. C. (1992b) The Stein-Chen method, point processes and compensators. *Ann. Prob.* **20**, 1504–1527.

[12] BARBOUR, A. D., CHEN, L. H. Y. AND CHOI, K. P. (1995) Poisson approximation for unbounded functions, I: independent summands. *Statist. Sinica* **5**, 749–766.

[13] BARBOUR, A. D., CHEN, L. H. Y. AND LOH, W. L. (1992) Compound Poisson approximation for nonnegative random variables via Stein's method. *Ann. Prob.* **20**, 1843–1866.

[14] BARBOUR, A. D. AND GREENWOOD, P. E. (1993) Rates of Poisson approximation to finite range random fields. *Ann. Appl. Prob.* **3**, 91–102.

[15] BARBOUR, A. D. AND GRÜBEL, R. (1995) The first divisible sum. *J. Theor. Prob.* **8**, 39-47.

[16] BARBOUR, A. D. AND HALL, P. (1984) Stein's method and the Berry-Esseen theorem. *Austral. J. Statist.* **26**, 8–15.

[17] BARBOUR, A. D., HOLST, L. AND JANSON, S. (1992) *Poisson approximation.* Oxford Studies in Probability **2**, Clarendon, Oxford.

[18] BARBOUR, A. D. AND UTEV, S. (1996) Solving the Stein equation in compound Poisson approximation. Preprint.

[19] BOLTHAUSEN, E. (1984) An estimate of the remainder in a combinatorial central limit theorem. *Z. Wahrsch. Verw. Gebiete* **66**, 379–386.

[20] BOLTHAUSEN, E. AND GÖTZE, F. (1993) The rate of convergence for multivariate sampling statistics. *Ann. Statist.* **21**, 1692–1710.

[21] BROWN, T. C. AND GREIG, D. (1996) Poisson approximation for point processes via monotone coupling. *Ann. Appl. Prob.* **6**, 545–560.

[22] BROWN, T. C. AND XIA, A. (1995a) On metrics in point process approximation. *Stochastics Rep.,* **52**, 247–263.

[23] BROWN, T. C. AND XIA, A. (1995b) On Stein-Chen factors for Poisson approximation. *Statist. Prob. Lett.* **23**, 327–332.

[24] CHEN, L. H. Y. (1975) Poisson approximation for dependent trials. *Ann. Prob.* **3**, 534–545.

[25] CHEN, L. H. Y. (1986) The rate of convergence in a central limit theorem for dependent random variables with arbitrary index set. IMA Preprint Series #243, Univ. Minnesota.

[26] CHEN, L. H. Y. AND CHOI, K. P. (1992) Some asymptotic and large deviation results in Poisson approximation. *Ann. Prob.* **20**, 1867–1876.

[27] CHEN, L. H. Y. AND ROOS, M. (1995) Compound Poisson approximation for unbounded functions on a group, with application to large deviations. *Prob. Theory Rel. Fields* **103**, 515–528.

[28] DEMBO, A. AND KARLIN, S. (1992) Poisson approximation for r-scan processes. *Ann. Appl. Prob.* **2**, 329–357.

[29] DEMBO, A. AND KARLIN, S. AND ZEITOUNI, O. (1994) Limit distribution of maximal non-aligned two-sequence segmental score. *Ann. Prob.* **22**, 2022–2039.

[30] DEMBO, A. AND RINOTT, Y. (1996) Some examples of normal approximations by Stein's method. In *Random Discrete Structures*, IMA vol 76 (D. Aldous and R. Pemantle, eds.), 25–44, Springer, New York.

[31] DIACONIS, P. (1977) The distribution of leading digits and uniform distribution mod 1. *Ann. Prob.* **5**, 72–81.

[32] DIACONIS, P. (1989) An example for Stein's method. Stanford Stat. Dept. Tech. Rep.

[33] EHM, W. (1991) Binomial approximation to the Poisson binomial distribution. *Statist. Prob. Lett.* **11**, 7–16.

[34] GESKE, M. X., GODBOLE, A. P., SCHAFFNER, A. A., SKOLNICK, A. M. AND WALLSTROM, G. L. (1995). Compound Poisson approximations for word patterns under Markovian hypotheses. *J. Appl. Prob.* **32**, 877–892.

[35] GOLDSTEIN, L. AND REINERT, G. (1996) Stein's method and the zero bias transformation with applications to simple random sampling. Preprint.

[36] GOLDSTEIN, L. AND RINOTT, Y. (1996) On multivariate normal approximations by Stein's method and size bias couplings. *J. Appl. Prob.* **33**, 1–17.

[37] GÖTZE, F. (1991) On the rate of convergence in the multivariate CLT. *Ann. Prob.* **19**, 724–729.

[38] HO, S. T. AND CHEN, L. H. Y. (1978) An L_p bound for the remainder in a combinatorial central limit theorem. *Ann. Prob.* **6**, 231–249.

[39] LOH, W. L. (1992). Stein's method and multinomial approximation. *Ann. Appl. Prob.* **2**, 536–554.

[40] LUK, M. (1994) *Stein's Method for the Gamma Distribution and Related Statistical Applications*. Ph.D. Thesis, Univ. Southern California.

[41] NEUHAUSER, C. (1994) A Poisson approximation for sequence comparisons with insertion and deletions. *Ann. Statist.* **22**, 1603–1629.

[42] PEKÖZ, E. (1995) Stein's method for geometric approximation. Tech. Rep. #225, Stat. Dept., Univ. California, Berkeley.

[43] RINOTT, Y. (1995) On normal approximation rates for certain sums of dependent random variables. *J. Comp. Appl. Math.* **55**, 135–143.

[44] RINOTT, Y. AND ROTAR, V. (1996) A multivariate CLT for local dependence with $n^{-1/2} \log n$ rate and applications to multivariate graph related statistics. *J. Multivar. Anal.* **56**, 333–350.

[45] ROOS, M. (1994) Stein's method for compound Poisson approximation: the local approach. *Ann. Appl. Prob.* **4**, 1177–1187.

[46] ROOS, M. AND STARK, D. (1996) Compound Poisson approximation of the number of visits to a small set in a Markov chain. Preprint.

[47] SCHBATH, S. (1995) Compound Poisson approximation of word counts in DNA sequences. *ESAIM: Prob. Statist.* **1**, 1–16.

[48] SOON, S. Y. T. (1996) Binomial approximation for dependent indicators. *Statist. Sinica* **6**, 703–714.

[49] STEIN, C. (1972) A bound for the error in the normal approximation to the distribution of a sum of dependent random variables. *Proc. Sixth Berkeley Symp. Math. Stat. Prob.* **2**, 583–602, Univ. California Press, Berkeley, Calif.

[50] STEIN, C. (1986) *Approximate Computation of Expectations.* Lecture Notes **7**, Inst. Math. Statist., Hayward, Calif.

[51] STEIN, C. (1992) A way of using auxiliary randomization. In *Probability Theory.* (L. H. Y. Chen, K. P. Choi, K. Hu and J.-H. Lou, eds.) 159-180. deGruyter, Berlin.

[52] STROOCK, D. W. (1993) *Probability Theory: an analytic view.* Cambridge Univ. Press, Cambridge, U. K.

[53] WATERMAN, M. S. (1995) *Introduction to Computational Biology.* Chapman &
     Hall, London.

[54] XIA, A. (1997) On the rate of Poisson process approximation to Bernoulli
     process. *J. Appl. Prob.*

TRILOGY OF COUPLINGS AND GENERAL FORMULAS FOR LOWER BOUND OF SPECTRAL GAP

MU-FA CHEN,* *Beijing Normal University*

Abstract

This paper starts from a nice application of the coupling method to a traditional topic: the estimation of the spectral gap (=the first non-trivial eigenvalue). Some new variational formulas for the lower bound of the spectral gap of Laplacian on manifold or elliptic operators in $\mathbb{R}^d$ or Markov chains are reported [10],[15],[16]. The new formulas are especially powerful for the lower bounds; they have no common points with the classical variational formula (which goes back to Lord Rayleigh (1877) or E. Fischer (1905)) and is particularly useful for the upper bounds. No analog of the new formulas has appeared before. The formulas not only enable us to recover or improve the main known results but also make a global change of the study on the topic. This will be illustrated by comparison of the new results with the known ones in geometry. Next, we will explain the mathematical tools for proving the results. That is, the trilogy of the recent development of the coupling theory: the Markovian coupling, the optimal Markovian coupling and the construction of distances for coupling. Finally, some related results and some problems for further study are also mentioned. It is hoped that the paper could be readable not only for probabilists but also for geometers and analysts.

DIFFUSIONS; MANIFOLD; MARKOV CHAINS; SPECTRAL GAP; COUPLINGS

AMS 1991 SUBJECT CLASSIFICATION: PRIMARY 35P15; 60H30
SECONDARY 60J27; 60J80

1. Background

1.1. *Definition* Consider a birth–death process with state space $E = \{0, 1, 2, \cdots\}$ and Q-matrix

$$
Q = (q_{ij}) = \begin{pmatrix}
-b_0 & b_0 & 0 & 0 & \cdots \\
a_1 & -(a_1 + b_1) & b_1 & 0 & \cdots \\
0 & a_2 & -(a_2 + b_2) & b_2 & \cdots \\
\vdots & \ddots & & \ddots & \ddots
\end{pmatrix}
$$

where $a_k,\ b_k > 0$. Since the sum of each row equals 0, we have $Q1 = 0 = 0 \cdot 1$. This means that the Q-matrix has an eigenvalue 0 with eigenvector 1. Next, consider the finite case, $E_n = \{0, 1, \cdots, n\}$. Then, the eigenvalues of $(-Q)$ are discrete: $0 = \lambda_0 < \lambda_1 \le \cdots \le \lambda_n$. Hence, there is a gap between λ_0 and λ_1: gap $(Q) := \lambda_1 - \lambda_0 = \lambda_1$. In

---

* Postal address: Department of Mathematics, Beijing Normal University, Beijing 100875, The People's Republic of China.
Research supported in part by NSFC and the State Education Commission of China.

the infinite case, the gap can be 0. Certainly, one can consider the self-adjoint elliptic operators in $\mathbb{R}^d$ or the Laplacian Δ on manifolds or an infinite-dimensional operator as in the study of interacting particle systems. In the last case, the operator depends on a parameter β. For different β, the system has completely different behavior.

1.2. *Applications* (1) Phase Transitions.

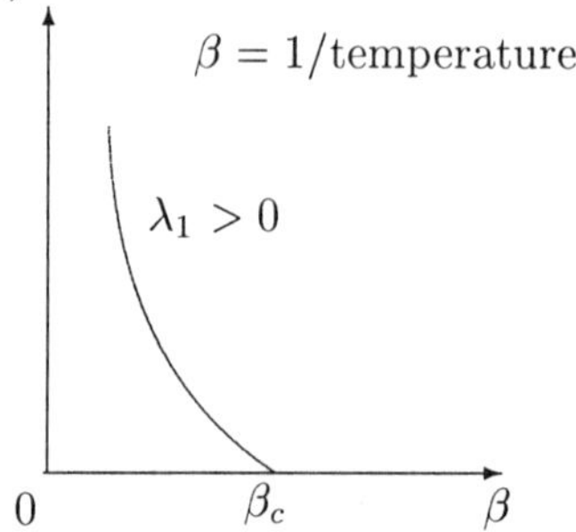

The picture means that at a higher temperature (small β), the corresponding semigroup $\{T_t\}_{t\geq 0}$ is exponentially ergodic in the L^2-sense: $\|T_t f - \pi(f)\| \leq \|f - \pi(f)\|e^{-\lambda_1 t}$, where $\pi(f) = \int f\,d\pi$, with the largest rate λ_1 and when the temperature goes to the critical value, the rate will go to zero. This provides a way to describe the phase transitions and it is now an active research field [6],[33],[39],[41],[43],[44],[51]. The next application we would like to mention is the
(2) Computer Science.

a) Complexity of randomized approximation algorithms. The existence of a spectral gap is used to prove a randomized approximation algorithm to be polynomial. See Jerrum and Sinclair (1989) for instance.

b) A fashionable application of the topic is the Markov chains Monte Carlo. There are too many publications to be listed here.

(3) Finally but not the last, the spectral gap has been used by Aldous and Brown (1993) and by Iscoe and McDonald (1994) for the asymptotics of the exit times.

1.3. *Difficulty* We have seen the importance of the topic but it is extremely difficult. To get some concrete feeling, let us look at the following examples.
a) Consider birth–death processes. Denote by g and $D(g)$ respectively the eigenfunction of λ_1 and the degree of g.

b_i	a_i	λ_1	$D(g)$
$i+1$	$2i$	1	1
$i+1$	$2i + 1.1$	2	2

The change of the death rate from $2i$ to $2i + 1.1$ leads to the change of λ_1 from one to two. More surprisingly, the order of g is changed from linear to quadratic. Next, for finite state spaces, it is trivial when $E = \{0,1\}$, $\lambda_1 = a_1 + b_0$. If we go one more step, $E = \{0,1,2\}$, then we have four parameters b_0, b_1 and a_1, a_2 only and

$$\lambda_1 = 2^{-1}\left[a_1 + a_2 + b_0 + b_1 - - - \sqrt{(a_1 - a_2 + b_0 - b_1)^2 + 4a_1 b_1}\right].$$

Now, the role for λ_1 played by the parameters becomes ambiguous.

b) Consider diffusions with operator $L = a(x)\dfrac{d^2}{dx^2} + b(x)\dfrac{d}{dx}$. The state space for the first row below is the full line and for the last two rows is the half line $[0, \infty)$ with reflection boundary.

$a(x)$	$b(x)$	λ_1	$D(g)$
1	$-x$	1	1
1	$-x$	2	2
1	$-(x+1)$	3	3

From these, one sees that the eigenvalue λ_1 is very sensitive and the relation between λ_1 and the coefficients (a_i, b_i) or $(a(x), b(x))$ can not be very simple.

One may think all these examples are rather special but the last one of birth–death process and the last two of diffusions are indeed new. Actually, we were unable in [13] to cover by our approach the general case of one-dimensional diffusions, for which an analytic approach was adopted.

2. Old Results and New Results

2.1. *Story of λ_1 in Geometry* The most well-developed subject of the first eigenvalue λ_1 is Riemannian geometry. For instance, a large part of each book [2], [4] and [40] is devoted to the problem. About 2000 references are included in [2]. For the latter use, we now review quickly some famous estimates obtained by geometers.

Let (M, g) be a compact and connected Riemannian manifold with Riemannian metric g. Denote by d and D respectively the dimension and the diameter of M. Assume that $\mathrm{Ricci}_M \geq Kg$ for some $K \in \mathbb{R}$. The main aim of the study is to use the geometric quantities d, D and K to estimate λ_k's of Laplacian Δ. The main lower bounds of λ_1 obtained by different authors are listed in the following table:

$$
\begin{array}{lll}
\text{Lichnerowicz (1958)} & \dfrac{d}{d-1}K, & K \geq 0 \\[2ex]
\text{Li and Yau (1980)} & \dfrac{\pi^2}{2D^2}, & K \geq 0 \\[2ex]
\text{Zhong and Yang (1984)} & \dfrac{\pi^2}{D^2}, & K \geq 0 \\[2ex]
\text{Li and Yau (1980)} & \dfrac{1}{D^2(d-1)\exp\left[1+\sqrt{1-4D^2K(d-1)}\,\right]}, & K \leq 0 \\[2ex]
\text{Cai (1991)} & \dfrac{\pi^2}{D^2} + K, & K \leq 0 \\[2ex]
\text{Yang (1989) and Jia (1991)} & \dfrac{\pi^2}{D^2}e^{-\alpha/2}, \quad \text{if } d \geq 5, & K \leq 0 \\[2ex]
\text{Yang (1989) and Jia (1991)} & \dfrac{\pi^2}{2D^2}e^{-\alpha'/2}, \quad \text{if } 2 \leq d \leq 4, & K \leq 0
\end{array}
$$

where $\alpha = D\sqrt{-K(d-1)}$ and $\alpha' = D\sqrt{-K((d-1)\vee 2)}$.

The first estimate is very good since it is optimal for the sphere in any dimension $d \geq 2$. The third one is optimal when $K = 0$. The last line but one (for all $d \geq 2$) is called the Yau's conjecture (mentioned in Yang (1989)). The importance of Li and Yau's work is that it introduced a new approach and made a deep influence to the subsequent study. No doubt, the results are very deep in geometry.

It was only recently that the coupling approach was introduced for the first time to study the estimate of λ_1 and produced the following estimates. Some of these are new in geometry.

2. Theorem. (Chen and Wang 1993).

$$\lambda_1 \geq \max\left\{\frac{\pi^2}{D^2}, \frac{d}{d-1}K, \frac{8}{D^2}+\frac{K}{3}\right\}, \qquad \text{if } K \geq 0$$

$$\lambda_1 \geq \max\left\{\frac{\pi^2}{D^2}+K, \frac{8}{D^2}+\frac{K}{3}, \frac{8}{D^2}\exp\left[\frac{D^2K}{8}\right], \frac{8}{D^2}\left(1+\frac{\alpha}{3}\right)e^{-\alpha/2}, \right.$$

$$\left. -\frac{K(d-1)}{4}\tanh^2\left(\frac{D}{2}\sqrt{\frac{-K}{d-1}}\right)\text{sech}^2\theta\right\}, \qquad \text{if } K \leq 0,$$

where θ is decreasing limit of θ_n: $\theta_1 = \frac{\alpha}{4}\tanh\left(\frac{D}{2}\sqrt{\frac{-K}{d-1}}\right)$, $\theta_n = \theta_1\tanh\theta_{n-1}$, $n \geq 2$. The last estimate is taken from [8] and it is sharp in some sense.

Clearly, all the sharp estimates in the previous table are included here. Moreover, the last two estimates in the table are also included:

$$\max\left\{\frac{\pi^2}{D^2}+K, \frac{8}{D^2}\left(1+\frac{\alpha}{3}\right)e^{-\alpha/2}\right\} \geq \frac{\pi^2}{D^2}e^{-\alpha/2}.$$

$$\Longrightarrow \lambda_1 \geq \frac{\pi^2}{D^2}e^{-\alpha/2}, \quad d \geq 2$$

$$\Longrightarrow \text{Yau's conjecture} \Longrightarrow \text{Yang and Jia's estimates.}$$

We have seen that in the past 40 years or so, geometers have made a series of hard efforts to improve the lower bounds step by step. The resulting bounds by different approaches are not comparable. On the other hand, several simple examples were in my mind for which I did not know how to handle by using our approach. Thus, I had a feeling for many years that each approach has its own advantage and there is no best one. I could not imagine, even half a year ago, that we could eventually find a general formula by using our approach.

2.2. New Results. Manifolds Define $\mathcal{F} = \{f \in C[0,D] : f > 0 \text{ on } (0,D)\}$, $C(r) = 1$ if $K = 0$ and

$$C(r) = \begin{cases} \cos^{d-1}\left[\frac{r}{2}\sqrt{\frac{K}{d-1}}\right], & \text{if } K > 0 \\ \cosh^{d-1}\left[\frac{r}{2}\sqrt{\frac{-K}{d-1}}\right], & \text{if } K < 0. \end{cases}$$

Then, our general formula is given as follows.

Theorem. (Chen and Wang (1997)[16]). For Laplacian on M, we have

$$\lambda_1 \geq 4\sup_{f\in\mathcal{F}}\inf_{r\in(0,D)} f(r)\left[\int_0^r C(s)^{-1}ds \int_s^D C(u)f(u)du\right]^{-1}.$$

Before moving further, let us recall the well-known classical variational formula, the Max–Min formula:*

$$\lambda_1 = \inf\left\{\mu(\|\nabla f\|^2)/\mu(f^2) : f \in C^1(M), \mu(f) = 0\right\},$$

where μ is the Riemannian measure on M. It is especially useful for the upper bound of λ_1 and is used in almost all of the literature on this topic. But it is much harder to handle the lower bound for which many approaches have been developed historically but no general formula ever appeared before. Comparing the formula with ours, one sees that there are no common points. To see the power of our formula, by setting $f = 1$, one can deduce all the bounds without underlines given in Theorem (1993)(cf. [12]). Of course, it should not be surprising that the new formula can produce a lot of new estimates since the test function f can be quite arbitrary. But it is surprising that the picture of the estimates of the first eigenvalue (given in [12] and [49] for instance) can be globally renewed, as illustrated by the following corollary.

Corollary. (Chen and Wang(1997)[16]).

$$\lambda_1 \geq \frac{\pi^2}{D^2} + \max\left\{\frac{\pi}{4d}, 1 - \frac{2}{\pi}\right\}K, \qquad K \geq 0$$

$$\lambda_1 \geq \frac{dK}{d-1}\left\{1 - \cos^d\left[\frac{D}{2}\sqrt{\frac{K}{d-1}}\right]\right\}^{-1}, \qquad d > 1, \quad K \geq 0$$

$$\lambda_1 \geq \frac{\pi^2}{D^2} + \left(\frac{\pi}{2} - 1\right)K, \qquad K \leq 0$$

$$\lambda_1 \geq \frac{\pi^2}{D^2}\sqrt{1 - \frac{2D^2K}{\pi^4}}\cosh^{1-d}\left[\frac{D}{2}\sqrt{\frac{-K}{d-1}}\right], \qquad d > 1, \quad K \leq 0.$$

The corollary improves respectively the Zhong and Yang, the Lichnernowicz, the Cai and the Yang and Jia estimate. For instance, since $D\sqrt{K/(d-1)} \leq \pi$ for $K \geq 0$ and usually the strict inequality holds, the second one above improves the Lichnerowicz's estimate.

2.3. *New Result. Diffusions in Half Line* For diffusions in $\mathbb{R}^d$ or for Markov chains, we have a similar story as for geometry but this is not discussed here. We mention only some general results. Consider the diffusions in half line with operator $L = a(x)\dfrac{d^2}{dx^2} + b(x)\dfrac{d}{dx}$ and reflecting boundary. Define

$$C(x) = \int_0^x \frac{b(y)}{a(y)}dy, \qquad \pi(dx) = \frac{1}{Z}\frac{e^{C(x)}}{a(x)}dx,$$

where Z is a normalizing constant. Set $\mathcal{F} = \{f \in L^1(\pi) : \pi(f) \geq 0 \text{ and } f'|_{(0,\infty)} > 0\}$.

---

* Historical Note: In [4], it is named the Rayleigh's formula, goes back to Lord S. J. W. Rayleigh (1877). On the other hand, in the book 'Inequalities' by E. F. Beckenback and R. Bellman (§25 and §26), it says that the formula goes back to E. Fischer (1905) and generalized by R. Courant (1924) (cf. [17]) and the original Rayleigh's result means $\lambda_0 = 0$ rather than λ_1. The relation of λ_1 and the L^2-exponential convergence mentioned above was studied in [33] and [5].

Theorem. (Chen and Wang(1997)[15]).

$$\lambda_1 \geq \sup_{f \in \mathcal{F}} \inf_{x>0} \left[\frac{e^{-C(x)}}{f'(x)} \int_x^\infty \frac{f(u)e^{C(u)}}{a(u)} du \right]^{-1}.$$

Moreover, in the regular case, the equality holds.

2.4. *New Result. Birth-Death Processes* Recall that

$$\pi_i = \frac{\mu_i}{\mu}, \quad \mu_0 = 1, \quad \mu_i = \frac{b_0 b_1 \cdots b_{i-1}}{a_1 a_2 \cdots a_i}, \quad i \geq 1, \quad \mu = \sum_i \mu_i.$$

Let $\mathcal{W} \subset L^1(\pi)$ be the set of all strictly increasing sequences $(w_i : i \geq 1)$ with $\sum_{i \geq 1} \mu_i w_i > 0$. Define

$$I_i(w) = b_i \mu_i (w_{i+1} - w_i) \bigg/ \sum_{j=i+1}^\infty \mu_j w_j, \; i \geq 1, \quad I_0(w) = b_0 \left(1 + w_1 \bigg/ \sum_{j=1}^\infty \mu_j w_j \right).$$

Theorem. (Chen(1996)[10]).

$$\lambda_1 = \sup_{w \in \mathcal{W}} \inf_{i \geq 0} I_i(w)$$

$$\lambda_1 = \sup_{(v_i > 0)} \inf_{i \geq 0} \{ a_{i+1} + b_i - a_i/v_{i-1} - b_{i+1} v_i \}.$$

We remark that here the equalities hold. In other words, we have complete dual variational formulas for λ_1. Furthermore, the formulas remain if λ_1 is replaced by the exponentially ergodic rate $\hat{\alpha}$ which is a traditional topic in the study of Markov chains [5] or [6]. The first formula above is a summation form which is similar to the integration form used previously. The second one is a differential form. An analog of the last form also works for the cases of manifolds or diffusions but is omitted here.

3. Trilogy of Couplings

3.1. *Markovian Couplings* We now turn to discussing the trilogy of couplings: The Markovian coupling, the optimal Markovian coupling and the construction of distances for couplings. We will also sketch the main proof of our results reported above. Since the story for Markov chains is similar, we concentrate on diffusions. Given an elliptic operator in $\mathbb{R}^d$

$$L = \sum_{i, j=1}^d a_{ij}(x) \frac{\partial^2}{\partial x_i \partial x_j} + \sum_{i=1}^d b_i(x) \frac{\partial}{\partial x_i}.$$

An elliptic (possibly degenerate) operator $\widetilde{L}$ on the product space $\mathbb{R}^d \times \mathbb{R}^d$ is called a **coupling of** L if it satisfies the following **marginality**:

$$\widetilde{L}f(x,y) = Lf(x) \text{ (resp. } \widetilde{L}f(x,y) = Lf(y)), \quad f \in C_b^2(\mathbb{R}^d), \; x \neq y,$$

where on the left-hand side, f is regarded as a bivariate function.

It is clear that the coefficients of any coupling operator $\widetilde{L}$ should be of the form

$$a(x,y) = \begin{pmatrix} a(x) & c(x,y) \\ c(x,y)^* & a(y) \end{pmatrix}, \qquad b(x,y) = \begin{pmatrix} b(x) \\ b(y) \end{pmatrix}.$$

This condition and the non-negative definite property of $a(x, y)$ consist of the **marginality** of $\widetilde{L}$ in the context of diffusions. Obviously, the only freedom is the choice of $c(x, y)$.

Three examples:

- **Classical coupling.** $c(x, y) \equiv 0, x \neq y$.

- **March coupling** (Chen and Li 1989). Let $a(x) = \sigma(x)^2$. Take $c(x, y) = \sigma(x)\sigma(y)$.

- **Coupling by reflection.** Set $\bar{u} = (x - y)/|x - y|$ and take

$$c(x, y) = \sigma(x)\left[\sigma(y) - 2\frac{\sigma(y)^{-1}\bar{u}\bar{u}^*}{|\sigma(y)^{-1}\bar{u}|^2}\right], \quad \det \sigma(y) \neq 0, \ x \neq y \ \text{[Lindvall and Rogers 1986]}$$

$$c(x, y) = \sigma(x)\left[I - 2\bar{u}\bar{u}^*\right]\sigma(y), \quad x \neq y \quad \text{[Chen and Li 1989]}.$$

The last coupling was extended to manifolds by W. S. Kendall (1986). See also M. Cranston (1991). In the case that $x = y$, the first and the third ones are defined to be the same as the second one.* Each coupling has its own `character`. A nice way to interpret the first coupling is to use a `Chinese idiom: fall in love at first sight`. The word 'march' is a `Chinese command to soldiers to start marching`. We are now ready to talk about

3.2. Sketch of the Main Proof Here we adopt the analytic language. Given a self-adjoint elliptic operator L, denote by $\{T_t\}_{t \geq 0}$ the semigroup determined by L: $T_t = e^{tL}$. Corresponding to $\widetilde{L}$, we have $\{\widetilde{T}_t\}_{t \geq 0}$. The coupling simply means that

$$(1) \qquad \widetilde{T}_t f(x, y) = T_t f(x) \ (\text{resp. } \widetilde{T}_t f(x, y) = T_t f(y))$$

for all $f \in C_b^2(\mathbb{R}^d)$ and all (x, y) $(x \neq y)$, where on the left-hand side, f is regarded as a bivariate function.

Step 1. Let g be the eigenfunction of $-L$ corresponding to λ_1. We have

$$\frac{d}{dt}T_t g(x) = T_t L g(x) = -\lambda_1 T_t g(x).$$

Hence

$$(2) \qquad T_t g(x) = g(x)e^{-\lambda_1 t}.$$

Step 2. Consider compact space. Since g is Lipschitz with respect to Riemannian distance ρ, g is a Lipschitz function. Denote by c_g the Lipschitz constant. Now, the main condition we need is the following:

$$(3) \qquad \widetilde{T}_t \rho(x, y) \leq \rho(x, y)e^{-\alpha t}.$$

---

* For this reason, the term 'basic coupling' was used in [11] for the second coupling. We now use the term 'march' rather than 'basic' since it is more intrinsic and consistent with the one used for Markov chains by the author years ago.

This condition is implied by

$$(4) \qquad \qquad \widetilde{L}\rho(x,y) \leq -\alpha\rho(x,y), \quad x \neq y$$

Setting $g_1(x,y) = g(x)$ and $g_2(x,y) = g(y)$, we obtain

$$\begin{aligned}
e^{-\lambda_1 t}|g(x) - g(y)| &= |T_t g(x) - T_t g(y)| \quad \text{(by (2))} \\
&= |\widetilde{T}_t g_1(x,y) - \widetilde{T}_t g_2(x,y)| \quad \text{(by (1))} \\
&\leq \widetilde{T}_t |g_1 - g_2|(x,y) \\
&\leq c_g \widetilde{T}_t \rho(x,y) \quad \text{(Lipschitz property)} \\
&\leq c_g \rho(x,y) e^{-\alpha t} \quad \text{(by (3))}.
\end{aligned}$$

Step 3. Choose $\{(x_n, y_n)\}$ so that $\dfrac{|g(x_n) - g(y_n)|}{\rho(x_n, y_n)} \to c_g$. We obtain $\lambda_1 \geq \alpha$. This completes the proof.

The proof is unbelievably straightforward. It is universal in the sense that it works for general Markov processes. A good point in the proof is the use of the eigenfunction so that we can achieve the sharp estimates. On the other hand, it is crucial that we do not need too much knowledge about the eigenfunction. Otherwise, there is no hope to work things out since the eigenvalue and its eigenfunction are either known or unknown simultaneously. Except the Lipschitz property of g with respect to the distance, which can be avoided by using a localizing procedure for the non-compact case, the key of the proof is clearly the condition (4). For this, one needs not only a good coupling but also a good choice of the distance.

3.3. *Optimal Markovian Coupling* Since there are infinitely many choices of coupling operators, it is natural to ask the following questions. Does there exist an optimal one? In what sense of optimality we are talking about?

*Definition 1 Let $(E, \rho, \mathcal{E})$ be a metric space. A coupling operator $\overline{L}$ is called ρ-***optimal*** if $\overline{L}\rho(x_1, x_2) = \inf_{\widetilde{L}} \widetilde{L}\rho(x_1, x_2)$ for all $x_1 \neq x_2$, where $\widetilde{L}$ varies over all coupling operators.*

To construct an optimal Markovian coupling is not an easy job even though there is often no problem for existence. Here, we mention a special case only.

Theorem 1 [Chen 1994] Let $f \in C^2(\mathbb{R}_+; \mathbb{R}_+)$ with $f(0) = 0$ and $f' > 0$. Suppose that $a(x) = \varphi(x)\sigma^2$ for some positive function φ, where σ is a constant matrix with $\det \sigma > 0$.

- *If $\rho(x,y) = f(|\sigma^{-1}(x - y)|)$ with $f' \leq 0$, then the coupling by reflection is ρ-optimal. That is, $c(x,y) = \sqrt{\varphi(x)}\left[\sigma^2 - 2\bar{u}\bar{u}^*/|\sigma^{-1}\bar{u}|^2\right]\sqrt{\varphi(y)}$.*

- *If $\rho(x,y) = f(|\sigma^{-1}(x - y)|)$ with $f' \geq 0$, then the march coupling is ρ-optimal. That is, $c(x,y) = \sqrt{\varphi(x)}\sigma^2\sqrt{\varphi(y)}$.*

- *If $d = 1$ and $\rho(x,y) = |x - y|$, then all the three couplings mentioned above are ρ-optimal.*

Part (2) of the theorem is newly added but it is an analog of the birth–death processes and its proof is similar to that of part (1). Note that in case (2), ρ may not be a distance but the definition of ρ-optimal coupling is still meaningful.

3.4. *Construction of Distances* In view of the above theorem, one sees that the optimal coupling depends heavily on ρ and furthermore, even for a fixed optimal coupling, there is still a large class of ρ that can be chosen, for which the resulting estimate of α given in (4) may be completely different. For instance, the sharp estimates for the Laplacian on manifolds can not be achieved if one is restricted to the Riemannian distance only. Thus, the construction of the distances plays a key role in the application of our coupling approach. However, we now have a unified construction for the distance ρ used for the three classes of processes discussed in the paper. Here, we write down the answer for the case of diffusions in the half line only.

$$g(r) = \int_0^r e^{-C(s)} ds \int_s^\infty \frac{f(u)e^{C(u)}}{a(u)} du, \quad f \in \mathcal{F}, \qquad \rho(x,y) = |g(x) - g(y)|.$$

This construction of distances consists of the last part of the trilogy.

4. Related Results and Problems

The new results presented in section 2 are particular ones of [10], [15] and [16]. Actually, in [16], we deal with the operator $\Delta + \nabla V$ for some $V \in C^2(M)$ on manifolds (maybe non-compact) with Neumann boundary or without boundary. In [15], we deal with self-adjoint elliptic operators in $\mathbb{R}^d$. It is not difficult to go to the full line from the half line but in the higher dimensional case one needs more work. Our new formulas are also meaningful for gradient estimates, Dirichlet eigenvalues, the mixed eigenvalues and many others. The recent papers, based on or closely related to the coupling approach, are partially collected in the references.

For a long period, the coupling method has been mainly used for the convergence in total variation which then deduces the study on success of couplings. The impression in one's mind is often that the coupling method is useful only if it is successful and it can only provide rough estimates. However, from the illustrations above, one sees how large a change has been made recently. Actually, the study of the spectral gap is only the most recent topic of various applications of the coupling method. One may refer to [6], [7] and [34] for other applications. For instance, coupling by reflection is a good choice for the present purpose. But when one looks for the order-preserving coupling, the march coupling is clearly better than the previous one. A nice application of a geometric generalization of the march coupling is given in [24] and [56]. Now, what coupling is the best for the order-preserving one? For this, we now have only a partial answer (see [61] for instance). Next, a fundamental problem for the couplings of time-continuous Markov processes is the uniqueness (well-posed) one. For Markov chains (more generally, for jump processes), this problem was solved completely (cf. [6; Theorem 5.16, Theorem 5.17], [7] and [30]). For diffusions, the same conclusion is conjectured to be true but only a partial solution is known now. Finally, what is the optimal coupling for weak convergence? How to construct 'good' couplings for semimartingales? These are only a few questions we mention here randomly. In conclusion, the theory of couplings is still too young. There is a lot to be done and the subject should have a nice future.

Acknowledgements

Most of the materials in the paper were or will be talked about at several conferences and universities or institutes: The 60th Anniversary Conference of Chinese Mathematical Society (May 1995, Beijing), the 23rd Conference on Stochastic Processes and Their Applications (June 1995, Singapore), the Symposium on Probability Towards the Year 2000 (October, 1995, New York), Stochastic Differential Geometry and Infinite-Dimensional Analysis (April 1996, Hangzhou), Workshop on Interacting Particle Systems and Their Applications (June 1996, Haifa), The Third Gaussian Symposium (August 1996, Beijing), Cornell University, University of Illinois, Institute of Applied Mathematics (Chinese Academy), Institute of Advanced Mathematics (Hangzhou), University of Biefeled, Bar-Ilan University and Technion-Israel Institute of Technology. The author would like to thank the following mathematicians for their hospitality and financial support: Prof. Louis H. Y. Chen, Dr. J. H. Lou and Dr. K. P. Choi at Singapore U., Prof. L. Accardi at U. of Roma II, Profs. C. Heyde, K. Sigman and Y. Z. Shao at Columbia U., Profs. R. Durrett, L. Gross and Z. Q. Chen at Cornell U., Prof. D. L. Burkholder at U. of Illinois, Prof. X. W. Zhuang at Fujian Normal U., Profs. Z. M. Ma and J. A. Yan at Applied Inst. Chin. Acad., Prof. D. Elworthy at Warwick U., Profs. F. Götze and M. Röckner at U. of Bielefeld, Prof. K. J. Hochberg at Bar-Ilan U. and Prof. B. Granovsky at Technion-Israel Inst. of Tech. The author also acknowledges Prof. F. Y. Wang for fruitful cooperation.

References

[1] Aldous, D. J. and Brown, M. (1993) Inequalities for rare events in time-reversible Markov chains, *IMS Lecture Notes-Monograph Series.* **22**, Stochastic Inequalities, 1–16.

[2] Bérard, P. H. (1986) *Spectral Geometry: Direct and Inverse Problem* LNM. vol. 1207, Springer-Verlag.

[3] Cai, K. R. (1991) Estimate on lower bound of the first eigenvalue of a compact Riemannian manifold *Chin. Ann. of Math.* **12(B)**, 267–271.

[4] Chavel, I. (1984) *Eigenvalues in Riemannian Geometry* Academic Press.

[5] Chen, M. F. (1991) Exponential L^2-convergence and L^2-spectral gap for Markov processes *Acta Math. Sin. New Ser.* **7**, 19–37.

[6] Chen, M. F. (1992) *From Markov Chains to Non-Equilibrium Particle Systems* World Scientific.

[7] Chen, M. F. (1994) Optimal Markovian couplings and application to Riemannian geometry, in *Prob. Theory Math. Statist.* Eds. B. Grigelionis et al. VPS/TEV, 121–142.

[8] Chen, M. F.(1994) Optimal Markovian couplings and applications *Acta Math. Sin. New Ser.* **10**, 260–275 .

[9] Chen, M. F. (1995) On ergodic region of Schlögl's model *Dirichlet Forms and Stoch. Proc.* Edited by Z. M. Ma, M. Röckner and J. A. Yan, Walter de Gruyter, 87–102.

[10] Chen, M. F. (1996) Estimation of spectral gap for Markov chains *Acta Math. Sinica New Ser.***12**, 337–360.

[11] Chen, M. F. and Li, S. F. (1989) Coupling methods for multi-dimensional diffusion processes *Ann. of Probab.* **17**, 151–177.

[12] Chen, M. F. and Wang, F. Y. (1993) Application of coupling method to the first eigenvalue on manifold *Sci. Sin.* (A), **23** (1993) (Chinese Edition), 1130–1140, **37** (1994) (English Edition), 1–14.

[13] Chen, M. F. and Wang, F. Y. (1995) Estimation of the first eigenvalue of second order elliptic operators *J. Funct. Anal.* **131**, 345–363.

[14] Chen, M. F. and Wang, F. Y. (1997) Estimates of logarithmic Sobolev constant – An improvement of Bakry–Emery criterion *J. Funct. Anal.* **144**, 287–300.

[15] Chen, M. F. and Wang, F. Y. (1997) Estimation of spectral gap for elliptic operators *Trans. Amer. Math. Soc.* **349**, 1209– 1237.

[16] Chen, M. F. and Wang, F. Y. (1997) General formula for lower bound of the first eigenvalue on Riemannian manifolds *Sci. Sin.* **27** (Chinese Edition), 34–42, **40** (English Edition), 384–394.

[17] Courant, R. and Hilbert, D. (1953) *Methods of Mathematical Physics* Interscience Publishers.

[18] Cranston, M. (1991) Gradient estimates on manifolds using coupling *J. Funct. Anal.* **99**, 110–124.

[19] Cranston, M. (1992) A probabilistic approach to gradient estimates *Canad. Math. Bull.* **35**, 46–55.

[20] Deuschel, J.-D. and Stroock, D. W. (1990) Hypercontractivity and spectral gap of symmetric diffusion with applications to the stochastic Ising models *J. Funct. Anal.* **92**, 30–48.

[21] Diaconis and Stroock (1991) Geometric bounds for eigenvalues of Markov chains *Ann. Appl. Prob.* **1**, 36–61.

[22] Doeblin, W. (1938) Exposé de la théorie des chaines simples constantes de Markov à un nombre dini d'etats *Rev. Math. Union Interbalkanique* **2**, 77–105.

[23] Fischer, E. (1905) Über quadratische Formen mit reellen Koeffizienten, *Monatsh. Math. Phys.* **16**, 234–249.

[24] Hsu, E. P. (1994) Logarithmic Sobolev inequality on path spaces *C.R. Acad. Sci. Paris* **320**, 1209–1214.

[25] Iscoe, I. and McDonald, D. (1994) Asymptotics of exit times for Markov jump processes (*I*) *Ann. Prob.* **22**, 372–397.

[26] Jerrum, M. R. and Sinclair, A. J. (1989) Approximating the permanent *SIAM J. Comput.* **18**, 1149–1178.

[27] Jia, F. (1991) Estimate of the first eigenvalue of a compact Riemannian manifold with Ricci curvature bounded below by a negative constant (In Chinese) *Chin. Ann. Math.* **12(A)**, 496–502.

[28] Kendall, W. (1986) Nonnegative Ricci curvature and the Brownian coupling property *Stochastics* **19**, 111–129.

[29] Kendall, W. S. (1994) Probability, convex, and harmonic maps *II*: smoothness via probabilistic gradient inequalities *J. Funct. Anal.* **124**.

[30] Lawler, G. F. and Sokal, A. D.(1988) Bounds on the L^2 spectrum for Markov chain and Markov processes: a generalization of Cheeger's inequality *Trans. Amer. Math. Soc.* **309**, 557–580.

[31] Li, P. and Yau, S. T. (1980) Estimates of eigenvalue of a compact Riemannian manifold *Ann. Math. Soc. Proc. Symp. Pure Math.* **36**, 205–240.

[32] Lichnerowicz, A. (1958) *Geometrie des Groupes des Transformationes* Dunod, Paris.

[33] Liggett, T. M. (1989) Exponential L_2 convergence of attractive reversible nearest particle systems*Ann. Probab.* **17**, 403-432.

[34] Lindvall, T. (1992) *Lectures on the Coupling Method* Wiley, New York.

[35] Lindvall, T. and Rogers, L. C. G. (1986) Coupling of multidimensional diffusion processes *Ann. of Probab.* **14**, 860–872.

[36] Lu, Y. G. (1994) An estimate on non-zero eigenvalues of Laplacian in non-linear version *preprint.*

[37] Lu, Y. G. (1994) Estimate of the first non-zero eigenvalue of Laplace-de Rahm and the Laplace-Beltrami operators *preprint.*

[38] Lü, J. S. (1995) Optimal coupling for single birth processes and applicaation to a class of infinite-dimensional reaction-diffusion processes (In Chinese) *J. Beijing Normal Univ.* **33**, 10–17.

[39] Minlos, R. A. and Trisch, A. (1994) Complete spectral decomposition of the generator for one-dimensional Glauber dynamics (In Russian) *Uspekhi Matem. Nauk,* **209–210**.

[40] Schoen, R. and Yau, S. T. (1988) *Differential Geometry* (In Chinese) Science Press, Beijing, China .

[41] Schonmann, R. H. and Shlosman, S. B. (1994) Complete analyticity for 2D Ising completed *Comm. Math. Phys.* **170,** 453–482 .

[42] Sinclair, A. J. and Jerrum, M. R. (1989) Approximate counting, uniform generation, and rapidly mixing Markov chains *Inform. and Comput.* **82,** 93–133

[43] Sokal, A. D. and Thomas, L. E.(1988) Absence of mass gap for a class of stochastic contour models *J. Statis. Phys.* **51,** 907–947 .

[44] Stroock, D. W. and Zegarlinski, B. (1992), The equivalence of the logarithmic Sobolev inequality and the Dobrushin-Shlosman mixing condition, *Comm. Math. Phys.* **144,** 303-323.

[45] Sullivan, W. G.(1984) The L^2 spectral gap of certain positive recurrent Markov chains and jump processes *Z. Wahrs.* **67,** 387–398.

[46] Wang, F. Y. (1994) Gradient estimates on $\mathbb{R}^d$ *Canad. Math. Bull.* **XX(2),** 1–11.

[47] Wang, F. Y. (1994) Gradient estimates for generalized harmonic function on manifold (In Chinese) *Chin. Sci. Bull.* **39** 492–495 .

[48] Wang, F. Y. (1994) Ergodicity for infinite-dimensional diffusion processes on manifolds *Sci. Sin. Ser A* **37,** 137–146.

[49] Wang, F. Y. (1994) Application of coupling method to the Neumann eigenvalue problem *Prob. Th. Rel. Fields* **98,** 299–306.

[50] Wang, F. Y. (1994) Estimate of the first Dirichlet eigenvalue by using the diffusion processes *Prob. Th. Rel. Fields* **101,** 363–369.

[51] Wang, F. Y. (1995) Uniqueness of Gibbs states and the L^2-convergence of infinite-dimensional reflecting diffusion processes *Sci. Sin. Ser A* **32,** 908–917 .

[52] Wang, F. Y. (1994) On estimates of logarithmic Sobolev constant (In Chinese) *J. Beijing Normal Univ.* **30,** 448–452.

[53] Wang, F. Y. (1996) Estimates of logarithmic Sobolev constant for finite volume continuous spin systems *J. Stat. Phys.* **84,** 277–293.

[54] Wang, F. Y. (1994) A Probabilistic approach to the first Dirichlet eigenvalue on non-compact Riemannian manifold *Acta Math. Sin. New Series* **13,** 116–126.

[55] Wang, F. Y. (1995) Spectral gap for diffusion processes on non-compact manifolds *Chin. Sci. Bull.* **40,** 1145–1149.

[56] Wang, F. Y. (1996) Logarithmic Sobolev inequalities for diffusion processes with application to path space *Chin. J. Appl. Prob. Stat.* **12,** 255–264.

[57] Wang, F. Y. and Xu, M. P. (1997) On order-preservation of couplings for multi-dimensional diffusion processes *Chin. J. Prob. Stat.* **13**, 142–148.

[58] Yang, H. C. (1989) Estimate of the first eigenvalue of a compact Riemannian manifold with Ricci curvature bounded below by a negative constant (In Chinese) *Sci. Sin.(A)* **32**, 698–700.

[59] Yuan, X. B. (1995) Gradient estimates and the first mixed eigenvalue *Master's thesis at Beijing Normal Univ.*

[60] Zhang, Y. H. (1994) Conservativity of couplings for jump processes (In Chinese) *J. Beijing Normal Univ.* **30**, 305–307.

[61] Zhang, Y. H.(1995) The construction of order-preserving coupling for one-dimensional Markov chains *Chin. J. Appl. Prob. Stat.* **12**, 376–382.

[62] Zhong, J. Q. and Yang, H. C. (1984) Estimates of the first eigenvalue of a compact Riemannian manifolds *Sci. Sin.* **27**, 1251–1265.

STOCHASTICITY AND CHAOS

P. COLLET,[*] *Ecole Polytechnique*

Abstract

We discuss some results concerning stochastic perturbations of chaotic systems. In particular stochastic stability of SRB measures, asymptotic laws for entrance and exit times in small sets and rates of leaking due to noise.

1. INTRODUCTION

Chaotic behavior of simple dynamical systems has recently been the object of intense investigation. There is a strong parallel with the theory of stochastic processes, and indeed at a very general and abstract level a stochastic process can be considered as a dynamical system. It is therefore not surprising that many ideas methods and questions from the theory of stochastic processes have been successfully transferred in the realm of dynamical systems. We will explain below some of these developments concerning stochastic perturbations of chaotic systems.

In order to define a dynamical system, one has first to give the phase space (denoted below by Ω), that is to say the space of all the possible states of the system. Then one has to introduce a time evolution of the states, namely of the points of the phase space. In the case of continuous time evolution, this is given by a semiflow on the phase space (for example if Ω is a manifold, this can be the semi flow which integrates a vector field). We recall that a semiflow is a map ϕ from $\mathbb{R}^+ \times \Omega$ to Ω satisfying for any $x \in \Omega$ $\phi_0(x) = x$ and for any t and $s \in \mathbb{R}^+$

$$\phi_{s+t}(x) = \phi_t \circ \phi_s(x) = \phi_t(\phi_s(x)) \ .$$

The interpretation is of course that if at time zero the system is in the state x, then under the evolution at time t it is in the state $\phi_t(x)$. A discrete time dynamical system is simply given by a map T from Ω into itself and the time evolution is given by iterations of T. Namely, if at time zero the system is in the state x, at time n (> 0) it will be in the state

$$T^n(x) = \underbrace{T \circ \cdots \circ T}_{n \text{ times}}(x) \ .$$

Chaos corresponds to erratic wandering of the state of the system during time evolution.

The above definitions are for dynamical systems which are time independent (holonomous), it is easy to generalize these definition to time dependent evolutions. We will

---

[*] Postal address: Centre de Physique Théorique, Laboratoire CNRS UPR 14, Ecole Polytechnique, F-91128 Palaiseau Cedex (France).

come back to this point below when defining random dynamical systems. Also we have not discussed the regularity of the map (flow) defining the dynamical system. This is of course a question of generality and context. For example, in ergodic theory one is often satisfied with only measurability hypothesis. In most cases below we will consider concrete dynamical systems and we will assume (unless otherwise stated) that Ω is a regular manifold and the map (flow) is also regular (that is to say with enough regularity so that the needed differential manipulations can be performed without questioning).

As remarked before, the analogy with stochastic processes is striking and in fact a stochastic process can be viewed as a dynamical system. It is therefore not surprising that the two theories have many relations. For example many ideas and results of ergodic theories which are usually formulated for dynamical systems apply as well to stochastic processes. Note however that dynamical systems are usually considered with compact phase spaces Ω whereas many interesting stochastic processes have unbounded trajectories or more precisely non compact phase spaces (as is the case for example of brownian motion).

Since both stochastic processes and dynamical systems produce complicated (erratic) behavior, it is natural to ask if interesting phenomena can occur through the interaction of two such systems. This is also a natural question from a Physical point of view. If we have a dynamical system modeling the evolution of some natural object, there is often some perturbations of a stochastic nature not taken into account by the model, for example thermal fluctuations. A major problem is to determine whether an interesting dynamical behavior is stable under these random fluctuations. For example if one considers a dissipative mechanical system with a stable equilibrium which is not of minimum energy, even a very tiny stochastic perturbation may allow the system to jump over a potential barrier (after eventually a very long time) and to reach states with lower energy (see [F.W.]). In this case we see that the asymptotic time behavior of the system is strongly perturbed even by a very small noise. There are a large number of interesting problems in this area of stochastic perturbations of dynamical systems ranging from purely theoretical to technological applications. In fact the denomination stochastic perturbations is even not well adapted since by relaxing the mixing properties of the stochastic perturbations one is led to perturbations by other deterministic systems, for example periodic or quasi periodic perturbations have a lot of interesting applications.

One can give general formulations of stochastic dynamical systems parallel to the two previous definitions with discrete and continuous time. For discrete time, one can for example pick at random a sequence of mappings of the phase space Ω and consider the composition of these maps. For the case of continuous time, there is the whole theory of stochastic differential equations. We refer to [F.W.] for this last case which will be only discussed occasionally below. We will often consider below the simple case of independent perturbations of maps. These are constructed as follows. First assume that Ω is a (real) vector space. Let ξ be a random variable with values in Ω. We consider a sequence of i.i.d. random variables (ξ_n) which are independent copies of ξ and a real number ϵ. If T is a map of our phase space Ω, we can construct a random sequence of maps defined by

$$T_n = T + \epsilon\, \xi_{n-1}\,. \tag{1}$$

We can now define a random dynamical system as follows. If the initial state of the system is x, we define $x_0 = x$. The (random) state of the system at time n (> 0) denoted by x_n is given by the recursion

$$x_n = T_n(x_{n-1}) = T(x_{n-1}) + \epsilon \, \xi_{n-1} \ .$$

Or in other words,

$$x_n = T_n \circ \cdots \circ T_1(x) \ .$$

It is easy to verify that we have defined in this way a Markov process with state space Ω. Note also that if the parameter $\epsilon = 0$ all the maps are equal to T and we recover the deterministic iteration. Note that if $\epsilon \neq 0$ we have a time dependent (non holonomous) system.

There is another interesting class of problems with random dynamical systems where one picks an evolution at random once for all and considers its large time behavior. We will not discuss these questions here.

We now consider in more detail the situation where the random dynamical system is a small perturbation of a deterministic one. The fundamental question is attributed to Kolmogorov and will be formulated below in the case of a map T (discrete time systems; the case of continuous time is similar). We will also assume a compact phase space. Assume μ is an invariant (Borel) probability measure for the time evolution, namely for any measurable set A we have

$$\mu(T^{-1}(A)) = \mu(A)$$

where T^{-1} is the set theoretic inverse of T, or in other words, for any continuous function g we have

$$\int g \circ T \, d\mu = \int g \, d\mu \ .$$

Assume also μ is ergodic, that is to say any measurable invariant function u ($u \circ T = u$ μ almost surely) is almost surely constant. If we introduce a stochastic perturbation of T we obtain a stochastic process which may have an invariant measure μ_ϵ depending of the amplitude of the perturbation ϵ. Of course the stochastic process may have several invariant measures or none. Under quite natural conditions on T and on the stochastic perturbations, the invariant measure is unique (see [F.W.]) and we will only consider this situation.

QUESTION. Is is true that for the weak topology of measures (*i.e.* after integration against any continuous function)

$$\lim_{\epsilon \to 0} \mu_\epsilon = \mu \ .$$

If this is the case this means that the dynamical system associated to the map T is rather stable against the small stochastic perturbation, at least as far as the measure μ is concerned. Of course stochastic stability may depend on the kind of stochastic perturbation which is used, and although stochastic stability is a desirable property in many situations, stochastic instability has also very interesting consequences as in the case of metastability mentioned above. It often happens that a dynamical system has many ergodic invariant measures, and a positive answer to the above question leads to a selection of a particular invariant measure.

As a simple example of stochastic stability let us consider the one dimensional potential flow

$$\frac{dx}{dt} = -x \ .$$

There is only one stable fixed point namely $x = 0$ and only one ergodic invariant measure, namely the Dirac measure at the origin. Consider now the stochastic differential equation

$$dX = -X \ dt + \epsilon \ dW(t) \tag{2}$$

where W is the one dimensional Brownian motion (in more Physical notations this is also $dX/dt = -X + \epsilon \ \xi(t)$ where $\xi(t)$ is the white noise). Formally when ϵ is set to zero we recover the potential flow. It turns out that the invariant measure of (2) is unique and known explicitly. It is absolutely continuous with respect to the Lebesgue measure with a density given by

$$\frac{d\mu_\epsilon}{dx} = \frac{e^{-x^2/\epsilon^2}}{\epsilon\sqrt{\pi}} \ .$$

It is an easy exercise to verify that when ϵ tends to zero, μ_ϵ converges weakly to the Dirac measure at the origin. In this case the answer to Kolmogorov's question is positive.

This is of course a very elementary example and one would like to analyze situations where the dynamical system has more interesting behavior, in particular chaotic situations.

There is an interesting class of chaotic dynamical systems which is now rather well understood and which is called axiom A systems (see [S.] [B.] and [R.] for some detailed analysis). For these systems the phase space Ω is a smooth (compact) manifold and the dynamics is given by a regular map T. Moreover, the tangent space E_x at every point x of the manifold can be written as a direct sum of two subspaces

$$E_x = E_x^u \oplus E_x^s$$

with the following properties

(i) $DT_x E_x^s = E_{T(x)}^s$.

(ii) $DT_x^{-1} E_x^u = E_{T^{-1}(x)}^u$ (we assume for simplicity T invertible but this is not necessary, see [H.P.S]).

(iii) There is a number $0 < \lambda < 1$ such that

$$\sup_x \max \left\{ \|DT_x|_{E_x^s}\| , \|DT_x^{-1}|_{E_x^u}\| \right\} \leq \lambda \ .$$

The last hypothesis means that we have uniform hyperbolicity of the tangent map. We will also assume that there is a unique attractor $\mathcal{A}$ (roughly speaking a set which attracts all the neighboring trajectories) which is irreducible and topologically mixing (see [B.] for these technical definitions which ensure finally uniqueness of the interesting measure). In fact one needs only assume properties i)-iii) above in a neighborhood of the attractor.

One can prove that there is for any $x \in \mathcal{A}$ a regular local stable manifold $W_{\text{loc}}^s(x)$ passing through x. This is a piece of submanifold tangent to E_x^s at x (with the same dimension) and such that if y and z belong to $W_{\text{loc}}^s(x)$ the distance between $T^n(y)$ and $T^n(z)$ tends to zero exponentially fast with n. Note that although the two trajectories approach each other very rapidly they may well wander around very erratically on the attractor. There is an equivalent construction of local unstable manifolds which can be defined in the same way using the inverse of T (we refer to [H.P.S.] for the non invertible case).

When the attractor $\mathcal{A}$ is not a simple object it is called a strange attractor (see [R.T.]). It has often a Hausdorff dimension smaller than the dimension of the phase space. The local structure of such an object is the (topological) product of a disk by a Cantor set, where the disks are the local unstable manifolds and the Cantor sets extend in the stable directions. Note that the local unstable manifolds are contained in the attractor.

For an axiom A system satisfying the above hypothesis there is a particularly interesting invariant measure called an SRB measure (after Sinai, Ruelle and Bowen), we formulate two of its properties in the following Theorem (see [E.R.]).

Theorem I.1. Under the above hypothesis, there is a unique invariant (ergodic and mixing) measure μ supported by the attractor $\mathcal{A}$ with the following properties.

1. There is a neighborhood U of $\mathcal{A}$ and a measurable subset V of U of positive Lebesgue measure such that for any continuous function g and for any $x \in U \backslash V$ we have

$$\lim_{n \to \infty} \frac{1}{n} \sum_{j=0}^{n-1} g(T^j(x)) = \int_{\mathcal{A}} g \, d\mu \, .$$

2. The measure μ conditioned to the local unstable manifolds is absolutely continuous with respect to the Lebesgue measure. In other words, if we denote by $\mathcal{W}$ the set of all unstable manifolds, there is a (probability) measure ν on $\mathcal{W}$ (the transverse measure) and for each $w \in \mathcal{W}$ there is an absolutely continuous probability measure $d\mu_w$ supported by w such that for any continuous function g

$$\int_{\mathcal{A}} g \, d\mu = \int_{\mathcal{W}} d\nu(w) \int_w g \, d\mu_w \, .$$

Note that the first condition is in some sense an extension of the ergodic theorem. Indeed, one could only apply Birkhoff's Ergodic Theorem to points x on the attractor which supports the measure. Property 1) tells us that time averages also converge for a large set of initial conditions outside the attractor.

There are several other beautiful properties of the SRB measures for which we refer to [E.R.]. From the point of view of the present paper the important one is the following.

Theorem I.2. The SRB measure is (the unique measure) stable by stochastic perturbations.

In other words, for these measures the answer to Kolmogorov's question is positive. This means in a sense that strong chaotic behaviors are stable against small stochastic perturbations. We refer to [Ki.], [Y.] and [L.Q.] for the conditions on the noise and the proof.

If the noise is additive as in (1), with for example a uniform distribution with a compact support, it turns out that the invariant measure μ_ϵ is absolutely continuous with respect to the Lebesgue measure of the phase space Ω. One can then try to understand how it collapses to the SRB measure which is singular with respect to the Lebesgue measure of the phase space. One can indeed determine the dominant contribution to the density of μ_ϵ for ϵ small. It turns out that this density can be written as an integral over the transverse measure ν of a family of local densities each associated to a local unstable manifold w. The local density ρ_w associated to w looks as follows. It is essentially the product of two functions which can be easily described using a system of coordinates given by w and its transverse directions. There is first the density $d\mu_w/dx_w$ of the measure $d\mu_w$ of 2) in the above Theorem. This function depends only on the coordinates on w. There is then a second function which depends on the normal coordinates to w but which has in those directions a width (size of the support) which tends to 0 when ϵ tends to zero. In other words this part tends to a Dirac measure and this is how one recovers a function supported by the attractor in the limit. We refer to [C1] for the detailed results.

The notion and properties (in particular stochastic stability) of SRB measures has been extended to other type of dynamical systems, for example to expanding unimodal maps [Be.Y1], Hénon map [Be.Y2], and piecewise expanding maps of the interval [Ba.Y.] and [B.K.]. We will come back to this last case in the next section.

From what we have seen above, it seems that strongly chaotic systems are rather stable against mild stochastic perturbations. It is therefore natural to ask if this is also the case for dynamical situations which are only mildly chaotic (for example zero topological entropy) or even for simple bifurcations. We will only mention some results in this direction. For the influence of noise on bifurcation we refer to [O.R.] and references therein. It has long been known that the accumulation of period doubling is unstable against stochastic perturbations, we refer to [C.L.] for a detailed study of this question and references to the original work.

2. HOLES AND LEAKING IN PIECEWISE EXPANDING MAPS

Piecewise expanding maps of the interval form another interesting class of chaotic dynamical systems for which a large number of results have been proven (see [L.M]). In some sense this class extends the class of axiom A systems described in the previous section at a reasonable cost in technical details of the proofs. This class of maps is of course less natural than unimodal maps of the interval or Hénon maps of the plane but their properties are far simpler to establish. This is therefore an interesting laboratory to develop ideas and test conjectures.

A map f of the interval $[0,1]$ is said to be piecewise expanding if there is a finite partition of the interval with boundary points $0 = a_0 < a_1 < \cdots < a_l = 1$ and a number $\rho > 1$ such that the restriction of f to any of the intervals $]a_i, a_{i+1}[$ $(0 \leq i < l)$ is C^2 and can be extended to a C^2 function in the closed interval (f, f', and f'' have limits at the boundary), and moreover there is an integer m such that

$$\min_{0 \leq i < l} \inf_{x \in]a_i, a_{i+1}[} |f^{m\prime}(x)| \geq \rho \, .$$

We will only deal below with the case $m = 1$, the case $m > 1$ being analogous. In the particular case where the finite set $\{a_0, \cdots, a_l\}$ is invariant, the map f is called a piecewise expanding Markov map of the interval, and the associated dynamical systems are essentially the same as the Axiom A systems at the topological level (except of course for invertibility). It is easy to show that any finite state Markov chain can be realized as a piecewise expanding Markov map of the interval. The simplest example being the map $f(x) = 2x$ mod 1 which corresponds to the flipping of a fair coin.

The most important result on piecewise expanding maps of the interval is a Theorem of Lasota and Yorke which asserts that there is at least one absolutely continuous invariant measure (see [L.Y.], [H.K.], [C.T]). Moreover if the dynamical system has a dense orbit, this measure is unique and mixing. We therefore have an analog of SRB measures, if we consider the attractor as the whole phase space.

Stochastic perturbations of piecewise expanding maps of the interval where studied in particular in [Ba.Y.] and [B.K.]. Their main result is the following.

Theorem II.1. If the map f is mixing and none of the points $a_1, \cdots, a_{l-1}$ is periodic, the absolutely invariant measure is stable against stochastic perturbations.

We refer to the original paper for the assumptions on the noise and details of the proof, and to [B.K.S.] for results on the correlations.

A counter example to stochastic stability had been previously constructed in [Ke.] in a situation where the above aperiodicity hypothesis is not satisfied. In fact the difficulty is already present at the deterministic level. Take a piecewise expanding map of the interval and perturb it (to another piecewise expanding map of the interval). Is it true that when the perturbation goes to zero the absolutely continuous invariant measure of the perturbed map converges to the absolutely continuous invariant measure of the unperturbed one ? This is not necessarily so. We first mention that expanding maps of the interval have in general many invariant ergodic probability measures singular with respect to the Lebesgue measure. A simple example of (deterministic) instability can be constructed as follows. The initial map is the map $f_0(x) = 2x$ mod 1, and its SRB measure is the normalized Lebesgue measure of the interval. Note that the Dirac measure at the point 1 is also an invariant ergodic probability measure. For $1/2 > \epsilon > 0$ one defines a map f_ϵ as follows.

$$f_\epsilon(x) = \begin{cases} f_0(x) & \text{if } 0 \leq x < 1 - \epsilon, \\ 1 - 2|x - 1 + \epsilon/2| & \text{if } 1 - \epsilon \leq x \leq 1. \end{cases}$$

All these maps are piecewise expanding with slope of modulus 2 and if we let ϵ tend to zero in the above formula we recover f_0. However for $\epsilon > 0$ small the dynamics of f_ϵ is very different from that of f_0. Indeed the interval $[1 - \epsilon, 1]$ is invariant for f_ϵ. Moreover, it can be proven that for Lebesgue almost every initial condition in the interval $[0, 1 - \epsilon[$ the corresponding orbit will penetrate into the interval $[1 - \epsilon, 1]$ after a finite time (and then stay there for ever). A simple example where this fact is obvious is obtained by taking $\epsilon = 2^{-k}$ where k is an integer larger than 1, in this case the map is Markov. Coming back to the general ϵ, we conclude that the SRB measure of f_ϵ is the (normalized) Lebesgue measure on the interval $[1 - \epsilon, 1]$ (it is easy to verify directly that this is an invariant measure). When ϵ tends to zero, this measure converges weakly to the Dirac measure of the point 1 which is clearly not the SRB measure of f_0.

This example shows in particular that even apparently small perturbations of a map can have strong effects on the dynamics. It turns out that the above instability phenomenon is related to another important question where one modifies the phase space instead of modifying the map by considering a smaller phase space. For example if H is a subset of the phase space Ω, one can consider the set of initial conditions whose orbit will never penetrate into H. If not empty, this is an invariant subset for the dynamics. One can also ask for a given initial condition not in H how long it takes for the orbit to reach H. This is an old question dating back to the early foundations of Statistical Mechanics and connected with the ergodic Theorem and Poincaré's recurrence time Theorem (*i.e.* how long would it take for all the molecules of oxygen in a room to cluster into the left half ?). If we denote by $\tau(x)$ the first time larger than 0 where the orbit of x enters the set H, one would like to obtain information about the statistical properties of τ. One suspects from Birkhoff's ergodic Theorem that the inverse of the measure of H should be an order of magnitude for τ. M.Kac proved [Ka.] that this is indeed the case if the initial condition itself belongs to the set H. Namely if the dynamical system defined by the map T has an ergodic invariant probability measure μ, the expectation of τ with respect to μ conditioned to start in H is exactly $1/\mu(H)$. In the case of expanding maps of the interval, one can investigate in more detail the behavior of τ. In this case there is a very useful technical tool to study the statistical behavior of the system which is called the transfer operator. For a piecewise expanding map f of the interval, the transfer operator which acts on functions on the interval is given by

$$Pg(x) = \sum_{f(y)=x} \frac{g(y)}{|f'(y)|} \, .$$

Among the simple but important properties of the transfer operator we have the identity

$$\int_0^1 u(x)Pg(x)dx = \int_0^1 g(x)\, u(f(x)) \, dx \, ,$$

where u and g are for example bounded measurable functions. In other words, the transfer operator is the adjoint with respect to the Lebesgue measure of the Koopman operator of composition with f. If we have a random variable X (belonging to $[0,1]$) with probability density g, the random variable $f(X)$ has probability density Pg. This implies in particular that the integral of Pg with respect to the Lebesgue measure is equal to the integral of g, and that the densities of the absolutely invariant measures are the (non negative) eigenvectors of eigenvalue 1 of P. It is this last property which is the starting point of the proof of the Theorem of Lasota and Yorke mentioned above, namely there is always a non zero non negative eigenvector for the eigenvalue one. Moreover many other statistical properties of the dynamical system can be derived from a spectral analysis of P: mixing, central limit theorem, decay of correlations, large deviations, etc. (see [H.K.] or [C2]). Of course when one speaks of spectral analysis one should say precisely in which Banach space one is working. A main discovery of Lasota and Yorke was that the space of functions of bounded variation is particularly well adapted to the spectral analysis. In this space, if the system has a dense orbit and is mixing, the spectrum consists of the simple eigenvalue 1, and the rest is contained in a disk centered at the origin of radius $\eta < 1$.

We now come back to the entrance time τ assuming that we have a unique ergodic absolutely continuous invariant measure $h(x)dx$ as above and a hole H. For any integer $n > 0$ we obviously have

$$\mathbb{P}(\tau > n) = \int_0^1 \prod_{j=1}^n \emptyset_{H^c}(f^j(x))h(x)\ dx\ ,$$

where $\emptyset_{H^c}$ is the characteristic function of the complement of the hole H. Using the basic properties of the operator P recursively, we obtain at once

$$\mathbb{P}(\tau > n) = \int_0^1 \emptyset_{H^c}(x)\big[P\emptyset_{H^c}\big]^{n-1}(h)(x)\ dx\ .$$

We see immediately that we can get information at least for large n if we have some control on the spectral theory of the operator $P\emptyset_{H^c}$. In particular, if H is a small interval one may expect that $P\emptyset_{H^c}$ is a small perturbation of P. Technically, in the space of functions of bounded variation this is not the case (recall that the total variation of the characteristic function of an interval is 2). Nevertheless, using adequate changes of norms and resummations the following result was proven in [Co.G.] under some weak hypotheses (see also [P] and [H]).

Theorem II.2. The distribution of the random variable $\tau/\mu(H)$ converges in law to a Poisson distribution when the length of the interval H converges to zero.

It has also been proved in [Co.G.] that the process of successive visits to H when viewed at the natural (Birkhoff's) time scale $1/\mu(H)$ converges in distribution to a Poisson point process. The main idea of the proof is to control the non small perturbation, proving that the expected result from a naive application of perturbation theory holds, namely the operator $P\emptyset_{H^c}$ has a dominant eigenvalue $1 - \mu(H)$ (plus some higher order corrections).

Similar results were obtained in [C.C.] for the close encounters of two trajectories, namely for $\epsilon > 0$ and x and y two initial conditions, one defines τ_ϵ as the smallest positive time such that the orbits of x and y come at a distance less than ϵ. The natural time scale is now $1/\epsilon$ and when ϵ tends to zero the distribution of τ_ϵ converges to a Poisson distribution. The process of successive ϵ close encounters leads in the limit to a Poisson point process. On the other hand, for the case of a diffeomorphism of the circle there is no convergence (see [C.F.]).

With these results we can now come back to the question of instability of the measure. In our example of the one parameter family f_ϵ of perturbations of $2x \bmod 1$ we have already mentioned the existence of an invariant segment $[1 - \epsilon, 1]$ which absorbs Lebesgue almost every trajectory. This implies that the transfer operator P_ϵ of f_ϵ, can be block triangularized into two pieces. The two operators which appear on the diagonal are respectively $\varnothing_{[1-\epsilon,1]} P_\epsilon \varnothing_{[1-\epsilon,1]}$ and $\varnothing_{[0,1-\epsilon]} P_\epsilon \varnothing_{[0,1-\epsilon]}$. The first one is simply, modulo a change of scale, the transfer operator for a fixed map and we have a simple eigenvalue one corresponding to the absolutely continuous invariant measure (the Lebesgue measure on $[1 - \epsilon, 1]$). The second operator is exactly of the form studied in [Co.G.]. For ϵ small it has an eigenvalue $1 - \mathcal{O}(1)\epsilon$ and a positive eigenvector h_ϵ. There is also a corresponding eigenvector ν_ϵ of the adjoint which is a positive measure supported by the Cantor set K_ϵ of initial conditions whose trajectory will never enter $[1 - \epsilon, 1]$ (K_ϵ has Lebesgue measure zero, and in fact it can be shown that it has a Hausdorff dimension smaller than 1). The measure $d\rho_\epsilon = h_\epsilon \, d\nu_\epsilon$ turns out to be an invariant measure. Now if we let ϵ tend to zero we have the following consequences of [Co.G.]. The Cantor set K_ϵ becomes fatter and fatter (its Hausdorff dimension converges to one) and in some sense it converges to the whole interval. The measure ν_ϵ converges weakly to the Lebesgue measure, and h_ϵ converges to 1. That is to say the measure $d\rho_\epsilon$ converges weakly to the SRB measure, and we have recovered stability of the measure if we consider for $\epsilon > 0$, not the SRB measure, but a measure which is more evenly distributed (although singular continuous). We refer to [C.M.S.2] for more details.

Coming back to the proof of stochastic stability in [Ba.Y.] and [B.K.] one can see that the condition of aperiodicity of the points $a_1, \cdots, a_{l-1}$ is used for the exclusion of small invariant segments in the perturbed map. Technically small invariant segments lead to bad estimates in a fundamental bound of Lasota and Yorke (estimates which diverge when the width of the segment goes to zero). The technique of [Co.G.] allows one to circumvent this difficulty and we conjecture that the above phenomenon of "exchange" of SRB measure occurs in all cases where perturbations create small invariant segments.

There is another interesting class of problems associated with stochastic perturbations which has to do with the escape from phase space. On a compact set, when introducing a stochastic perturbation one is very careful to ensure that the noise somehow disappears at the boundary so that the stochastic process stays for ever in the domain. In the opposite situation, namely when the trajectory can leave the phase space one has of course a loss of the total probability. If one imagines simulating the system by looking at the motion of (many) independent particles, this means that a particle dies when it jumps outside the phase space. Two main questions are the rate of decay of the number of particles (rate of decay of the total mass), and the distribution of the particles which have survived up to a given time. There are similar questions and results in a deterministic setting for the escape of the neighborhood of a repeller (see [C.M.S.1] and references therein), but the present problem is more analogous to the above problem of hole (with the hole at the exterior). For piecewise expanding maps of the interval and simple independent noise this problem was studied in details in [C.M.S.2]. In the case of small stochastic perturbations, one can write a transfer operator which describes the time evolution of the probability density and for small noise one can get information by perturbation theory from the zero noise case (this is also the idea in [Ba.Y.] and [B.K.]). One of course meets all the problems mentioned above of non smallness of the perturbation in the natural norm, but all the techniques developed previously can be applied to this case and one gets the following result.

Theorem II.3. Under some mild hypothesis (see [C.M.S.2]) the probability that a trajectory has not left the interval up to time n decays (asymptotically) exponentially fast with a decay rate $a\epsilon + o(\epsilon)$ where $a > 0$ depends only on the map and the noise distribution.

In other words if one starts a simulation with a large number of independent particles, one has to wait a time of order ϵ^{-1} to see a substantial depletion of the system. This result can be compared with a purely diffusive situation (which corresponds to the identity map) and where it can be shown that the typical time scale is at least of order ϵ^{-2}. In other words, this is a situation where chaos enhances diffusion.

These two results above can be explained in a simple way. For the case of diffusion, particles starting away form the boundary of the interval which perform a random walk with step size of order ϵ can only travel, after time n, a distance of order $n\epsilon^2$ (from the central limit theorem, at least with large probability). Hence we get a time of order ϵ^{-2} to travel a distance of order one. In the case with chaos, we know from [Co.G.] that the typical time to reach an ϵ neighborhood of the boundary is of order ϵ^{-1} (Birkhoff's time scale). On the other hand, when the trajectory is at a distance of order ϵ of the boundary it has a substantial probability to jump out and we get finally the time scale of order ϵ^{-1}. We see that it is really the mixing of phase space provided by the chaotic system which is responsible for the acceleration of the death rate. We refer to [C.M.S.2] for technical details.

3. OPEN PROBLEMS

We have already mentioned the intimate relations between the study of chaotic dynamical systems and the theory of stochastic processes. This is a general idea that will probably give other fruitful achievements. For example there does not seem to be up to now an equivalent of martingale theory on the dynamical systems side. The statistical methods used in the study of stochastic processes should also have very useful analogues for dynamical systems.

It is of course an important question to generalize the above results to broader classes of dynamical systems or stochastic perturbations, and it is to be expected that this will lead to an intense activity in the near future. In particular up to now the results have been obtained under assumptions of rapidly decaying correlations of the noise (or simply independence). Some interesting phenomena may occur with slowly decaying correlations of the stochastic perturbations and for quasi periodic perturbations.

Applications up to now are rather scarce but should develop in the near future. An example is the dynamo problem in astrophysics. The problem (in its simple form) is to understand how transport by a turbulent flow can amplify a magnetic field and to compute the exponential growth rate in the limit of zero (small) magnetic resistivity. This can be formulated as a problem of small stochastic perturbations of a dynamical system through the relation between white noise and laplacian. It is believed that a noise, however small, can conspire with the large deviations of the system to produce a non trivial result for the growth rate. We refer to [Ch.G.] for a review. Other interesting questions on transport are discussed in [B.W.Z.].

REFERENCES

- [B.] R.Bowen. *Equilibrium States and the Ergodic Theory of Anosov Diffeomorphisms*. Lecture Notes in Mathematics 470. Springer, Berlin Heidelberg New York 1975.

- [Ba.Y.] V.Balady, L.-S.Young. On the spectra of randomly perturbed expanding maps. Commun. Math. Phys. **156**, 355-385 (1993).

- [Be.Y1] M.Benedicks, L.-S.Young. Absolutely continuous invariant measures and random perturbations for some one dimensional maps. Ergod. Theor. & Dyn. Syst. **12**, 13-37 (1992).

- [Be.Y1] M.Benedicks, L.-S.Young. SRB measures for certain Hénon maps. Invent. Math. **112**, 541-576 (1993).

- [B.K.] M.Blank, G.Keller. Stochastic stability versus localization in chaotic dynamical systems. Preprint 1996.

- [B.K.S.] V.Baladi, A.Kondah, B.Schmitt. Random correlations for small perturbations of expanding maps. Preprint 1995.

- [B.W.Z.] M.N.Bussac, R.B.White, L.Zuppiroli. Particle and heat transport in a partially stochastic magnetic filed. Physics Letters A, **190**, 101-105 (1994).

- [Ch.G.] S.Childress, A.Gilbert. *Stretch, twist, fold: the fast dynamo*. Springer, Berlin, Heidelberg, New York 1995.

- [C.C.] Z.Coelho, P.Collet. Asymptotic Limit Law for the Close Approach of Two Trajectories in Expanding Maps of the Circle. Prob. Theor. and Related Fields **99**, 237-250 (1994)

- [C1] P.Collet Stochastic perturbations of the invariant measure of some hyperbolic dynamical systems. In *Nonlinear Evolution and Chaotic Phenomena*, G.Gallavotti and P.Zweifel editors, Plenum, New York London 1988.

- [C2] P.Collet. Some Ergodic Properties of Maps of the Interval. In "Dynamical Systems & Frustrated Systems", R.Bamon, J.-M.Gambaudo and S.Martinez editors, to appear.

- [C.F.] Z.Coelho, E.de Faria. Limit laws of entrance times for homeomorphisms of the circle. To appear in the Israel Journal of Maths.

- [C.G.] P.Collet, A.Galves. Asymptotic distribution of entrance times for expanding maps of the interval. *Dynamical Systems and Applications.* R.P. Agarwal editor, World Scientific 1995.

- [C.L.] P.Collet, A.Lesne. Renormalization group analysis of dynamical systems with noise. Journ. Stat. Phys. **57**, 967 (1989).

- [C.M.S1] P.Collet, S.Martinez, B.Schmitt. The Yorke-Pianigiani measure and the asymptotic law on the limit Cantor set of expanding systems. Nonlinearity **7**, 1437-1443 (1994).

- [C.M.S2] P.Collet, S.Martinez, B.Schmitt. In preparation.

- [E.R.] J.-P. Eckmann, D.Ruelle. Ergodic theory of chaos and strange attractors. Rev. Mod. Phys. **57**, 617-656 (1985).

- [F.W.] M.Freidlin, A.Wentzell. *Random perturbations of dynamical systems.* Springer, Berlin Heidelberg New York 1984.

- [H.] M.Hirata. Poisson law for axiom A diffeomorphisms. Ergod. Theor. & Dyn. Syst. **13**, 533-556 (1993).

- [H.K.] F.Hofbauer, G.Keller. Ergodic properties of invariant measures for piecewise monotonic transformations. Math. Z. **180**, 119-140 (1982).

- [H.P.S.] M.Hirsch, C.Pugh, M.Shub. Invariant Manifolds. Lecture Notes in Mathematics 583. Springer, Berlin Heidelberg New York 1977.

- [Ka.] M.Kac. Bull. Amer. Math. Soc.**53**, 1002 (1947).

- [Ke.] G.Keller. Stochastic stability in some chaotic dynamical systems. Mh. Math. **94**, 313-333 (1982).

- [Ki.] Y.Kifer. *Random perturbations of dynamical systems.* Birkhauser, Boston 1988.

- [L.Q.] P.D.Liu, M.Qian. Smooth ergodic theory of random dynamical systems. Lecture Notes in Mathematics **1606**, Springer Verlag 1995.

- [L.Y.] A.Lasota, J.Yorke. On the existence of invariant measures for piecewise monotone transformations. Trans. Amer. Math. Soc. **186**, 481-488 (1973).

- [L.M.] A.Lasota, M.Mackey. *Probabilistic Properties of deterministic Systems.* Cambridge University Press, Cambridge 1985.

- [O.R.] J.Olarrea, F.J.Rubia. Stochastic Hopf Bifurcation. Phys. Rev. **E 53**, 268-271 (1996).

- [P.] B.Pitskel. Ergod. Th. & Dynam. Sys. **11**, 501 (1991).

- [R.] D.Ruelle. *Thermodynamic Formalism.* Addison-Wesley, Reading 1978.

- [R.T.] D.Ruelle, F.Takens. On the nature of turbulence. Commun. Math. Phys. **20**, 167-192 (1971) and **21**, 64 (1971).

- [S.] S.Smale. Differentiable dynamical systems. Bull. Amer. Math. Soc. **73**, 747-817 (1967).

- [Y.] L.-S.Young. Stochastic stability of hyperbolic attractors. Ergod. Theor. & Dyn. Syst. **6**, 311-319 (1986).

DECOUPLING INEQUALITIES: A SECOND GENERATION OF MARTINGALE INEQUALITIES

VICTOR H. DE LA PEÑA,* *Columbia University*

Abstract

The theory of martingale inequalities has been central in the development of modern probability theory. Recently this theory has been expanded widely through the introduction of decoupling inequalities, which provide natural extensions in cases where the variables take values in general spaces or when a martingale structure is not available. Typically, decoupling inequalities are used to transform problems involving sums of dependent random variables into problems involving sums of (conditionally) independent random variables. This transformation particularly permits the use of traditional results when dealing with sums of dependent variables. In this paper an account of the theory of decoupling inequalities is given with emphasis on its relations to the theory of martingale inequalities, and its applications and extensions to a wide range of problems, including best constants on martingale inequalities, stopping time problems, U-statistics, random graphs, quadratic forms and stochastic integration.

1. Introduction

In this paper we provide a brief account of the development of the theory of decoupling inequalities. Given a sequence of random variables adapted to an increasing sequence of sigma algebras, a decoupling inequality aims at reducing the level of dependence present in a given function of the sequence by means of inequalities relating the expectation of the function of the sequence of interest to the expectation of the same function of another sequence

The earliest reference to a general theoretical result in the area is an inequality of McConnell and Burkholder found in Burkholder (1983). Their decoupling inequality dealt with comparing the L_p-norm of a martingale transform of Rademacher variables to the L_p-norm of a sum of conditionally independent variables. The motivation for this inequality was a study of integral operators on Lebesgue–Bochner spaces. Its relation to square function martingale inequalities can also be traced back to that paper. As explained by Burkholder (personal communication) his motivation for this 'first' decoupling inequality was a desire to extend the concept of square function inequalities to more general spaces.

In the general framework, decoupling inequalities have been mainly derived for comparing the L_p-norms of sums of pairs of sequences of random variables adapted to a common filtration and having similar conditional marginals. Stronger results are obtained when the second sequence involved is restricted to have a conditional

---

* Postal address: Department of Statistics, Columbia University, New York, NY 10025, USA.

independence property. In this case one can obtain exponential inequalities (cf. de la Peña (1994)) and improve on the constants involved in these inequalities (cf. Hitczenko (1994b)). A key result of the theory, found in Kwapien and Woyczynski (1989; 1992), roughly states that for any sequence of dependent random variables $\{d_i\}$, there is a second sequence $\{e_i\}$ such that e_i has the same conditional distribution as d_i for all i, but the second sequence is in addition conditionally independent, therefore making the above cited inequalities widely applicable.

The Principle of Conditioning introduced in Jakubowski (1986) has been derived parallel to the theory of decoupling. This principle, by means of a decoupling-type argument, reduces the proof of the central limit theorem for martingales due to Brown and Eagleson (cf. Kwapien and Woyczynski (1992)) to the proof of the central limit theorem for sums of independent random variables.

A typical application of the theory of decoupling consists in a re-interpretation of Wald's (1945) equation for randomly stopped sums of independent random variables. As will be seen later, Wald's first equation can be viewed as implying that, on the average, the stopping time and the sum involved can be considered as being independent of one another, hence 'decoupled'. This re-interpretation has provided a natural framework for obtaining useful extensions of Wald's equation to the case of high moments and Banach-valued variables. In particular it has been used in a series of papers including Klass (1988), de la Peña and Govindarajulu (1992), and Hitczenko (1994a).

A second important application of decoupling involves showing that U-statistics and, more generally, quadratic forms of independent random variables can be viewed as sums of conditionally independent random variables. Examples of such applications may be found in the seminal paper of McConnell and Taqqu (1986), followed by those of Kwapien (1987), Bourgain and Tzafriri (1987), de la Peña (1992a), Kwapien and Woyczynski (1992), Gine and Zinn (1994), and finally de la Peña and Montgomery-Smith (1994,1995), where important results are derived comparing the tail probabilities of the processes.

As statistical applications of decoupling we can cite results in the theory of density estimation through the development of symmetrization inequalities. Arcones and Giné (1993) used and developed symmetrization and decoupling inequalities found in de la Peña (1992a) in obtaining limiting results for a functional version of a U-statistic known as a U-process. Their study of this type of process was motivated by problems of the theory of density estimation introduced in Nolan and Pollard (1987).

Finally, applications of decoupling to problems in mathematics may be found in Bourgain and Tzafriri (1987) and Janson and Nowicki (1991). The first paper derives conditions for the invertibility properties of sub-matrices by applying a decoupling inequality for quadratic forms of independent random variables. The second paper introduces a generalized version of a U-statistic to deal with problems in random graphs, making the known decoupling results for U-statistics readily applicable in this context.

The paper is organized as follows. In Section 2 Wald's equation for randomly stopped sums is introduced and extensions are derived using a new formulation. Applications are given concerning the first exit time of a random walk on the sphere. Section 3 focuses on the study of the properties of a generalized version of U-statistics which includes as special cases U-statistics and quadratic forms of independent random variables. More specifically, we introduce a tail probability inequality that compares the tail probability of generalized U-statistics to their decoupled counterpart, and discuss applications of this result to the theory of random graphs, stochastic integration, density estimation and the invertibility of matrices. Section 4 provides a brief introduction to general concepts and inequalities that relate to the theory of decoupling and in Section 5 we make a comparison between decoupling inequalities and martingale inequalities showing some important cases when decoupling inequalities are to be preferred to martingale inequalities. Section 6 relates the general theory presented in Section 4 to the results of Section 2 by means of solving the longstanding open problem of extending Wald's equation to the case of U-statistics. Finally, we give concluding remarks in Section 7.

2. Wald's Equation and a Re-Formulation

In this section we introduce Wald's equation for randomly stopped sums of independent random variables and use it to motivate the theory of decoupling inequalities. Wald's (1945) equation, established in its full generality by Blackwell (1946), is stated next along with its second moment analogue.

Theorem 2.1 (Wald 1945). *Let $\{X_i\}$ be i.i.d. random variables with finite mean, $S_n = X_1 + X_2 + ... + X_n$, and T a stopping time adapted to $\{X_i\}$. Then,*

$$(2.1) \qquad\qquad ES_T = ET \cdot EX_1, \qquad \textit{(First equation)}$$

whenever $ET < \infty$.
If in addition $EX_1 = 0$, $EX_1^2 < \infty$, then

$$(2.2) \qquad\qquad ES_T^2 = ET \cdot EX_1^2, \qquad \textit{(Second equation)}$$

whenever $ET < \infty$.

Wald's first equation and, more generally, the foundations of the theory of sequential analysis were developed by Wald during World War II to deal with problems related to destructive sampling. As pointed out by Rubin (1995), a problem of great importance during the war was that of optimizing resources. The army wanted to provide a rule to its soldiers that would help them in cases of great danger, when their rifles might not be working properly. The rule was as follows. If a rifle misfiers, try again. If it misfiers twice in a row, this is strong indication that the rifle is out of order so you might want to change the rifle. In order for this policy to be enforced successfully, the army needed to make sure that the quality of its bullets was superb. Under the Neyman Pearson model, in order to insure the quality of bullet batches, high numbers of bullets had to be used so that the statistical tests could attain both a small type one error and a high power as desired. It seemed wasteful to require the destruction of a large number of bullets in cases where there was overwhelming initial evidence supporting the hypothesis that the quality of the bullets was high. It was therefore suggested that a statistical methodology allowing for the decision of accepting or rejecting a bullet batch should be based on a sequential procedure which would permit the taking of a decision based on information gathered in a sequential fashion. The decision to reject or accept the hypothesis would be made repeatedly and would be based on the information available at every stage of the sampling plan. Wald was confronted with the general problem in 1944 and devised the theoretical foundations of the theory of sequential analysis over a period of three months. The cornerstone of this theory has come to be known as Wald's first equation.

Next, we will present a re-formulation of the statements of Wald's equations that implies that the stopping time can be considered as being independent from the original sequence and hence decoupled.

Theorem 2.2. *Let $\{X_i\}$ and $\{\tilde{X}_i\}$ be two independent sequences of i.i.d. r.v.'s with $\mathcal{L}(X_i) = \mathcal{L}(\tilde{X}_i)$ for all $i \geq 1$. Set $\tilde{S}_n = \tilde{X}_1 + ... + \tilde{X}_n$. Then,*

$$(2.3) \qquad\qquad ES_T = E\tilde{S}_T,$$

whenever $|EX_1| < \infty$ and $ET < \infty$, and

$$(2.4) \qquad\qquad ES_T^2 = E\tilde{S}_T^2,$$

whenever $EX_1 = 0, EX_1^2 < \infty$ and $ET < \infty$.

Equation (2.3) is obtained through the chain of equalities. $ES_T = ET \cdot EX_1 = ET \cdot E\tilde{X}_1 = E\tilde{S}_T$, where the first corresponds to Wald's equation and the second one uses the fact that the X's and $\tilde{X}'$s have the same distribution and the third uses Fubini's theorem only, due to the independence of $\{\tilde{X}_i\}$ and T. Equation (2.4) can be obtained similarly. It is to be noted that (2.3) and (2.4) are examples of 'decoupling' equalities. The term indicates the fact that the dependence between the stopping time and the X's has been decoupled (disposed of) through the introduction of the $\tilde{X}'$s. The power of this heuristic has been exploited in a series of papers beginning with Klass (1988). Recently, de la Peña and Govindarajulu (1992) treated the case of (2.4) when the mean of the variables is not zero. In that case one gets the result $ES_T^2 \leq 2E\tilde{S}_T^2$ and the result is sharp. Later on Hitczenko (1994a) extended this result to p^{th} powers of S_T. Klass' (1988) paper deals with the general case of Banach valued random variables. The main result of his paper can be easily extended by making use of Montgomery-Smith (1993) (for details see de la Peña and Giné (1996)) as follows:

Theorem 2.3. *Let $\{X_i\}$ be i.i.d. random variables with values in a Banach Space, $(B, ||\cdot||)$ with T a stopping time adapted to this sequence. Let $\{\tilde{X}_i\}$ be an independent copy of $\{X_i\}$, and S_n and $\tilde{S}_n$ the respective sums. Then, for all $p > 0$, there are universal constants $c_1, c_2, \ 0 < c_1, c_2 < \infty$, not depending on the distribution of the random variables involved, the norm $||\cdot||$, or p such that*

$$(2.5) \qquad\qquad E||S_T||^p \leq c_1 c_2^p E||\tilde{S}_T||^p.$$

The key element in the proof of the above result is the independence found between X_i and $1(T \geq i)$ and the observation that $S_T = \sum_{i=1}^{T} X_i = \sum_{i=1}^{\infty} X_i 1(T \geq i)$. The independence between $\{\tilde{X}_i\}$ and T permits the re-writing of the right hand side of (2.5) as follows:

$$(2.6) \qquad E||\tilde{S}_T||^p = \sum_{n=1}^{\infty} E||\tilde{S}_n||^p P(T=n) = \sum_{n=1}^{\infty} E||S_n||^p P(T=n).$$

Using this fact, we will show that (2.5) gives information on the first time T_r, a random walk exits a sphere $W = \{x\epsilon B : ||x|| \le r\}$, where $||\cdot||$ stands for the Euclidean norm. Following Griffin and McConnell (1992) (where the case when $B = R^k$ is considered), for $0 < p \le q$, we seek conditions under which,

$$(2.7) \qquad \sup_{r \ge 1} \frac{E||S_{T_r}||^p}{r^q} < \infty.$$

From Theorem 2.3, it is easy to see that

$$\sup_{r \ge 1} \frac{E||S_{T_r}||^p}{r^q} \le c_1 c_2^p \sup_{r \ge 1} \frac{E||\tilde{S}_{T_r}||^p}{r^q}.$$

Therefore, we obtain conditions for the finiteness of the left hand side of the above in terms of the distribution of T. In particular, if $p = q = 1$, and $E||S_n|| \le c'\sqrt{n}$, then a sufficient condition for (2.7) is that $\sup_r \frac{E\sqrt{T_r}}{r} < \infty$.

3. U-Statistics and Quadratic Forms

In this section we introduce decoupling inequalities for a generalized class of U-statistic which includes quadratic forms.

Let $\{X_i\}$ be independent random variables and define D_n by the equation,

$$(3.1) \qquad D_n = \sum_{1 \le i \ne j \le n} f_{ij}(X_i, X_j) = \sum_{j=1}^{n} \sum_{i=1, i \ne j}^{n} f_{ij}(X_i, X_j).$$

Let $\{\tilde{X}_i\}$ be an independent copy of $\{X_i\}$ and set

$$(3.2) \qquad \tilde{D}_n = \sum_{1 \le i \ne j \le n} f_{ij}(X_i, \tilde{X}_j) = \sum_{j=1}^{n} \sum_{i=1, i \ne j}^{n} f_{ij}(X_i, \tilde{X}_j).$$

Fixing $\{X_i\}$, $\tilde{D}_n = \sum_{j=1}^{n} \{\sum_{i=1, i \ne j}^{n} f_{ij}(X_i, \tilde{X}_j)\} = \sum_{j=1}^{n} G_j(\tilde{X}_j)$ shows that $\tilde{D}_n$ is a sum of independent random variables.

It is possible to give a geometric interpretation that connects D_n to $\tilde{D}_n$. If we let $f_{ij}(X_i, X_j) = d(X_i, X_j)$ where $d(X, Y)$ stands for the distance from X to Y, then D_n gives a measure of the within distance for the points $\{X_1, ..., X_n\}$ and $\tilde{D}_n$ is a measure of the distance between the two sets of points $\{X_1, ..., X_n\}$ and $\{\tilde{X}_1, ..., \tilde{X}_n\}$. The following theorem shows that in a very broad sense, both 'random' distances are equivalent.

Theorem 3.1 (de la Peña and Montgomery-Smith (1994)). *Let $\{X_i\}$ be a sequence of independent random variables with values in a measurable space $(S, \mathcal{S})$. Let $\{\tilde{X}_i\}$ be an independent copy of $\{X_i\}$. Let $\{f_{ij}\}$ be a family of functions taking $(S \times S) \to B$ where $(B, \|\cdot\|)$ is a Banach Space. Then, for all $t \geq 0$,*

$$(3.3) \qquad P(\|\sum_{j=1}^{n}\sum_{i=1,j\neq i}^{n} f_{ij}(X_i, X_j)\| \geq t) \leq cP(\|\sum_{j=1}^{n}\sum_{i=1,j\neq i}^{n} f_{ij}(X_i, \tilde{X}_j)\| \geq \frac{t}{c}),$$

where $0 < c < \infty$ is a universal constant.

The reverse inequality to (3.3) holds whenever $f_{ij}(X_i, X_j) = f_{ji}(X_j, X_i)$ for all $1 \leq i, j \leq n$. Moreover, the case of multilinear forms and higher order U-statistics has also been proved and can be found in de la Peña and Montgomery-Smith (1995).

One of the key tools in the proof of (3.3) which allowed the introduction of the $\{\tilde{X}_i\}$ sequence is the following elementary inequality (possibly of independent interest). Let X, Y be i.i.d. Then, for all $t > 0$,

$$(3.4) \qquad\qquad P(\|X\| \geq t) \leq 3P(\|X + Y\| \geq \frac{2}{3}t).$$

It is easy to see that the case $f_{ij}(X_i, X_j) = a_{ij}X_iX_j$ gives quadratic forms while $f_{ij}(X_i, X_j) = f(X_i, X_j)/\binom{n}{2}$ gives U-statistics, Since the sample variance is a U-statistic, Theorem 3.1 gives insight into the behaviour of the sample variance as shown by the following example.

Example 3.1. Let $\{X_i\}$ be a sequence of independent random variables. Let $\{\tilde{X}_i\}$ be an independent copy of $\{X_i\}$. Let Var_n denote its sample variance, and $\bar{X} = (X_1 + ... + X_n)/n$. Then

$$(3.5) \qquad\qquad Var_n = \frac{\sum_{i=1}^{n}(X_i - \bar{X})^2}{(n-1)} = \sum_{j=1}^{n}\sum_{i=1;i\neq j}^{n}\frac{(X_i - X_j)^2}{\binom{n}{2}}.$$

Further, for all $n, t > 0$ there is a numerical finite constant $c > 0$ such that

$$(3.6) \qquad \frac{1}{c}P(\sum_{j=1}^{n} G_j(\tilde{X}_j) \geq ct) \leq P(Var_n \geq t) \leq cP(\sum_{j=1}^{n} G_j(\tilde{X}_j) \geq \frac{t}{c}),$$

where $G_j(Y) = \sum_{i=1;i\neq j}^{n}\frac{(X_i - Y)^2}{\binom{n}{2}}$.

The best constant c in this example is not known. The results found in de la Peña (1992c) suggest $c = 4$ as a possible choice.

Theorem 3.1 builds upon (and generalizes) works of several researchers including decoupling inequalities found in McConnell and Taqqu (1986), Kwapien (1987), Bourgain and Tzafriri (1987), Kwapien and Woyczynski (1992), de la Peña (1992a), de la Peña, Montgomery-Smith and Szulga (1994), Giné and Zinn (1994)

An important application of decoupling inequalities for quadratic forms to the study of the invertibility properties of sub-matrices can be found in Bourgain and

Tzafriri (1987). Another key application of results like Theorem 3.1 is to obtaining symmetrization inequalities by using the fact that the decoupled variables are conditionally independent. In the special case of U-statistics, Nolan and Pollard (1987) obtained symmetrization inequalities that are closely linked to our results. The motivation behind their work was the development of limiting theorems for a functional version of a U-statistic known as a U-process. Their motivation for the study of this type of process were problems on the theory of density estimation. Arcones and Giné (1993) extended the work of Nolan and Pollard (1987) to the case of multivariate U-statistics and more generally U-processes. Their papers deal with limit theorems of the type $\frac{\sum_{1 \le i < j \le n} f(X_i, X_j)}{A_n} \to \gamma_f$ where γ_f is a Gaussian process indexed by $f \epsilon \Pi$, with Π an appropriate family of functions. Their approach included a use of a special case of Theorem 3.1 found in de la Peña (1992a) to symmetrize the U-processes involved.

There are potential applications of our results to the study of random graphs. This connection is more apparent in a study of Janson and Nowicki (1991), which deals with the properties of random graphs by introducing a generalized form of U-statistics. One example taken from their work will help illustrate the connection with decoupling inequalities.

Example 3.2. Let $\{X_i\}, i = 1, .., n$ be i.i.d., $P(X_i = 1) = p_1$, $P(X_i = 0) = 1 - p_1$. Let Y_{ij} be i.i.d., $P(Y_{ij} = 1) = p_2$, $P(Y_{ij} = 0) = 1 - p_2$. Then, $Y_{ij} = 1$ indicates that there is an edge between vertices i and j and $X_i = 1$ when vertex i is red while $X_i = 0$ when vertex i is blue. Finally, let $f(x_1, x_2, y) = yI(x_1 \ne x_2)$. Then,

$$(3.7) \qquad S_{n,2}(f) = \sum_{1 \le i < j \le n} f(X_i, X_j, Y_{ij}),$$

counts the edges with one red and one blue end-point. It is easy to see how to extend Theorem 3.1 to the case of this type of statistic since, using the assumed independence between $\{X_i\}$ and $\{y_{ij}\}$, conditioning on $\{y_{ij}\}$ we have a random variable for which Theorem 3.1 can be applied. An application of this theorem reduces the problem to one involving sums of conditionally independent random variables.

Inequalities like the one presented in Theorem 3.1 are also important in the study of stochastic integrals of the form

$$(3.8) \qquad \int_{t=0}^{\infty} \int_{s=0}^{t} f(s,t) dY_s dY_t,$$

where Y_s is a process with independent increments. The reason for this is that stochastic integrals as the one introduced above arise as limits of quadratic forms of the type $\sum_{1 \le i, j \le n} a_{ij} Y_i Y_j$, where the $\{Y_i\}$ are independent random variables (see Kwapien and Woyczynski (1992)).

4. General Theory of Decoupling

In what follows we present a brief introduction to the general theory of decoupling. We begin with the key defintions and examples which will be followed by a series of inequalities.

Definition 4.1. *Let $\{d_i\}$ be a sequence of random variables adapted to $\{\mathcal{F}_i\}$ (increasing). An $\{\mathcal{F}_i\}$-adapted sequence $\{e_i\}$ is $\{\mathcal{F}_i\}$-tangent to $\{d_i\}$ if for all i,*

$$(4.1) \qquad \mathcal{L}(d_i|\mathcal{F}_{i-1}) = \mathcal{L}(e_i|\mathcal{F}_{i-1}).$$

We will illustrate this definition, let $\{X_i\}$ be a sequence of independent random variables. Let $\{\tilde{X}_i\}$ be an independent copy of $\{X_i\}$. Set $\mathcal{F}_n = \sigma(\{X_i, i = 1, ..., n; \tilde{X}_i, i = 1, ..., n\})$. In Section 2, we studied sums of the form $S_T = \sum_{i=1}^{\infty} X_i 1(T \geq i)$, where T is a stopping time adapted to $\mathcal{F}_n$. If we let

$$d_i = X_i 1(T \geq i), \qquad \text{then,} \qquad S_T = \sum_{i=1}^{\infty} X_i 1(T \geq i) = \sum_i d_i$$

and also

$$e_i = \tilde{X}_i 1(T \geq i), \qquad \text{with,} \qquad \tilde{S}_T = \sum_{i=1}^{\infty} \tilde{X}_i 1(T \geq i) = \sum_i e_i.$$

Likewise in Section 3 we studied the case of U-statistics which can be dealt with by letting

$$f_j = \sum_{i=1}^{j-1} f_{ij}(X_i, X_j),$$

$$(4.2) \qquad \frac{D_n}{2} = \sum_{j=2}^{n}\{\sum_{i=1}^{j-1} f_{ij}(X_i, X_j)\} = \sum_{j=2}^{n} f_j$$

$$g_j = \sum_{i=1}^{j-1} f_{ij}(X_i, \tilde{X}_j),$$

$$(4.3) \qquad \frac{\tilde{D}'_n}{2} = \sum_{j=2}^{n}\{\sum_{i=1}^{j-1} f_{ij}(X_i, \tilde{X}_j)\} = \sum_{j=2}^{n} g_j.$$

It is easy to see that, the above definitions make the sequences $\{d_i\}$ and $\{e_i\}$ $\{\mathcal{F}_n\}$-tangent since

$$(4.4) \qquad \mathcal{L}(d_j|X_1, ..., X_{j-1}; \tilde{X}_1, ..., \tilde{X}_{j-1}) = \mathcal{L}(e_j|X_1, ..., X_{j-1}; \tilde{X}_1, ..., \tilde{X}_{j-1}).$$

The same is true for the sequences $\{f_i\}$ and $\{g_i\}$.

In what follows we will introduce a series of inequalities that can be used to compare tangent sequences. The first general inequality appeared in Zinn (1985). Hitczenko (1988) provided an extension of Zinn's result which we state next.

Theorem 4.1 (Zinn (1985), Hitczenko (1988)). *Let $\{d_i\}$, $\{e_i\}$ be $\{\mathcal{F}_i\}$-tangent. Assume that either $d_i \geq 0$ for all i or $\{d_i\}$ is a sequence of conditionally symmetric random variables $(\mathcal{L}(d_i|\mathcal{F}_{i-1}) = \mathcal{L}(-d_i|\mathcal{F}_{i-1}))$. Then, for all $p > 0$,*

$$(4.5) \qquad c_p E \sup_n |\sum_{i=1}^n e_i|^p \leq E \sup_n |\sum_{i=1}^n d_i|^p \leq C_p E \sup_n |\sum_{i=1}^n e_i|^p.$$

In de la Peña (1993) it is shown that one can use one proof for both (4.5) and square-function inequalities in the case of conditionally symmetric martingale difference sequences.

A different set of conditions to the ones stated in Theorem 4.1 is necessary in order to obtain stronger results. In particular, one can obtain exponential-type inequalities and one can also remove the dependence on p of C_p in (4.5) by assuming that the bounding random variables satisfy the independence property introduced next.

Definition 4.2. *A sequence $\{e_i\}$ of random variables adapted to an increasing sequence of σ-fields $\{\mathcal{F}_i\}$ contained in $\mathcal{F}$ is said to satisfy condition CI if there exists a σ-algebra $\mathcal{G}$ contained in $\mathcal{F}$ such that $\{e_i\}$ is conditionally independent given $\mathcal{G}$ and*

$$(4.6) \qquad \mathcal{L}(d_i|\mathcal{F}_{i-1}) = \mathcal{L}(e_i|\mathcal{F}_{i-1}) = \mathcal{L}(e_i|\mathcal{G})$$

for all i. The sequence $\{e_i\}$ is said to be DECOUPLED.

It is easy to see that the CI condition is satisfied by the examples presented above, taking $\mathcal{G} = \sigma\{X_i\}$, $i = 1, ..., n$.

Using a conditioning argument that relies heavily on the conditional independence properties of the variables, the following exponential decoupling inequality was obtained in de la Peña (1994). An example of Hitczenko found in that paper shows that this result is sharp.

Theorem 4.3 (de la Peña (1994)). *Let $\{d_i\}$, $\{e_i\}$ be $\{\mathcal{F}_i\}$-tangent. Assume that $\{e_i\}$ is decoupled (CI). Then for all finite t,*

$$(4.7) \qquad E \exp(t \sum_{i=1}^n d_i) \leq \sqrt{E \exp(2t \sum_{i=1}^n e_i)}.$$

As a consequence of ideas related to Theorem 4.3, a one-sided L_p inequality is possible.

This inequality has the important feature that the constant involved does not depend on p and hence can be used to develop exponential inequalities as well.

Theorem 4.4 (Hitczenko (1994b)). *Let $\{d_i\}$, $\{e_i\}$ be $\{\mathcal{F}_i\}$-tangent. Assume that $\{e_i\}$ is decoupled (CI). Then, for all $p \geq 1$,*

$$(4.8) \qquad E|\sum_{i=1}^n d_i|^p \leq (C)^p E|\sum_{i=1}^n e_i|^p.$$

The key feature of this result is the form of the constant C that does not depend on p and this permits the transfer of results involving sums of independent random variables to the case of martingales. In particular, Hitczenko (1994b) (see also Hitczenko (1990)) shows how to use Theorem 4.4 to obtain the best (asymptotic) value for the constants in Rosenthal's inequality for martingales using the known results for sums of independent random variables.

A result that underlines the importance of Theorems 4.3 and 4.4 by showing their applicability in dealing with sums of arbitrarily dependent random variables states that for any sequence of random variables there is a tangent sequence $\{e_i\}$ which in addition satisfies the condition CI (Kwapien and Woyczynski (1989), Jakubowski (1986)).

As shown in de la Peña (1994), one approach for constructing a decoupled (CI) sequence is to proceed sequentially. Assuming that at the jth stage in the process producing $\{d_i\}$ it is possible to obtain a conditionally independent copy e_j of d_j given $d_1, ..., d_{j-1}$. The following diagram taken from de la Peña (1994) illustrates the idea. Let e_1 be an independent copy of d_1. For each $j \geq 2$,

$$d_1 \to d_2 \to d_3 \to d_4 \to .... \to d_{j-1} \to d_j$$
$$\searrow$$
$$e_1 \quad e_2 \quad e_3 \quad e_4 \quad .... \: e_{j-1} \quad e_j.$$

At the jth stage, given $\{d_1, ..., d_{j-1}\}$, d_j and e_j are drawn from independent repetitions of the same random mechanism. Then, it is easy to see that taking $\mathcal{G} = \sigma\{d_i\}$, $\mathcal{L}(d_i|\mathcal{F}_{i-1}) = \mathcal{L}(e_i|\mathcal{F}_{i-1}) = \mathcal{L}(e_i|\mathcal{G})$, and it also follows that the constructed variables $\{e_i\}$ are conditionally independent given $\mathcal{G}$.

5. Decoupling vs Square Function Inequalities

As discussed in the introduction, the strong connection between martingale inequalities and decoupling inequalities can be traced back to Burkholder (1983). Moreover, Kwapien and Woyczynski (1992) develops the theory of stochastic integration by using a decoupling approach. As discussed earlier, in de la Peña (1993) it is shown that the same proof can be given to obtain general decoupling inequalities and square function inequalities. The following corollary illustrates the way in which decoupling inequalities can be used to advantage in the case where a martingale is present.

Corollary 5.1. *Let $\{d_i\}$ be a martingale difference sequence adapted to $\{\mathcal{F}_i\}$. Let $\{e_i\}$ be its associated decoupled (CI), $\{\mathcal{F}_i\}$-tangent sequence. Then, for all $p \geq 1$, there exist universal constants $0 < c_p < C_p < \infty$ depending on p only, such that,*

$$(5.1) \qquad c_p E \sup_n |\sum_{i=1}^{n} d_i|^p \leq E \sup_n |\sum_{i=1}^{n} e_i|^p \leq C_p E (\sum_{i=1}^{\infty} d_i^2)^{\frac{p}{2}}.$$

For details of this see de la Peña and Giné (1996).

Example 5.1. Let $\{X_i\}$ be a sequence of independent mean zero random variables. Let T be a stopping time adapted to $\{X_i\}$, and $\{\tilde{X}_i\}$ an independent copy of $\{X_i\}$. Then,

$$(5.2) \qquad c_p E \sup_{n \le T} |\sum_{i=1}^{n} X_i|^p \le E \sup_{n \le T} |\sum_{i=1}^{n} \tilde{X}_i|^p \le C_p E(\sum_{i=1}^{T} X_i^2)^{\frac{p}{2}}.$$

In the special case the X's are symmetric α-stable with $1 < p < \alpha \le 2$, decoupling and Lévy's inequality give the bound

$$(5.3) \qquad E \sup_{n \le T} |\sum_{i=1}^{n} \tilde{X}_i|^p \le c^p E |\sum_{i=1}^{T} \tilde{X}_i|^p = c^p ET^{\frac{p}{\alpha}} E|X_1|^p$$

while the square function inequality is less informative.

Finally, we remark that a more general result than Corollary 5.1 is possible by applying the results known as the Burkholder-Davis-Gundy inequalities. (See e.g. Chow and Teicher (1988) for this result.)

6. Wald's Equation for U-statistics

In this section we return to our original problem of the study Wald's equation and its extension to the case of U-statistics. This problem, which remained open for several years, was solved in de la Peña and Lai (1997) and de la Peña and Lai (1995). In obtaining the proofs, general decoupling results for tangent sequences and U-statistics were used and developed. We refer the reader to the above papers for the proofs. Special cases of the results were obtained in de la Peña (1992b), and Chow, de la Peña and Teicher Chow, de la Peña and Teicher (1993).

Theorem 6.1 (de la Peña and Lai (1997)). *Let $\{X_i\}$ be a sequence of i.i.d. random variables with values in a measurable space $(S, \mathcal{S})$. Let $f : (S \times S) \to R$ be any function satisfying the following conditions:*
i) $E(f(X_i, X_j)|X_i) = E(f(X_i, X_j)|X_j) = 0$
ii) $E|f(X_i, X_j)|^p < \infty$ for $1 < p \le 2$. Then,

$$(6.1) \qquad E \sum_{1 \le i < j \le T} f(X_i, X_j) = 0,$$

whenever $ET^{\frac{1}{p-1}} < \infty$.

The optimality of the above result can be checked since by letting $p = 2$, $f(x, y) = xy$ we get Wald's second equation from Theorem 6.1 and Wald's first equation. That is, applying Theorem 6.1 we have that whenever $ET < \infty$ and $EX_1^2 < \infty$, $\frac{1}{2}E(S_T^2 - \sum_{i=1}^{T} X_i^2) = E \sum_{1 \le i < j \le T} X_i X_j = 0$, which along with Wald's first equation gives, $ES_T^2 = E \sum_{i=1}^{T} X_i^2 = EX_1^2 ET$.

In obtaining the proof of Theorem 6.1, one uses general decoupling inequalities, square function inequalities along with a symmetrization argument designed to 'neutralize' the effect the stopping time has on the variables. The main difficulty in deriving te result stems form the lack of independence present in the sum

$$\sum_{1 \le i < j \le T} f(X_i, X_j) =$$

$$\sum_{j=2}^{T} \sum_{i=1}^{j-1} f(X_i, X_j) = \sum_{j=2}^{\infty} \{ \sum_{i=1}^{j-1} f(X_i, X_j) \} 1(T \ge j),$$

since $\sum_{i=1}^{j-1} f(X_i, X_j)$ is not independent from $1(T \ge j)$ and this prevents the direct use of the techniques introduced by Klass (1988) in dealing with the proof of Theorem 2.3 (see the remarks immediately following Theorem 2.3 for more details on this).

7. Concluding Remarks

In this paper we have provided a survey of the theory and applications of decoupling inequalities. A key feature of decoupling inequalities as shown in this account is that they always hold for dealing with sums of arbitrarily dependent variables.

They have been a very powerful device in solving several important problems in sequential analysis, density estimation through its connection to the theory of U-statistics, stochastic integration, the invertibility properties of matrices and several other areas. Moreover, when specializing to the study of martingales, they are very frequently comparable to square function inequalities and are often more informative than typical martingale inequalities.

Currently, work is underway to extend decoupling inequalities to randomly stopped continuous time processes like Bessel processes, and to developed central limit theorems for a sequence of two by two tables (cf. Kou and Ying (1995)). It is expected that in the near future, new decoupling inequalities would be derived for such specialized processes as branching processes. It is also envisioned that decoupling would find several applications in statistics as is the case now with martingale inequalities.

References

[1] Arcones, M. and Giné, E. (1993). Limit theorems for U-processes. Ann. Probab. **21** 1494-1542.

[2] Blackwell, D. (1946). On an equation of Wald. *Ann. Math. Statist.* **17** 84-87.

[3] Bourgain, J. and Tzafriri, L. (1987). Invertibility of 'large' submatrices with applications to the geometry of Banach spaces and harmonic analysis. *Israel J. Math.* **57** 137-224.

[4] Burkholder, D. L. (1973). Distribution function inequalities for martingales. *Ann. Probab.* **1** 19-42.

[5] Burkholder, D. L. (1983). A geometric condition that implies the existence of certain singular integrals of Banach-space-valued functions. *Conference on harmonic Analysis in Honor of A. Zygmund* (W. Beckner, A.P. Calderon, R. Fefferman, and P. Jones, eds.), Wadsworth, Belmont, CA. pp. 270-286.

[6] Chow, Y. S. and Teicher, H. (1988). *Probability Theory: Independence, Interchangeability, Martingales.* Second Ed. Springer, New York.

[7] Chow, Y. S., de la Peña, V. H. and Teicher, H. (1993). Wald's lemma for a class of de-normalized U-statistics. *Ann. Probab.* **21** 1151-1158.

[8] de la Peña, V. H. (1992a). Decoupling and Khintchine's inequalities for U-statistics. *Ann. Probab.*, Vol. **20** 1877-1892.

[9] de la Peña, V. H. (1992b). Sharp bounds on the L_p norm of a randomly stopped multilinear form with an application to Wald's equation. In *Probab. in Banach Spaces VIII* (R. M. Dudley et al. eds.). Birkhauser, New York.

[10] de la Peña, V. H. (1992c). Nuevas desigualdades para U-estadísticas y gráficas aleatorias. *Proceedings of the fourth Latin American Congress of Probab. and Math. Statist. (CLAPEM)*, México City, September 1990.

[11] de la Peña, V. H. (1994). A bound on the moment generating function of a sum of dependent variables with an application to simple random sampling without replacement. *Ann. Inst. Henry Poincaré. Probabilités et Statistiques*, **30** 197-211.

[12] de la Peña, V. H. (1995). Correction to: A bound on the moment generating function of a sum of dependent variables with an application to simple random sampling without replacement. *Ann. Inst. Henry Poincaré. Probabilités et Statistiques*, **31** 703-704.

[13] de la Peña, V. H. and Giné, E. (1996). Preliminary version of a book on decoupling.

[14] de la Peña, V. H. and Govindarajulu, Z. (1992). Bounds for second moment of a randomly stopped sum of independent variables. *Statist. Probab. Letters* **14** 275-281.

[15] de la Peña, V. H. and Lai, T. L. (1997). Wald's equation and asymptotic bias of randomly stopped U-statistics. *Proc. Amer. Math. Soc.* **125**, 917-925.

[16] de la Peña, V. H. and Lai, T. L. (1995). Moments of randomly stopped U-statistics. To appear in *Ann. Probab.*

[17] de la Peña, V. H., Montgomery-Smith, S. J. and Szulga, J. (1994). Contraction and decoupling inequalities for multilinear forms and U-statistics. *Ann. Probab.* **22** 1745-1765.

[18] de la Peña, V. H. and Montgomery-Smith, S. J. (1994). Bounds on the tail probability of U-statistics and quadratic forms. *Bull. Amer. Math. Soc.* **31** 223-227.

[19] de la Peña, V. H. and Montgomery-Smith, S. J. (1995). Decoupling inequalities for the tail probabilities of multivariate U-statistics. *Ann. Probab.* **23** 806-816.

[20] Giné, E. and Zinn, J. (1994). A remark on convergence in distribution of U-statistics. *Ann. Probab.* **22** 117-125.

[21] Griffin, P. S. and McConnell, T. R. (1992). On the position of a random walk at the time of first exit from a sphere. *Ann. Probab.* **20** 825-854.

[22] Hitczenko, P. (1988). Comparison of moments for tangent sequences of random variables. *Probab. Theory Related Fields.* **78** 223-230.

[23] Hitczenko, P. (1990). Best constants in martingale version of Rosenthal's inequality. *Ann. Probab.* **18** 1656-1668.

[24] Hitczenko, P. (1994a). Sharp inequality for randomly stopped sums of independent non-negative random variables. *Stochastic Process. Appl.* **51** 63-73.

[25] Hitczenko, P. (1994b). On a domination of sums of random variables by sums of conditionally independent ones. *Ann. Probab.* **22** 453-468.

[26] Janson, S. and Nowicki, K. (1991). The asymptotic distributions of generalized U-statistics with applications to random graphs. *Probab. Theory Related Fields.* **90** 341-375.

[27] Jakubowski, A. (1986). Principle of conditioning in limit theorems for sums of random variables. *Ann. Probab.* **14** 902-915.

[28] Klass, M. J. (1988). A best possible improvement of Wald's equation *Ann. Probab.* **16** 840-853.

[29] Klass, M. J. (1990). Uniform lower bounds for randomly stopped Banach space valued random sums. *Ann. Probab.* **18** 790-809.

[30] Kou, S. G., and Ying, Z. (1995). A central limit theorem for a sequence of 2×2 tables. Preprint.

[31] Kwapien, S. and Woyczynski, W. (1992). Random series and stochastic integrals: simple and multiple. Birkhauser, NY.

[32] McConnell, T.R. and Taqqu, M.S. (1986). Decoupling inequalities for multilinear forms in independent symmetric random variables. *Ann. Probab.* **14** 943-954.

[33] Montgomery-Smith, S. J. (1993). Comparison of sums of independent identically distributed random variables. Probab. Math. Statist. **14** 281-285.

[34] Nolan, D. and Pollard D. (1987). U-processes: rates of convergence. *Ann. Statist.* **15** 780-799.

[35] Wald, A. (1945). Some generalizations of the theory of cumulative sums of random variables. *Ann. Math. Statist.* **16** 287-293.

[36] Zinn, J. (1985). Comparison of Martingale differences. *Probab. in Banach Spaces V, Lecture Notes in Math.* **1153** 453-457. Springer-Verlag, Berlin.

STOCHASTIC DIFFERENTIAL EQUATIONS ON MANIFOLDS

K. D. ELWORTHY,* *University of Warwick*

1. Introduction

A. The title is designed to indicate those particular aspects of stochastic differential equations which will be considered here: these are almost equally valid when the manifold in question is $\mathbb{R}^n$ (although compactness is often a useful simplifying assumption). In fact one of the main themes here will be that stochastic differential equations, even on $\mathbb{R}^n$, induce non-trivial differential-geometric structures and these structures are an important tool in analyzing the behaviour of the solutions of the s.d.e.

The equations will be Stratonovich equations driven by white noise: i.e. of the form

$$(1) \qquad dx_t = X(x_t) \circ dB_t + A(x_t) dt$$

with state space a C^∞, n-dimensional, manifold M e.g. $M = \mathbb{R}^n$ or $M = S^n$, the n-sphere. The coefficients are a smooth vector field A on M, so $A(x)$ lies in the tangent space $T_x M$ to M at x, for each x in M, and a smooth X which gives a linear map $X(x) : \mathbb{R}^m \to T_x M$ for each x in M. This injects the noise, which lies on $\mathbb{R}^m$, derived from $(B_t : t \geq 0)$ a Brownian motion on $\mathbb{R}^m$. In general $m \neq n$.

Such an equation is often considered as a random perturbation of the ordinary dynamical system

$$\frac{d\sigma}{dt} = A(\sigma(\cdot)).$$

However some caution is needed in doing this: see §4C below.

Despite the general sounding title we will not be considering equations with jumps, e.g. see [64], [17], [5], [39], [42], or more general equations as in [10], [37], [53]. We could allow $(B_t : t \geq 0)$ to be infinite dimensional and this would give a more complete picture, see the discussion in [30] and [33].

B. Historically we must start by referring to Itô's 1950 paper [49], while his remarkable ICM article [50] foretold what geometry there might be if a good theory of stochastic analysis on manifolds could be developed. Dynkin followed this up in 1968 [24]. Another early remarkable paper was that of Daletskii and Shnaiderman (now Belapolskya) [20], in 1969 using s.d.e. on infinite dimensional Lie groups to construct measures on the groups (but see also [28]). However, it was not until the Stratonovich

---

* Postal address: Mathematics Institute, University of Warwick, Coventry CV4 7AL, UK.

approach became current in the early 70's, [16] that the geometry and analysis could be coupled together in an efficient way (e.g. as in [61]).

C. Associated to (1) is the Markov process of its solutions, generator $\mathcal{A}$, Markov semigroup $(P_t : t \geq 0)$, and local solution flow $\{\xi_t(x) : 0 \leq t < \zeta(x), x \in M\}$ e.g. see [26] [48], [53]. In terms of the vector fields $X_i, i = 1$ to m, given by $X_i = X(\cdot)e_i$ for $e_1, ..., e_m$ an orthonormal base of $\mathbb{R}^n$ we have, on sufficiently smooth functions

$$(2) \qquad\qquad \mathcal{A} = L_{X_i} L_{X_i} + L_A$$

where L_A refers to Lie differentiation, i.e. differentiation in the direction of A: for $f : M \to \mathbb{R}$ and $x \in M$

$$L_A(f)(x) = df(A(x)).$$

Recall that the equation is *non-degenerate* if $X(x)$ is surjective for each x in M or equivalently if $\mathcal{A}$ is elliptic. If so we can (and always will in this article) give $T_x M$ the quotient inner product $<, >$ induced by $X(x)$. This makes M a Riemannian manifold (the first differential geometric structure induced by (1)). The generator can then be written

$$(3) \qquad\qquad \mathcal{A} = \frac{1}{2}\Delta + L_{A^X}$$

where Δ is the Laplace-Beltrami operator of the Riemannian structure

$$\Delta f = \text{ div grad } f = \text{trace } df$$

and A^X is the vector field

$$A^X(x) = \frac{1}{2}\sum_{i=1}^{m} \nabla X_i\left(X_i(x)\right) + A(x)$$

using ∇ to refer to covariant differentiation using the Levi-Civita connection (see below).

There is also the Itô map for each $x_0 \in M$

$$\mathcal{I} \equiv \mathcal{I}^{x_0} : \Omega \to C_{x_0}(M)$$

where $(\Omega, \mathcal{F}, P)$ is the probability space of $(B_t : t \geq 0)$ and $C_{x_0}(M)$ is the space of continuous paths with values in M (or its compactification if there is explosion) starting at x_0 at time $t = 0$ and $\mathcal{I}^{x_0}(\omega)_t = \xi_t(x_0, \omega)$.

Acknowledgements. Joint work, some as yet unpublished, with Y. Le Jan and X.-M. Li has been heavily drawn upon, as have conversations with Z. Ma. Research has been supported by EPSRC grant GR/H67263.

2. Three examples

A. Gradient systems. Suppose M is a submanifold of the Euclidean space $\mathbb{R}^n$. Then the tangent space $T_x M$ at any point x of M can be identified with a subspace of $\mathbb{R}^m$. We get an s.d.e. by taking $X(x) : \mathbb{R}^m \to T_x M$ to be the orthogonal projection and setting $A \equiv 0$. This is non-degenerate and the Riemannian metric induced on M is just the one it inherits as a submanifold. The generator $\mathcal{A}$ turns out to be $\frac{1}{2}\Delta$ so the solutions of (1) are Brownian motions on M, e.g. see [26], [29]. This is called a *gradient Brownian system* because the vector fields X_i, $i = 1$ to m are gradients (of the coordinate functions).

In [67] Itô remarks that it was when Stroock showed him this method of constructing Brownian motion on the sphere S^n, (using its standard embedding, so $m = n + 1$) that he became convinced of the usefulness of Stratonovich equations.

B. Invariant equations on Lie groups. Let M = G, a Lie group, with *id* its identity element and let L_a and R_a denote left and right multiplication by an element a of G. Take $m = n$ and let $X(id) : \mathbb{R}^n \to T_{id}G$ be some linear isomorphism. Define

$$X(g) : \mathbb{R}^n \to T_g G$$

by left invariance i.e. $X(g)e = TL_g(X(id)e)$. The induced Riemannian metric on G is the left invariant metric determined by that induced on the Lie algebra $T_{id}G$ by $X(id)$. If this is bi-invariant, and $A \equiv 0$, then the generator is again $\frac{1}{2}\Delta$ and the solutions are therefore Brownian motions, e.g. [26], [29].

If $(g_t : t \geq 0)$ is the solution with $g_0 = id$ then the solution flow can be written as right translation by $(g_t : t \geq 0)$, for a in M

$$\xi_t(a) = a g_t.$$

C. The canonical equation on the frame bundle. Now suppose that M has a given Riemannian metric. Let $\pi : OM \to M$ be the corresponding bundle of orthonormal frames. Thus u is in $\pi^{-1}(x)$ if u is an isometry $u : \mathbb{R}^n \to T_x M$ of the standard inner product of $\mathbb{R}^n$ with $<,>_x$ on $T_x M$. The Levi-Civita connection for the Riemannian structure determines a *horizontal lifting* $\mathbf{h}_u : T_{\pi(u)}OM \to T_u OM$ for each u in OM and hence an
$X(u) : \mathbb{R}^n \to T_u OM$ by

$$X(u)e = \mathbf{h}_u(u(e)), \qquad u \in OM, e \in \mathbb{R}^n.$$

The s.d.e.

$$(4) \qquad\qquad du_t = X(u_t) \circ dB_t$$

now on OM is canonically determined by the Riemannian structure. For smooth, rather than Brownian paths it goes back to Kobayashi's analysis of E. Cartan's development map ("rolling without slipping") [52] and for Brownian paths was similarly used by Eells and Elworthy, e.g. see [28] to obtain the 'stochastic development' construction of Brownian motion on M stimulated by an earlier version by Gangolli [43], [25]. The point of this construction is that if $(u_t : t \geq 0)$ is a solution of (4) then its projection $x_t = \pi(u_t)$ is a Brownian motion on M. For details see [48], [26], [29], [21]. When $\Omega = C_0(\mathbb{R}^n)$ with $(B_t : t \geq 0)$ this gives an essentially canonical map $\mathcal{D} : C_0(\mathbb{R}^n) \to C_{x_0}(M)$ mapping Wiener measure to the Brownian motion measure on $C_{x_0}(M)$. (It depends on the choice of $u_0 \in \pi^{-1}(x_0)$ but if $u_0 : \mathbb{R}^n \to T_{x_0}M$ is used as an identification it is canonical on $C_0(T_{x_0}M)$.)

D. For $M = S^n$ with n $= 1$ or 3 there is a Lie group structure as well as a natural embedding. We therefore have 3 ways of constructing Brownian motions on M. The first two have solution flows (for the third the flow lies on OM not M) but it is easy to see that though the solutions have the same laws the solution flows behave quite differently, [15]. One question which we will address is on the limitations of this difference.

E. For other examples of stochastic flows see [54], where there is infinite dimensional noise, and [60].

3. Solution Flows

A. When M is compact or $M = \mathbb{R}^n$ and the coefficients of (1) together with their derivatives are bounded there is a global solution flow consisting of diffeomorphisms see [53] [26], [14]. However, even if the Markov process is conservative i.e. $P_t 1 \equiv 1$, so (1) has no explosion, or is *complete*, equivalently $\zeta(x_0) = \infty$ almost surely for each $x_0 \in M$, it may be that no continuous flow exists for all time. The standard example is the equation $dx_t = dB_t$ on M where $M = \mathbb{R}^2 - 0$. If a continuous flow exists (so $\zeta(x_0) = \infty$ for all $x_0 \in M$ almost surely) (1) is said to be *strongly complete* or *strictly conservative*, [26], [53]. The mechanisms by which this property fails do not seem easy to analyse. Examples such as the equation $dx_t = dB_t$ on $M = \mathbb{R}^n - L$ where L is a q-dimensional subspace or submanifold led to a more refined definition [59]: the equation is *strongly p-complete* if, loosely speaking, the flow can be chosen continuous on any p-dimensional subset of M. It is shown in [59] that this is implied by certain moment conditions on the (formal) derivatives $T_{x_0}\xi_t : T_{x_0}M \to T_{\xi_t(x_0)}M, x_0 \in M, t \geq 0$, using any complete Riemannian metric for M.

B. Another important concept is that of moment stability. For a given Riemannian metric on M we say (1) is *strongly p-moment stable* if $\mu_k(p) < 0$ for all compact subsets K of M where

$$\mu_k(p) = \limsup_{t \to \infty} \sup_{x \in K} \log \mathbb{E}|T_x \xi_t|^p$$

for any $p \in \mathbb{R}^1$, e.g. see [6], [29], [8].

It is shown in [35] that strong p-moment stability has strong topological consequences for compact M e.g. if $p \geq \frac{1}{2}(n+1)$ it can only occur if M is homotopically equivalent to a sphere. This holds for any s.d.e. (1) and so is an example where the state space, rather than just the generator, restricts the possible behaviour of the flow.

For non-compact manifolds the situation is not so well understood. However strong 1-completeness implies that the flow moves smooth closed curves on M continuously while strong 1-moment stability implies that their lengths are shrunk on average; recurrence of the underlying Markov process will imply that such curves cannot escape out to infinity. Recurrence is also implied by moment conditions on $T\xi_t$ for elliptic symmetrisable $\mathcal{A}$, provided the induced metric is complete [58]. These are the ingredients used by X.-M. Li [58] to show that such generators cannot arise from an s.d.e. (1) which is non-explosive, strongly moment stable and with $\mathbb{E}\sup_{s\leq t}|T_x\xi_s|$ bounded on compact subsets of M, unless M is simply connected.

This is relevant to the case $M = \mathbb{R}^n$ when (1) has some periodicity. For a very detailed analysis of the behaviour of curves moved by a special class of flows see [18].

C. As an example where the geometry induced by the generator more directly restricts the behaviour of the flow, from [36] we have: For M compact if the generator $\mathcal{A}$ associated to (1) is $\frac{1}{2}\Delta$ for the induced metric and that metric has everywhere non-positive Ricci curvature then the flow cannot be 1-moment stable.

D. In one direction it is possible to apply these results to particular cases, such as gradient systems, to find purely geometric conditions which imply that (1) is strongly p–moment stable (for example) and so obtain differential geometric applications [35], [58]. From the point of view of the qualitative behaviour of s.d.e. it is the other direction which is important e.g. what conditions on X, A (or even better on M or $\mathcal{A}$) imply particular stability properties. For other purposes in stochastic analysis, e.g. see §5B below, it is important to know about the moments of the H-derivative of the Itô map $\mathcal{I}^{x_0}$. For all of these the study of the derivative $T\xi_t$ of the flow is a major ingredient.

E. Although the law of the flow ξ (e.g. as a process with values in a suitable space of maps from M to M) is not determined by $\mathcal{A}$ it was shown by Baxendale [7], that it is determined by the law of the diffusion process $\{(\xi_t(x), \xi_t(y)) : t \geq 0, x, y \in M\}$ on $M \times M$: this is essentially to say that it is determined by a covariance tensor and drift vector or equivalently by a Gaussian measure on the space of vector fields on M, see also [57], [53], or the discussion in [30].

4. LeJan-Watanabe connections induced by non-degenerate s.d.e.

A. A connection (on M or more specifically but confusingly, on the tangent bundle TM) is a rule for differentiating sections Z of TM i.e. vector fields. For such Z since $Z(x)$ is in a different space T_xM for each x in M there is no way of doing this determined only by the differentiable structure of M, apart from using the derivative $TZ : TM \to TTM$ which, especially after iteration, is rather clumsy and unmanagable because of the essentially redundant terms and high dimension of TTM (consider the case when $M = \mathbb{R}^n$). A choice of connection gives a 'covariant derivative' $\nabla Z : TM \to TM$ for differentiable Z and $\nabla Z(v) \in T_xM$, often written

$\nabla_v Z$, is considered as the derivative of Z in the direction v, for v any tangent vector at x. When M has a given Riemannian metric $<,>_x \colon x \in M$ the connection is said to be *metric* if for any two vector fields Z_1, Z_2 the derivative of the function $x \mapsto < Z_1(x), Z_2(x) >_x$ is given by

$$d < Z_1, Z_2 > . (v) = < \nabla Z_1(v), Z_2(x) >_x + < Z_1(x), \nabla Z_2(v) >_x$$

for any tangent vector v at x.

If $\sigma : (a,b) \to M$ is a piecewise differentiable path then $\frac{DZ}{\partial t} \in T_{\sigma(t)}M$ can be defined by

$$\frac{DZ}{\partial t} = \nabla Z(\dot{\sigma}(t))$$

for a given connection. It depends only on $\{Z(\sigma(s)) : s \in (a,b)\}$ and so makes sense when $Z : (a,b) \to TM$ is a differentiable function with $Z(s) \in T_{\sigma(s)}M$, i.e. a vector field along σ. Interpreting the relevant derivatives of σ which occur as Stratonovich differentials allows extensions of this construction to differentiation along the sample paths of solutions of (1), [26], [29], or more generally along semi-martingales in M, [37].

To fit in with the usual conventions in this situation the differential notation DZ, etc., is used.

B. In general there is a torsion tensor $T : TM \times TM \to TM$ with

$$(5) \qquad \frac{D}{\partial s}\frac{\partial}{\partial t}\sigma = \frac{D}{\partial t}\frac{\partial}{\partial s}\sigma + T\left(\frac{\partial\sigma}{\partial s}, \frac{\partial\sigma}{\partial t}\right)$$

for a smooth 2 parameter mapping $\sigma : (a,b) \times (c,d) \to M$.

The *Levi-Civita connection* for a given Riemannian metric is the unique metric connection with T identically zero. In future we will reserve the notation for this connection. For a general connection $\tilde{\nabla}$ there is the *adjoint* connection $\tilde{\nabla}'$ introduced by Driver [21] and given by

$$\tilde{\nabla}'Z(v) = \tilde{\nabla}Z(v) + T(Z(x), v) \qquad\qquad v \in T_xM.$$

When $\tilde{\nabla}$ is metric this may or may not be metric (for the same metric). If it is, $\tilde{\nabla}$ is said to be *torsion skew symmetric* [21]. In terms of adjoint connections equation (5) can be written

$$(6) \qquad \frac{\tilde{D}'}{\partial s}\frac{\partial}{\partial t}\sigma = \frac{\tilde{D}}{\partial t}\frac{\partial}{\partial s}\sigma.$$

B. Now suppose (1) is non-degenerate and give M the induced Riemannian metric. From [33] we can define a metric connection $\check{\nabla}$ on M by

$$(7) \qquad \check{\nabla} Z(v) = X(x_0) d[Y(\cdot)Z(\cdot)](v).$$

for $v \in T_{x_0} M$, where on the right hand side we are taking the derivative in the direction v of the function $x \mapsto Y(x)Z(x)$ for Y the adjoint of X:

$$Y(x) = X(x)^* : T_x M \to \mathbb{R}^m.$$

This agrees with the connection defined by Le Jan and Watanabe [57] using the covariance tensor of the flow of (1) mentioned in §3E above. Geometrically it is the push forward by X of the trivial connection on $M \times \mathbb{R}^m$. It has the characterising property that for each $u \in T_{x_0} M$

$$(8) \qquad \check{\nabla} Z^u(v) = 0, \qquad v \in T_{x_0} M$$

where Z^u is the vector field

$$(9) \qquad Z^u(x) = X(x)Y(x_0)u.$$

For a gradient system, as in §2A, this connection is just the Levi-Civita connection. For a left invariant s.d.e. on a Lie group G, as in §2B, this connection is the flat left invariant connection (all left invariant vector fields have vanishing covariant derivatives), and the adjoint connection is the flat right invariant connection on G. The latter is metric for a right invariant metric and so our connection is only torsion skew symmetric if the induced left invariant metric is also right invariant.

In general the torsion is given [33] by the Lie bracket

$$\check{T}(u,v) = -[Z^u, Z^v](x_0) \qquad u, v \in T_{x_0} M.$$

As observed in [57] the generator $\mathcal{A}$ can be written as

$$(10) \qquad \mathcal{A}(f) = \frac{1}{2} \mathrm{trace} \check{\nabla} \, \mathrm{grad} f + L_A(f).$$

C. For a given elliptic generator $\mathcal{A}$, Ikeda and Watanabe constructed a metric connection with torsion (in general) so that if the stochastic development construction described in §2C is applied using the horizontal lift coming from this connection rather than the Levi-Civita connection then the resulting process $x_t \equiv \pi(u_t)$ on M is a diffusion with generator $\mathcal{A}$, [48]. Now any metric connection can be shown to be the LeJan-Watanabe connection for some X. In particular this is true for the Ikeda-Watanabe connection. From this it follows [33] that for any elliptic $\mathcal{A}$ (with smooth coefficients, no zero order term) there is an X such that the solution diffusion process to

$$dx_t = X(x_t) \circ dB_t$$

has generator $\mathcal{A}$, i.e.

$$\mathcal{A} = \sum_{j=1}^{m} L_{X_j} L_{X_j}.$$

D. For any $\tilde{\nabla}$ with adjoint $\tilde{\nabla}'$ if $v_t = T\xi_t(v_0)$, from (6) we have the covariant Stratonovich equation

$$(11) \qquad \tilde{D}'v_t = \tilde{\nabla}X(v_t) \circ dB_t + \tilde{\nabla}A(v_t)dt$$

(By $\tilde{\nabla}X(v_t)e$ we mean $\tilde{\nabla}_{v_t}(X(\cdot)e)$).

When we use $\tilde{\nabla} = \check{\nabla}$ and let $\check{\nabla}$ etc. refer to the adjoint of $\check{\nabla}$ it is possible to decompose $(B_t : t \geq 0)$ into independent components one of which lies in the kernel of $X(x_t)$ for each t with the other being orthogonal to this kernel. (Strictly speaking they are independent after parallel translation to $T_{x_0}M$). The σ-algebra generated by the latter is just $\mathcal{F}^{x_0}$, the σ-algebra generated by $\{\xi_t(x_0) : t \geq 0\}$. However the characterising property (8) of the Le Jan-Watanabe connection ensures that when we decompose B_t in (11) the term involving the latter process vanishes and so the only noise which is left in (11) comes from what is a Brownian motion independent of $\mathcal{F}^{x_0}$. Using this together with a conversion of (11) to an Itô equation it was shown by Elworthy, Le Jan and Li [33] extending the result for gradient systems by Elworthy and Yor [36] that if $\bar{v}_t = \mathbb{E}\{v_t|\mathcal{F}^{x_0}\}$ then $\{\bar{v}_t : t \geq 0\}$ satisfies the (ordinary) covariant differential equation along the paths of $\{x_t : t \geq 0\}$

$$(12) \qquad \frac{\hat{D}\bar{v}_t}{\partial t} = -\frac{1}{2}\check{\mathrm{Ric}}^{\#}(\bar{v}_t) + \check{\nabla}A(\bar{v}_t)$$

where $\check{\mathrm{Ric}}^{\#} : TM \to TM$ comes from the Ricci curvature of the Le Jan-Watanabe connection. Here the conditional expectation $\mathbb{E}\{v_t|\mathcal{F}^{x_0}\}$ is given by parallel translating v_t back to $T_{x_0}M$, taking conditional expectations in the usual sense and then parallel translating back to $T_{x_t}M$. The idea is that $\bar{v}_t$ is just $T\xi_t(v_0)$ with the "extraneous noise" (coming from the fact that $X(x)$ may have a kernel) filtered out. Of course some bounds on X, A are needed for $\bar{v}_t$ to be even defined; to be safe we can assume M compact here.

Formula (12) or more generally the method which was used to obtain it is very useful for getting moment estimates on $|T\xi_t|$. For example if $\check{\nabla}$ is metric for some Riemannian metric $<,>'_x$, and M is compact then it is immediate from (12) that (1) can not be strongly 2-moment stable unless

$$< \check{\mathrm{Ric}}^{\#}(v), v >'_x -2 < \check{\nabla}A(v), v >'_x \,> 0$$

for some tangent vector v at some point x. [33].

E. Many of these results go over to certain classes of degenerate equations [31]. In general these connections appear to be a very useful tool in the study of stochastic flows, also giving insight into how the flows act on differential forms [33]. There is also somewhat related work in [1].

5. Estimates for Itô maps

A. Now suppose that $(B_t : t \geq 0)$ is the canonical version of Brownian motion, so that $(\Omega, F, \mathbb{P})$ is $C_0(\mathbb{R}^m)$ with Wiener measure. Let H be the corresponding Cameron-Martin space $L_0^{2,1}(\mathbb{R}^m)$. Given h in H the formal derivative of $\mathcal{I}^{x_0} : C_0(\mathbb{R}^m) \to C_{x_0}(M)$ in the direction h is seen, via (6), to satisfy

$$(13) \qquad \frac{\hat{D}}{\partial t}\Lambda_t = \check{\nabla}X(\Lambda_t) \circ dB_t + \check{\nabla}A(\Lambda_t)dt + X(x_t)(\dot{h}_t)dt$$

with $\Lambda_0 = 0$, which, by 'variation of constants' has solution

$$(14) \qquad \Lambda_t = T\xi_t\left(\int_0^t T\xi_s^{-1}(X(x_s)\dot{h}_s)ds\right).$$

Indeed $\{\Lambda_t : t \geq\}0$ is the H-derivative of $\mathcal{I}^{x_0}$

$$\Lambda_t(\omega) = (D\mathcal{I}^{x_0}(\omega)(h.))_t \in T_{x_t}M$$

as in Malliavin calculus, e.g. see [11], [48].

Let $\theta_s : C_0(\mathbb{R}^m) \to C_0(\mathbb{R}^m)$ be the shift, given by

$$\theta_s(\sigma)(t) = \sigma(t+s) - \sigma(s) \qquad \text{for} \ \ 0 \leq s, t.$$

The cocycle property for stochastic flows gives

$$T_{x_0}\xi_t(\omega) \circ T_{x_s}\xi_s^{-1} = T_{x_s}\xi_{t-s}(\theta_s\omega).$$

It follows [32] from §4D that Λ_t conditioned on the solution $\{x_t : t \geq 0\}$ is given by

$$(15) \qquad \mathbb{E}\{\Lambda_t|\mathcal{F}^{x_0}\} = W_t^A \int_0^t (W_s^A)^{-1}(X(x_s)\dot{h}_s)ds$$

where $W_t^A : T_{x_0}M \to T_{x_t}M$ is the random linear map solving

$$(16) \qquad \frac{DW_t^A}{\partial t} = -\frac{1}{2}\check{\mathrm{Ric}}^{\#}(W_t^A-) + \check{\nabla}A(W_t^A-)dt,$$

the "Hessian flow" of [30].

B. The main import of (15), qualitatively, is that the conditional expectation is almost surely bounded when $\hat{\nabla}$ is a metric connection for a metric for which $-\frac{1}{2}\check{\mathrm{Ric}}+\check{\nabla}A$ is bounded. The basic technique can be used to get estimates on higher moments of Λ_t. This enabled the logarithmic Sobolov inequality on $C_{x_0}M$ to be deduced from that on $C_0(\mathbb{R}^m)$ in [3], with extensions to the loop space of M in [2].

6. Integration by parts formulae

A. Infinite dimensional integration by parts formulae have a long history e.g. see [13]. In essence they are equations of the form

$$(17) \qquad \int_M df(V(\sigma))\mu(d\sigma) = -\int_M f(\sigma) \ \mathrm{div} \ V(\sigma)\mu(d\sigma)$$

for μ a measure on a possibly ∞-dimensional manifold $\mathbb{M}$ (e.g. Wiener measure on paths on $\mathbb{R}^n$) with a vector field V and function $f : \mathbb{M} \to \mathbb{R}^1$ and some 'divergence operator', div, which takes suitable vector fields to functions. Of course there need to be modifications if $\mathbb{M}$ has boundary, e.g. $\mathbb{M}$ a domain in Wiener space, [28] [66]. From the beginning [13] it was clear that in infinite dimensions there were restrictions on the directions $V(x)$ in which one could differentiate in order to have such formulae. The notion of H-differentiability systematically formulated by Gross [45] was a key to proofs of change of variable formulae successively by Gross, Kuo, Ramer, with consequent integration by parts formulae [44], [28], for $V(x)$ in some Hilbert 'tangent space', e.g. the Cameron-Martin space. This notion was shown to be of great practical importance with the advent of Malliavin calculus, [62], with integration by parts on classical Wiener space being a key to the probabilistic approach to Hörmander type hypoellipticity results via Malliavin's integration by parts formula, [62], [11], [9], [68]. The results in 1982 by Driver [21] on integration by parts formulae for Brownian motion measures on path spaces of Riemannian manifolds provided impetus for a lot of activity in this area, mainly concerned with analysis on path or loop spaces. One avowed aim is to obtain a good theory of operators on such spaces, e.g. see [55], and hopefully to tie in with work of a similar nature by geometers and topologists [51]. However, despite some success, as in [56], the frameworks often seem very different [41]. On the other hand classical Wiener measure has such a wealth of structure, e.g. the Fock space structure of its L^2 space, that the analysis of more general measures such as Wiener measures on paths and loops on Riemannian manifolds (or just those measures on paths in $\mathbb{R}^n$ arising from diffusion processes) has a great deal of intrinsic interest. The case of Lie groups has already created some nice structure [23], [46] following earlier work by Gross.

On a more abstract level there is the theory of differential measures introduced by Fomin. See for example [12], [65], [19], [9]. The sort of problem which can be addressed here is, given that there is an integration by parts formula as (17), when can one conclude that the vector field V has a flow which leaves the measure μ quasi-invariant? Also what differentiability or quasi-invariance properties do the laws of solutions of certain stochastic evolution equations possess, or the same questions for equilibrium measures of infinite dimensional systems.

B. There is a general procedure to 'push forward' integration by parts formulae by sufficiently regular maps. Informally suppose (17) holds for a class $\mathbb{C}$ of functions f on $\mathbb{M}$ and a given V, and suppose $\Phi : \mathbb{M} \to \mathbb{N}$ is a map which can be differentiated in the directions of V to give a measurable map $\sigma \mapsto T_\sigma \Phi(V(\sigma))$ from $\mathbb{M}$ to the tangent bundle $T\mathbb{N}$. Let $f : \mathbb{N} \to \mathbb{R}^1$ be differentiable with $f \circ \Phi$ in $\mathbb{C}$. Then (17) gives

$$\int_M df(T_\sigma \Phi(V(\sigma)))\mu(d\sigma) = -\int_M f(\Phi(\sigma)) \text{ div } V(\sigma)\mu(d\sigma)$$

from which comes

$$(18) \qquad \begin{aligned} &\int_N df\{\mathbb{E}T_\sigma\Phi(V(\sigma))|\Phi(\sigma) = y\}\Phi_*(\mu)(dy) \\ &= \int_N f(y)\mathbb{E}\{divV(\sigma))|\Phi(\sigma) = y\}\Phi_*(\mu)(dy) \end{aligned}$$

where $\Phi_*(\mu)$ is the image measure of μ by Φ , i.e. the law of μ . This shows that there is an integration by parts formula for the measure $\Phi_*(\mu)$ using vector fields on N of the form $y \mapsto \mathbb{E}\{T_\sigma\Phi(V(\sigma))|\Phi(\sigma) = y\}$.

C. Starting from the integration by parts formula on Wiener space $C_0(\mathbb{R}^m)$:

$$\mathbb{E} \, df(h\cdot) = \mathbb{E}f \int_0^1 < \dot{h}_s, dB_s >$$

where $h.$ is in the Cameron-Martin space H, and using the notation of §5A, equation (18) for Φ the Itô map $\mathcal{I}^{x_0}$ gives

$$(19) \qquad \begin{aligned} &\int_{C_{x_0}(M)} df\left(\mathbb{E}\left\{\Lambda_t \,|\mathcal{I}^{x_0} = y\right\}\right)\mu_{x_0}(dy) \\ &= \int_{C_{x_0}(M)} f(y)\,\mathbb{E}\left\{\int_0^1 < \dot{h}_s, dB_s > |\mathcal{I}^{x_0} = y\right\}\mu_{x_0}(dy) \end{aligned}$$

for μ_{x_0} the law on $C_{x_0}(M)$ of $\{x_t : 0 \le t \le 1\}$.

Applying equation (15) we see we have integration by parts formulae for vector fields on $C_{x_0}(M)$ of the form $W_t^A \int_0^1 (W_s^A)^{-1}(X(x_s)\dot{h}_s)ds, 0 \le t \le 1$ obtaining Driver's formulae [21], [47] in a slightly different form: for details and extensions to degenerate S.D.E. see [32] with an elementary derivation in [34] for the non-degenerate case. (It is simpler to use the Girsanov-Maruyama formula rather than Malliavin calculus once it is known what the formulae should be.)

Note that our S.D.E. (1) has now determined an S.D.E. type coefficient on $C_{x_0}(M)$, namely

$$X : C_{x_0}(M) \times H \to TC_{x_0}(M)$$

$$(20) \qquad X(y)(h.)_t := W_t^A \int_0^t (W_s^A)^{-1}(X(y_s)\dot{h}_s)ds, \qquad 0 \le t \le 1$$

which could be used to inject the canonical cylindrical Gaussian measure of H into the tangent spaces to $TC_{x_0}(M)$ and so (hopefully) integrated to obtain (hopefully) interesting infinite dimensional processes on $C_{x_0}(M)$. There are different versions of X in use. e.g. with [40], [4], [34], [21]. Since X is only measurable, approaches using Dirichlet form theory have generally proved more successful to obtain such processes see [22], [27], but also see [38], [63]. However the calculus on $C_{x_0}(M)$ determined by X (in the same sort of way that our original X determined a calculus on M) does seems to be of fundamental importance: after all a special case of (1) is the equation $dx_t = dB_t$ on $\mathbb{R}^n$ with $B.$ a BM($\mathbb{R}^n$).

References

[1] ACCARDI, L. AND MOHARI, A. (1996). On the structure of classical and quantumn flows. *J. Funct. Anal.* **135**, 421–455.

[2] AIDA, S. (1997). On the irreducibility of certain Dirichlet forms on loop spaces over compact homogeneous spaces. In *New Trends in stochastic Analysis, Proc. Taniguchi Symposium, Sept. 1995, Charingworth.* World Scientific Press, Singapore.

[3] AIDA, S. AND ELWORTHY, K. (1995). Differential calculus on path and loop spaces. 1. Logarithmic Sobolev inequalities on path spaces. *C. R. Acad. Sci. Paris, t. 321, série I* 97–102.

[4] AIRAULT, H. AND MALLIAVIN, P. (1996). Integration by parts formulas and dilation vector fields on elliptic probability spaces. *Probab. Theory Relat. Fields* **106**, 447–494.

[5] APPLEBAUM, D. (1995). A horizontal Levy process on the bundle of orthonormal frames over a complete Riemannian manifold. In *Sém. de Probabilitś XXIX, Lecture Notes in Maths 1613.* Springer-Verlag, Berlin. pp. 166–180.

[6] ARNOLD, L. AND WIHSTUTZ, V. E. (1986). *Lyapunov Exponents. Proceedings, Bremen 1984. Lecture Notes in Mathematics 1186.* Springer-Verlag, Berlin.

[7] BAXENDALE, P. (1984). Brownian motions in the diffeomorphism groups I. *Compositio Math.* **53**, 19–50.

[8] BAXENDALE, P. AND STROOCK, D. W. (1988). Large deviations and stochastic flows of diffeomorphisms. *Probab. Theory Relat. fields* **81**, 521–554.

[9] BELL, D. (1987). *The Malliavin Calculus.* Pitman Monographs and Surveys in Pure Maths. 34. Longman, London.

[10] BELOPOLSKAYA, Y. I. AND DALECKY (1989). *Stochastic equations and differential geometry.* Kluwer, Dordrecht.

[11] BISMUT, J. M. (1981). Martingales, the Malliavin calculus and Hoemander's theorems. In *Stochastic Integrals, Lecture Notes in Maths. 851.* ed. D. Williams. Springer-Verlag, Berlin. pp. 85–109.

[12] BOGACHEV, V. (1995). Differentiable measures and Malliavin calculus. Preprints di Matematica no. 16, Scuola Normale Superiore, Pisa.

[13] CAMERON, R. (1951). The first variation of an indefinite Wiener integral. *Proc. Amer. Math Soc.* **2**, 914–924.

[14] CARVERHILL, A. AND ELWORTHY, D. (1983). Flows of stochastic dynamical systems: the functional analytic approach. *Z. Wahrscheinlichkeitstheorie Verw. Gebiete* **65**, 245–267.

[15] CARVERHILL, A. P., CHAPPELL, M. AND ELWORTHY, K. (1986). Characterisitic exponents for stochastic flows. In *Stochastic processes-Mathematics and physics. Proceedings, Bielefeld 1984, pp. 52–72. Lecture Notes in Mathematics 1158.* ed. S. Alberverio and et al. Springer Verlag, Berlin.

[16] CLARK, J. (1973). An introduction to stochastic differential equations on manifolds. In *Geometric methods in systems theory.* ed. D. M. R. Brockett. pp. 131–149.

[17] COHEN, S. (1995). Some Markov properties of stochastic differential equations with jumps. In *Sém. de Probabilitiés XXIX. Lecture Notes in Maths 1613.* ed. J. A. et al. Springer-Verlag, Berlin. pp. 181–193.

[18] CRANSTON, M. AND LEJAN, Y. (1997). To appear in the Proceedings of 1997 Bremen conference on Random Dynamical Systems.

[19] DALESTKII, Y. L. AND FOMIN, S. (1993). *Measures and differential equations in infinite dimensional spaces.* Kluwer, Dordrecht.

[20] DALETSKII, Y. L. AND SHNAIDERMAN, Y. I. (1969). Diffusions and quasi-invariant measures on infinite dimensional Lie groups. *Funktsional'nyi Analiz i Ego prilozheniya* **3**, 88–90. English translation: Funct. Anal. Appl., 3 156–8.

[21] DRIVER, B. (1992). A Cameron–Martin type quasi-invariance theorem for Brownian motion on a compact Riemannian manifold. *J. Funct. Anal.* **100**, 272–377.

[22] DRIVER, B. AND ROECKNER, M. (1992). Construction of diffusions on path and loop spaces of compact Riemannian manifolds. *C.R. Acad. Sci Paris Sér 1 Math* **315**, 603–608.

[23] DRIVER, B. K. (1995). On the Kakutani–Itô–Segal–Gross and Segal–Bargmann–Hall isomorphisms. *J. of Funct. Anal.* **133**, 69–128.

[24] DYNKIN, E. (1968). Diffusion of tensors. *Dokl. Akad. Nank SSSR* **179**, 1264–1267. English translations: Soviet Math. Dokl. 9 (1968), no 532–535.

[25] EELLS, J. AND ELWORTHY, K. (1971). On Fredholm manifolds. *Actes Congress Intern. Maths. 1970* **Tome 2**, 215–220.

[26] ELWORTHY, K. (1982). *Stochastic Differential Equations on Manifolds, Lecture Notes Series 70*. Cambridge University Press, Cambridge.

[27] ELWORTHY, K. AND MA, Z. (1997). Vector fields on mapping spaces and related Dirichlet forms and diffusions. To appear in Osaka J. Math.

[28] ELWORTHY, K. D. (1975). Measures on infinite dimensional manifolds. In *Functional Integration and its Applications*. Oxford University Press, Oxford. pp. 60–68.

[29] ELWORTHY, K. D. (1988). Geometric aspects of diffusions on manifolds. In *Ecole d'Eté de Probabilités de Saint-Flour XV–XVII, 1985–1987. Lecture Notes in Mathematics*. ed. P. L. Hennequin. vol. 1362. Springer-Verlag, Berlin. pp. 276–425.

[30] ELWORTHY, K. D. (1992). Stochastic flows on Riemannian manifolds. In *Diffusion processes and related problems in analysis, volume II. Birkhauser Progress in Probability*. ed. M. A. Pinsky and V. Wihstutz. Birkhauser, Boston. pp. 37–72.

[31] ELWORTHY, K. D., LEJAN, Y. AND LI, X.-M. (1996). Integration by parts formulae for degenerate diffusion measures on path spaces and diffeomorphism groups. *C. R. Acad. Sci. t. 323 série 1* **921–926,**.

[32] ELWORTHY, K. D., LEJAN, Y. AND LI, X.-M. (1997). In preparation.

[33] ELWORTHY, K. D., LEJAN, Y. AND LI, X.-M. (1997). Concerning the geometry of stochastic differential equations and stochastic flows. In *New Trends in Stochastic Analysis, Proc. Taniguchi Symposium, Sept. 1995, Charingworth*. ed. K. D. Elworthy and I. S. S. Kusuoka. World Scientific Press, Singapore.

[34] ELWORTHY, K. D. AND LI, X.-M. (1996). A class of integration by parts formulae in stochastic analysis I. In *Itô's Stochastic Calculus and Probability Theory*. Springer-Verlag, Berlin. pp. 15–30.

[35] ELWORTHY, K. D. AND ROSENBERG, S. (1996). Homotopy and homology vanishing theorems and the stability of stochastic flows. *Geometric and Functional Analysis* **6**, 51–78.

[36] ELWORTHY, K. D. AND YOR, M. (1993). Conditional expectations for derivatives of certain stochastic flows. In *Sem. de Prob. XXVII. Lecture Notes in Maths. 1557*. ed. J. Azéma, P. Meyer, and M. Yor. Springer-Verlag, Berlin. pp. 159–172.

[37] EMERY, M. (1989). *Stochastic Calculus in Manifolds*. Springer-Verlag, Berlin.

[38] ENCHEV, O. AND STROOCK, D. (1995). Towards a Riemannian geometry on the path space over a Riemannian manifold. *J. Funct. Anal.* **134**, 392–416.

[39] ESTRADE, A. (1992). Exponentielle stochastique et intégrale multiplicative discontinues. *Ann. Inst. H. Poincaré (Prob. Stat.)* **28**, 107–129.

[40] FANG, S. AND MALLIAVIN, P. (1993). Stochastic analysis on the path spaces of a Riemannian manifold. *J. Funct. Anal.* **118**, 249–274.

[41] FEHÉR, L., STIPSICZ, A. AND SZENTHE, J. E. (1992). *Lecture Notes of the Conference on Topological Quantum Field Theories and Geometry of Loop Spaces*. World Scientific, Singapore.

[42] FUJIWARA, T. (1991). Stochastic differential equations of jump type on manifolds and Levy flows. *J. Math. Kyoto Univ.* **31**, 99–119.

[43] GANGOLLI, R. (1964). On the construction of certain diffusions on a differentiable manifold. *Z.F. Wahrscheinlichkeitstheorie. Verw. Greb.* **2**, 402–419.

[44] GOODMAN, V. (1972). A divergence theorem for Hilbert spaces. *Trans. Amer, Math Soc.* **164**, 411–.

[45] GROSS, L. (1967). Potential theory on Hilbert space. *J. Funct. Anal.* **1**, 123–181.

[46] HIJAB, O. (1995). Hermite functions on compact Lie groups, ii. *J. Funct. Anal.* **133**, 41–49.

[47] HSU, E. P. (1995). Quasi-invariance of the Wiener measure on the path space over a compact Riemannian manifold. *J. Funct. Anal.* **134**, 417–450.

[48] IKEDA, N. AND WATANABE, S. (1981). *Stochastic Differential Equations and Diffusion Processes*. North-Holland, Amsterdam.

[49] ITÔ, K. (1950). On stochastic differential equations on a differentiable manifold. *Nagoya Math. J.* **1**, 35–47.

[50] ITÔ, K. (1963). The Brownian motion and tensor fields on a Riemannian manifold. *Proc. Internat. Congr. Math. Stockholm, 1962, Djursholm: Inst. Mittag-Leffler* 536–539.

[51] JONES, J. D. S. AND LEANDRE, R. (1991). L^p chen forms on loop spaces. In *Stochastic Analysis*. ed. M. Barlow and N. Bingham. Cambridge University Press, Cambridge. pp. 104–162.

[52] KOBAYASHI, S. (1957). Theory of connections. *Annali di Mat.* **43,** 119–194.

[53] KUNITA, H. (1990). *Stochastic Flows and Stochastic Differential Equations.* Cambridge Studies in Advanced Mathematics, 24. Cambridge University Press, Cambridge.

[54] LE JAN, Y. (1985). On isotropic Brownian motions. *Z. Wahrscheinlichkeitstheorie Verw Geb* **70,** 609–720.

[55] LÉANDRE, R. (1995). Hilbert space of spinor fields over the free loop space. Preprint 95/no 25. Institut Elie Cartan, Universit H. Poincaré, Nancy 1.

[56] LÉANDRE, R. (1997). Brownian cohomology of a homogeneous manifold. In *New trends in Stochastic Analysis, Proc. Taniguchi Workshop (Charingworth, Sept. 1994).* ed. K. Elworthy, S. Kusuoka, and I. Shigekawa. World Scientific, Singapore. pp. 305–347.

[57] LEJAN, Y. AND WATANABE, S. (1984). Stochastic flows of diffeomorphisms. In *Stochastic Analysis, Proc. of the Taniguchi Int. Symp., Katata and Kyoto, 1982.* ed. K. Itô. North-Holland, Amsterdam. pp. 307–332.

[58] LI, X.-M. (1994). Stochastic differential equations on noncompact manifolds: moment stability and its topological consequences. *Probab. Theory Relat. Fields* **100,** 417–428.

[59] LI, X.-M. (1994). Strong p-completeness and the existence of smooth flows on noncompact manifolds. *Probab. Theory Relat. Fields* **100,** 485–511.

[60] LIAO, M. (1992). Stochastic flows on the boundaries of Lie groups. *Stochastics and Stochastic Reports* **39,** 213–238.

[61] MALLIAVIN, P. (1978). Geométrie differentielle stochastique. Sem. de Math. Sup. 64. Université de Montréal.

[62] MALLIAVIN, P. (1978). Stochastic calculus of variation and hypoelliptic operators. In *Proc. Intern. Symp. S.D.E., Kyoto 1976.* ed. K. Itô. Wiley-Interscience, New York. pp. 195–263.

[63] NORRIS, J. (1967). Twisted sheets. *J. Funct. Anal.* **132,** 273–334.

[64] ROGERSON, S. J. (1981). Stochastic differential equations with discontinuous sample paths and differential geometry. Control Theory Centre Report No 101, University of Warwick, Coventry, England. Submitted to 'Stochastics'.

[65] SMOLYANOV, O. AND WEIZSÄKER, H. (1993). Differentiable families of measures. *J. Funct. Anal.* **118,** 454–476.

[66] STENGLE, G. (1968). A divergence theorem for Gaussian stochastic process expectations. *J. Math. Anal. Appl.* **21,** 537–546.

[67] STROOCK, D. AND VARADHAN, S. ʳ. (1987). *Kiyosi Itô Selected Papers.* Springer-Verlag, Berlin.

[68] WATANABE, S. (1984). *Lectures on stochastic differential equations and Malliavin calculus.* Bombay: Tata Institute of Fundamental Research. Springer-Verlag, Berlin.

EXTENDING FLOWS OF CLASSICAL MARKOV PROCESSES TO QUANTUM FLOWS IN FOCK SPACE

F. FAGNOLA,* *Università di Genova*

Abstract

We describe a rather general scheme for extending the flow of a classical Markov process to a quantum flow on the algebra of all bounded operator on an L^2 space of the state space. We discuss open problems and possible developements.

1. Introduction

The notion of quantum stochastic process on a *-algebra $\mathcal{A}$ with values in a *-algebra $\mathcal{B}$ as a family $j = (j_t)_{t>0}$ of *-homomorphisms $j_t : \mathcal{B} \to \mathcal{A}$ was introduced by Accardi, Frigerio and Lewis in [2]. It generalizes the concept of a classical stochastic process $(x(t))_{t\geq 0}$ on a probability space $(\Omega, \mathcal{F}, \mathbb{P})$ with values in a measurable space $(E, \mathcal{E})$. In fact the process x can be defined through the homomorphisms

$$j_t : \mathcal{L}^\infty(E, \mathcal{E}) \to \mathcal{L}^\infty(\Omega, \mathcal{F}, \mathbb{P}), \qquad j_t(f) = f(x(t)).$$

We shall call stochastic flow a stochastic process in the sense of Accardi, Frigerio and Lewis with homomorphisms j_t satisfying a stochastic differential equation. In this note we suppose that $E = \mathbb{R}^d$ for simplicity. In many interesting cases the homomorphisms j_t of the flow can be written in the form

$$(1.1) \qquad j_t(f) = V(t)(M_f \otimes I)V(t)^*$$

where M_f is the multiplication operator by f acting on $L^2(\mathbb{R}^d)$ space, I the identity operator on $L^2(\Omega, \mathcal{F}, \mathbb{P})$ and $(V(t))_{t\geq 0}$ is a family of isometries or unitaries acting on $L^2(\mathbb{R}^d) \otimes L^2(\Omega, \mathcal{F}, \mathbb{P})$.

Flows of the form (1.1) can be constructed in a natural way starting either from a deterministic or a stochastic evolution. We give first an example associated to a deterministic evolution. Let $b : \mathbb{R} \to \mathbb{R}$ be a Lipschitz function. The unique solution of the Cauchy problem

$$
\begin{aligned}
x'(t) &= b(x(t)) \\
x(0) &= x.
\end{aligned}
$$
(1.2)

defines a one-to-one map $x \to x(t)$ for all $t \in \mathbb{R}$. Let $V(t)$ be the unitary operator on $L^2(\mathbb{R})$

$$(V(t)u)(x) = \left| \frac{b(x(t))}{b(x)} \right|^{\frac{1}{2}} u(x(t))$$

---

* Postal address: Dipartimento di Matematica, Università di Genova, Via Dodecaneso 35, I - 16146 Genova, Italia.

(with the understanding $0/0 = 1$). Then $V(t)^* = V(-t)$ and $(V(t))_{t\geq 0}$ is a group of unitary operators. Identifying f with the multiplication operator on $L^2(\mathbb{R})$ by the function $x \to f(x)$, it is easy to check that $V(t)fV(t)^*$ agrees with the multiplication operator on $L^2(\mathbb{R})$ by the function $x \to f(x(t))$. Thus the flow associated with the dynamical system (1.2) can be represented in the form (1.1) and $(V(t))_{t\geq 0}$ is a unitary group.

A stochastic example can be constructed by considering an homogeneous classical diffusion x on $\mathbb{R}$ with covariance σ and drift b regular enough so that it is the unique solution of a stochastic differential equation

$$(1.3) \qquad dx(t) = \sigma(x(t))dB(t) + b(x(t))dt,$$

where B is a brownian motion on a given probability space $(\Omega, \mathcal{F}, \mathbb{P})$. For every function $f : \mathbb{R} \to \mathbb{R}$ twice differentiable the process $(f(x(t))_{t\geq 0}$ satisfies the stochastic differential equation

$$(1.4) \qquad df(x(t)) = (Lf)(x(t))dB(t) + (Af)(x(t))dt, \qquad x(0) = x$$

where A, the infinitesimal generator of the diffusion, and L are given by

$$(Af)(x) = \frac{1}{2}(\sigma(x))^2 f'(x) + b(x)f'(x)$$
$$(1.5) \qquad (Lf)(x) = \sigma(x)f'(x)$$

Suppose for simplicity that σ is constant and $b = 0$. Thinking of the random variables $B(t)$ as multiplication operators on $L^2(\Omega, \mathcal{F}, \mathbb{P})$ and denote by ∂ the usual derivation operator on $L^2(\mathbb{R})$. The operator $i\partial \otimes B(t)$ acting on the tensor product Hilbert space $L^2(\mathbb{R}) \otimes L^2(\Omega, \mathcal{F}, \mathbb{P})$ is selfadjoint. Therefore the operator

$$V(t) = \exp\left(i\partial \otimes B(t)\right)$$

is unitary. Moreover, using the classical Itô formula, it is easy to check that $V(t)$ satisfies the stochastic differential equation

$$dV(t) = V(t)\left(i\sigma\partial dB(t) - \frac{\sigma^2}{2}\partial^2 dt\right)$$

Define $j_t(f) = V(t)(f \otimes I)V(t)^*$ where f denotes the multiplication operator by a twice differentiable bounded function f. Using again the Itô formula, one shows that $j_t(f)$ satisfies the equation (1.4). Therefore the uniqueness of the solution of (1.4) implies that $j_t(f)$ coincides with the multiplication operator by the function $x \to f(x(t))$. In this case, however, the family of operators $(V(t))_{t\geq 0}$ is neither a group nor a semigroup in fact it is a Markov cocycle in the sense of [1].

Note that, whenever the homomorphisms have the form (1.1), the multiplication operator f can be replaced by an arbitrary bounded operator. Thus in both the above examples they admit an extension to the algebra of all bounded operators on $L^2(\mathbb{R})$ with values in the algebra of all bounded operators on $L^2(\mathbb{R}) \otimes L^2(\Omega, \mathcal{A}, \mathbb{P})$. By construction the restriction to the algebra of multiplication operators of the new flow coincides with the original classical commutative flow.

In general whenever the flow $(k_t)_{t\geq 0}$ on the algebra of all bounded operators on $L^2(\mathbb{R})$ (or $L^2(\mathbb{R}^d)$) has the property that, when restricted to the multiplication operators $((fu)(x) = f(x)u(x)$ for all $f \in L^\infty(\mathbb{R})$, $u \in L^2(\mathbb{R}))$ it induces a classical flow (in the sense that $k_t(f) = f(x(t))$ for some classical stochastic process $(x(t))_{t\geq 0}$ we shall say that $(k_t)_{t\geq 0}$ *is a quantum extension of the classical flow* $(j_t)_{t\geq 0}$.

In quantum physics there arise many quantum extensions of classical flows; therefore the following question is quite natural from the mathematical point of view: *which classical flows admit a quantum extension?* At the moment a satisfactory solution of this problem has not been found mainly because of analytical difficulties. However, for a large class of stochastic processes it has been solved affirmatively. This class includes:

- homogeneous diffusions with smooth coefficients and strictly elliptic covariance,

- some diffusion with non-smooth drift (e.g. the transient Bessel processes),

- countable state Markov chains,

- pure jump processes on a locally compact group.

This problem is also interesting from a probabilistic point of view because it is a natural generalization of the theory of classical stochastic processes (at least of those processes which satisfy a stochastic differential equation). In the following we shall illustrate the basic ideas involved in the extension of a classical flow to a quantum one in the case of diffusions and birth-and-death processes. Moreover we describe some rigorous results on this problem.

2. Some results in quantum stochastic calculus

Let $\Gamma\left(L^2\left(\mathbb{R}; \mathbb{C}^N\right)\right)$ be the boson Fock space over $L^2\left(\mathbb{R}; \mathbb{C}^N\right)$ and denote by $e(f)$ the exponential vector corresponding to an element f of $L^2\left(\mathbb{R}; \mathbb{C}^N\right)$. We denote by $\mathcal{H}$ the tensor product space

$$\mathcal{H} = h \otimes \Gamma\left(L^2(\mathbb{R}; \mathbb{C}^N)\right)$$

where h is another complex separable Hilbert space. The algebra of all bounded operators on h (resp. $\mathcal{H}$) will be denoted by $\mathcal{B}(h)$ (resp. $\mathcal{B}(\mathcal{H})$) and the identity operator of both will be denoted by I. Let $\{e_l\}_{l=1}^N$ be the canonical basis of $\mathbb{C}^N$ and let $|e_m\rangle\langle e_l|$ denote the projection operator on $\mathbb{C}^N$

$$x \to \langle e_l, x\rangle\, e_m.$$

The basic quantum noises $\{\Lambda_m^l\}_{0\leq l,m\leq N}$ in $\Gamma\left(L^2(\mathbb{R};\mathbb{C}^N)\right)$ are defined by

$$\Lambda_0^l(t)e(f) = A\left(\chi_{(0,t)}e_l\right) = \left\langle \chi_{(0,t)}e_l, f\right\rangle e(f) \qquad\qquad \text{for } l \neq 0,$$

$$\Lambda_m^l(t)e(f) = \Lambda\left(\chi_{(0,t)}|e_m\rangle\langle e_l|\right)e(f) = -i\frac{d}{d\varepsilon}e\left(e^{i\varepsilon\chi_{(0,t)}|e_m\rangle\langle e_l|}f\right)\Big|_{\varepsilon=0} \qquad \text{for } l,m \neq 0,$$

$$\Lambda_m^0(t)e(f) = A^+\left(\chi_{(0,t)}e_m\right)e(f) = \frac{d}{d\varepsilon}e(f + \chi_{(0,t)}e_m)\Big|_{\varepsilon=0} \qquad\qquad \text{for } m \neq 0,$$

$$\Lambda_0^0(t) = tI.$$

We assume that the reader is familiar with quantum stochastic calculus in the boson Fock space and refer to the books of [18] and [21], for a detailed exposition. Let $\mathcal{A}$ denote a sub*algebra of $\mathcal{B}(h)$ and let $\mathcal{D}$ denote a sub*algebra of $\mathcal{A}$ which is dense in $\mathcal{A}$ for the σ-weak topology.

Definition 2.1. A quantum flow on $\mathcal{A}$ with structure maps

$$\theta_m^l : \mathcal{D} \to \mathcal{A}, \qquad 0 \leq l,m \leq N$$

is a family $\{j_t\}_{t\geq 0}$ of *homomorphisms $j_t : \mathcal{A} \to \mathcal{B}(\mathcal{H})$ satisfying the q.s.d.e. in h

$$(2.1) \qquad\qquad dj_t(X) = \sum_{l,m=0}^{N} j_t(\theta_m^l(X))d\Lambda_l^m(t), \qquad j_0(X) = X$$

for all $X \in \mathcal{D}$. A quantum flow $\{j_t\}_{t\geq 0}$ on $\mathcal{A}$ is identity preserving if the identity operator I belongs to $\mathcal{A}$ and $j_t(I)$ is the identity operator on $\mathcal{B}(\mathcal{H})$ for all $t \geq 0$.

Computing the stochastic differential of the processes $(j_t(YX))_{t\geq 0}$ and $(j_t(Y) j_t(X))_{t\geq 0}$ (see, for instance, ([18] Ch. VI, 3.5), ([21] Prop. 28.1)) it can be shown that the structure maps of a quantum flow satisfy the structure relations

$$(2.2) \qquad\qquad \theta_m^l(YX) = Y\theta_m^l(X) + \theta_m^l(Y)X + \sum_{k=1}^{N}\theta_k^l(Y)\theta_m^k(X)$$

for all $X, Y \in \mathcal{D}$ and all $l, m \in \{0, 1, \ldots, N\}$.

Definition 2.2. A quantum flow $\{j_t\}_{t\geq 0}$ on $\mathcal{A}$ is commutative if $\mathcal{A}$ is commutative and

$$[j_s(Y), j_t(X)] = 0$$

for all $X, Y \in \mathcal{D}$ and all $s, t \geq 0$.

A large class of quantum flows of the form (1.1) where $V = \{V(t)\}_{t\geq 0}$ is a unitary cocycle satisfying an Hudson–Parthasarathy q.s.d.e.

$$(2.3) \qquad\qquad dV(t) = \sum_{l,m=0}^{N} V(t)L_m^l d\Lambda_l^m(t), \qquad V(0) = I$$

and $(L_m^l)_{0 \le l, m \le N}$ belong to $\mathcal{B}(h)$ and satisfy the following conditions

$$(2.4) \qquad L_m^l + (L_l^m)^* + \sum_{k=1}^{N} (L_l^k)^* L_m^k = 0$$

$$(2.5) \qquad L_m^l + (L_l^m)^* + \sum_{k=1}^{N} L_k^l (L_k^m)^* = 0$$

for all $l, m \in \{0, 1, \ldots, N\}$. The conditions (2.4) (resp. (2.5)) are necessary and sufficient in order the operators $V(t)$ to be isometries (resp. coisometries) (see, for instance, [18] p.143, [21] p.227). When (2.4) holds the family of homomorphisms (1.1) satisfies the q.s.d.e. (2.1) and the structure maps $\left(\theta_m^l\right)_{0 \le l, m \le N}$ are related to the operators L_m^l by

$$(2.6) \qquad \theta_m^l(X) = L_m^l X + X(L_l^m)^* + \sum_{k=1}^{N} L_k^l X (L_k^m)^*.$$

Moreover the flow is identity preserving if and only if (2.5) holds and the flow is commutative if and only if $\mathcal{A}$ is commutative and invariant under the action of the structure maps (see [22]).

3. Extension of one-dimensional homogeneous diffusions

If we want to construct the quantum flow extending the flow of a classical Markov process, the discussion in the former section and formula (1.4) suggest that the quantum generator θ_0^0 must agree with the infinitesimal generator A of the classical Markov process on a large class of bounded functions. Therefore the problem of the quantum extension of a classical Markov process can be studied as follows: given the generator A of a classical Markov semigroup find the generator θ_0^0 of a quantum Markov semigroup which, when restricted to the operators of multiplication by a function f, coincides with the multiplication operators by $A(f)$ and the structure maps θ_m^l ($l + m > 0$) satisfying the structure relations (2.2) of a quantum flow $(j_t)_{t \ge 0}$ which, when restricted to the operators of multiplication by a function f, gives $f(x_t)$ where $(x_t)_{t \ge 0}$ is the classical Markov process generated by A. This problem, because of (2.6), can often be reduced to the problem of finding a suitable form for the operators L_m^l depending on the classical generator. The existence of a quantum extension of a classical process is not always guaranteed. Moreover, even if the existence holds, uniqueness usually does not.

A first example of a recipe for finding a possible choice of the operators L_m^l in the q.s.d.e. (2.3) from the given classical generator is given by a classical homogeneous diffusion with infinitesimal generator A given by (1.5). The natural candidate as commutative *algebra $\mathcal{A}$ is $\mathcal{C}_b^0(\mathbb{R}; \mathbb{IC})$, the algebra of complex-valued bounded continuous functions on the real line, and $\mathcal{D}$ will be an appropriate sub*algebra of $\mathcal{C}_b^0(\mathbb{R}; \mathbb{IC})$ containing regular functions. Let $N = 1$. The structure relation (2.2) with $l = m = 0$

yields

$$\theta_1^0(g)\theta_0^1(f) = \theta_0^0(gf) - g\theta_0^0(f) - \theta_0^0(g)f$$
$$= A(gf) - gA(f) - A(g)f$$
$$= \sigma^2 g' f'$$

for all $g, f \in \mathcal{D}$. Therefore a possible choice for $\theta_1^0(f)$ and $\theta_0^1(f)$ is

$$(3.1) \qquad \left(\theta_1^0(f)\right)(x) = \left(\theta_0^1(f)\right)(x) = \sigma(x)f'(x)$$

Note that these structure maps act as derivations on $\mathcal{C}_b^1(\mathbb{R}; \mathbb{C})$ therefore the other structure relations (2.2) are fulfilled if beyond (2.7) we also chose

$$(3.2) \qquad \theta_1^1(f) = 0.$$

From the special form of the above structure maps we can easily find a possible choice for the operators L_m^l. In fact (3.2) holds if we choose $L_1^1 = 0$. Now we want to extend maps (3.1), given on multiplication operators by bounded functions, to arbitrary bounded operators on $L^2(\mathbb{R})$. To this end we use the identity

$$[\partial, f] = \partial f - f\partial = f'$$

(where ∂ is denotes the derivation operator on $L^2(\mathbb{R})$) and the fact that, for any multiplication operator ρ one has

$$[\partial + \rho, f] = [\partial, f]$$

to write the right-hand side of (3.1) in the form

$$\theta_0^1(f) = \left[L_0^1, f\right]$$

with

$$(3.3) \qquad L_0^1 = \sigma f' + \rho$$

where ρ is a multiplication operator to be determined later. Now, from condition (2.4) with $l = m = 0$, we obtain

$$L_0^0 + (L_0^0)^* = 2\mathrm{Re}L_0^0 = -(L_0^1)^* L_0^1$$

therefore the operator L_0^0 can be chosen of the form

$$(3.4) \qquad L_0^0 = -\frac{1}{2}(L_0^1)^* L_0^1 - iH$$

where $H = \mathrm{Im}L_0^0$ is a symmetric operator to be determined. In order to determine ρ and H we use again the identities (2.6) with $l = m = 0$ and $X = f$. Replacing L_0^0 by the right-hand side of (3.4) in this identity we find:

$$-i[H, f] = A(f) + \frac{1}{2}\left((L_0^1)^* L_0^1 f - 2(L_0^1)^* f L_0^1 + f(L_0^1)^* L_0^1\right)$$
$$(3.5) \qquad\qquad = (b + \sigma\rho - \sigma\sigma')f'$$

Therefore, in order θ_0^0 to agree with the infinitesimal generator of the diffusion process, we can either consider the symmetric operator

$$-iHu(x) = \eta(x)u'(x) + \frac{1}{2}\eta'(x)u(x)$$

where $\eta : \mathbb{R} \to \mathbb{R}$ is the function multiplying f' in (3.5) or, when the function $1/\sigma$ is bounded, we can choose the function ρ in such a way that $b + \sigma\rho - \sigma\sigma'$ vanishes. In any case there are infinitely many possible choices.

This discussion is far from being rigorous because one must deal very careful with domain problems even in the simplest and cases (see, for example [6]).

4. Extension of birth-and-death-processes

The state space of the process is the set of natural numbers $\mathbb{N}$, however, for a reason that will be clear in a while, we consider the Hilbert space $l^2(\mathbb{Z})$. Denoting by $(\lambda^2(n))_{n\geq0}$ (resp. $(\mu^2(n))_{n\geq0}$) the birth (resp. death) intensities with $\mu(0) = 0$ and letting $\lambda(n) = 0$, $\mu(n) = 0$ for $n < 0$, the classical generator A of the flow is

$$(4.1) \qquad (A(f))(n) = \lambda^2(n)(f(n+1) - f(n)) + \mu^2(n)(f(n-1) - f(n))$$

for all f in the domain of the infinitesimal generator of the classical process. Our problem is to find structure maps θ_m^l, defined on a suitable subspace of all bounded operators on $l^2(\mathbb{Z})$, which satisfy (2.2) and $\theta_0^0(f) = A(f)$ for every $f \in l^\infty(\mathbb{Z})$. Now, since the generator (4.1) is the sum of a pure-birth generator and a pure-death generator, it is quite natural to look for a solution of (2.2) with $N = 2$. The structure relations (2.2) with $l = m = 0$ and $N = 2$ becomes

$$\theta_1^0(g)\theta_0^1(f) + \theta_2^0(g)\theta_0^2(f) = \theta_0^0(gf) - g\theta_0^0(f) - \theta_0^0(g)f$$
$$= \lambda^2(n)(g(n+1) - g(n))(f(n+1) - f(n))$$
$$+ \mu^2(n)(g(n-1) - g(n))(f(n-1) - f(n))$$

This is satisfied if we choose

$$(4.2) \qquad (\theta_0^1(f))(n) = (\theta_1^0(f))(n) = \lambda(n)(f(n+1) - f(n))$$
$$(4.3) \qquad (\theta_0^2(f))(n) = (\theta_2^0(f))(n) = \mu(n)(f(n-1) - f(n))$$

where $\lambda(n)$ and $\mu(n)$ are supposed to be non-negative. The structure relations (2.2) with $l = 1$ and $m = 0$ yield

$$\theta_1^1(g)\lambda(n)(f(n+1) - f(n)) + \theta_2^1(g)\mu(n)(f(n-1) - f(n))$$
$$= \lambda(n)(g(n+1) - g(n))(f(n+1) - f(n))$$

for all integer n. This condition is satisfied if we choose

$$(4.4) \qquad (\theta_1^1(g))(n) = g(n+1) - g(n), \qquad \theta_2^1(g) = 0.$$

A similar computation with $l = 2$ and $m = 0$ shows that the other structure relations are satisfied by choosing

$$(4.5) \qquad (\theta_2^2(g))(n) = g(n - 1) - g(n), \qquad \theta_1^2(g) = 0.$$

Let S and the right shift operator on $l^2(\mathbb{Z})$ and note that

$$(SfS^*)(n) = f(n + 1), \qquad (S^*fS)(n) = f(n - 1).$$

Then, defining

$$L_1^1 = S - I, \quad L_2^2 = S^* - I, \quad L_2^1 = L_1^2 = 0,$$

it is easy to check that the conditions (2.6), (4.4) and (4.5) for $l, m > 0$ are fulfilled. Note that, by taking $l^2(\mathbb{IN})$ instead of $l^2(\mathbb{Z})$, the condition (2.5) would not be satisfied. Considering the number operator N on $l^2(\mathbb{Z})$, defined by

$$D(N) = \left\{ x \in l^2(\mathbb{Z}) \ \Big| \ \sum_{n \in \mathbb{Z}} n^2 |x_n|^2 < +\infty \right\}, \qquad Nx = (nx_n)_{n \in \mathbb{Z}}$$

we note that

$$\theta_0^1(f) = \theta_1^0(f) = \lambda(N)(f(N + 1) - f(N))$$
$$\theta_0^2(f) = \theta_2^0(f) = \mu(N)(f(N - 1) - f(N))$$

Therefore the relations (2.6), (4.2) and (4.3) are satisfied with

$$L_0^1 = -\lambda(N), \quad L_1^0 = \lambda(N)S, \quad L_0^2 = -\mu(N), \quad L_2^0 = \mu(N)S^*.$$

Finally we can choose

$$L_0^0 = -\frac{1}{2}\left((L_0^1)^* L_0^1 + (L_0^2)^* L_0^2\right) = -\frac{1}{2}\left(\lambda^2(N) + \mu^2(N)\right)$$

and check that (2.6) with $l = m = 0$ turns out to be fulfilled and θ_0^0 acts on f as the given infinitesimal generator.

In the following section we will give an account of the results allowing to deal with q.s.d.e. of the form (2.3) with unbounded operators L_m^l and q.s.d.e. of the form (2.1) with unbounded structure maps θ_m^l. These results enable us to construct several quantum flows as (1.1) and, in particular, some quantum flows whose restrictions to a suitable abelian algebras correspond to the flow of a prescribed classical process.

5. Some results on quantum flows with unbounded structure maps

Let D be a dense linear submanifold of h and let $(L_m^l)_{l,m=0}^N$, be linear operators on h satisfying the following conditions:

(**A1**) L_0^0 is the infinitesimal generator of a strongly continuous contraction semigroup on h and D is a core for L_0^0

(**A2**) for all $l, m \in \{0, 1, \ldots, N\}$ the domain of the operators L_m^l contains D and we have

$$\langle v, L_m^l u \rangle + \langle L_l^m v, u \rangle + \sum_{k=1}^{N} \langle L_l^k v, L_m^k u \rangle = 0 \tag{5.1}$$

for all $v, u \in D$.

Note that (**A1**) and (**A2**) are always satisfied whenever the above operators are bounded and (2.4) holds. In the sequel we shall denote the operator L_0^0 by G. By a simple argument (see, for instance, [7] Prop. 2.5) we can show that the assumption (**A2**) implies that the operators L_0^l with $l \in \{1, \ldots, N\}$ admit an extension to the domain of G. A standard argument (see, for example, [21] p.143) shows also that the same assumption forces the operators L_m^l with $l, m \in \{1, \ldots, N\}$ to be bounded. For all integer $n \geq 1$ let

$$R(n; G) = (nI - G)^{-1} \tag{5.2}$$

and consider the operators with domain D

$$G_n = n^2 R(n; G^*) G R(n; G), \quad (L_0^l)_n = n L_0^l R(n; G), \quad (L_l^0)_n = n R(n; G^*) L_l^0.$$

The operator G_n has a bounded extension by the well-known properties of the Yosida approximations (5.2). Using the assumptions (**A1**) and (**A2**) one can show that $(L_0^l)_n$, $(L_l^0)_n$ have also. The properties of the Yosida approximations (5.2) yield

$$\lim_{n \to \infty} G_n u = G u, \qquad \lim_{n \to \infty} (L_0^l)_n u = L_0^l u, \tag{5.3}$$

for all $u \in D(G)$ and, for all $u \in D$,

$$\lim_{n \to \infty} (L_l^0)_n u = (L_l^0)_n u.$$

We can prove the following result (see, for example, [13] Prop. 3.4).

Theorem 5.1. Suppose that the assumptions (**A1**), (**A2**) hold. Then the sequence of isometric cocycles satisfying the q.s.d.e. in h

$$dV_n(t) = \sum_{l,m=0}^{N} V_n(t) (L_m^l)_n d\Lambda_l^m(t) \tag{5.4}$$

with the initial condition $V_n(0) = I$ converges weakly to a contractive cocycle V satisfying the q.s.d.e. in D

$$dV(t) = \sum_{l,m=0}^{N} V(t) L_m^l d\Lambda_l^m(t), \tag{5.5}$$

with the initial condition $V(0) = I$. The solution of the q.s.d.e. (5.5) is unique.

The following theorem gives necessary and sufficient conditions in order the cocycle V to be isometric. We refer to [13] Th. 5.3 for the proof (see also [14] Th. 3.2).

Theorem 5.2. Let V be the unique contractive cocycle satisfying the q.s.d.e. (5.5) in D. Let $\mathcal{Q}_\lambda$ ($\lambda > 0$) be the linear maps in $\mathcal{B}(h)$ defined by

$$(5.6) \qquad \langle v, \mathcal{Q}_\lambda(X)u \rangle = \sum_{l=1}^{N} \int_0^\infty e^{-\lambda s} \langle L_0^l P(s)v, X L_0^l P(s)u \rangle \, ds$$

for all $v, u \in D$ and all $X \in \mathcal{B}(h)$. For every $X \in \mathcal{B}(h)$ let $\mathcal{L}(X)$ be the bilinear form on h with domain $D \times D$ defined by

$$\langle v, \mathcal{L}(X)u \rangle = \langle Gv, Xu \rangle + \sum_{l=1}^{N} \langle L_0^l v, X L_0^l u \rangle + \langle v, XGu \rangle.$$

The following conditions are equivalent:

1) the contraction $V(t)$ is an isometry for all $t \geq 0$,

2) for all $t \geq 0$ the sequence $\{V_n(t)\}_{n \geq 1}$ of isometric cocycles satisfying the q.s.d.e. (5.4) converges strongly to $V(t)$,

3) there exists no positive nonzero element X of $\mathcal{B}(h)$ satisfying the equation $\lambda X = \mathcal{L}(X)$ for a fixed $\lambda > 0$,

4) there exists no positive nonzero element X of $\mathcal{B}(h)$ satisfying the equation $\lambda X = \mathcal{Q}_\lambda(X)$ for a fixed $\lambda > 0$,

5) the decreasing sequence $\{\mathcal{Q}_\lambda^n(I)\}_{n \geq 0}$ of bounded operators on h converges strongly to 0 for a fixed $\lambda > 0$.

We recall now Journé's notion of *dual cocycle* [16] which allows to apply the Theorem 5.2 also to check whether the operators $V^*(t)$ are isometries. For each $t \geq 0$ let ρ_t be the unitary time reversal on the interval $[0, t]$ defined on $L^2\left(\mathbb{R}_+ ; \mathbb{C}^N\right)$ by

$$(\rho_t f)(x) = \begin{cases} f(t-x) & \text{if } 0 \leq x \leq t, \\ f(x) & \text{if } x > t. \end{cases}$$

Let $\mathcal{R}_t$ be the second quantisation of ρ_t. The family $\widetilde{V}$ of operators $\left\{\widetilde{V}(t)\right\}_{t \geq 0}$ defined by

$$\widetilde{V}(t) = \mathcal{R}_t V^*(t) \mathcal{R}_t$$

is called the *dual cocycle* of V. The following proposition gives a version of Journé's time reversal principle.

Let us introduce two more assumptions:

(A3) the linear manifold D is contained in the domain of the operators $(L_l^m)^*$ for all $l, m \in \{0, 1, \ldots, N\}$ and we have

$$\langle v, L_m^l u \rangle + \langle L_l^m v, u \rangle + \sum_{k=1}^{N} \langle (L_k^l)^* v, (L_k^m)^* u \rangle = 0$$

for all $v, u \in D$,

(A4) for all $u \in D$ and all $n \geq 1$ the vector $R(n; G)u$ belongs to the domain of G^* and the sequence $(nG^* R(n; G)u)_{n \geq 1}$ converges.

Note that **(A3)** is equivalent to (2.5) when the operators involved are bounded.

Proposition 5.3. Suppose that the assumptions **(A1)**, ... ,**(A4)** hold and let V be the unique contractive cocycle satisfying the q.s.d.e. (5.5). Then the dual cocycle $\widetilde{V}$ of V satisfies the q.s.d.e. in D

$$(5.7) \qquad d\widetilde{V}(t) = \sum_{l,m=0}^{N} \widetilde{V}(t)(L_m^l)^* d\Lambda_m^l(t),$$

with the initial condition $\widetilde{V}(0) = I$.

We refer to [13] Prop. 5.4 for the proof. By Proposition 5.3, in order to prove that a cocycle V is a coisometry it suffices to apply the Theorem 5.2 to the dual cocycle $\widetilde{V}$ considering the map $\widetilde{Q}_\lambda$ defined by

$$\left\langle v, \widetilde{Q}_\lambda(X)u \right\rangle = \int_0^\infty e^{-\lambda s} \left\langle (L_0^l)^* P^*(s)v, X(L_0^l)^* P^*(s) \right\rangle ds$$

Next result allows to obtain the q.s.d.e. satisfied by the process $\{j_t(X)\}_{t \geq 0}$ where

$$j_t(X) = V(t)XV^*(t)$$

and X is an appropriate element of $\mathcal{B}(h)$. Let $j^{(n)}$ be the approximating quantum flow on $\mathcal{B}(h)$ defined by

$$j_t^{(n)}(X) = V_n(t)XV_n^*(t)$$

where the structure maps $\{^{(n)}\theta_m^l\}_{0 \leq l,m \leq N}$ are determined by the coefficients of the q.s.d.e. (5.4).

Theorem 5.4. Suppose that the assumptions **(A1)**,...,**(A4)** hold. Moreover suppose that the cocycle V is a coisometry. Let $\mathcal{A}$ be a sub*algebra of $\mathcal{B}(h)$ and let $\mathcal{D}$ be a sub*algebra of $\mathcal{A}$ with the following properties:

1) for all $X \in \mathcal{D}$ and all $u \in D$ the strong limits

$$s-\lim_{n \to \infty} {}^{(n)}\theta_m^l(X)u \qquad\qquad (0 \leq l, m \leq N)$$

and define bounded operators $\theta_m^l(X)$ satisfying, for all $u, v \in D$

(5.8)

$$\langle v, \theta_m^l(X)u \rangle = \langle (L_m^l)^*v, Xu \rangle + \langle v, X(L_m^l)^*u \rangle + \sum_{k=1}^{N} \langle (L_k^l)^*v, X(L_k^m)^*u \rangle$$

2) for all $X \in \mathcal{D}$ and all $l, m \in \{0, 1, \ldots, N\}$ we have

$$\sup_{n \geq 1} \left\| {}^{(n)}\theta_m^l(X) \right\|_\infty < +\infty.$$

Then, for every $X \in \mathcal{D}$, the quantum process $\{j_t(X)\}_{t \geq 0}$ satisfies the q.s.d.e. in D

$$dj_t(X) = \sum_{l,m=0}^{N} j_t\left(\theta_m^l(X)\right) d\Lambda_l^m(t), \qquad j_0(X) = X,$$

where the structure maps $\{\theta_m^l\}_{0 \leq l,m \leq N}$ mapping $\mathcal{D}$ to $\mathcal{A}$ are given by (5.8).

We refer to [14], Theorem 4.3 for the proof.

Theorem 5.5. Under the assumptions of Theorem 5.4, suppose that $\mathcal{A}$ is commutative and $\mathcal{D}$ is w^*-dense in $\mathcal{A}$. Then, for all $X, Y \in \mathcal{A}$ and all $s, t \geq 0$, the operators $j_t(X), j_s(Y)$ commute.

The proof can be easily adapted from [14], Proposition 4.5 with the first step of the induction argument as in [6] Theorem 2.4.

The above results can be applied to construct the quantum flow extending the following classes of classical processes:

1. pure birth and pure death processes [12]

2. countable state Markov chains [19]

3. multidimensional diffusion with 'regular' covariance and drift [14]

6. Conclusions

The boson Fock quantum stochastic calculus developed by Hudson and Parthasarathy [17] allows to construct several quantum flows of the form (1.1) which extend the flow of a classical Markov process to the algebra of all bounded operators on a Hilbert space which is often an L^2 space of the state space. The family of unitaries $V = \{V(t)\}_{t \geq 0}$ is a cocycle with respect to time shift satisfying an appropriate quantum stochastic differential equation. Since q.s.d.e. are driven by brownian motions and Poisson processes which are mutually non-commuting, in this way we can realise a classical Markov process as a non-commutative functional of brownian motions and Poisson processes. Moreover the quantum flow (1.1) enjoys the quantum Markov property in the sense of Accardi [1] and satisfies a q.s.d.e..

A class of quantum flows which are not of the form (1.1) has been studied by Evans and Hudson in [11]. A detailed account on this subject for classical Markov processes with bounded generator can be found in the recent books by Meyer [18] and Parthasarathy [21].

The generalisation to unbounded generators is necessary to construct a non-commutative extension of flows of most interesting classical Markov processes. In this note we described the general scheme developed in [14] for constructing such extension. The main steps are the following:

1. from the infinitesimal generator of a classical process find a possible choice of the structure maps of the q.s.d.e. satisfied by the flow j and the coefficients of the q.s.d.e. satisfied by the cocycle V,

2. show the existence of V by an approximation procedure,

3. show that V is unitary,

4. study a 'regularity' property of a sequence $\{j^{(n)}\}_{n \geq 1}$ of approximating quantum flows with bounded structure maps.

5. show that the homomorphisms $\{j_t\}_{t \geq 0}$ are also homomorphism of a smaller commutative algebra.

This scheme can be successfully followed to construct, for example, the quantum flows extending the classical flow of strongly elliptic multidimensional diffusion processes with 'regular" covariance and drift (see [14]), discrete state quantum Markov chains (see [18], [19], [20]), and some other simple classical stochastic processes.

The success of the scheme we outlined depends essentially on the form of infinitesimal generator. Some techniques remind constructions of classical Markov processes done in [9]. The most difficult step usually is to show that the cocycle V is unitary. The necessary and sufficient conditions given by Theorem 5.2 in practice can be checked only in the case of the infinitesimal generator of a classical countable state Markov chain. When the infinitesimal generator is a second order differential operator the only way to prove unitarity is by checking a sufficient condition obtained in [7]. Unfortunately this method seems very hard to apply when dealing with pseudo differential operators.

The most difficult problem, however, arises when it turns out that the cocycle V is not isometric (see, for example, [6] and [8]). This always occurs when the classical process has some 'singular state', for example a boundary point that can be reached in finite time, so that the infinitesimal generator depends on boundary conditions. In this case the condition (**A2**) might not be satisfied therefore we expect that the cocycle V satisfies a very singular q.s.d.e.. Moreover the choice of the operators L_m^l seems to involve the study of a quantum analogue of the classical boundary theory.

The attempts to study directly the equation (2.1) have been successful only when the structure maps are bounded [10] or satisfy and analyticity condition which is very hard to check in practice [15]. A remarkable result of Accardi and Mohari [4] states that, under a simple regularity condition, every quantum flow must satisfy an equation of the form (2.1). However there are no known results (besides those in the classical commutative cases and [15] in the noncommutative one) allowing to prove that given unbounded structure maps satisfying the structure relations the $(j_t)_{t\geq 0}$ solving (2.1) are homomorphisms.

The direct study of the equation for $(j_t)_{t\geq 0}$ however seems important because, even in the simplest cases, it might not be possible to find a suitable choice of the coefficients of equation (2.3). In fact consider, for example, the classical Markov process on $\mathbb{R}$ with infinitesimal generator

$$(Af)(x) = f(0) - f(x).$$

The arguments of sections 3 and 4 yields

$$-(\theta_0^1(f))(x) = -(\theta_1^0(f))(x) = (\theta_1^1(f))(x) = f(0) - f(x)$$

and it is not possible to find a unitary operator on $L^2(\mathbb{R})$ such that $(SfS^*)(x) = f(0)$.

Several results could be easily rewritten in the representation free quantum stochastic calculus of [3]; we restricted ourselves to boson Fock space for simplicity.

We think that the analytical methods described here can be used also to construct quantum diffusions on manifold studied by Applebaum in [5] and Sauvageot in [23] mainly from the algebraic point of view.

References

[1] L. Accardi: On the quantum Feynman–Kac formula. *Rend. Sem. Mat. Fis. Milano* **XLVIII** 135–179 (1978).

[2] L. Accardi, A. Frigerio, J.T. Lewis: Quantum stochastic processes. *R.I.M.S. Kyoto Univ.* **18**, 97–133, (1982).

[3] L. Accardi, F. Fagnola, J. Quaegebeur: A representation free quantum stochastic calculus. *J. Funct. Anal.*, **104** No.1, 149–197 (1992).

[4] L. Accardi, A. Mohari: On the Structure of Classical and Quantum Flows. To appear in: *J. Funct. Anal.*

[5] D. Applebaum: Towards a Quantum Theory of Classical Diffusions on Riemannian Manifolds. *Quantum Probability and Related Topics* **VI**, 93–109, (1991).

[6] B.V.R. Bhat, F. Fagnola, K.B. Sinha: On quantum extensions of semigroups of brownian motions on an half-line. *Russ. J. Math. Phys.* **4** n.1, (1996).

[7] A.M. Chebotarev, F. Fagnola: Sufficient conditions for conservativity of quantum dynamical semigroups. *J. Funct. Anal.* **118**, 131–153, (1993).

[8] A.M. Chebotarev, F. Fagnola: On quantum extensions of the Azéma martingale semigroup. *Sém. Prob.* **XXIX**, 1–16, Springer LNM 1613, (1995).

[9] S.N. Ethier, T.G. Kurtz: *Markov Processes. Characterization and Convergence.* John Wiley & Sons, New York, Brisbane, Singapore (1986).

[10] M. Evans: Existence of Quantum Diffusions. *Probab. Th. Rel. Fields* **81**, 473–483 (1989).

[11] M. Evans, R.L. Hudson: Multidimesional quantum diffusions. *Proc. Oberwolfach 1987 (Lecture Notes in Math.* **1303**) Springer-Verlag, Berlin, Heidelberg, New York 69–88, (1988).

[12] F. Fagnola: Pure birth and pure death processes as quantum flows in Fock space. *Sankhyā*, **53**, Serie **A** (1991).

[13] F. Fagnola: Characterisation of isometric and unitary weakly differentiable cocycles in Fock space. Preprint UTM **358** (1991), *Quantum Probability and Related Topics* **VIII**, 143–164, (1993).

[14] F. Fagnola: Diffusion processes in Fock space. Preprint UTM **379** (1992), *Quantum Probability and Related Topics* **IX** 189–214 (1994).

[15] F. Fagnola, Kalyan B. Sinha: Quantum flows with unbounded structure maps and finite degrees of freedom. *J. London Math. Soc.* (2) **48**, 537–551, (1993).

[16] J.-L. Journé,: Structure des cocycles markoviens sur l'espace de Fock. *Probab. Th. Rel. Fields* **75**, 291–316, (1987).

[17] R.L. Hudson, K.R. Parthasarathy: Quantum Ito's formula and stochastic evolutions. *Commun. Math. Phys.* **93**, 301–323, (1984).

[18] P.-A. Meyer: *Quantum Probability for Probabilists.* Springer Verlag, Berlin Heidelberg, New York (Lecture Notes in Math. **1538**) (1993).

[19] A. Mohari, Kalyan B. Sinha: Quantum flows with infinite degrees of freedom and countable state Markov processes. *Sankhyā*, **52**, Serie **A** 43–57 (1990).

[20] A. Mohari, K.B. Sinha: Stochastic dilation of minimal quantum dynamical semigroup. *Proc. Indian Acad. Sciences (Math.)* **102** (3), 159–170, (1993).

[21] K.R. Parthasarathy: *An Introduction to Quantum Stochastic Calculus.* Birkhäuser Verlag, Basel (1992).

[22] K.R. Parthasarathy, Kalyan B. Sinha: Markov chains as Evans-Hudson diffusions in Fock space. *Sém. Prob.* **XXIV**, 362–369, (1990).

[23] J.L. Sauvageot: Towards a Quantum Theory of Classical Diffusions on Riemanian Manifolds. *Quantum Probability and Related Topics* **VII**, 299–316, (1992).

QUANTUM STOCHASTIC ANALYSIS AFTER FOUR DECADES

R L HUDSON,* *University of Nottingham*

Abstract

This is a personal view of the development of quantum stochastic analysis from early days to the present time, with particular emphasis on quantum stochastic calculus.

1. Introduction

The notion of an explicitly quantum or noncommutative noise was introduced into the physical literature by Senitzky [Se]. Much earlier Wiener had discovered, in his homogeneous chaos [Wi], what was subsequently recognised as the decomposition into multiparticle subspaces of the physical concept of Fock space [Fo]. This connection was developed and abstracted by Segal [Sg]. The advantages of using the language of stochastic integrals was recognised by Nelson [Ne] in the context of quantum field theory, and urged by my student Cockroft and myself in [CH] when we initiated the systematic mathematical study of noncommutative Brownian motions arising from quantum formalism. In [H_1] it was shown that such quantum Brownian motions enjoyed a strong Markov property leading [PS_1] to a probabilistic extension of the continuous tensor product structure of Fock space. This paper included an abstract probabilistic characterization of quantum Brownian motion (called here quantum Wiener process), such that a 'cyclic' Brownian motion lives in a Fock space, and a general one in the tensor product of a Fock with an 'initial' Hilbert space. The need for a comprehensive theory of quantum stochastic integration was foreshadowed by the description of the second quantized time orthogonal dilation [HIP] as a 'stochastic product integral'. Two such theories were developed simultaneously [HSt, HP_1, HP_2, BSW]. Barnett, Streater and Wilde took as their starting point, instead of the Wiener–Fock formalism of 'Bose–Einstein' second quantization, which allows Brownian motion to be expressed as the sum of mutually noncommuting 'creation' and 'annihilation' processes which however individually commute with themselves at different times, the corresponding sum of creation and annihilation processes from the physically parallel concept of 'Fermi-Dirac' second quantization, in which commutation relations are replaced by anticommutation relations. In particular the sum, called the Ito–Clifford process, was now a noncommutative process which they regarded as a quantum analogue of Brownian motion. Hudson and Parthasarathy emphasised instead the rôle of the separate creation and annihilation processes as independent integrators in a quantum stochastic calculus which thus naturally extended noncommutatively, rather than paralleled, the classical Ito calculus of Brownian motion.

---

* Postal address: University of Nottingham, Nottingham NG7 2RD, UK.

Later [HP$_3$] we showed that, remarkably, the 'Fermionic' Ito–Clifford theory could be unified with the 'Bosonic' theory by a simple stochastic differential description.

In this paper I shall be mainly concerned to review some mathematical developments with which I have been involved of the Hudson–Parthasarathy calculus. The choice of topics is thus subjective and the inclusion or exclusion of any particular work may well prove to be uncorrelated with its eventual significance. However a brief mention of some of the physical applications of quantum stochastic calculus should also be made now. These include the dilation of irreversible evolutions in quantum statistical physics [HP$_1$, HSh], the quantum theory of continual measurement in [BL, BS, BHH], the dynamical Stark effect [MR], the quantum Hall effect [HR$_1$], and the interaction of matter and radiation [A]. More recently the recognition of scaling limits as functional central limits [AV] has led to the inclusion within limiting forms of dynamical evolution, such as the weak coupling limit, quantum stochastic descriptions of evolution of the whole system in perturbation theories [AL, AFL, AGL]. Remarkably these developments have led to the application in physics of the 'free' quantum noise [Sp] based on the notion of free (noncommutative) independence introduced by Voiculescu [VDN].

The remainder of this paper is organised as follows. In Section 2 we give a brief introduction to the notion of a Fock space emphasising its continuous tensor product structure and to classes of operators in Fock space of probabilistic importance, before describing so-called 'probabilistic interpretations', which are Hilbert space isomorphisms from Fock space to the L^2 spaces of probability spaces carrying a classical stochastic process such as Brownian motion and Poisson processes, under which certain operator valued processes transform to multiplication by the classical process and the natural filtration of Fock space determined by the continuous tensor product structure corresponds to the filtration of the classical process. In Section 3 we introduce, from this point of view, the Hudson–Parthasarathy quantum stochastic calculus as a noncommutative generalization and unification of the Ito calculus of Brownian motion and of the stochastic calculus of the Poisson process. In Section 4 stochastic evolutions and flows are constructed as solutions of quantum stochastic differential equations and a perturbation theory of the latter by the former is described. In Section 5 we show that the generators of evolutions and flows and the perturbation of the latter can be described by theories of cohomology for involutive algebras of various degrees of sophistication, including cyclic cohomology. Finally, in Section 6, we describe connections between multidimensional quantum stochastic calculus and representations of Lie algebras and Lie groups, and describe some recent work extending these to Lie superalgebras which generalises the Boson–Fermion unification scheme of [HP$_3$].

2. Fock space and its probabilistic interpretations

Let $\mathfrak{h}$ be a Hilbert space; usually we shall take

$$\mathfrak{h} = L^2(\mathbb{R}_+; \mathcal{K})$$

All Hilbert spaces are complex, with inner product $\langle\ ,\rangle$ linear on the right.

to be the Hilbert space of square-integrable (with respect to Lebesgue measure) vector-valued functions on the non-negative half-line $\mathbb{R}_+$, taking values in a finite-dimensional Hilbert space $\mathcal{K}$, whose dimension may for example eventually be identified with that of a multi-dimensional Brownian motion. $L^2(\mathbb{R}_+; \mathcal{K})$ is canonically identified with the Hilbert space tensor product $L^2(\mathbb{R}_+) \otimes \mathcal{K}$ by the extension of the identifications

$$f \otimes u = fu \qquad (f \in L^2(\mathbb{R}_+), \ u \in \mathcal{K}).$$

The *Fock space* over $\mathfrak{h}$ is another Hilbert space $\mathcal{F}(\mathfrak{h})$, conveniently defined axiomatically as follows; $\mathcal{F}(\mathfrak{h})$ is generated as a Hilbert space by a family of so-called *exponential vectors* $e(f)$, $f \in \mathfrak{h}$ satisfying

$$\langle e(f), e(g) \rangle_{\mathcal{F}(\mathfrak{h})} = \exp \langle f, g \rangle_{\mathfrak{h}}.$$

$\mathcal{F}(\mathfrak{h})$ is unique in so far as given two candidate Fock spaces, there is a unique Hilbert space isomorphism between them which exchanges exponential vectors labelled by the same vector in $\mathfrak{h}$; its existence follows from the fact that the map

$$\mathfrak{h} \times \mathfrak{h} \ni (f, g) \mapsto \exp \langle f, g \rangle$$

is a non-negative definite kernel over $\mathfrak{h}$.

For each direct sum decomposition $\mathfrak{h} = \mathfrak{h}_1 \oplus \mathfrak{h}_2$ we can identify $\mathcal{F}(\mathfrak{h})$ with the Hilbert space tensor product $\mathcal{F}(\mathfrak{h}_1) \otimes \mathcal{F}(\mathfrak{h}_2)$ by extending the identifications

$$e(f) = e(f_1) \otimes e(f_2) \qquad (f = (f_1, f_2) \in \mathfrak{h} = \mathfrak{h}_1 \oplus \mathfrak{h}_2).$$

In particular when $\mathfrak{h} = L^2(\mathbb{R}_+, \mathcal{K})$ the *continuous tensor product structure* of $\mathcal{H} =: \mathcal{F}(\mathfrak{h})$ corresponds to the continuum of direct sum decompositions

$$(2.1) \qquad \mathfrak{h} = \mathfrak{h}_t \oplus \mathfrak{h}^t, \qquad \mathfrak{h}_t = L^2([0, t]; \mathcal{K}), \qquad \mathfrak{h}^t = L^2((t, \infty); \mathcal{K}).$$

We write

$$(2.2) \qquad \mathcal{H} = \mathcal{H}_t \otimes \mathcal{H}^t, \qquad e(f) = e(f_t) \otimes e(f^t), \qquad f = (f_t, f^t).$$

We also introduce the *exponential domain* $\mathcal{E}$, that is the linear span, dense in $\mathcal{H}$, of the exponential vectors $f \in \mathfrak{h}$ (sometimes, according to context which will be made clear, the labels f of the exponential vectors will be restricted to belong to a dense subspace of $\mathfrak{h}$ chosen so that whenever it contains f it also contains f_t and f^t, for example the subspace $L^\infty_{\mathrm{loc}}(\mathbb{R}_+; \mathcal{K})$ of locally bounded elements of $\mathfrak{h}$) and write

$$(2.3) \qquad \mathcal{E} = \mathcal{E}_t \widehat{\otimes} \mathcal{E}^t$$

where $\widehat{\otimes}$ denotes the algebraic tensor product. In this paper all operators in the Fock space $\mathcal{H}$ will be defined on the exponential domain $\mathcal{E}$ (possibly restricted as above) and will have adjoints whose domains include $\mathcal{E}$; for such an operator T, $T^\dagger$ denotes the restriction to $\mathcal{E}$ of the full adjoint T^*. For example the *annihilation* and *creation* operators $a(f)$ and $a^\dagger(f)$ $(= a(f)^\dagger)$ corresponding to $f \in \mathfrak{h}$ are defined by

$$a(f)e(g) = \langle f, g \rangle \, e(g), a^\dagger(f)e(g) = \frac{d}{d\epsilon} e(g + \epsilon f)\Big|_{\epsilon=0}.$$

The *second quantization* $\Gamma(S)$ of a bounded operator $S \in B(\mathfrak{h})$ is defined by

$$\Gamma(S)e(f) = e(Sf)$$

and the *differential second quantization* $\gamma(S)$ by

$$\gamma(S)e(f) = \frac{d}{d\epsilon} e(\exp(\epsilon S)f).$$

We also introduce the operators $e^{a(f)}, e^{a^\dagger(f)}$, $f \in h$, where actions are

$$e^{a(f)}e(g) = e^{\langle f, g \rangle}e(g), \qquad e^{a^\dagger(f)}e(g) = e(g + f).$$

The operators $\Gamma(S)$, $e^{a^\dagger(f)}$, $e^{a(g)}$, $S \in B(\mathfrak{h})$, $f, g \in \mathfrak{h}$ leave $\mathcal{E}$ invariant and can be composed according to the rule

$$e^{a^\dagger(f_1)}\Gamma(S_1)e^{a(g_1)}e^{a^\dagger(f_2)}\Gamma(S_2)e^{a(g_2)}$$

(2.4)
$$= e^{\langle g_1, f_2 \rangle}e^{a^\dagger(f_1 + S_1 f_2)}\Gamma(S_1 S_2)e^{a(S_2^* g_1 + g_2)}.$$

When $\mathfrak{h} = L^2(\mathbb{R}_+, \mathcal{K})$ we define a *process* to be a family of operators

$$E = (E(t), t \in \mathbb{R}_+)$$

and say that it is *adapted* if each $E(t)$ takes the form $E(t) = E_t \widehat{\otimes} 1^t$ in the decomposition (2.3). The *adjoint process* $E^\dagger = \left(E^\dagger(t), t \in \mathbb{R}_+\right)$ of an adapted process is evidently adapted. Given vectors u and $v \in \mathcal{K}$ and a bounded operator S on $\mathcal{K}$, the *creation*, *annihilation* and *number* processes of strengths u, v and s respectively are the adapted processes $A_u^\dagger$, A_v and Λ_S defined by

$$A_u^\dagger = a\left(\chi_{[0,t]}u\right), \qquad A_v^\dagger = a^\dagger\left(\chi_{[0,t]}v\right), \qquad \Lambda_s(t) = \gamma\left(M_{\chi_{[0,t]}} \otimes S\right)$$

where $\chi_{[0,t]}$ denotes the indicator function of the set $[0, t]$ and M_f denotes multiplication by the function f as an operator on $L^2(\mathbb{R}_+)$. It is also convenient to introduce the time process $T(t) = t1$ where 1 is the identity operator. From (2.4) we have

$$e^{A_{u_1}^\dagger(t)}e^{\Lambda_{s_1}(t)}e^{A_{u_1}(t)}e^{A_{u_2}^\dagger(t)}e^{\Lambda_{s_2}(t)}e^{A_{v_2}(t)}$$

(2.5)
$$= e^{\langle v_1, u_2 \rangle t}e^{A_{u_1 + e^{S_1}u_2}^\dagger(t)}e^{\Lambda_{S_1}(t)}e^{\Lambda_{S_2}(t)}e^{A_{e_2^{S^*}v_1 + v_2}(t)}.$$

Now let $\mathcal{K}$ be one-dimensional, and note that the algebra $\mathbb{C}$ is canonically identified with $B(\mathcal{K})$, but the Hilbert space $\mathbb{C}$ is identified with the Hilbert space $\mathcal{K}$ only after choosing a unit basis vector $u \in \mathcal{K}$, $\|u\| = 1$ and identifying $z \in \mathbb{C}$ with zu. Thus the number process $\Lambda = \Lambda_1$ is canonical, but the creation and annihilation processes

$A^\dagger = A_u^\dagger$, $A = A_u$ are basis dependent; choosing a different basis vector corresponds to the 'gauge transformation'

$$(2.6) \qquad A^\dagger \mapsto e^{i\theta} A^\dagger, \qquad A \mapsto e^{-i\theta} A.$$

Having made such a choice of basis, we may verify from (2.5) that, for each $t \in \mathbb{R}_+$, the family of operators

$$(2.7) \qquad U_x(t) = e^{\frac{1}{2}x^2 t} e^{-x A^\dagger(t)} e^{x A(t)} = e^{-\frac{1}{2}x^2 t} e^{x A(t)} e^{-x A^\dagger(t)}$$

satisfy

$$U_x(t)U_y(t) = U_{x+y}(t), \qquad U_{-x}(t) = U_x(t)^\dagger, \qquad U_0(t) = 1$$

and thus extends uniquely to a continuous unitary representation $\mathbb{R} \ni x \mapsto U_x(t)$ of the group $(\mathbb{R}, +)$. The infinitesimal generator $\frac{d}{dx}U_x(t)\big|_{x=0}$ acts on $\mathcal{E}$ as $A(t) - A^\dagger(t)$; in this sense we may write

$$U_x(t) = e^{x\left(A(t) - A^\dagger(t)\right)}, \qquad x \in \mathbb{R}.$$

Furthermore it follows from (2.5) that for all $x, y \in \mathbb{R}$ and $s, t \in \mathbb{R}_+$

$$U_x(t)U_y(s) = U_y(s)U_x(t);$$

in this sense the adapted process

$$P(t) = -i\left(A(t) - A^\dagger(t)\right), \qquad t \in \mathbb{R}_+$$

consists of mutually commuting essentially self-adjoint operators. Using (2.5), we may calculate the joint characteristic functions of this process in the 'vacuum state' $e(0)$ as

$$(2.8) \qquad f_{t_1,\ldots,t_n}(x_1,\ldots,x_n) = \left\langle e(0), \prod_{j=1}^{n} e^{i x_j P(t_j)} e(0) \right\rangle$$

$$= \left\langle e(0), \prod_{j=1}^{n} U_{x_j}(t_j) e(0) \right\rangle$$

$$= \exp\left(\frac{1}{2} \sum_{j,k=1}^{n} t_j \wedge t_k x_j x_k\right),$$

where $n \in \mathbb{N}$, $t_1,\ldots,t_n \in \mathbb{R}_+$, $x_1,\ldots,x_n \in \mathbb{R}$ We may recognise these as the joint characteristic functions of Brownian motion and declare that the *momentum* process P is a Brownian motion in the vacuum state. More precisely, using the fact that the vectors $U_{x_1}(t_1) \cdots U_{x_n}(t_n)e(0)$ form a generating family in $\mathcal{H}$, it can be shown that there is a unique Hilbert space isomorphism D_P from $\mathcal{H}$ to the L^2-Hilbert space $L^2(\mathbb{W})$ of Wiener measure such that

---

Here $t \wedge s$ is the minimum of $\{t, s\}$.

(a) $D_P e(0) \equiv 1$,

(b) the self-adjoint closure of $D_P P(t) D_P^{-1}$ is the operator $M_{X(t)}$ of multiplication by the canonical Brownian motion $X(t)$.

We thus obtain a *probabilistic interpretation* of Fock space by diagonalising the *momentum process* P as a Brownian motion.

To obtain another probabilistic interpretation we replace (2.7) by

$$U_x(t) = e^{-(1-e^{ix})|z|^2 T(t)} e^{iz(1-e^{ix})A^\dagger(t)} e^{ix\Lambda(t)} e^{-i\bar{z}(-e^{ix}+1)A(t)}.$$

Then it may be verified that, once again, for each fixed t this defines a continuous unitary representation of $(\mathbb{R}, +)$, and that these representations commute for different values of t. In this case the infinitesimal generator acts on exponential vectors as

$$\Pi_z(t) = -i\frac{d}{dx}U_x(t)\Big|_{x=0} = \Lambda(t) - i\left(zA^\dagger(t) - \bar{z}A(t)\right) + |z|^2 T(t)$$

and the joint vacuum state characteristic functions (2.8) are now

$$f_{t_1,\ldots,t_n}(x_1,\ldots,x_n) = \prod_{j=1}^{n} \exp\left(e^{i\sum_{k=j}^{n} x_k} - 1\right)(t_j - t_{j-1}),$$

$$n \in \mathbb{N}, \; t_1,\ldots,t_n \in \mathbb{R}_+, \; x_1,\ldots,x_n \in \mathbb{R}.$$

Recognising these as the joint characteristic functions of the Poisson process of intensity measure $|z|^2\, dt$ and again verifying that the set of vectors $U_x(t_1)\cdots U_{x_n}(t_n)e(0)$, $n \in \mathbb{N}$, $t_1,\ldots,t_n \in \mathbb{R}_+$, $x_1,\ldots,x_n \in \mathbb{R}$ generates $\mathcal{H}$ we obtain a probabilistic interpretation of Fock space as the L^2-space of the Poisson process. Notice that, for a given intensity $|z|^2\, dt$, the Fock space realisations

$$\Pi_z = \Lambda - i(zA^\dagger - \bar{z}A) + |z|^2 T$$

do not commute for different choices of $\arg z$ and correspond to different probabilistic interpretations of Fock space. Similarly, if we make a different choice of basis vector $u \in \mathcal{K}$ leading to the gauge transformation (2.6), we obtain a different probabilistic interpretation of Fock space based on the Brownian motion $ie^{i\theta}A^\dagger - ie^{-i\theta}A$. The choice $\theta = \frac{\pi}{2}$ leads to the *position process* $Q = A^\dagger + A$ expressing Brownian motion as the sum of creation and annihilation processes.

As emphasized by Attal and Meyer [AM], in all such probabilistic interpretations the natural filtration of Fock space is transformed into the filtration of the classical process, that is, for each $t \in \mathbb{R}_+$, the diagonalising transformation D maps the subspace $\mathcal{H}_t \otimes e^t(0) \subseteq \mathcal{H}$ onto the subspace of square-integrable complex functions on the probability space of the interpretation which are measurable with respect to the σ-field generated by the classical process of the interpretation up to time t.

The Brownian and Poissonian interpretations of the Fock space $\mathcal{F}\left(L^2\left(\mathbb{R}_+, \mathcal{K}\right)\right)$ with $\dim \mathcal{K} = 1$ are special cases of the old result of Parthasarathy and Schmidt [PSc] that an arbitrary classical process with independent increments can be realised in a Fock space. In general the space $\mathcal{K}$ needed to achieve this is infinite dimensional; it is not difficult to see that, with $\dim \mathcal{K} = 1$, essentially only Brownian motion and Poisson processes can be realised.

3. Quantum stochastic calculus

It is convenient to denote by K_Σ the linear combination

$$(3.1) \qquad K_\Sigma = xT + A_u^\dagger + A_v + \Lambda_T, \quad x \in \mathbb{C},\ u, v \in \mathcal{K},\ T \in B(\mathcal{K})$$

of time, creation and annihilation processes, where Σ is the operator on the Hilbert space $\mathbb{C} \oplus \mathcal{K}$ given by

$$\Sigma(y, w) = (xy + \langle v, w \rangle, yu + Tw), \quad y \in \mathbb{C},\ w \in \mathcal{K}.$$

Evidently $K_\Sigma^\dagger = K_{\Sigma^*}$. For $f \in \mathfrak{h} = L^2(\mathbb{R}_+, \mathcal{K})$, we write

$$\tilde{f}(t) = (1, f(t)) \in \mathbb{C} \oplus \mathcal{K},\ t \in \mathbb{R}_+.$$

Then it is easily verified that, for $f, g \in \mathfrak{h}$, $t \in \mathbb{R}_+$

$$(3.2) \qquad \langle e(f), K_\Sigma(t)e(y) \rangle = \int_{0<s<t} \left\langle \tilde{f}(s), \Sigma\tilde{g}(s) \right\rangle ds\, \langle e(f), e(g) \rangle$$

and that

$$\langle K_{\Sigma_1}(t)^\dagger e(f), K_{\Sigma_2}(t)e(g) \rangle$$

$$= \left\{ \iint_{0<s_1<s_2<t} \left\langle \tilde{f} \otimes \tilde{f}(s_1, s_2), (K_{\Sigma_1} \otimes K_{\Sigma_2} + K_{\Sigma_2} \otimes K_{\Sigma_1})\, \tilde{g} \otimes \tilde{g}(s_1, s_2) \right\rangle ds_1 ds_2 \right.$$

$$(3.3)$$

$$\left. + \int_{0<s<t} \left\langle \tilde{f}(s), K_{\Sigma_1} \triangle K_{\Sigma_2}\tilde{g}(s) \right\rangle ds \right\} \langle e(f), e(g) \rangle$$

where $\tilde{f} \otimes \tilde{f}$ denotes the $(\mathbb{C} \oplus \mathcal{K}) \otimes (\mathbb{C} \oplus \mathcal{K})$-valued function

$$\tilde{f} \otimes \tilde{f}(s_1, s_2) = \tilde{f}(s_1) \otimes \tilde{f}(s_2)$$

and $\triangle$ is the projection operator $0 \oplus 1$ in $\mathbb{C} \oplus \mathcal{K}$

The definition of stochastic integral processes

$$I(t) = \int_0^t E(s)dK_\Sigma(s)$$

follows the traditional route, beginning with *elementary* adapted processes E of the form

$$E(t) = \begin{cases} 0 & 0 \le t < t, \\ E(t_1) & t_1 \le t < t_2, \\ 0 & t_2 \le t, \end{cases}$$

for which

$$I(t) = E(t_1)\left(K_\Sigma(t_2 \wedge t) - K_\Sigma(t_1 \wedge t)\right).$$

Here the product of unbounded operators of the form $E(t_1)\,(K_\Sigma(s) - K_\Sigma(t_1))$ is meaningful as an algebraic tensor product operator of the form $E_{t_1}\widehat{\otimes K^{t_1}}(s)$. Evidently I is an adapted process and

$$(3.4) \qquad I^\dagger(t) = \int_0^t E^\dagger(s)dK_\Sigma^\dagger(s).$$

From (3.2) we deduce the *first fundamental formula* (for an elementary integrand)

$$(3.5) \qquad \left\langle e(f), \int_0^t E(s)dK_\Sigma(s)e(g) \right\rangle = \int_0^t \left\langle \tilde{f}(s), \Sigma\tilde{g}(s) \right\rangle \left\langle e(f), E(s)g(s) \right\rangle ds$$

and from (3.3) the *second fundamental formula*, for

$$I_j(t) = \int_0^t E_j(s)dK_{\Sigma_j}(s), \quad j = 1, 2,$$

$$\langle I_1(t)e(f), I_2(t)e(g) \rangle$$
$$= \int_0^t \Big\{ \left\langle \tilde{f}(s), \Sigma_2\tilde{g}(s) \right\rangle \langle I_1(s)e(f), E_2(s)g(s) \rangle$$
$$+ \left\langle \Sigma_1\tilde{f}(s), \tilde{g}(s) \right\rangle \langle E_1(s)e(f), I_2(s)e(g) \rangle$$
$$(3.6) \qquad + \left\langle \Sigma_1\tilde{f}(s), \Delta\Sigma_2\tilde{g}(s) \right\rangle \langle E_1(s)e(f), E_2(s)e(g) \rangle \Big\} ds.$$

Stochastic integrals of simple processes, that is, of finite sums of elementary processes are defined by additivity:

$$\int_0^t \sum_j E_j(s)dK_\Sigma(s) = \sum_j \int_0^t E_j(s)dK_\Sigma(s).$$

Evidently (3.4), (3.5) and (3.6) hold for the extended integral. To extend the integral to a wider class of integrands, we first put $f = g$ and $I_1 = I_2 = I = \int EdK_\Sigma$ in (3.6) to obtain

$$\left\| \int_0^t E(s)dK_\Sigma(s)e(f) \right\|^2$$
$$= \int_0^t \Big\{ 2\,\mathrm{Re}\left\langle \Sigma\tilde{f}(s), \tilde{f}(s) \right\rangle \langle E(s)e(f), I(s)e(f) \rangle$$
$$+ \|\Delta\Sigma f(s)\|^2 \|E(s)e(f)\|^2 \Big\} ds.$$

Writing this equation in differential form, setting

$$\phi_1 = \Sigma\tilde{f}(t) \otimes E(t)e(f), \qquad \phi_2 = \tilde{f}(t) \otimes I(t)e(f)$$

and using the inequality

$$2\,\mathrm{Re}\langle \phi_1, \phi_2 \rangle \leq \|\phi_1\|^2 + \|\phi_2\|^2$$

we obtain

$$\frac{d}{dt}\left\|\int_0^t E(s)dK_\Sigma(s)e(f)\right\|^2$$
$$\leq \left(\left\|\Sigma\tilde{f}(t)\right\|^2 + \left\|\Delta\Sigma\tilde{f}(t)\right\|^2\right)\|E(t)e(f)\|^2 + \left\|\tilde{f}(t)\right\|^2\|I(t)e(f)\|^2.$$

Multiplying by the integrating factor $\exp\left(-\int_0^t\left\|\tilde{f}(s)\right\|^2 ds\right)$ and integrating the resulting inequality, we get

$$\left\|\int_0^t E(s)dK_\Sigma(s)e(f)\right\|^2$$
$$(3.7)\qquad \leq \int_0^t \exp\left(\int_s^t \|f(s)\|^2 ds\right)\left(\left\|\Sigma\tilde{f}(s)\right\|^2 + \left\|\Delta\Sigma\tilde{f}(s)\right\|^2\right)\|E(s)e(f)\|^2 ds.$$

Let us now restrict the exponential domain $\mathcal{E}$ to be the span of the vectors $e(f)$ where $f \in \mathfrak{h}$ is locally bounded (in fact this restriction is needed only if number-integrals are present, that is if $T \neq 0$ in (3.1)). Then from (3.7) we obtain the *fundamental estimate* in the following form. For arbitrary $T \in \mathbb{R}_+$ there exists a positive constant $c(f,T)$ such that, for all $t \in [0,T]$,

$$(3.8)\qquad \left\|\int_0^t E(s)dK_\Sigma(s)e(f)\right\|^2 \leq c(f,T)\int_0^t \|E(s)e(f)\|^2 ds.$$

Extension of stochastic integration now proceeds along standard lines; we say that an adapted process E is *integrable* if there exists a sequence $(E_n,\ n \in \mathbb{N})$ of simple adapted processes such that, for arbitrary $t \in \mathbb{R}_+$ and $f \in \mathfrak{h}$,

$$\lim_n \int_0^t \|(E(s) - E_n(s))e(f)\|^2 ds = 0$$

together with the corresponding relation for the adjoint processes. Using the fundamental estimate it can then be shown that the sequence of vectors $\int_0^t E_n(s)dK_\Sigma(s)e(f)$ is Cauchy, hence convergent for each $t \in \mathbb{R}$, moreover the convergence is uniform for t belonging to finite intervals and the limit is independent of the choice of approximating sequence (E_n) and defines the action of $\int_0^t E(s)dK_\Sigma(s)$ on the (restricted) exponential domain. Because the convergence is uniform the two fundamental formulas and the fundamental estimate continue to hold for the extended integrable as does the relation (3.4). Adapted processes E which are continuous, in the sense that each $t \mapsto E(t)e(f)$ is continuous from $\mathbb{R}_+$ to $\mathfrak{h}$, can be shown to be integrable. The fundamental estimate shows that stochastic integrals are continuous; hence iterated integrals are well defined.

Using (3.5), the second fundamental formula can be used to express the product of two stochastic integrals in differential form as

$$(3.9)\qquad d(I_1 I_2) = dI_1 I_2 + I_1 dI_2 + dI_1 dI_2.$$

Here, to overcome problems of multiplying operators which do not in general leave the exponential domain invariant, products of operators are in the weak sense defined by

$$\langle e(f), I_1(t)I_2(t)e(g)\rangle =: \left\langle I_1^\dagger(t)e(f), I_2(t)e(g)\right\rangle.$$

The right hand side of (3.9) is reduced to the form EdK_Σ by the rule that the differentials dK_{Σ_j} commute with adapted processes:

$$dK_{\Sigma_j}E = EdK_{\Sigma_j}$$

and are multiplied by the rule

$$(3.10) \qquad dK_{\Sigma_1}dK_{\Sigma_2} = dK_{\Sigma_1\triangle\Sigma_2}.$$

In terms of the time, creation, annihilation and number processes (3.10) becomes the *quantum Ito table*

$$(3.11)$$

	$d\Lambda_T$	$dA_u^\dagger$	dA_v	dT
$d\Lambda_S$	$d\Lambda_{ST}$	$dA_{Su}^\dagger$	0	0
$dA_u^\dagger$	0	0	0	0
dA_v	dA_{T^*v}	$\langle v,u\rangle\, dT$	0	0
dT	0	0	0	0

in particular in the case $\dim\mathcal{K} = 1$ we obtain

$$(3.12)$$

	$d\Lambda$	$dA^\dagger$	dA	dT
$d\Lambda$	$d\Lambda$	$dA^\dagger$	0	0
$dA^\dagger$	0	0	0	0
dA	dA	dT	0	0
dT	0	0	0	0

(3.12) contains the familiar Ito tables for the Brownian motion P and Poisson process Π_z

$$(3.13)$$

	dP	dT
dP	dT	0
dT	0	0

	$d\Pi_z$
$d\Pi_z$	$d\Pi_z$

In applications of quantum stochastic calculus in physics and elsewhere a supplementary Hilbert space $\mathcal{H}_0$, called the *initial space*, is adjoined to the Fock space and the action takes place in the Hilbert space tensor product $\mathcal{H}_0\otimes\mathcal{F}(\mathfrak{h})$. A process E of operators on the domain $\mathcal{H}_0\widehat{\otimes}\mathcal{E}$ is adapted if each $E(t)$ is of form $E_t\otimes 1$ on $\mathcal{H}_0\widehat{\otimes}\mathcal{E} = \left(\mathcal{H}_0\widehat{\otimes}\mathcal{E}_t\right)\widehat{\otimes}\mathcal{E}^t$. The integrator processes K_Σ are extended to $\mathcal{H}_0\widehat{\otimes}\mathcal{E}$ by algebraic ampliation, identifying such $K_\Sigma(t)$ with $1\widehat{\otimes}K_\Sigma(t)$. The two fundamental formulas and fundamental estimate become:

$$\left\langle u \otimes e(f), \int_0^t E(s) dK_\Sigma(s) v \otimes e(g) \right\rangle$$

(3.14)
$$= \int_0^t \left\langle \tilde{f}(s), \Sigma \tilde{g}(s) \right\rangle \langle u \otimes e(f), E(s) v \otimes g(s) \rangle \, ds$$

$$\langle I_1(t) u \otimes e(f), I_2(t) v \otimes e(g) \rangle$$

$$= \int_0^t \left\{ \left\langle \tilde{f}(s), \Sigma_2 \tilde{g}(s) \right\rangle \langle I_1(s) u \otimes e(f), E_2(s) v \otimes e(g) \rangle \right.$$

$$+ \left\langle \Sigma_1 \tilde{f}(s), \tilde{g}(s) \right\rangle \langle E_1(s) u \otimes e(f), I_2(s) v \otimes e(g) \rangle$$

(3.15)
$$\left. + \left\langle \Sigma_1 \tilde{f}(s), \Delta \Sigma_2 \tilde{g}(s) \right\rangle \langle E_1(s) u \otimes e(f), E_2(s) v \otimes e(g) \rangle \right\} ds$$

(3.16)
$$\left\| \int_0^t E(s) dK_\Sigma(s) u \otimes e(f) \right\|^2 \le c(f,T) \int_0^t \| E(s) u \otimes e(f) \|^2 \, ds.$$

4. Quantum stochastic evolutions and flows

Let $(u_1, \ldots, u_N)$ be an orthonormal basis of $\mathcal{K}$ and define component integrator processes Λ_β^α, $\alpha, \beta = 0, 1, \ldots, N$ by

(4.1)
$$\Lambda_0^0 = T, \qquad \Lambda_j^0 = A_{u_j}^\dagger, \qquad \Lambda_0^k = A_{u_k}, \qquad \Lambda_j^k = \Lambda_{|u_j\rangle\langle u_k|}, \qquad j, k = 1, \ldots, N.$$

We consider the stochastic differential equation

(4.2)
$$dU = U l_\beta^\alpha d\Lambda_\alpha^\beta, \qquad U(0) = 1.$$

Here the l_β^α are bounded operators on the initial space $\mathcal{H}_0$, identified with their ampliations $l_\beta^\alpha \otimes 1$ to $\mathcal{H}_0 \otimes \mathcal{F}(\mathfrak{h})$, and we use the convention that repeated Greek suffices (one lower and one upper) are summed over from 0 to N, whereas repeated Roman suffices are summed from 1 to N.

Using the fundamental estimate we can use the usual iterative method to construct a solution, defining U_n inductively by $U_0 \equiv 1$,

$$U_n(t) = 1 + \int_0^t U_{n-1}(s) l_\beta^\alpha d\Lambda_\alpha^\beta(s)$$

and showing that $U_n(t)$ converges on the domain $\mathcal{H}_0 \widehat{\otimes} \mathcal{E}$ to a solution of (4.2).

Necessary conditions for (4.2) to have a unitary valued solution are found from the second fundamental formula; differentiating with respect to t the relations

$$\langle U(t) u \otimes e(f), U(t) v \otimes e(g) \rangle = \langle U(t)^\dagger u \otimes e(f), U(t)^\dagger v \otimes e(g) \rangle$$
$$= \langle u \otimes e(f), v \otimes e(g) \rangle$$

we find, using the assumed unitarity of $U(t)$ and the independence of the basic stochastic differentials [Li] that

$$(4.3) \qquad l_\beta^\beta + \left(l_\alpha^\beta\right)^* + l_j^\alpha \left(l_j^\beta\right)^* = \left(l_\alpha^\beta\right)^* + l_\beta^\alpha + \left(l_\alpha^j\right)^* l_\beta^j = 0.$$

Moreover the first of these conditions is also sufficient for U to be co-isometry valued, that is $UU^\dagger \equiv 1$, and in particular bounded. If this is satisfied a variant of the usual uniqueness argument based again on the fundamental estimate shows that (4.2) has a unique solution. Furthermore it can then be shown [HP$_2$], by appealing to the uniqueness theorem for an ordinary differential equation for a $B(\mathcal{H}_0 \otimes \mathcal{F}(\mathfrak{h}))$-valued function with bounded coefficients, that the second of the two conditions (4.3) is sufficient for unitarity.

The vacuum conditional expectation $T(t)$ of such a unitary process is defined by

$$\langle u, T(t)v \rangle = \langle u \otimes e(0), U(t)v \otimes e(0) \rangle$$

and is a uniformly continuous semigroup of contraction operators on $\mathcal{H}_0$ of which l_0^0 is the infinitesimal generator.

When $\dim \mathcal{K} = 1$, the unitarity conditions (4.2) become

$$(4.4) \qquad \left(l_0^0, l_0^1 l_1^0, l_1^1\right) = \left(ih - \tfrac{1}{2}l^*l, l, -t^*w, w - 1\right)$$

where $h, l, w \in B(\mathcal{H}_0)$ with h self-adjoint and w unitary. In this case, by using the quantum martingale representation theorem of [PSi$_2$] it can be shown [HL] that such unitary processes are characterized by the following three properties.

(a) $t \mapsto U(t)$ is a unitary-valued adapted process and, for $t < s$, $U(t)^*U(s)$ is the ampliation of an operator on $\mathcal{H}_0 \otimes \mathcal{F}(L^2((s,t), \mathcal{K}))$ to $\mathcal{H}_0 \otimes L^2(\mathbb{R}_+); \mathcal{K})$.

(b) $t \mapsto T(t)$ defined by (4.4) is uniformly continuous.

(c) For all $s, t \in \mathbb{R}_+$

$$U(t) = \Gamma_s^\dagger U^\dagger(s) U(s+t)\Gamma_s$$

where Γ_s is the second quantized forward shift defined by

$$\Gamma_s(u \otimes e(f)) = u \otimes e(f_s), \qquad f_s(t) = \begin{cases} 0 & t < s, \\ f(t-s) & t \geq s. \end{cases}$$

A similar characterization of multi-dimensional unitary evolutions requires a multi-dimensional version of the Parthasarathy–Sinha martingale representation theorem. A more interesting but evidently [Jo] technically difficult generalization would be a 'stochastic Stone's theorem' in which the driving coefficients (4.4) would become unbounded, and the unphysical uniform continuity of (b) above relaxed to strong continuity.

Let $\mathcal{A}$ be a unital C^*-subalgebra of $\mathcal{B}(\mathcal{H}_0)$ called the *initial algebra*. A *quantum stochastic flow* over $\mathcal{A}$ is a family $J = (J_t : t \in \mathbb{R}_+)$ of unital C^*-algebra morphisms $J_t : \mathcal{A} \to \mathcal{B}(\mathcal{H}_0 \otimes \mathcal{F}(h))$ such that

---

In earlier literature the term *quantum diffusion* was used.

(a) for each $x \in \mathcal{A}$, $t \mapsto J_t(x)$ is an integrable adapted process,

(b) there are maps $\lambda_\beta^\alpha : \mathcal{A} \to \mathcal{A}$, $\alpha_1 \beta = 0, 1, \ldots, N$ such that for all $x \in \mathcal{A}$ and $t \in \mathbb{R}_+$

$$(4.5) \qquad J_t(x) = x \otimes 1 + \int_0^t J_s(\lambda_\beta^\alpha(x)) d\Lambda_\alpha^\beta(s).$$

Here the repeated suffix convention remains in force. The maps λ_β^α are called the *structure maps* of the flow.

Each unitary evolution U determines a flow by

$$(4.6) \qquad J_t(x) = U(t) x \otimes 1 U(t)^*;$$

if U satisfies (4.2) then the structure maps are given by

$$(4.7) \qquad \lambda_\beta^\alpha(x) = l_\beta^\alpha x + x(l_\alpha^\beta)^* + l_j^\alpha x(l_j^\beta)^*.$$

Such a flow is called *inner*. That not every flow is inner is shown by the following example. We take $\mathcal{A} = \mathcal{B}(\mathcal{H}_0)$ where the initial space $\mathcal{H}_0$ is infinite dimensional, and let σ be a non-inner endomorphism of $\mathcal{B}(\mathcal{H}_0)$, for example we could define

$$(4.8) \qquad \sigma(x) = W(x \otimes 1)W^*$$

where W is a Hilbert space isomorphism from $\mathcal{H}_0 \otimes \mathcal{H}_1$ to $\mathcal{H}_0$ and the auxiliary Hilbert space $\mathcal{H}_1$ is of dimension greater than 1. Using one-dimensional quantum stochastic calculus, $\dim \mathcal{K} = 1$, we use the spectral decomposition

$$\Lambda(t) = \sum_{n=0}^\infty n E_n(t)$$

to define

$$(4.9) \qquad J_t(x) = \sigma^{\Lambda(t)}(x) =: \sum_{n=0}^\infty \sigma^n(x) \otimes 1 \, E_n(t), \quad x \in \mathcal{B}(\mathcal{H}_0).$$

It may be verified that

$$J_t(x) = x \otimes 1 + \int_0^t J_s(\sigma(x) - x) d\Lambda(s)$$

and hence that J is a quantum stochastic flow over $\mathcal{B}(\mathcal{H}_0)$.

In consequence of the linearity, unitality and $*$-map properties of each J_t together with independence of the stochastic differentials [Li], the structure maps of a flow J are linear, vanish on the unit element and satisfy

$$(4.10) \qquad \lambda_\beta^\alpha(x^*) = (\lambda_\alpha^\beta(x))^*, \quad x \in \mathcal{A}, \; \alpha, \beta = 0, 1, \ldots, N.$$

By equating the stochastic differentials of both sides of the multiplicativity condition

$$J(xy) = J(x)J(y),$$

we find similarly that

$$(4.11) \qquad \lambda_\beta^\alpha(xy) = x\lambda_\beta^\alpha(y) + \lambda_\beta^\alpha(x)y + \lambda_j^\alpha(x)\lambda_\beta^j(y).$$

It can be shown conversely [Ev] that, given maps $\lambda_\beta^\alpha : \mathcal{A} \to \mathcal{A}$ satisfying all these conditions which are in addition bounded as operators on $\mathcal{A}$, then there exists a unique flow satisfying (4.5). This is the basic existence theorem of the subject of quantum stochastic flows. Another existence theorem [EH] is for inner perturbations of flows. Consider a flow J having bounded structure maps satisfying (4.5), and a unitary evolution U satisfying (4.5) with bounded l_β^α, $\alpha, \beta = 0, 1, \ldots, N$. Then the stochastic differential equation for a process U^J

$$(4.12) \qquad dU^J = U^J J(l_\beta^\alpha)d\Lambda_\alpha^\beta, \quad U_0^J(0) = 1$$

has a unique unitary solution. Moreover $\tilde{J}$ defined by

$$\tilde{J}_t(x) = U^J(t)J_t(x)U^J(t)^*$$

is a flow whose structure maps are given by

$$\begin{aligned}
\tilde{\lambda}_\beta^\alpha(x) = &\lambda_\beta^\alpha(x) + l_\beta^\alpha x + x(l_\alpha^\beta)^* + l_j^\alpha x(l_j^\beta)^* \\
(4.13) \qquad &+ l_j^\alpha \lambda_\beta^j(x) + \lambda_j^\alpha(x)(l_j^\beta)^* + l_j^\alpha \lambda_k^j(x)(l_k^\beta)^*.
\end{aligned}$$

Notice that (4.12) is not of the form (4.2) in so far as the driving coefficients $J(l_\beta^\alpha)$ are adapted process rather than fixed elements of the initial space. We call $\tilde{J}$ the perturbation of J by U. Perturbations of inner flows are inner; if J consists of conjugation by V, the unitary evolution satisfying

$$V(t) = 1 + \int_0^t V(s)m_\beta^\alpha d\Lambda_\alpha^\beta(s),$$

then $\tilde{J}$ consists of conjugation by the evolution $U^J V$, which satisfies

$$dU^J V = U^J V \tilde{l}_\beta^\alpha d\Lambda_\alpha^\beta, \quad U^J V(0) = 1$$

where

$$(4.14) \qquad \tilde{l}_\beta^\alpha = l_\beta^\alpha + m_\beta^\alpha + l_j^\alpha m_\beta^j.$$

In the one-dimensional case $\dim \mathcal{K} = 1$ the structure maps of the flow

$$\begin{aligned}
dJ_t(x) = x \otimes 1 + &\int_0^t (J_s(\lambda(x))d\Lambda(s) + J_s(\alpha(x))dA^\dagger(s) \\
(4.15) \qquad &+ J_s(\alpha^\dagger(x))dA(s) + J_s(\tau(x))dT(s))
\end{aligned}$$

as well as being linear and vanishing on the identity operator satisfy

$$(4.16) \qquad \lambda(x^*) = \lambda(x)^*, \quad \alpha(x^*) = \alpha^\dagger(x)^*, \quad \tau(x^*) = \tau(x)^*$$

and

$$(4.17\mathrm{a}) \qquad \lambda(xy) = \lambda(x)y + x\lambda(y) + \lambda(x)\lambda(y)$$

$$(4.17\mathrm{b}) \qquad \alpha(xy) = \alpha(x)y + x\alpha(y) + \lambda(x)\alpha(y)$$

$$(4.17\mathrm{c}) \qquad \alpha^\dagger(xy) = \qquad\qquad\qquad \alpha^\dagger(x)y + x\alpha^\dagger(y) + \alpha^\dagger(x)\lambda(y)$$

$$(4.17\mathrm{d}) \qquad \tau(xy) = \tau(x)y + x\tau(y) + \alpha^\dagger(x)\alpha(y)$$

in view of (4.10) and (4.11). Adding xy to both sides of (4.17a) we see that $\lambda = \sigma - id$ where σ is an endomorphism of $\mathcal{A}$. Making this substitution in (4.17b) and (4.17c) we find that

$$\alpha(xy) = \alpha(x)y + \sigma(x)\alpha(y), \quad \alpha^\dagger(xy) = \alpha^\dagger(x)\sigma(y) + x\alpha^\dagger(y),$$

that is α and $\alpha^\dagger$ are respectively left and right σ-derivations of $\mathcal{A}$. If α (and hence $\alpha^\dagger$) is an *inner* σ-derivation, that is,

$$(4.18) \qquad \alpha(x) = lx - \sigma(x)l, \qquad \alpha^\dagger(x) = xl^* - \sigma(x)l^*,$$

then a map τ_0 satisfying (4.17d) is given by

$$(4.19) \qquad \tau_0(x) = -\tfrac{1}{2}(l^* lx - l^*\sigma(x)l + xl^*l)$$

and the general such map τ differs from τ_0 by a $*$-derivation of $\mathcal{A}$. In the case $\mathcal{A} = \mathcal{B}(\mathcal{H}_0)$ every σ-derivation and every derivation is inner, and it follows that every quantum stochastic flow over $\mathcal{B}(\mathcal{H}_0)$ is an inner perturbation of one of form (4.9).

By taking vacuum conditional expectations, we can construct from a flow J a semigroup of maps from the initial algebra to itself defined by

$$\langle u, \mathcal{T}_t(x)v \rangle = \langle u \otimes e(0), J_t(x)v \otimes e(0) \rangle, \quad u, v \in \mathcal{H}_0, \; x \in \mathcal{H}_0, \; t \in \mathbb{R}_+.$$

Being the composition of a conditional expectation and a $*$-morphism, each $\mathcal{T}_t$ is completely positive, that is to say it maps positive elements to positive elements and, for each natural number n, $\mathcal{T}_t \otimes id_{M_n}$ has the same property on the tensor product $\mathcal{A} \otimes M_n$ of $\mathcal{A}$ with the $n \times n$ matrix $*$-algebra M_n. Moreover $\mathcal{T}_t$ evidently maps the identity to itself; thus we have a *quantum dynamical semigroup*, a one-parameter semigroup of unital completely positive. If the structure map $\lambda_0^0 : \mathcal{A} \to \mathcal{A}$ is bounded then $\mathcal{T}_t = e^{t\lambda_0^0}$ is uniformly continuous. A natural question is whether, conversely, every uniformly continuous quantum dynamical semigroup can be obtained in this way by conditioning a flow, that is, admits a *stochastic dilation*.

In the case $\mathcal{A} = \mathcal{B}(\mathcal{H}_0)$ it does [HSh]. Indeed, adapting the characterization of Lindblad [Ld], we may write the infinitesimal generator of the semigroup in the form

$$\tau(x) = i[h, x] + \tau_0(x)$$

This property, sometimes called *conservativity*, corresponds to unitality of J; recent work [BP,Ch] relaxes both conditions.

where τ_0 is given by (4.19). But then the semigroup is obtained by conditioning the perturbation of the flow (4.9) by the evolution U given by

$$dU = U(ldA^\dagger - l^\dagger dA + (ih - \tfrac{1}{2}l^\dagger l)dT), \qquad U(0) = 1.$$

An alternative construction of a stochastic dilation in the case $\mathcal{A} = \mathcal{B}(\mathcal{H}_0)$ was given in [HP$_4$]. The flow is inner in this case but in general an infinite dimensional extension of quantum stochastic calculus is needed.

5. Cohomology of generators of quantum stochastic flows

The structure relations satisfied by structure maps of quantum stochastic flows can be analysed using Hochschild cohomology of associative algebras [Ho]. In this theory, given an algebra $\mathcal{A}$ and a *module* for $\mathcal{A}$, that is a complex vector space $\mathcal{M}$ equipped with bilinear maps called the *left* and *right actions*

$$\mathcal{A} \times \mathcal{M} \ni (a, m) \mapsto am \in \mathcal{M}, \qquad \mathcal{M} \times \mathcal{A} \ni (m, a) \mapsto ma \in \mathcal{M}$$

satisfying

$$(5.1) \qquad a_1(a_2 m) = (a_1 a_2)m, \qquad a_1(ma_2) = (a_1 m)a_2, \qquad (ma_1)a_2 = m(a_1 a_2),$$

we define, for $n = 0, 1, 2, \ldots$, the space $C_n = C_n(\mathcal{A}, \mathcal{M})$ of n-cochains to consist of all n-linear maps $\xi : \mathcal{A} \times \cdots \times \mathcal{A} \to \mathcal{M}$, and the coboundary map $b : C_n \to C_{n+1}$ for each n by

$$b\xi(x_0, x_1, \ldots, x_n) = x_0\xi(x_1, \ldots, x_n) + \sum_{j=1}^{n-1}(-1)^j\xi_n(x_0, x_1, \ldots, x_{j-1}x_j, x_{j+1}, \ldots, x_n)$$

$$+ (-1)^n\xi(x_0, \ldots, x_{n-1})x_n.$$

The space Z_n of n-cocycles is the kernel of $b : C_n \to C_{n+1}$ and the space $\mathcal{B}_n$ of n-coboundaries is the range of $b : C_{n-1} \to C_n$. Because $b^2 = 0$, $\mathcal{B}_n \subseteq Z_n$ and we can define the nth *cohomology space* H_n to be the quotient $Z_n/\mathcal{B}_n$. In the $*$-algebra version of the theory $\mathcal{M}$ carries an involution $\dagger$ and (5.1) is supplemented by

$$(5.2) \qquad\qquad\qquad (am)^\dagger = m^\dagger a^*.$$

n-cochains ξ are required to satisfy the relation

$$\xi(x_1, \ldots, x_n)^\dagger = \xi(x_1^*, \ldots, x_n^*).$$

Infinitesimal generators of deterministic quantum flows, in the form of uniformly continuous one-parameter groups of automorphisms of a C^*-algebra $\mathcal{A}$, are derivations of $\mathcal{A}$ commuting with the involution $*$, that is, they are 1-cocycles in the $*$-algebra Hochschild cohomology in which $\mathcal{A}$ itself is the module and the actions are given by multiplication. Such flows are inner, given by conjugation by one parameter unitary groups, if the derivation is inner, equivalently it is a 1-coboundary. Thus the first cohomology space represents the obstruction to flows being inner. To achieve a similar analysis for quantum stochastic flows we shall need a generalisation, due to Lue [Lu], of Hochschild cohomology.

Before describing this, let us use the Hochschild theory to analyse the structure relations (4.16), (4.17) in the case of a one-dimensional quantum stochastic flow [H$_2$]. (A similar analysis is possible in the many dimensional case [EH$_2$].) As before we write $\lambda = \sigma - id$ when $\sigma \in \text{End } \mathcal{A}$. Ignoring the involution for the moment, we introduce Hochschild cohomologies on $\mathcal{A}$ as module in which the left (respectively right) action is given by

$$(a, m) \to \sigma(a)m, \qquad ((m, a) \to m\sigma(a)),$$

the remaining action being by multiplication. Then the structure maps α and $\alpha^\dagger$ are respectively elements of the spaces $^\sigma Z_1$ and Z_1^σ of 1-cocycles in these cohomologies. On the other hand the relation (4.17d) requires the 2-cochain $\eta_{\alpha^\dagger,\alpha}$, where

$$\eta_{\alpha,\beta}(x, y) = \alpha(x)\beta(y)$$

to be an element of the space $\mathcal{B}_2$ of 2-coboundaries in the cohomology based on the module $\mathcal{A}$ with both actions given by multiplication. It is not difficult to prove the implications

$$\alpha \in {}^\sigma Z_1, \qquad \beta \in Z_1^\sigma \Rightarrow \eta_{\alpha,\beta} \in Z_2,$$
$$\alpha \in {}^\sigma B_1, \qquad \beta \in B_1^\sigma \Rightarrow \eta_{\alpha,\beta} \in B_2.$$

However in cases where both first and second cohomology spaces are non-trivial, it is possible that, for given $\alpha, \alpha^\dagger$ and σ satisfying (4.16), (4.17a,b,c) there does not exist a map τ satisfying (4.17d). An example [HR$_1$] is provided by the irrational rotation algebra; the 'obstruction' is overcome by enlarging the algebra and plays a rôle in the theory of the quantum Hall effect [HR$_2$].

In the $*$-algebraic version of Lue's 'nonabelian' generalisation of Hochschild cohomology, the module $\mathcal{M}$, carrying left and right actions of $\mathcal{A}$ satisfying (5.1) and (5.2), is itself equipped with a multiplication. n-cochains are no longer linear maps (thus 'nonlinear' rather than 'nonabelian' might be a better description). Instead the space of 0-chains consists of elements $u \in \mathcal{M}$ satisfying

$$(5.3) \qquad l + l^\dagger + ll^\dagger = l^\dagger + l + ll^\dagger = 0.$$

Notice that if multiplication in $\mathcal{M}$ is trivial, all products vanishing, this condition reduces to that for a Hochschild 0-chain. The 0-chains form a group C_0 under the composition

$$(5.4) \qquad l \circ m = l + m + lm.$$

The space of 1-cocycles Z_1 is defined to consist of linear $*$-maps $\lambda : \mathcal{A} \to \mathcal{M}$ satisfying

$$(5.5) \qquad \lambda(xy) = \lambda(x)y + x\lambda(y) + \lambda(x)\lambda(y).$$

The coboundary map $b : C_0 \to Z_1$ is defined by

$$(5.6) \qquad bl(x) = lx + xl^\dagger + lxl^\dagger;$$

(5.3) ensures that bl satisfies (5.5). More generally there is a right action of the group C_0 on Z, given by

$$(5.7) \quad (\lambda, l) \mapsto \lambda^l; \qquad \lambda^l(x) = \lambda(x) + l(x + \lambda(x)) + (x + \lambda(x))l^\dagger + l(x + \lambda(x))l^\dagger.$$

To apply this theory to describe generators of quantum stochatic flows, we take the module $\mathcal{M}$ to be the tensor product $*$-algebra $\mathcal{A} \otimes \mathcal{I}$, where $\mathcal{I}$ is the *Ito algebra*, that is the associative $*$-algebra of differentials dK_Σ, equipped with the multiplication and involution

$$dK_{\Sigma_1} dK_{\Sigma_2} = dK_{\Sigma_1 \triangle \Sigma_2}, \qquad dK_\Sigma^\dagger = dK_{\Sigma^*};$$

and is equipped with the actions defined by

$$a(b \otimes dK_\Sigma) = ab \otimes dK_\Sigma = (a \otimes dK_\Sigma)b.$$

The *generator* of the unitary evolution (4.2) is defined to be the element of $\mathcal{A} \otimes \mathcal{I}$

$$l = l_\beta^\alpha \otimes d\Lambda_\alpha^\beta;$$

the *generator* of the flow (4.5) is the map from $\mathcal{A}$ to $\mathcal{A} \otimes \mathcal{I}$

$$\lambda : x \mapsto \lambda_\beta^\alpha(x) \otimes d\Lambda_\alpha^\beta.$$

Then it may be verified that the unitarity condition (4.3) is the condition (5.3) that l be a 0-cochain, the structure relation (4.11) is the condition (5.5) that λ be a 1-cocycle, the formula (4.7) for the structure maps of an inner flow is just (5.6) and the perturbation formula (4.13) is the action (5.7). The formula (4.14) corresponds to the group structure (5.4) of 0-chains.

The relation (5.5) satisfied by the generator of a flow also gives a connection with cyclic cohomology [CH]. Cyclic cocycles can be characterized by traces on the *Cuntz algebra* $\mathcal{C}(\mathcal{A})$ [CC], which may be defined as the $*$-algebra freely generated by elements $x \in \mathcal{A}$ and $q(x)$, where q is a linear $*$-map satisfying

$$q(xy) = q(x)y + xq(y) + q(x)q(y).$$

Replacing q by $-q$ we obtain the relation (5.5). The number of cyclic cocycles found directly this way is rather small because of the paucity of taces on subalgebras of $\mathcal{I}$. Fermionic or $\mathbb{Z}_2$-graded theories are more promising [HSj].

6. Lie algebraic structure in quantum stochastic calculus

A simple functional Ito formula for one-dimensional Brownian motion B is

$$(6.1) \qquad dF(B, T) = F(B + dB, T + dT) - F(B, T).$$

Here F is a polynomial in two indeterminates. (6.1) can be deduced from the more familiar Ito formula

$$dF(B, T) = F'(B, T)dB + \left(\tfrac{1}{2}F''(B, T) + \dot{F}(B, T)\right) dT,$$

where $'$ and $\cdot$ denote differentiation with respect to the first and second indeterminate respectively, using the multiplication table

(6.2)

$$
\begin{array}{c|cc}
 & dB & dT \\
\hline
dB & dT & 0 \\
dT & 0 & 0
\end{array}.
$$

More generally, given any closed Ito multiplication table

(6.3)
$$dA_j d\Lambda_k = c^l_{jk} d\Lambda_l, \quad j,k = 1,\dots,n$$

we may form the stochastic differential of a polynomial function

$$F(\Lambda_1,\dots,\Lambda_n) = F(\Lambda)$$

as

(6.4)
$$dF(\Lambda) = F(\Lambda + d\Lambda) - F(\Lambda).$$

Writing (6.4) in integral form

$$F(\Lambda(t)) = F(0) + \int_0^t \left(F(\Lambda(s) + d\Lambda(s)) - F(\Lambda(s)) \right)$$

and iterating we obtain a chaotic decomposition (in general incuding time as an integrator)

(6.5)
$$F(\Lambda(t)) = \sum_m I_m^t \left(\sum_{s \subseteq \{1,\dots,m\}} (-1)^{m-|s|} F\left(\sum_{j \in s} d\Lambda(s_j) \right) \right).$$

Here I_m^t is the iterated integral $\int_{0 < s_1 < \cdots < s_m < t}$ and the summation over m terminates at the degree of F. The right hand sides of (6.1) and (6.4) are reduced to linear combinations (with polynomial coefficients) of the basic differentials by the rule that the latter commute with polynomial coefficients, and are themselves multiplied by the rules (6.2), (6.3) to remove terms of degree more than one. The subtraction of F avoids the embarrassment of a degree 0 term (the Ito algebra has no identity). Similarly the integrand in (6.5) is reduced to linear combinations of the m-fold integrators $d\Lambda_{j_1}(t_1) \cdots d\Lambda_{j_m}(t_m)$ by applying the Ito multiplication to products of the $d\Lambda$ corresponding to the same time t_j. The differencing ensures that the embarrassment of a multi-fold integrator of order less than m is avoided, so that I_m^t can be applied meaningfully.

These formulae remain valid in theories of quantum stochastic calculus in which the multiplication table (6.3) becomes noncommutative [HPu]. That they make sense at all involves some subtlety. Let us introduce the abstract Lie algebra $\mathcal{L}$ corresponding to equipping the Ito algebra $\mathcal{I}$ with the commutator Lie bracket. Thus, for each $L \in \mathcal{L}$ we have a corresponding Ito differential $d\Lambda_L$ and the map $L \mapsto d\Lambda_L$ is a Lie algebra morphism from $\mathcal{L}$ to $\mathcal{I}_{\mathrm{Lie}}$. It is a cardinal feature of quantum stochastic calculus that the processes Λ_L also form representations of the Lie algebra $\mathcal{L}$, but only in the weak sense that, for exponential vectors $e(f), e(g)$ and $t \in \mathbb{R}_+$,

$$\langle e(f), \Lambda_{[L,M]}(t)e(g)\rangle = \left\langle \Lambda_L^\dagger(t)e(f), \Lambda_M(t)e(g)\right\rangle - \left\langle \Lambda_M^\dagger(t)e(f), \Lambda_L(t)e(g)\right\rangle.$$

In fact these are the differentials of the Lie group representations implicit in (2.5). Notice that the associative algebra $\mathcal{P}$ generated (with weak multiplication) by the integrator processes Λ_L possesses a natural unit, the constant process 1.

We may now interpret (6.4) in the noncommutative case as follows. For each element U of the universal enveloping algebra $\mathcal{U}$ of $\mathcal{L}$, there is a corresponding proces $F(\Lambda)$ got by extending the Lie algebra morphism $L \mapsto \Lambda_L$ to a unital associative algebra morphism π_0 from $\mathcal{U}$ to $\mathcal{P}$. The differential $dF(\Lambda)$ is found as follows. We extend the Lie algebra morphism $L \mapsto d\Lambda_L$, to an associative algebra morphism π_1, not from $\mathcal{U}$ which is impossible since $\mathcal{I}$ has no identity, but from the ideal $\mathcal{J}$ in $\mathcal{U}$ generated by $\mathcal{L}$, to the Ito algebra $\mathcal{I}$. The right hand side $F(\Lambda + d\Lambda) - F(\Lambda)$ of (6.4) is then the image under

$$\pi_0 \otimes \pi_1 : \mathcal{U} \otimes \mathcal{J} \to \mathcal{P} \otimes \mathcal{I}$$

of the element $\gamma(U) - U \otimes 1$ of $\mathcal{U} \otimes \mathcal{J}$, where γ is the coproduct of $\mathcal{U}$, that is, the unital associative algebra morphism which extends the Lie algebra morphism

$$\mathcal{L} \ni L \mapsto L \otimes 1 + 1 \otimes L \in (\mathcal{U} \otimes \mathcal{U})_{\mathrm{Lie}}.$$

That this is meaningful depends on the easily verified fact that $\gamma(U) - U \otimes 1$ does indeed belong to $\mathcal{U} \otimes \mathcal{J} \subset \mathcal{U} \otimes \mathcal{U}$.

The right hand side of (6.5) admits a similar interpretation using the higher-order coproducts which extend the Lie algebra morphisms

$$\mathcal{L} \ni L \mapsto L \otimes 1 \otimes \cdots \otimes 1 + 1 \otimes L \otimes \cdots \otimes 1 + \cdots + 1 \otimes \cdots \otimes L \in (\mathcal{U} \otimes \cdots \otimes \mathcal{U})_{\mathrm{Lie}}.$$

The proofs of (6.4) and (6.5) are based on a higher order generalization of the Ito product formula. For details see [HPu].

This theory admits an amusing analogue for Lie superalgebras [EyH, Ey] which is at the same time a multidimensional generalization of the Boson–Fermion unification scheme of [HP$_3$]. Let us recall from Section 3 that the Ito algebra $\mathcal{I}$ of all differentials dK_Σ is realised as the associative $*$-algebra of operators Σ on $\mathbb{C} \oplus \mathcal{K}$ with the multiplication $\Sigma_1 \triangle \Sigma_2$. Now let us introduce a $\mathbb{Z}_2$-grading of the Hilbert space $\mathbb{C} \oplus \mathcal{K}$, that is an orthogonal decomposition, of form

$$(6.6) \qquad\qquad \mathbb{C} \oplus \mathcal{K} = (\mathbb{C} \oplus \mathcal{K}_+) \oplus \mathcal{K}_-$$

where $\mathcal{K}_\pm$ are mutually orthogonal subspaces of $\mathcal{K}$. Correspondingly, $\mathcal{I}$ becomes a $\mathbb{Z}_2$-graded algebra

$$\mathcal{I} = \mathcal{I}_+ \oplus \mathcal{I}_-$$

where $\mathcal{I}_+$ and $\mathcal{I}_-$ are respectively the subspaces of elements of $\mathcal{I}$ which commute and anticommute with the grading operator which is 1 on $\mathbb{C} \oplus \mathcal{K}_+$ and -1 on $\mathcal{K}_-$.

Now we may form the corresponding Lie superalgebra $\mathcal{I}_{\text{sLie}}$ by equipping $\mathcal{I}$ with the supercommutator bracket

$$\{\Sigma_1, \Sigma_2\} = \Sigma_1 \bigtriangleup \Sigma_2 - (-1)^{\sigma(\Sigma_1)\sigma(\Sigma_2)} \Sigma_2 \bigtriangleup \Sigma_1$$

where $\sigma(\Sigma)$ denotes the grade (0 if Σ is even, 1 if it is odd) of $\Sigma \in \mathcal{I}$.

Introducing for each $t \in \mathbb{R}_+$ a grading of the Fock space $F\left(L^2(\mathbb{R}_+) \otimes \mathcal{K}\right)$ by means of the grading operator

$$R(t) = \Gamma\left(M_{\chi_{[0,t]}} \otimes (1_+ \oplus (-1_-)) + M_{\chi_{(t,\infty)}} \otimes 1\right)$$

where $1_\pm$ are the identity operators on $\mathcal{K}_\pm$, we may now $\mathbb{Z}_2$-grade processes according to whether they commute or anticommute with the process R. Now define processes $\tilde{K}_\Sigma$, $\Sigma \in \mathcal{I}$ by the stochastic integral prescription

$$(6.7) \qquad\qquad \tilde{K}_\Sigma(t) = \int_0^t R(s)^{\sigma(\Sigma)} dK_\Sigma.$$

Then it can be shown [EyH] that the processes $\tilde{K}_\Sigma$ form a realisation of the Lie superalgebra $\mathcal{I}_{\text{sLie}}$, just as the K_Σ realise $\mathcal{I}$, in the sense of weak multiplication. The analogues of (6.4) and (6.5) hold [Ey] when interpreted using the natural coproduct structure of the universal enveloping algebra of the abstract Lie superalgebra $\mathcal{L}_s \approx \mathcal{I}_{\text{sLie}}$.

When $\dim \mathcal{K} = 1$ there is only one non-trivial choice of grading (6.6) which is got by taking $\mathcal{K}_+ = \{0\}$, $\mathcal{K}_- = \mathcal{K}$ and the process R is then given by $R = (-1)^\Lambda$. In this case the odd processes (6.7) are Fermionic annihilation and creation processes $B, B^\dagger$ satisfying

$$BB^\dagger + B^\dagger B = T.$$

This is the Boson–Fermion unification schcme of [HP$_3$]; (6.7) takes the simple differential form

$$dB = (-1)^\Lambda dA, \qquad dB^\dagger = (-1)^\Lambda dA^\dagger.$$

Acknowledgement

Part of this paper was written while the author visited the Centro V Volterra, Roma whose warm hospitality is acknowledged.

References

[A]   D B Applebaum, Stochastic dilations of the Bloch Equation in Boson and Fermion noise, *J Phys A* **19** (1986) 937–959.

[AFL] L Accardi, A Frigerio and Y G Lu, The weak coupling limit as a quantum central limit, *Commun Math Phys* **131** (1991) 537–576.

[AGL] L Accardi, J Gough and Y G Lu, On the stochastic limit for quantum theory, *Rep Math Phys* **36** (1995) 155–187.

[AL]   L Accardi and Y G Lu, On the weak coupling limit for quantum electrodynamics, pp 16–22, in *Probabilistic methods in mathematical physics*, ed F Guerra *et al*, World Scientific (1992).

[AM]   S Attal and P-A Meyer, Interpretation probabiliste et extension des integrales stochastiques noncommutatives, Strasbourg preprint (1994).

[AV]   L Accardi and I Volovich, The stochastic limit of quantum field theory, Rome II preprint (1994).

[BHH]   V P Belavkin, O Hirota and R L Hudson (eds) *Quantum Communication and Measurement*, Plenum (1995).

[BL]   A Barchielli and G Lupieri, Quantum stochastic calculus, operation valued stochastic processes and continual measurements in quantum mechanics, *J Math Phys* **26** (1985) 2222–2230.

[BS]   V P Belavkin and P Staszewski, Nondemolition measurement of a free quantum particle, *Phys Rev A* **45** (1992) 1347–1356.

[1]   C Barnett, R F Streater and I Wilde, The Ito Clifford integral, *J Funct Anal* **48** (1982) 172–212.

[CC]   A Connes and J Cuntz, Quasi-homomorphismes, cohomologie cyclique et positivité, *Commun Math Phys* **114** (1988) 515–526.

[CF]   A Chebotarev and F Fagnola, Sufficient conditions for conservativity of quantum dynamical semigroups, *J Funct Anal* **118** (1993) 113–153.

[CH]   A M Cockroft and R L Hudson, Quantum mechanical Wiener processes, *J Multivariate Anal.* **7** (1977) 107–124.

[CoH]   P Beazley Cohen and R L Hudson, Generators of quantum stochastic flows, Cuntz morphisms and cyclic cohomology, Nottingham preprint (1994).

[EH$_1$]   M P Evans and R L Hudson, Perturbations of quantum diffusions, *J London Math Soc* (2) **41** (1990) 373–384.

[EH$_2$]   M P Evans and R L Hudson, Multidimensional quantum diffusions, pp 69–88, in *Quantum Probability IV*, ed L Accardi *et al*, Springer LNM **1303** (1988).

[Ev]   M P Evans, Existence of quantum diffusions, *Prob Theory and Related Fields* **81** (1989) 473–483.

[Ey]   T M W Eyre, Chaotic expansion for Lie superalgebras in quantum stochastic calculus, Nottingham preprint (1996).

[EyH]   T M W Eyre and R L Hudson, Representation of Lie superalgebras and generalized Boson–Fermion equivalence in quantum stochastic calculus, Nottingham preprint (1996), to appear in *Commun Math Phys*.

[Fo]   V A Fock, Konfigurationsraum und zweite Quantelung, *Z Physik* **75**, 622–647 (1932).

[H$_1$]   R L Hudson, The strong Markov property for canonical Wiener processes, *J Funct Anal* **37** (1980) 68–87.

[H$_2$]   R L Hudson, Quantum diffusions and cohomology of algebras, pp 479–485, *Proceedings of 1st World Congress of Bernoulli Society*, Tashkent 1986, Vol 1, ed Yu Prohorov *et al*, VNU (1987).

[HIP]   R L Hudson, P D F Ion and K R Parthasarathy, Time orthogonal unitary dilations and noncommutative Feynman Kac formulae, *Commun Math Phys* **83** (1982) 261–280.

[HLi]   R L Hudson and J M Lindsay, On characterizing quantum stochastic evolutions, *Math Proc Camb Phil Soc* **102** (1987) 363–369.

[Ho]   G Hochschild, On the cohomology groups of an associative algebra, *Ann of Math* **46** (1945) 58–67.

[HP$_1$] R L Hudson and K R Parthasarathy, Quantum diffusions, pp 111–121 in *Theory and applications of random fields*, proceedings, Bangalore 1982, ed V Kallianpur, Springer LN Control theory and IS **49** (1983).

[HP$_2$] R L Hudson and K R Parthasarathy, Quantum Ito's formula and stochastic evolutions, *Commun Math Phys* **93** (1984) 301–323.

[HP$_3$] R L Hudson and K R Parthasarathy, Unification of Boson and Fermion quantum stochastic calculus, *Commun Math Phys* **104** (1986) 457–470.

[HP$_4$] R L Hudson and K R Parthasarathy, Stochastic dilations of uniformly continuous quantum dynamical semigroups, *Acta Applicandae Math* **2** (1984) 457–470.

[HPu]   R L Hudson and S Pulmannova, Chaotic expansion of elements of the universal enveloping algebra of a Lie algebra associated with a quantum stochastic calculus, Nottingham preprint (1996).

[HR$_1$] R L Hudson and P Robinson, Quantum diffusions and the noncommutative torus, *Lett Math Phys* **15** (1988) 47–53.

[HR$_2$] R L Hudson and P Robinson, Quantum diffusions on the noncommutative torus and solid state physics, pp 338–345, in *Differential geometric methods in solid state physics*, Chester, 1988, ed A Solomon, World Scientific 1989.

[HSh]   R L Hudson and P Shepperson, Stochastic dilations of quantum dynamical semigroups using one-dimensional quantum stochastic calculus, pp 216–218 in *Quantum Probability V*, ed L Accardi *et al*, Springer LNM **1442** (1990).

[HSj]   R L Hudson and V R Struleckaja, Nonabelian cohomology and Fermionic flows over Z_2-graded *-algebras, *Lett Math Phys* **38** (1996) 13–22.

[HSt]   R L Hudson and R F Streater, Noncommutative martingales and stochastic integrals in Fock space, pp 216–227 in *Stochastic processes in quantum theory and statistical physics*, proceedings, Marseilles 1982, ed Albeverio, Springer LN Physics **173** (1983).

[Jo]   J L Journé, Structure des cocycles Markoviens sur l'espace de Fock, *Prob Theor Rel Fields* **75** (1987) 291–316.

[Ld]   G Lindblad, On the generators of quantum dynamical semigroups, *Commun Math Phys* **48** (1976) 119–130.

[Li]   J M Lindsay, Independence for quantum stochastic integrators, pp 325–332, in *Quantum probability VI*, ed L Accardi *et al*, World Scientific (1991).

[Lu]   A S Lue, Nonabelian cohomology of associative algebras, *Quart Jour Math* **19** (1968) 159–180.

[MR]   H Maassen and P Robinson, Quantum stochastic calculus and the dynamical Stark effect, *Rep Math Phys* **30** (1992) 185–203.

[Ne]   E Nelson, The free Markoff field, *J Funct Anal* **12** (1973) 211–227.

[PM]   A Mohari and K R Parthasarathy, On a class of generalized Evans–Hudson flows related to classical Markov processes, pp 221–249, in *Quantum Probability and Applications VII*, ed L Accardi *et al*, World Scientific (1992).

[PSc]   K R Parthasarathy and K Schmidt, Factorizable representations of current groups and the Araki–Woods embedding theorem, *Acta Mathematica* **128** (1972) 53–71.

[PSi$_1$]   K R Parthasarathy and K B Sinha, Stop times in Fock space stochastic calculus, pp 495–498, in *Proceedings of 1st World Congress of Bernoulli Society*, Tashkent 1986 vol 1, ed Yu Prohorov *et al*, VNU (1987).

[PSi$_2$]   K R Parthasarathy and K B Sinha, Stochastic integral representations of bounded quantum martingales in Fock space, *J Funct Anal* **67** (1986) 126–151.

[Se]   I R Senitzky, Dissipation in quantum mechanics. The harmonic oscillator, *Phys Rev* **119** (1960) 670–679.

[Sg]   I E Segal, Tensor algebras over Hilbert spaces, *Trans Amer Math Soc* **81** (1956) 106–134.

[Sp]   R Speicher, A new example of 'Independence' and 'White noise', *Prob Theor Rel Fields* **84** (1990) 141–154.

[VDN]   D Voiculescu, K J Dykema and A Nica, *Free random variables*, American Mathematical Society CRM Monographs (1992).

[Wi]   N Wiener, The homogeneous chaos, *Amer J Math* **60** (1930) 897–936.

PERFECT SIMULATION FOR THE AREA-INTERACTION POINT PROCESS

WILFRID S. KENDALL,* *University of Warwick*

Abstract

Because so many random processes arising in stochastic geometry are quite intractable to analysis, simulation is an important part of the stochastic geometry toolkit. Typically, a Markov point process such as the area-interaction point process is simulated *(approximately)* as the long-run equilibrium distribution of a (usually reversible) Markov chain such as a spatial birth-and-death process. This is a useful method, but it can be very hard to be precise about the length of simulation required to ensure that the long-run approximation is good. The splendid idea of Propp and Wilson [17] suggests a way forward: they propose a coupling method which delivers exact simulation of equilibrium distributions of (finite-state-space) Markov chains. In this paper their idea is extended to deal with perfect simulation of attractive area-interaction point processes in bounded windows. A simple modification of the basic algorithm is described which provides perfect simulation of the repulsive case as well (which being non-monotonic might have been thought out of reach). Results from simulations using a C computer program are reported; these confirm the practicality of this approach in both attractive and repulsive cases. The paper concludes by mentioning other point processes which can be simulated perfectly in this way, and by speculating on useful future directions of research. Clearly workers in stochastic geometry should now seek wherever possible to incorporate the Propp and Wilson idea in their simulation algorithms.

PERFECT SIMULATION; SPATIAL BIRTH-AND-DEATH PROCESSES; AREA-INTERACTION POINT PROCESS; STRAUSS POINT PROCESS; EXCLUDED-AREA POINT PROCESS; POISSON POINT PROCESS; BOOLEAN MODEL; COUPLING; MARKOV CHAIN MONTE CARLO

AMS 1991 SUBJECT CLASSIFICATION: PRIMARY 62M30
SECONDARY 60G55; 60K35

1. Perfect simulation

Conventional Markov chain Monte Carlo techniques (MCMC) consist of long-run simulations of Markov chains: it is hoped that the long runs are sufficiently long that the Markov chain distribution is close to equilibrium. This vastly increases the range of applicability of Bayesian methods, as one can avoid computationally intractable issues relating to evaluation of the normalization constant of the posterior density. This is because simulating (approximately) the density as a long-run equilibrium distribution of a (typically reversible) Markov chain will be based ultimately on ratios of various posterior densities, thus avoiding reference to the normalization constant.

---

* Postal address: Statistics, University of Warwick, Coventry CV4 7AL, UK.

For a discussion of these issues in a spatial context see [3] (other papers in the same journal issue provide useful discussion of MCMC in other statistical contexts).

There is however a catch. How can one tell whether one is close enough to equilibrium? There are various empirical techniques and theoretical bounds, but the problem remains rather intransigent. This is a vexed question even in the simpler situation of simulating large-state-space Markov chains possessing substantial symmetry (arising in statistical mechanics and as discrete approximations to models in stochastic geometry).

A recent paper by Propp and Wilson [17] shows how one can solve the problem completely in several cases, in which one can modify the simulation algorithm using coupling techniques so that the simulation itself provides a certificate stating when equilibrium has been achieved exactly! This allows *perfect simulation* of the equilibrium distribution of suitable finite-state-space Markov chains. This paper demonstrates the possibilities for stochastic geometry and spatial statistics, by showing how to exploit Propp and Wilson's ideas to develop an algorithm for perfect simulation of the equilibrium distribution of a certain spatial birth-and-death process, thus obtaining perfect simulation of the area-interaction point processes of [1].

The reader is referred to the Propp–Wilson paper for full details of their application of coupling techniques to simulation. Here we briefly discuss the simplest possible interesting case (though in fact the notation for the general case is identical).

Suppose that X is a two-state discrete-time Markov chain, with state-space $\mathcal{S}$ and stochastic matrix

$$P = \begin{pmatrix} 0 & 1 \\ \frac{1}{2} & \frac{1}{2} \end{pmatrix}.$$

We can view $\{X_t : t \geq 0\}$ as produced by a (discrete-time) *stochastic flow*: if $X_0 = i$ then $X_t = F_{0,t}(i)$, where the random maps $F_{s,t} : \mathcal{S} \to \mathcal{S}$ ($s \leq t$) are defined by the composition $F_{s,t} = H_{s+1} \circ \ldots \circ H_t$, using an independent and identically distributed sequence $\{H_k : k \in \mathbb{Z}\}$ of transition maps $H_k : \mathcal{S} \to \mathcal{S}$ with distribution

$$\mathbb{P}\left[H_k(1) = 2 \right] = 1,$$
$$\mathbb{P}\left[H_k(2) = 1 \right] = \mathbb{P}\left[H_k(2) = 2 \right] = \frac{1}{2}.$$

Of course everything can be calculated exactly in this simple case, but consider how one might obtain the equilibrium distribution ($\frac{1}{3}$, $\frac{2}{3}$) of X by simulation. Orthodox simulation methods would approximate it by sampling $X_t(i) = F_{0,t}(i)$ for large t. Propp and Wilson point out that $Y_t(i) = F_{-t,0}(i)$ has the same distribution as $X(i)$ (though $\{Y_t : t \geq 0\}$ no longer forms a Markov chain) but also $Y_t(i)$ does not depend on i when $t \geq T_c$, for some (random, almost surely finite) T_c. Consequently $Y_\infty = \lim_{t \to \infty} Y_t$ exists and has the equilibrium distribution. Better yet, the random variable T_c can be defined as a *coalescence time* which can be viewed as a stopping time in a reverse filtration:

$$T_c = \inf\{t : F_{-t,0} \text{ has singleton image}\}.$$

Hence $F_{-T_c,0}(i) = Y_\infty$ gives an algorithm for sampling from the equilibrium distribution in finite (but random) time.

In our simple example $T_c = \inf\{t : H_{-t}(2) = 2\}$ has a Geometric distribution. More generally T_c is distributed as the first time to complete coupling of copies of X started at all possible initial positions and allowed to coalesce when they meet. This will work whenever the Markov chain is irreducible and ergodic. There is a relationship to work on estimation of the second eigenvalue using coupling techniques (see Chen's article in this volume, also [5]) as well as to other coupling ideas, as discussed in Propp and Wilson.

The essence of the above approach is as follows. When we can show that

(a) a Markov chain X_t on a state-space $\mathcal{S}$ is realized by a stochastic flow $F_{s,t} : \mathcal{S} \to \mathcal{S}$ ($s \leq t$), in the sense that $X_t = F_{s,t}(X_s)$; and
(b) there is *coalescence* of $F_{-T,0}$ for all sufficiently negative $-T$, in the sense that with probability one for all sufficiently negative $-T$ the result $F_{-T,0}(x)$ does not depend on the initial value $x \in \mathcal{S}$;

then an equilibrium distribution for the chain X exists and is given by

$$Y_\infty = \lim_{T \to \infty} F_{-T,0}(x),$$

where the limit Y_∞ does not depend on the initial value x. We describe this situation as one where *perfect simulation* of the equilibrium distribution is possible, because in principle it then suffices to sample from $F_{-t,0}(x)$ for some $t > T$ (in which case $F_{-t,0}(x) = Y_\infty$).

For perfect simulation to be practical it must be computationally feasible to check whether or not coalescence has happened by (parsimonious!) inspection of the flow $F_{-T,-t}$ over a given simulation interval ($-t \in [-T, 0]$). Propp and Wilson [17] observe that in case of monotonicity (as for example in the case of attractive interacting particle systems) this can be done at a cost which is comparable to conventional simulation techniques: if there are states $\mathbf{0}$ and $\mathbf{1}$ in $\mathcal{S}$ such that $F_{-T,0}(x)$ is sandwiched between $F_{-T,0}(\mathbf{0})$ and $F_{-T,0}(\mathbf{1})$ for all $x \in \mathcal{S}$ then coalescence has occurred exactly when $F_{-T,0}(\mathbf{0}) = F_{-T,0}(\mathbf{1})$.

Note that the *Principle* described by Letac [11, p164] is closely related to the above. However Letac's arguments apply under the condition of almost-sure convergence of the Y_t (rather than the stronger condition of coalescence in finite time) and Letac applies the principle to deliver powerful theoretical characterization arguments rather than simulation algorithms. Letac's ideas have found application in actuarial science as well as theoretical probability: see [7].

I was introduced to Propp and Wilson's idea by Julian Besag at a conference at CWI Amsterdam in November 1995, organized by Ilya Molchanov and Michael Keane. After discussion with Adrian Baddeley, I set myself the challenge of considering how one might make these general ideas work in the context of simulation of point processes, and the following is the first fruits of this investigation. There are two main issues to deal with: firstly that the state-space $\mathcal{S}$ is now uncountable, and secondly that monotonicity does not apply for many interesting point process models. We see below how *both* these issues can be dealt with effectively in the case of perfect simulation for the area-interaction point process. (Indeed the methods described here also generalize to repulsive Strauss point processes, including the limiting case of the excluded-area or hard-core point process.) So this paper records a successful response to this challenge. Clearly stochastic geometers must now be more ambitious about their simulation objectives: after the ideas of Propp and Wilson we can no longer be content merely with long-run approximations to equilibrium distributions but instead must strive after perfection.

It should be mentioned that spatial birth-and-death process equilibrium theory is no stranger to coupling techniques, which figure in the seminal work of Preston [15], also in [13] (for hard-core processes) and [14] (where a geometric ergodicity result is established). A recent paper [9] establishes similar results by direct arguments about point processes. What is new here is the application of coupling to provide exact simulation applicable to several varieties of point process.

It is appropriate in the setting of this volume to note that this work was greatly assisted by the World-Wide Web, which enabled rapid and easy access to a preprint version of [17]. Coupling ideas of this kind typically spread extremely rapidly because of their profoundly intuitive nature and (*post hoc!*) simplicity: for example the fundamental reflection Brownian coupling idea due to Lindvall [12] spread so fast that by the time I heard of it the idea had already become detached from the name of its originator and it took some careful detective work to find out who deserved the credit! Propp and Wilson's idea has the same stamp of profound simplicity; the World-Wide Web has ensured that their names remain attached to it. May modern technology continue to be servant rather than master to scholarship.

In the following we shall use some conventional notation from stochastic geometry:

m_r denotes Lebesgue measure for the Euclidean space $\mathbb{R}^r$;

$\oplus$ denotes Minkowski addition applied to sets: thus $A \oplus B = \{x : x = a + b \text{ for } a \in A, b \in B\}$;

If X is a finite point process then $\#(X)$ is the number of points in X.

2. The area-interaction point process X

This process was introduced in a careful study by Baddeley and Van Lieshout [1]. It is intended to be a single point process model which for differing parameters exhibits either clustering (*attraction*) or dispersion (*repulsion*) between points, relative to the Poisson point process. This is achieved by weighting the Poisson point process distribution according to an exponential of the area (in the planar case) of the union of congruent planar bodies centered at the points of the process. This produces a delightfully simple and appealing model. It should be contrasted with the Strauss point process, commonly used in stochastic geometry, which weights according to an exponential of the number of point pairs closer than some fixed threshold: while the Strauss model is successful in modelling repulsion neither it nor obvious generalizations can produce successful models for attraction (indeed in the case of the Strauss model a divergence means the resulting point process fails to exist!). Thus the importance of the area-interaction point process for stochastic geometry lies exactly in the fact that it provides a simple model which for various parameter values can model either attraction or repulsion between points.

In this section we briefly summarize the main features of this point process, as described in [1].

In its simplest form, the area-interaction point process is a random process X of points in $\mathbb{R}^d$, with distribution having a Radon–Nikodým density or weight $p(X)$ with respect to the unit-rate Poisson process restricted to a compact window, where

$$p(X) = \alpha \lambda^{\#(X)} \gamma^{-m_d(X \oplus G)}.$$

Here α is a norming constant, λ, γ are positive parameters, and the *grain* G is a compact (typically convex) subset of $\mathbb{R}^d$. The set $X \oplus G$ is given by

$$X \oplus G = \bigcup \{x \oplus G : x \in X\},$$

so it would be the *Boolean model* formed by grain G and germs from X if X were a standard (i.e. unweighted) Poisson process. Thus the realization $X \oplus G$ can be thought of as a 'weighted Boolean model'.

As is usual in the theory of such point processes, the normalization parameter α is not computable in closed form except in trivial cases: the area-interaction means that it is related to exponential moments of coverage of Boolean models. Of course this intractability means simulation methods are of great importance.

The parameter λ can be thought of as controlling the intensity of X (though in fact the actual intensity of the point process X is a complicated function of λ, γ and G).

The parameter γ controls the *area-interaction* between the points of X: $\gamma > 1$ is the *attractive* case and $\gamma < 1$ is the *repulsive* case.

The attractive case of $\gamma > 1$ (and grain G a ball) was already known in statistical physics as the 'penetrable spheres' model of Widom and Rowlinson [21], important because (unusually for such models) it is possible to prove that phase-transition occurs in case $d > 1$. Further details of this connection also provide a simulation method for area-interaction point processes *via* accept-reject methods producing single-type margins of conditioned two-type Poisson processes, both for $\gamma > 1$ and $\gamma < 1$.

The area-interaction point process is well-defined for all values of the parameters λ, γ, and compact G, satisfies a stability condition of Ruelle type, and can therefore be extended to a well-defined point process over the whole of Euclidean space.

There are several possible generalizations. One which is relevant here is to replace Lebesgue measure by other Borel measures: replacing Lebesgue measure m_d by a suitable multiple of counting measure on a lattice is convenient for simulation purposes.

As well as methods related to conditioned two-type Poisson point processes as indicated above, one can simulate the area-interaction point process using the common method of simulating spatial birth-and-death processes till they are deemed to be close to equilibrium. In the remainder of this paper we will use this as our test case for perfect simulation of point processes: our objective is to describe a modification of spatial birth-and-death process simulation which does indeed give perfect (and practical!) simulation for an area-interaction point process. We will also see that this modification is susceptible to generalization, for example to Strauss point processes.

3. A coupling construction for perfect simulation of X

Consider the task of perfect simulation of an attractive area-interaction process X in a bounded window $W \subset \mathbb{R}^d$. Let $G \subset \mathbb{R}^d$ be the compact non-random grain on which the definition of X is based, let λ be the intensity of the underlying reference Poisson process, and let $\gamma > 1$ be the parameter of (attractive) area-interaction.

We shall use a space-time Boolean model and $[0,1]$-valued marks to build coalescing spatial birth-and-death processes whose common equilibrium distribution is the distribution of X. (These spatial birth-and-death processes provide the analogue of the discrete-time stochastic flow $F_{s,t}$ of Section 1.) The coalescence means that we are able to sample from the spatial birth-and-death processes as if they had been run from time $-\infty$, and consequently our sample must have the equilibrium distribution. A point of finesse is that the coalescence takes place *conditional on the realization of the marked grain process of the Boolean model*: nevertheless the Propp–Wilson argument goes through.

Our construction will assume empty boundary conditions, but non-trivial boundary conditions can be catered for by an easy modification.

The underlying space-time Boolean model is based on grains which are space-time cylinders

$$(G \oplus x) \times [s, s + \ell] \subset \mathbb{R}^d \times \mathbb{R}$$

with spatial cross-section $G \oplus x$. We say that the *birth time* of the above space-time cylinder is s, while its *death time* is $s + \ell$. We parametrize such space-time cylinders by the representing space built from coordinates x, s, ℓ:

$$(x, s, \ell) \in \mathcal{C} = \mathbb{R}^d \times \mathbb{R} \times (0, \infty).$$

We want to mark the grains with independent uniform $[0,1]$ marks, so let Z be the Poisson process on $\mathcal{C} \times [0,1]$ governed by the intensity measure

$$\left(\lambda e^{-\ell} \, \mathbb{I}_{[\, x \in W \,]}\right) \, m_d(dx) \, m_1(ds) \, m_1(d\ell) \, m_1(dp).$$

Hence the intensity measure factorizes as the product of the uniform measure of intensity λ on $W \times \mathbb{R}$, the exponential(1) distribution for length ℓ and the uniform $[0,1]$ distribution for mark p. From Z we derive a space-time Boolean model $\Psi \subset W \times \mathbb{R}$ by taking the union of the grains and forgetting the marks:

$$\Psi = \bigcup \left\{ (G \oplus x) \times [s, s + \ell] \; : \; [(x, s, \ell); \, p] \in Z \right\}.$$

Now fix $-T < 0$: for $u \in [-T, \, 0]$ we define time-evolving point processes

$$Y^{\max}(-T, u), \quad Y^{\min}(-T, u), \quad Y(-T, u)$$

on $W \subseteq \mathbb{R}^d$. They are grown recursively from initial configurations

$$Y^{\max}(-T, -T), \quad Y^{\min}(-T, -T), \quad Y(-T, -T),$$

and will be built using the sub-process of marked points representing those space-time cylinders of Z which overlap $[-T, \, 0]$:

$$Z^{[-T, 0]} = \left\{ [(x, s, \ell); p] \in Z \; : \; -T \le s + \ell, \, s \le 0 \right\}.$$

The initial configurations are defined as follows: set

$$Y^{\max}(-T, -T) = \Big\{ x \ : \ [(x, s, \ell); p] \in Z^{[-T, 0]}, \ s \leq -T \leq s + \ell \Big\}$$

$$Y^{\min}(-T, -T) = \Big\{ x \ : \ [(x, s, \ell); p] \in Z^{[-T, 0]}, \ s \leq -T \leq s + \ell, \ p \leq \gamma^{-m_d(G)} \Big\}$$

while $Y(-T, -T)$ is some arbitrary subset of $Y^{\max}(-T, -T)$ which we do not specify, except to require that it must be a superset of $Y^{\min}(-T, -T)$. We shall call such a $Y(-T, u)$ $(u \in [-T, 0])$ *appropriate*, since it is not necessary to consider any other Y in order to obtain the desired equilibrium.

It will be convenient to define three auxiliary spatial germ-grain models

$$\Xi^{\max}(-T, u), \quad \Xi^{\min}(-T, u), \quad \Xi(-T, u)$$

by

$$\Xi^{\max}(-T, u) = Y^{\max}(-T, u) \oplus G = \bigcup \Big\{ G \oplus x \ : \ x \in Y^{\max}(-T, u) \Big\},$$

$$\Xi^{\min}(-T, u) = Y^{\min}(-T, u) \oplus G = \bigcup \Big\{ G \oplus x \ : \ x \in Y^{\min}(-T, u) \Big\},$$

$$\Xi(-T, u) = Y(-T, u) \oplus G = \bigcup \Big\{ G \oplus x \ : \ x \in Y(-T, u) \Big\}.$$

Now let $-T < t_1 < ... < t_N < 0$ be the ordered list of space-time cylinder birth and death times lying in $[-T, 0]$, so that

$$\{ t_1, ..., t_N \} = [-T, 0] \cap \bigcup \Big\{ s, s + \ell \ : \ [(x, s, \ell); p] \in Z^{[-T, 0]} \Big\}.$$

By simple properties of Poisson processes we may argue that: no two birth/death times coincide; moreover corresponding to each birth/death time t_i there is a unique $[(x, s, \ell); \ p] \in Z$ with $s = t_i$ in case of birth, and $s + \ell = t_i$ in case of death.

We can now give the recursive definition of $Y(-T, u)$ for $u \in [-T, 0]$; the definitions of $Y^{\max}(-T, u)$, $Y^{\min}(-T, u)$ are entirely analogous, but in (c) below replace Ξ by $\Xi^{\max}$, $\Xi^{\min}$ respectively.

Proposition 1. The following algorithm simulates a spatial birth-and-death process.

(a) *Suppose that $Y(-T, u)$ is already defined, and that t_i is the least birth/death time greater than u. We set*

$$Y(-T, u + v) = Y(-T, u) \qquad for \ u \leq u + v < t_i \,.$$

(If there is no such time then we set $Y(-T, u + v) = Y(-T, u)$ for all $v \geq 0$, and the definition of $Y(-T, \cdot)$ is then complete!) What happens next depends on whether t_i is a death time (in which case we go to (b) below) or a birth time (in which case we go to (c) below).

(b) *If t_i is a death time, so $t_i = s + \ell$ for some $[(x, s, \ell); \ p] \in Z$, then we delete the point x: we set*

$$Y(-T, t_i) = Y(-T, t_i-) \setminus \{x\} \,.$$

(Here, of course, $Y(-T, t_i-)$ is the configuration of $Y(-T, \cdot)$ immediately prior to time t_i.) We then continue the construction using (a) above.

(c) If t_i is a birth time, so $t_i = s$ for some $[(x, s, \ell); \, p] \in Z$, then we test whether

$$p \leq \gamma^{-m_d((G \oplus x) \backslash \Xi(-T, t_i -))}.$$

(Recall that

$$\Xi(-T, t_i -) = \Xi(-T, t_i -) \oplus G$$

is given by the union of grains $\bigcup \{ G \oplus x \, : \, x \in Y(-T, t_i -) \}$.) If this is the case then we add the point x: we set

$$Y(-T, t_i) = Y(-T, t_i -) \cup \{x\}.$$

Otherwise we do nothing: we set

$$Y(-T, t_i) = Y(-T, t_i -).$$

In either case we then continue the construction using (a) above.

It is clear that this construction is complete and unambiguous (because to each birth/death time there corresponds a unique representing point of Z as mentioned above) and can be programmed on a computer. Indeed I have written a C program `Perfect`, currently in the experimental stage, which uses the above algorithm to conduct perfect simulations of area-interaction planar point processes, subject to an approximation of Lebesgue measure by a discrete measure as mentioned in Section 2 above. Further information about `Perfect` can be obtained from me at `w.s.kendall@warwick.ac.uk`.

In order to see how this provides perfect simulation it is necessary only to check a number of simple points, which we list below as propositions and whose proofs are largely left as exercises for the reader. (As is often the case with coupling arguments, to give heavily rigorous proofs would be rather like explaining a joke: it tends to spoil the fun! However the following propositions really deserve arguments demanding more space than can be spared here: I hope to deliver these arguments in a follow-up article dealing with further work.) As a consequence of these propositions one can obtain perfect simulations of area-interaction point processes from inspection of simulation runs of these spatial birth-and-death processes.

Proposition 2. The various spatial birth-and-death processes

$$Y^{\max}(-T, u), \quad Y^{\min}(-T, u), \quad Y(-T, u),$$

viewed as functions of $u \in [-T, 0]$, have the distribution of the target area-interaction process X as equilibrium distribution.

To check this is merely a matter of confirming that detailed balance holds and that simple regularity conditions apply, since in equilibrium these spatial birth-and-death processes are reversible. See [16] for the theory of spatial birth-and-death processes and [18], [19] for the use of spatial birth-and-death processes in simulation of point processes such as X. See [20, Chapter 5] for a summary exposition in book form. Section 3 of [1] gives details which apply to the processes $Y^{\max}(-T, \cdot)$, $Y^{\min}(-T, \cdot)$, $Y(-T, \cdot)$.

Proposition 3. The following simple modification of the algorithm in Proposition 1 deals with the repulsive case of $\gamma < 1$: increase the intensity of Z to

$$\left(\gamma^{-m_d(G)} \times \lambda e^{-\ell} \mathbb{I}_{[x \in W]}\right) m_d(dx)\, m_1(ds)\, m_1(d\ell)\, m_1(dp),$$

alter the initialization of $Y^{\min}(-T, T)$ to

$$Y^{\min}(-T, -T) = \left\{x \ : \ [(x, s, \ell); p] \in Z^{[-T, 0]}, \ s \leq -T \leq s + \ell, \ p \leq \gamma^{m_d(G)}\right\},$$

and alter the test in (c) to add x when

$$p \leq \gamma^{m_d(G) - m_d((G \oplus x) \setminus \Xi(-T, t_i-))}.$$

The result of the construction is then a spatial birth-and-death process whose equilibrium distribution is that of a repulsive area-interaction point process of parameter $\gamma < 1$ based on a Poisson point process of underlying intensity λ. (Again this is simply a matter of confirming that detailed balance holds and that simple regularity conditions apply.) As we see below, it is then less clear how one should confirm that exact simulation has taken place, since this is no longer an attractive point process. Nevertheless we will show that it can be done.

The construction of the processes $Y^{\max}(-T, u)$, $Y^{\min}(-T, u)$, $Y(-T, u)$ is based entirely on the randomness provided by $Z^{[-T, 0]}$. For $-S < -T$ the point process $Z^{[-S, 0]}$ can be produced by augmenting $Z^{[-T, 0]}$, and the resulting construction of $Y^{\max}(-S, u)$, $Y^{\min}(-S, u)$, $Y(-S, u)$ is strongly related to the construction over $[-T, 0]$: for example for all S the final patterns $Y^{\max}(-S, 0)$, $Y^{\min}(-S, 0)$, $Y(-S, 0)$ are subsets of a single fixed point pattern

$$\left\{x \ : \ [(x, s, \ell); p] \in Z, \ s \leq 0 \leq s + \ell\right\}.$$

In particular we can consider *all* constructions based on a single given realization of the underlying space-time cylinder process Z. In the attractive case ($\gamma > 1$) this leads to a crucial monotonicity property.

Proposition 4. In the attractive case $\gamma > 1$, if $-S \leq -T \leq u \leq 0$ then

$$Y^{\min}(-T, u) \subseteq Y^{\min}(-S, u) \subseteq Y(-S, u) \subseteq Y^{\max}(-S, u) \subseteq Y^{\max}(-T, u)$$

whenever $Y^{\min}(-S, -S) \subseteq Y(-S, -S) \subseteq Y^{\max}(-S, -S)$.

This is best seen by arguing recursively. Certainly the result holds for $u = -T$, since the algorithm described in Proposition 1 always adds points $[(x, s, \ell); p] \in Z$ for which $p \leq \gamma^{-m_d(G)}$. If the result holds up to time u then we have a similar inclusion for the corresponding random sets:

$$\Xi^{\min}(-T, u) \subseteq \Xi^{\min}(-S, u) \subseteq \Xi(-S, u) \subseteq \Xi^{\max}(-S, u) \subseteq \Xi^{\max}(-T, u).$$

But if $[(x, u, \ell); p] \in Z$ is a point to be added at time u then this monotonicity, together with the birth criterion

$$p \leq \gamma^{-m_d((G \oplus x) \setminus \Xi^*(-T, t_i-))},$$

for Ξ^* running through

$$\Xi^{\min}(-T, u), \quad \Xi^{\min}(-S, u), \quad \Xi(-S, u), \quad \Xi^{\max}(-S, u), \quad \Xi^{\max}(-T, u),$$

implies that x will be added in such a way as to maintain monotonicity. (Remember that we are dealing with the attractive case $\gamma > 1$ here.)

Thus appropriate $Y(-S, \cdot)$ form a subclass of appropriate $Y(-T, \cdot)$. Unfortunately in the repulsive case $\gamma < 1$ the above argument breaks down badly and this monotonicity does not hold. We shall see below (Proposition 7) how to fix this in a practical way. First we note how it is that coalescence allows us to do perfect simulation.

Suppose that we are given an initial point pattern $\{x_1, ..., x_k\}$ with which to start off the simulation at time $-S$. Point x_i lasts for an exponential time r_i (of unit mean), so all initial points die out by time $-T$ if $-S < -T - \max_i r_i$. Using the space-time cylinder process Z, we construct a spatial birth-and-death process $\tilde{Y}(-S, u)$ with initial condition $\{x_1, ..., x_k\}$ at time $-S$ as follows. We set $\tilde{\Xi}(-S, u) = \tilde{Y}(-S, u) \oplus G$ for $-S \leq u \leq 0$, and set $\tilde{Y}(-S, -S) = \{x_1, ..., x_k\}$. The point x_i is deleted from the pattern at time $-S + r_i$. Otherwise $\tilde{Y}(-S, \cdot)$ evolves according to the algorithm described in Proposition 1, using of course $\tilde{\Xi}(-S, \cdot)$ to compute the area bias and so to determine when points from Z are to be added to the pattern.

It follows from the algorithm that for $-T \in [-S + \max_i r_i, 0]$ the point pattern $\tilde{Y}(-S, -T)$ is *appropriate* in the sense defined above: that is to say $Y^{\min}(-T, -T) \subseteq \tilde{Y}(-S, -T) \subseteq Y^{\max}(-T, -T)$. Consequently if T is large enough for coalescence, so that $Y^{\min}(-T, 0) = Y^{\max}(-T, 0)$, then $Y^{\min}(-T, 0) = Y^{\max}(-T, 0) = \tilde{Y}(-S, 0)$ no longer depends on S (as long as $S > T + \max_i r_i$). Hence

(a) $\lim_{S \to \infty} \tilde{Y}(-S, 0)$ exists and has the equilibrium distribution, and
(b) by definition of coalescence being observable, we can choose a 'stopping time' T (i.e. $-T_0 \leq -T$ is observable from the realization $Z^{[-T_0, 0]}$) such that $Y(-T, 0) = \lim_{S \to \infty} \tilde{Y}(-S, 0)$ for all appropriate $Y(-T, 0)$, including $Y^{\min}(-T, 0)$.

Thus *in principle* we can sample from $\lim_{S \to \infty} \tilde{Y}(-S, 0)$ and thus from the equilibrium distribution, which is the desired distribution of the area-interaction point process. A simple percolation argument for the Boolean model Ψ in the bounded window W now shows that for sufficiently large T we have

$$Y^{\min}(-T, 0) = Y^{\max}(-T, 0),$$

and therefore we have coalescence and the possibility at least in principle of perfect simulation.

For suppose that $-\tau$ is the random time which is the supremum of times such that the fixed-time slice $\mathbb{R}^d \times \{0\}$ is *not* topologically connected to $\mathbb{R}^d \times \{-T\}$ by Ψ whenever $-T < -\tau$. (So $\mathbb{R}^d \times \{0\}$ and $\mathbb{R}^d \times \{0\}$ are in different connected components of $\mathbb{R}^d \times \{0\} \cup \Psi \cup \mathbb{R}^d \times \{1\} \subset \mathbb{R}^d \times \mathbb{R}$ whenever $-T < -\tau$.) This is almost surely finite since we have stipulated that the window W must be bounded. Notice also that τ is a 'stopping time', in the sense that the event $[-T < -\tau]$ is measurable with respect to $Z^{[-T, 0]}$, so that τ is observable based only on simulation over the range $[-T, 0]$. It follows easily from the construction that if $-T < -\tau$ then the final values $Y^{\max}(-T, -T), Y^{\min}(-T, -T), Y(-T, -T)$ must all agree and no longer depend on

T or initial conditions. Consequently the construction can be used to make a perfect simulation of an area-interaction process.

Proposition 5. Perfect simulation of an area-interaction point process may be obtained by the following algorithm.

Choose a $-T_1 < 0$, simulate and construct $Z^{[-T_1,\,0]}$, $Y^{\max}(-T_1,0)$. If $-T_1 < -\tau$ then $Y^{\max}(-T_1,0)$ $(= Y(-T_1,0)$ for all appropriate $Y(-T_1,0))$ is a sample from the desired equilibrium distribution. If not then choose $-T_2 < -T_1 < 0$, etc, and repeat till $-T_n < -\tau$. At each stage care must be taken to extend the fundamental space-time cylinder process Z so that $Z^{[-T_{n-1},0]} \subseteq Z^{[-T_n,0]}$. Then return $Y^{\max}(-T_n,0)$ as a sample from the desired equilibrium distribution.

The required extension property $Z^{[-T_{n-1},0]} \subseteq Z^{[-T_n,0]}$ was easily obtained in the C implementation `Perfect` by using pointers to build up a list of the space-time cylinders in $Z^{[-T_{n-1},0]}$, respectively $Z^{[-T_n,0]}$, deriving a list of potential birth/death times, then sorting the derived list in time order in order to conduct the actual simulation.

This provides perfect simulation for area-interaction point processes, whether they are attractive or repulsive. However τ may be large, and so we need more practical methods of determining when coalescence might have happened. Proposition 4 provides the key for the attractive case $\gamma > 1$, as monotonicity gives an easy way of avoiding checking for $-T_n < -\tau$.

Proposition 6. Perfect simulation of an area-interaction point process may be obtained by the following modified algorithm. Increase T in the construction given in Proposition 5 (always by extending previous constructions of the fundamental space-time cylinder process Z) until it is observed that

$$Y^{\max}(-T_n,0) = Y^{\min}(-T_n,0)\,.$$

Then $Y^{\max}(-T_n,0) = Y^{\min}(-T_n,0)$ is a sample from the desired equilibrium distribution.

For in such a case Proposition 4 implies that

$$Y(-S,0) = Y^{\min}(-T,0) = Y^{\max}(-T,0)$$

for all appropriate $Y(-S,0)$ where $-S < -T$, even if $-T_n < -\tau$ does not yet hold, and so coalescence has occurred. Simulation experiments have shown that this monotonicity idea (already used to great effect in the case of finite discrete state-space by [17]) provides a practical means of perfect simulation of an attractive area-interaction spatial birth-and-death process (in contrast the algorithm described in Proposition 5 is unusably slow): for example the pattern of 55 points in Figure 1, with parameters as described there, can be produced in 30 seconds by the (unoptimized) experimental version of `Perfect`, working in a slow multi-tasking environment using less than 640k RAM on my five-year-old personal computer (Acorn A440/1 `ARM3` with `FPA10` floating-point coprocessor). This simulation used 495 birth-or-death steps to produce the 55 points of $Y^{\min}(-T,0) = Y^{\max}(-T,0)$ (incidentally it is interesting to compare this with the *heuristic* recommendations which can be found

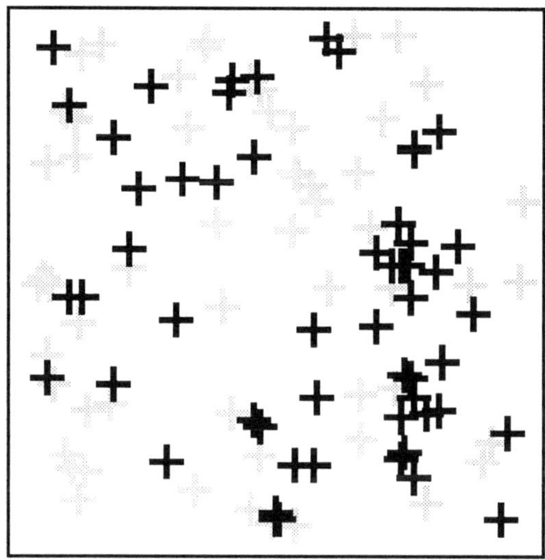

Figure 1. Perfect simulation of attractive area-interaction point process: 55 points, $\gamma = e^2 = 7.39$, $\lambda = 1$, square window of side-length 10, square grain side-length 0.7. Shaded crosses mark centres of sections of space-time cylinders in Z which intersect the time-slice 0.

in the literature, suggesting for example $4\times$ number of points for moderate parameter values for systematic single-point updating algorithms for Strauss point processes).

Monotonicity is not available for the repulsive case, and simulation experiments show that the percolation criterion using τ defined in Proposition 5 is not practical. Indeed my joint PhD student Elke Thönnes has observed that the probability $\mathbb{P}\left[-\tau > -T\right]$ is subject to a crude bound in terms of the distribution of residual busy period for an $M/M/\infty$ queue, but that this is unusably large in practical situations. Better bounds will be obtained from continuum percolation arguments: in particular even when the observation window is unbounded we still know that τ will be almost surely finite whenever the underlying intensity λ (respectively $\tilde{\lambda}\gamma^{-m_d(G)}$ for the repulsive case) is less than a percolation constant. Indeed rather similar percolation arguments have been used in [4] to give elegantly simple derivations of phase transition for the attractive area-interaction point process, in its older guise as the Widom–Rowlinson model. From the point of view of perfect simulation, however, theoretical bounds on $\mathbb{P}\left[-\tau < -T\right]$ *etc* are of lesser importance than the *observability* of $[-\tau < -T_n]$, which suggests that perfect simulation might be a feasible possibility.

It is delightful to report that there is a remarkably simple modification of the algorithm described in Propositions 1, 5, and 6 which is effective in determining equilibrium in the repulsive case.

Proposition 7. Modify the constructions of $Y^{\max}$, $Y^{\min}$ from the method described in Proposition 1 as follows. In (c) of the construction, when t_i is a birth time, the *corresponding point x is added in the case of $Y^{\max}$ when*

$$p \leq \gamma^{m_d(G)-m_d\left((G\oplus x)\setminus\Xi^{\min}(-T,t_i-)\right)},$$

and in the case of $Y^{\min}$ when

$$p \leq \gamma^{m_d(G)-m_d\left((G\oplus x)\setminus\Xi^{\max}(-T,t_i-)\right)}.$$

Then the algorithm which samples $Y^{\max}(-T,0) = Y^{\min}(-T,0)$ after coalescence produces a sample from the relevant repulsive area-interaction point process.

In this algorithm the constructions of $Y^{\max}$, $Y^{\min}$ are now coupled, and are no longer individually Markov. (In order to keep the Markov property we would have to consider the two-type spatial birth-and-death process $(Y^{\max}, Y^{\min})$.) But it is now simple to check by an inductive argument that

$$Y^{\min}(-T,u) \subseteq Y(-S,u) \subseteq Y^{\max}(-T,u)$$

whenever $-S \leq -T$, $u \in [-T,0]$ and

$$Y^{\min}(-T,-T) \subseteq Y(-T,-T) \subseteq Y^{\max}(-T,-T).$$

For the desired relation holds at time $u = -T$, deaths are the same, and the test for a $Y^{\max}$ birth (being based on $Y^{\min}$) is always no more stringent than the test for the corresponding $Y^{\min}$ birth (being based on $Y^{\max}$ which by induction contains $Y^{\min}$), nor indeed is it more stringent than the test for the corresponding Y birth. It therefore follows that if $Y^{\min}(-T,0) = Y^{\max}(-T,0)$ then $Y(-S,0) = Y^{\min}(-T,0) = Y^{\max}(-T,0)$ for all appropriate $Y(-S,0)$ where $-S < -T$, and so the common pattern of $Y^{\min}(-T,0) = Y^{\max}(-T,0)$ is a sample from the equilibrium distribution. In summary, we replace the lower and upper bounding spatial birth-and-death processes of the attractive case with well-behaved lower and upper bounding processes which are no longer spatial birth-and-death processes, but which are well-adapted to our problem!

Simulation experiments have shown that this provides a practical means of perfect simulation of a repulsive area-interaction spatial birth-and-death process: for example the pattern of 55 points in Figure 2, with parameters as described there, can be produced in 7 seconds by **Perfect**, running on the same configuration as above. (That the same number of points were obtained as in the Figure 1 is the result of an informal rejection-sampling undertaken in order to produce comparable figures!) This simulation used 1132 birth-or-death steps to produce the 55 points of $Y^{\min}(-T,0)$, equivalently $Y^{\max}(-T,0)$.

Implementation of the repulsive case required the addition of just three statements in **C** code!

The case of non-empty boundary conditions can be dealt with quite simply for all variants of the algorithm described above, just by adding the boundary points to both $Y^{\min}$ and $Y^{\max}$ for all time.

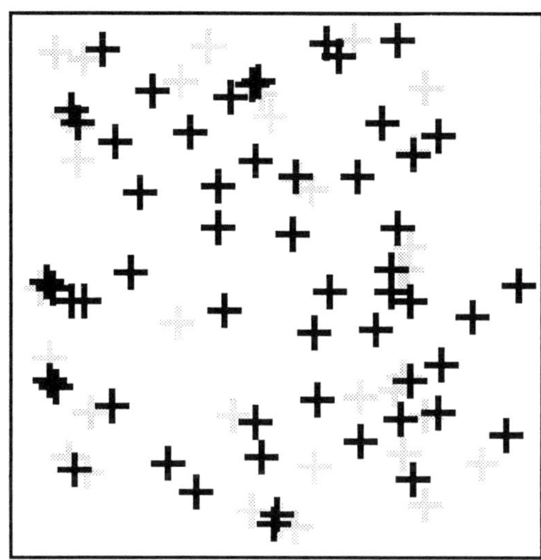

Figure 2.   Perfect simulation of repulsive area-interaction point process: 55 points, $\gamma =$ $e^{-2} = 0.135$, $\lambda = 0.25$, square window of side-length 10, square grain side-length 0.7. Shaded crosses mark centres of sections of space-time cylinders in the modified Z which intersect the time-slice 0.

4. Conclusion

The above shows that perfect simulation is possible for both attractive and repulsive area-interaction point processes. Of course this is just the beginning! Further questions which should now be addressed include the following.

(a) Can one establish good bounds on the run-times required by these perfect convergence algorithms? A combination of heuristics and crude simulation experiments suggests that if L is the linear dimension of (square) window size, then *for fixed parameters which are not too extreme* the time required for simulations over windows of increasing size should be about $O\left(L^2(\log(L))^2\right)$ (the dominant contribution here arising from sorting birth/death incidents arising from the underlying space-time cylinder process over the time interval required for coupling). Whatever the correct asymptotic, the constant multiplier will be extremely sensitive to the parameters. In particular it is interesting to note that the repulsive case $\gamma < 1$ actually results in faster convergence than the attractive case $\gamma > 1$, if one standardizes on observed intensity of points. It would be useful to have a mathematical explanation of this.

(b) The monotonicity of this construction can be exploited to prove monotonic relationships between various area-interaction point processes, and also limit theorems. The simplest example of this is that every area-interaction point process can be placed on a probability space such that it is contained in one Poisson point process, and contains another! Notice also that for fixed λ and for $\gamma > 1$ the construction gives a simple proof that the intensity of the area-interaction point process decreases as γ increases. Finally, for fixed γ the construction shows that for high λ (hence high intensity) the area-interaction point process is nearly Poisson. None of these results is particularly deep (indeed this monotonicity has already been exploited several times in the literature: see [4] for a recent example, and [9] for non-coupling proofs), but they illustrate the point that a good simulation method provides qualitative as well as quantitative insight into a process!

(c) As with the standard spatial birth-and-death simulation approach to point processes, there is an alternative algorithm for perfect simulation based on constant birth rate and variable death rate.

(d) Is there a way of doing perfect simulation for add-delete algorithms for a wide variety of point processes?

(e) Can one extend these ideas to other point processes, and more generally to the situation of Markov chain Monte Carlo sampling (MCMC) from posteriors obtained by conditioning point processes on observed discrete data? Certainly there are simple modifications of the algorithm described in Proposition 7 (which can be implemented by only minor changes in the C program **Perfect**) yielding perfect simulation of Strauss point processes with purely repulsive pairwise interactions (in other words, the amenable case typically arising in spatial statistics contexts), and even excluded volume point processes (but here theoretical considerations, which I have confirmed by simulation experiments, show that the method slows down rapidly as the underlying Poisson intensity approaches a certain directed continuum percolation threshold). More generally, any Markov point process with bounded local energy should be amenable to this approach: this includes (for appropriate parameter values) the connected-component interaction ('continuum random cluster model') proposed by Møller. I plan to follow up on this in a further paper. The question of what can be learned for MCMC seems much harder to me: maybe all one can obtain will be useful diagnostics based on the coupling ideas described above.

(f) A particularly pleasing aspect of such extensions is that purely repulsive Strauss point processes and excluded-area point processes can be simulated using the same underlying space-time cylinder process Z as is used for the area-interaction point processes. This allows for both theoretical and experimental comparisons between rather different kinds of point processes.

(g) It would be useful and possibly of theoretical interest to work out faster ways of determining when convergence has occurred for excluded-area point processes. When viewed in space-time the simulation algorithm above can be regarded as a feed-forward logical neural network, with threshold interaction and combining inputs using inclusive OR. Unfortunately this approach looks infeasible, because of links between coalescence-determination and the NP-complete SATISFIABILITY problem of computer science ([8, page 39]). However perhaps the intrinsic localization of the spatial birth-and-death process can come to the rescue here.

Such questions place a premium on examining and inventing further varieties of coupling methods. For example, consider the challenge, *could reflection coupling be useful in some cases where monotonicity does not apply?* It might be possible to modify the geometric ideas of [10] to produce practical algorithms which in suitable cases recognize (at least approximately) when the final result does not depend on the initial conditions. Experience in stochastic differential geometry suggests that some successful coupling strategies work in rather counter-intuitive situations; see [2]. There is some work on nonhomeomorphic coalescing stochastic flows ([6]) which may be relevant here. In any case, the success of the preliminary work reported above in providing perfect simulation of area-interaction point processes, and particularly the surprisingly minor modification which deals effectively with repulsive area-interaction point processes, encourages me in the view that creativity rather than pessimism is the appropriate response to this challenge.

Acknowledgements

I gratefully acknowledge the encouragement and helpful remarks of Professors J. Møller (Aalborg University) and O. Häggström (Chalmers University, Goteborg).

References

[1] BADDELEY, A. J. AND VAN LIESHOUT, M. N. M. (1995). Area-interaction point processes. *Ann. Inst. Statist. Math.* **47**, 601–619.

[2] BEN AROUS, G., CRANSTON, M. AND KENDALL, W. S. (1995). Coupling constructions for hypoelliptic diffusions: Two examples. In *Stochastic Analysis: Summer Research Institute July 11–30, 1993*. ed. M. Cranston and M. Pinsky. vol. 57. American Mathematical Society, Providence, RI. pp. 193–212.

[3] BESAG, J. AND GREEN, P. J. (1993). Spatial statistics and Bayesian computation (with discussion). *J. Roy. Statist. Soc.* **55**, 25–37.

[4] CHAYES, J. T., CHAYES, L. AND KOTECKÝ, R. (1995). The analysis of the Widom–Rowlinson model by stochastic geometric methods. *Comm. Math. Phys.* **172**, 551–569.

[5] CHEN, M.-F. (1994). Optimal couplings and application to Riemannian geometry. In *Probability Theory and Mathematical Statistics*. ed. B. Grigelionis et al. vol. 1. VPS/TEV. p. 15.

[6] DARLING, R. W. R. (1987). *Constructing Nonhomeomorphic Stochastic Flows*. vol. 376. American Mathematical Society, Providence, RI.

[7] DUFRESNE, D. (1991). The distribution of a perpetuity, with applications to risk theory and pension funding. *Scand. Actuarial J.* 39–79.

[8] GAREY, M. R. AND JOHNSON, D. S. (1979). *Computers and Intractability: A Guide to the Theory of NP-completeness*. Freeman, San Francisco, CA.

[9] GEORGII, H.-O. AND KÜNETH, T. (1996). Stochastic comparison of point random fields. (submitted).

[10] KENDALL, W. S. (1986). Nonnegative Ricci curvature and the Brownian coupling property. *Stochastics* **19**, 111–129.

[11] LETAC, G. (1986). A contraction principle for certain Markov chains and its applications. In *Random Matrices and their Applications*. ed. J. Cohen, H. Kesten, and C. Newman. vol. 50. pp. 263–273.

[12] LINDVALL, T. (1982). On coupling of Brownian motions. *Technical report*. University of Göteborg Department of Mathematics, Chalmers University of Technology & University of Göteborg.

[13] LOTWICK, H. W. AND SILVERMAN, B. W. (1981). Convergence of spatial birth-and-death processes. *Math. Proc. Camb. Phil. Soc.* **90**, 155–165.

[14] MØLLER, J. (1989). On the rate of convergence of spatial birth-and-death processes. *Ann. Inst. Statist. Math.* **41**, 565–581.

[15] PRESTON, C. J. (1976). *Random Fields*. vol. 534 of *Springer Lecture Notes in Mathematics*. Springer, Berlin.

[16] PRESTON, C. J. (1977). Spatial birth-and-death processes. *Bull. Inst. Int. Statist.* **46 (2)**, 371–391.

[17] PROPP, J. G. AND WILSON, D. B. (1996). Exact sampling with coupled markov chains and applications to statistical mechanics. *Random Structures and Algorithms* **9,** 223–252.

[18] RIPLEY, B. D. (1977). Modelling spatial patterns (with discussion). *J. Roy. Statist. Soc.* **39,** 172–212.

[19] RIPLEY, B. D. (1979). Simulating spatial patterns: dependent samples from a multivariate density. *Appl. Statist.* **28,** 109–112.

[20] STOYAN, D., KENDALL, W. S. AND MECKE, J. (1995). *Stochastic Geometry and its Applications.* 2nd edn. Wiley, Chichester.

[21] WIDOM, B. AND ROWLINSON, J. S. (1970). New model for the study of liquid–vapor phase transitions. *J. Chem. Phys.* **52,** 1670–1684.

QUANTUM PROBABILITY SEEN BY A CLASSICAL PROBABILIST

P.A. MEYER,[*] *Université de Strasbourg*

My first contact with QP took place in 1982 at the Bangalore conference on Random Fields, where I heard Hudson and Parthasarathy lecture on their recent definition of Boson Brownian motion. I found their talk fascinating, and I decided to learn this language. I will try to explain to classical probabilists the kind of pleasure I had with QP for the following ten years or so.

What is QP? In all cases, probability theory deals with random variables, with their laws, and what we call their 'dependence'. In classical probability, dependence is characterized by giving joint laws, and independence in particular is characterized by joint laws of product type. The difference between QP and classical probability lies in the definition of random variables. To avoid discussions, let us accept the definition of von Neumann: a random variable with values in a measurable space $(E, \mathcal{E})$ is a spectral measure X over E, that is, a countably additive mapping from measurable sets $A \in \mathcal{E}$ to *projections* X_A in some (given) complex Hilbert space $\mathcal{H}$, such that $X_E = I$. Given a unit vector $\mathbf{1}$ in $\mathcal{H}$, the *law of X in the (pure) state $\mathbf{1}$* is the probability measure $A \longmapsto <\mathbf{1}, X_A \mathbf{1}>$. In the case of real valued r.v.'s, the 1–1 correspondence between spectral measures and selfadjoint (possibly unbounded) operators, leads to the standard language of quantum mechanics.

To interpret in this way a classical E-valued random variable X on Kolmogorov's $(\Omega, \mathcal{F}, \mathcal{P})$, we take for $\mathcal{H}$ the Hilbert space $L^2(\mathcal{P})$, for X_A the *multiplication operator* by $I_{\{X \in A\}}$. In the pure state $\mathbf{1}$ corresponding to the vector $1 \in L^2$, the law of X is the same in the classical and the quantum sense.

In classical probability, any two random variables X, Y taking values in E, F define jointly a random variable $Z = (X, Y)$ taking values in $E \times F$ whose 'marginals' are X and Y. In QP, the existence of such a spectral measure implies that X *and Y commute,* and conversely, if they do we may define Z by the formula $Z_{A \times B} = X_A Y_B$. Thus *non-commuting r.v.'s have no joint law in QP.* On the other hand, the fun of QP comes from the positive statement: given a family, possibly uncountable, of commuting random variables X_t, they do have a joint law, and therefore give rise to a stochastic process in the classical sense. This is the quantum substitute for the Kolmogorov construction theorem, and applying it to get a variety of processes is the classical probabilist's reward.

A last remark: the space of all complex, finite linear combinations of indicator functions is an algebra $\mathcal{A}$, with involution $*$ given by complex conjugation, and we may extend by linearity a random variable X taking values in E to a *homomorphism*

---

[*] Postal address: Département de Mathématiques, Université de Strasbourg, 67 Strasbourg, France.

(preserving involution and unit) from $\mathcal{A}$ to the $*$-algebra $\mathcal{L}(\mathcal{H})$ of bounded operators on $\mathcal{H}$. Replacing $\mathcal{A}$ by a non-commutative algebra opens the way to a broader view of QP, in which it becomes a set of prescriptions to 'extract probability from algebra'.

This is not foreign to the spirit of classical probability: a few years ago, W. Kendall was testing whether the then prevalent symbolic program 'Reduce' could handle Stochastic Calculus computations*. He found that the only 'probabilistic' instructions that had to be fed to the computer were '$\mathbb{E}(dX_t) = 0$' and '$dX_t^2 = dt$'. He commented to me recently that the main mathematical problem in the passage from algebra to probability was the gap between a local martingale and a true martingale, somewhat similar to the gap between a symmetric and a self-adjoint operator.

Fock space. Let us take for $(\Omega, \mathcal{F}, \mathbb{P})$ the classical Wiener space for one-dimensional Brownian motion (X_t) starting at 0, for $\mathcal{H}$ the Hilbert space $L^2(\mathbb{P})$, for $\mathbf{1}$ the function 1. We are familiar with the σ–fields $\mathcal{F}_t$ generated by the process up to time t, and $\mathcal{F}^t$ generated by the *increments* of the process after time t; they are independent. The corresponding L^2 spaces are denoted by $\mathcal{H}_t$ and $\mathcal{H}^t$, and the independence property translates into the fact that $\mathcal{H} = \mathcal{H}_t \otimes \mathcal{H}^t$. $\mathcal{H}^t$ itself can be split again in this way, thus $\mathcal{H}$ has what is called a *continuous tensor product* structure.

We recall two important facts about Brownian motion. First, it is a 'normal martingale', by which we mean a square integrable martingale such that $<X, X>_t = t$ (that is, $X_t^2 - t$ is a martingale); this allows the definition of stochastic integrals $\int H_s dX_s$ provided H is adapted (more precisely, predictable) and $\mathbb{E}[\int |H_s^2|ds] < \infty$. Then every square integrable r.v. H has a representation

$$(1) \qquad\qquad H = \mathbb{E}[H] + \int_0^\infty H_s \, dX_s, \mathbb{E}[\int |H_s|^2 dX_s] < \infty \ .$$

This is called PRP, the *Predictable Representation Property* of Brownian motion. A more precise result is the celebrated *Wiener chaos expansion*,

$$H = \mathbb{E}[H] + \sum_{n \geq 1} \int_{s_1 < \cdots < s_n} h_n(s_1, \ldots, s_n) \, dX_{s_1} \ldots dX_{s_n}$$

where h_n is a L^2 function on the 'simplex' $\Sigma_n = \{s_1 < \cdots < s_n\}$, and the multiple integral (defined by iteration) defines an isometry between $L^2(\Sigma_n)$ and a subspace of $L^2(\Omega)$ called the *n-th Wiener chaos* $\mathcal{C}_n$ — by convention $\mathcal{C}_0$ consists of all (complex) constant r.v.'s. These subspaces make sense for an arbitrary normal martingale, and the spaces $\mathcal{C}_n$ turn out to be orthogonal. We say that X has the *Chaos Representation Property* (CRP) if their sum is dense in L^2 (of the natural σ-field of X). This chaos expansion formally arises from iterating the PRP: we express H_s in (1) as $\mathbb{E}[H_s] + \int_0^s H_{st} dX_t$ (integral extended up to time s by adaptation), separate the first term which becomes the $h_1(s)$ in the first Wiener integral, apply the PRP again to H_{st}, etc. Thus the proof of the CRP consists in showing that a remainder term tends to 0 in L^2.

Let now N_t be a Poisson process with unit jumps and intensity λ. Then $N_t - \lambda t$ is a martingale with square bracket N_t, and therefore $Y_t = (N_t - \lambda t)/\sqrt{\lambda}$ is normal. It

---

* He has since written a complete Stochastic Calculus package called 'Ito', usable with different programs.

was already known to Wiener that it has the CRP (and therefore the PRP). Thus *the Hilbert spaces for Brownian motion and Poisson processes have the same description in terms of multiple integrals.* This is more than all infinite dimensional Hilbert spaces being isomorphic: they have a lot of structure in common — the filtration of Hilbert spaces $\mathcal{H}_t$, the continuous tensor product structure, the distinguished curves in Hilbert space spanned by X_t or Y_t are the same (thus from now on we will use the same notation). Also the Ito integrals of adapted processes are the same. This structure is called *simple Fock space,* and it turns out to be the quantum probabilistic structure underlying Brownian motion and Poisson processes.

Multiple Fock spaces arise when we consider several (possibly infinitely many) independent Brownian motions or Poisson processes; X_t in this case is a vector process with components X_t^α, and we get large numbers of multiple integrals

$$\int_{\{s_1<\ldots<s_n\}} f(s_1,\ldots,s_n)\, dX_{s_1}^{\alpha_1}\ldots dX_{s_n}^{\alpha_n}$$

mutually orthogonal for different choices of n, $\alpha_1,\ldots,\alpha_n$.

Then where lies the difference between Brownian motion and Poisson processes? *It lies in the multiplication.* Two vectors in $\mathcal{H}$ are multiplied in different ways when considered as random variables in Wiener space or in Poisson space. This is a subtle matter since multiplication does not map $L^2\times L^2$ to L^2, but to L^1, but still the general idea is right: probability becomes closely tied with associative algebras, and the possibility opens to include as well *non-commutative* multiplications into probability. Why? Not only because 'it's nice to know it can be done', but because the stability of matter in the real world depends on the existence of *fermions.*

Creation, annihilation and number operators. We return to Brownian motion, consider a (deterministic) real-valued square integrable function $h(s)$ and denote by X_h the Gaussian r.v. $\int h(s)\,dX_s$. For $h = I_{[0,t]}$ we simply get X_t. We denote by Q_h the (unbounded self-adjoint) operator of multiplication by X_h. The multiplication formula for stochastic integrals says that Q_h acting on a vector $F \in \mathcal{C}_n$ (the n-th chaos) produces two terms, in $\mathcal{C}_{n+1}$ and $\mathcal{C}_{n-1}$. Let us denote them by $a_h^+ F$ and $a_h^- F$ respectively. These two mappings are called *creation and annihilation operators.* They are unbounded in the L^2 sense, like Q_h itself, but closable, and mutually adjoint in the precise sense of this term in Hilbert space theory.

Annihilation operators have a simple description: a_h^- is nothing but the derivative operator along the Cameron-Martin function $\eta(t) = \int_0^t h(s)\,ds$, a good friend of everyone interested in Wiener space analysis. The interpretation of creation operators is slightly more complicated (they are Skorohod integrals).

Creation and annihilation operators satisfy the so called CCR,

$$[a_h^-, a_k^+] = <h, k> I \ .$$

Now, given two mutually adjoint operators, they have another useful self-adjoint combination. Let us put

$$P_h = i\,(a_h^+ - a_h^-) \ .$$

This is only a formal definition, since we do not give explicitly the domain. To prove that an operator A is self-adjoint, a powerful way consists in showing that it generates an unitary group e^{itA}. It turns out that in the case of P_h, this unitary group is explicit, and much more is true: let T be the mapping $\omega \longmapsto \omega + \eta$ from Ω into itself (a Cameron-Martin translation) and let ϕ be the well-known Cameron-Martin density of $T(\mathbb{P})$ w.r.t. $\mathbb{P}$. Then the mapping $f \longmapsto (f\phi^{-1/2})\circ T$ is unitary, and all these mappings corresponding to all possible Cameron-Martin functions η commute. Then the spectral measures corresponding to their self-adjoint generators also commute, and *they can be considered as classical random variables*. In fact, there is a unitary mapping of $L^2(\mathbb{P})$, called the Fourier-Wiener transform, which intertwins Q_h and P_h and preserves the state $\mathbf{1}$, and therefore (P_h) is a Gaussian field like (Q_h), with the same law, and (P_t) is a new Brownian motion. The operators Q_h and P_h are related like position and momentum operators in classical quantum mechanics, and satisfy a 'canonical commutation relation' $[P_h, Q_k] = -2i < h, k > I$

Let us now return to Poisson processes. Every compensated Poisson martingale X_t, normalized so that $X_t^2 - t$ is a martingale, is characterized by its jump height c,and we may allow also a negative height; $c = 0$ is the limiting case of Brownian motion. It turns out — and this may be the most attractive result of the H–P paper for a classical probabilist — that the multiplication operator by X_t is given by

$$X_t = Q_t + cN_t \ .$$

N_t is a new self-adjoint operator, the *number operator* up to time t. For $t = \infty$ it is the standard number or Ornstein–Uhlenbeck operator on $\mathcal{H}$ from 'Malliavin's calculus'; for $t < \infty$ it is the same, but acting only on the $\mathcal{H}_t$ factor. Thus, contrary to $a_t^\pm$ it is a second order operator.

It turns out — and this had been somewhat overlooked by physicists — that N_t must be introduced as a basic piece in the building-up of quantum stochastic calculus, on the same footing as $a_t^\pm$. For this reason, I like denoting it a_t°. It is also interesting to note that in the multiple Fock space case, a^+ and a^- become a line and a column of operators, while a° becomes a square matrix.

Quantum stochastic calculus. Quantum stochastic calculus was developed by Hudson-Parthasarathy in close analogy with the classical Ito calculus. It consists of three main ideas.

1) To define 'stochastic integrals' of operators

$$A_t = A_0 + \int_0^t H_s \, da_s^- + J_s \, da_s^\circ + K_s \, da_s^+ + L_s \, ds \ ,$$

and to get at least a product formula (called somewhat improperly a quantum Ito formula), to express the operator product $A_t B_t$ of two stochastic integrals as a stochastic integral.

2) To investigate which 'processes of operators' admit such a representation. In particular, since it is easy to define families of operators in this Fock space set-up which generalize martingales, the question arises whether all reasonable martingales are representable as stochastic integrals.

3) To develop a theory of stochastic differential equations, linear and non-linear ('quantum flows').

These are not problems to be described in a short time, but there are a few remarks I would like to mention.

First of all, why just these three processes, which came out of the analysis of Wiener and Poisson multiplication operators? We got a neat answer from the brief passage of J.-L. Journi through Quantum Probability: Fock space has a discrete analogue ('quantum Bernoulli space'), which is simply the L^2 space of the Bernoulli random walk. This situation is so simple that it can be completely analyzed, the new information at each time being described by a (2,2) matrix, and the four processes correspond to the four arguments of that matrix. Thus we need no less and no more than 4 processes, and 3 basic martingales, coming exactly on the same footing: creation and annihilation have no privilege over number operators.

Second point, Journi gave a beautiful example of a bounded quantum martingale that *is not* representable as a stochastic integral, except in a very loose, distribution sense. In the positive direction, Parthasarathy and Sinha studied the representation problem, and gave simple criteria for representable martingales, which were recently transformed by Attal into simple criteria for representable 'semimartingales' of operators.

Third point, the Ito table. All the calculus of Brownian motion can be deduced from the formula $dX_t^2 = dt$ (and $dt\,dX_t = 0$) — the QP-minded probabilist will add the commutation of dX_s and dX_t for $s \neq t$. For the (normalized) compensated Poisson processes with jump size c, this formula becomes

$$dX_t^2 = dt + c\,dX_t \ .$$

Let me mention that, if the rule $dX_s^2 = dt$ is preserved, but for $s \neq t$ dX_s and dX_t are required to *anticommute,* we get the *fermion noise stochastic calculus,* which was developed in a series of papers by Barnett, Streater and Wilde. However, from the point of view of quantum stochastic calculus (not of physics) fermions can be reduced to bosons by a 'continuous Jordan-Wigner transformation', and therefore no separate theory seems necessary.

In the quantum case, the Ito table should involve all the products of the da_t^ε between themselves and with dt. It is shorter to write only the products which are different from 0:

$$da_t^-\,da_t^\circ = da_t^-,\, da_t^-\,da_t^+ = dt,\, da_t^\circ\,da_t^+ = da_t^+ \ .$$

For $s \neq t$, commutation is assumed to hold between all operators.

The search for the CRP. Brownian motion and compensated Poisson processes have independent increments. On the other hand, multiple integrals exist and are orthogonal for any normal martingale X_t. Thus the problem arises of finding new examples of normal martingales, in one or several dimensions, which possess the CRP. A necessary condition is that they should possess the PRP, a condition which had been investigated for purely probabilistic reasons (PRP is a natural condition in mathematical finance).

Our first remark will be that in discrete time, and with the appropriate version of 'normality', the PRP can be studied very explicitly, and a simple dimension argument shows that it always implies the CRP. This gave good hopes for continuous time, which have turned out to be frustrated.

In his main paper on closed random sets, Azima came across two beautiful normal martingales connected with the set of zeroes of Brownian motion (B_t), and proved they have the PRP. The first one is given by the explicit formula

$$X_t = \operatorname{sgn}(B_t)\,\sqrt{(2(t - G_t))}$$

where G_t is the last zero before time t. A sample function is easy to draw: it is deterministic (quadratic) in every excursion interval, and keeps the memory of the interval itself and of the excursion sign. The second martingale is $|X_t| - \mathcal{L}_t$, subtracting a local time to compensate the submartingale $|X_t|$. It keeps the memory of the excursion intervals, but no longer of the excursion signs. Both martingales are normal and have the PRP. Only the first Azima martingale has been studied from the point of view of CRP.

Let us remark that if a martingale (X_t) is normal, the process $[X, X]_t - t$ is a martingale, where $[X, X]$ is its 'square bracket' from martingale theory. If X has the PRP, this martingale is a stochastic integral, and therefore we have a relation which we write in differential form

$$d[X, X]_t = dt + \Phi_t\, dX_t$$

and call (following Emery) the *structure equation* of the martingale — or rather, it attempts to describe a martingale *law*. When Φ_t depends only on X_{t-} (the left limit is required by previsibility) we have a candidate for a Markov process, but usually Φ depends on the whole past. As a general subject, structure equations are disappointing, as they may have no solution at all, or too many, but concrete structure equations have interest, as the following examples show. The simplest equation is $d[X, X]_t = dt$, and describes Brownian motion. The next is $d[X, X]_t = dt + c\,dX_t$, and describes a compensated Poisson process with jump height c. The following level of complexity is

$$d[X, X]_t = dt + \beta X_{t-}\, dX_t$$

(we might have included a term $a\,dX_t$ on the right, but one can show it makes no essential difference). *The case $\beta = -1$, as discovered by Emery, is the first Azima martingale,* and it has become standard to call the whole family 'the Azima martingales'. For $\beta = -2$ another interesting process occurs, which has a deterministic absolute value $|X_t| = \sqrt{t}$ and was discovered by Protter-Sharpe as an illustration of Gilat's theorem. Between these values $\beta = -2$ and $\beta = 0$ lies the so called *good interval* for the Azima family, as shown by Emery:

Theorem 1 In the good interval, the solutions of the structure equation (starting from any initial value) are unique in law, define a Markov process, and possess the CRP.— Much less is known on the two outside intervals. Practically nothing on the right side, somewhat more on the left side. In this case Emery proved that uniqueness in law (and therefore the PRP) holds without restriction, and the CRP for all initial values except possibly 0. Also for the initial value 0, the conditional CRP after time ε holds for every $\varepsilon > 0$ — that is, the coefficients of a chaotic expansion are not required to be deterministic, but only to be $\mathcal{F}_\varepsilon$-measurable.

It is likely that simple processes of this kind do occur in nature, and one should keep a watchful eye to find their hiding place.

On the other hand, in a paper presented at a recent St Petersburgh symposium, Emery has given an astonishing counterexample of a normal martingale which has the PRP, the conditional CRP for any $\varepsilon > 0$, and for which the CRP fails.

Before leaving this subject, let me mention the following result of Parthasarathy. In the good interval, the L^2 space of the Azima martingale is Fock space, and therefore the multiplication operator by X_t can be interpreted as a self-adjoint (bounded) operator on this space. As such, it satisfies a quantum s.d.e.

$$dX_t = dQ_t + \beta X_t \, dN_t$$

for which existence and uniqueness can be proved. Also, Schürmann has been able to identify the Azima martingales as commutative components of vector valued (non-commutative) processes with independent increments, and this relates the theory to quantum groups.

One point must be clarified: how can it occur that the Azima martingale has a L^2 space which splits as $\mathcal{H}_t \otimes \mathcal{H}^t$ without having independent increments? The answer is that $\mathcal{H}^t$ is the space of all r.v.'s whose chaos expansion vanishes before time t, and this is *strictly smaller* than the L^2 space generated by the increments after time t.

Structure equations in several dimensions have been recently discussed by Attal and Emery, though their very fine results are purely commutative. But there is a much simpler result that deserves to be mentioned. It is due to Biane:

Theorem 2 Let (X_t) be a continuous time irreducible Markov chain with $n + 1$ states. Then the L^2 space it generates is isomorphic to Fock space with multiplicity n.— As for the chain itself, it can be shown to satisfy a simple quantum stochastic differential equation. Thus Kolmogorov's intuition that diffusions and Markov chains have much in common, has a precise justification in a non-commutative set-up.

Thus QP has led to the discovery of a new property of finite Markov chains, the simplest and oldest of stochastic processes!

Appendix

For the reader's convenience, I give now some information on more technical points: the way a classical s.d.e. or a Markov process can be interpreted in the non-commutative world. I will allow myself to be sketchy.

For simplicity, let us consider a diffusion process $X_t = (X_t^i)$ taking values in $\mathbb{R}^d$, satisfying the I to s.d.e.

$$X_t = X_0 + \sum_\alpha \int_0^t a_\alpha(X_s) \, dB_s^\alpha + \int_0^t b(X_s) \, ds \ .$$

Generally one takes $X_0 = x$, a fixed point. Taking expectations, one gets a semigroup on functions

$$P_t f(x) = \mathbb{E}[f \circ X_t(x, \omega)] \ ,$$

whose generator is a second order operator

$$L = \tfrac{1}{2} \sum_{\alpha i j} a_\alpha^i a_\alpha^j D_{ij} + \sum_i b^i D_i \ .$$

The function of three variables $X(x, t, \omega)$, suitably made precise, is called a *stochastic flow,* whence the name of *quantum flows* for the corresponding non-commutative object below.

The above s.d.e. is meaningful in a manifold E, taking the x^i to be local coordinates, but this raises delicate enough localisation and pasting problemes. It is therefore interesting to use the following global approach: apply to the equation an arbitrary $\mathcal{C}^\infty$ function f on E, and use Ito's formula. Then we have

$$f(X_t) = f(X_0) + \sum_\alpha \int_0^t L_\alpha f(X_s) \, dB_s^\alpha + \int_0^t Lf(X_s) \, ds$$

where $L_\alpha f = \sum_i a_\alpha^i D_i f$ and L is the generator as above.

A useful notation consists in defining $dB_t^0 = dt$, $L_0 = L$, and distinguishing two kinds of greek indices: $\alpha, \beta, \gamma \ldots$ which do not assume the value 0, and $\rho, \sigma, \tau \ldots$ which assume it.

Then the construction is transcribed into algebraic language: we have two 'quantum spaces', the manifold E and the probability space Ω, both represented by unital algebras, $\mathcal{A} = \mathcal{C}^\infty(E)$ for E and $\mathcal{B} = L^\infty(\Omega)$ for Ω — more precisely, the role of time is described by an increasing family $(\mathcal{B}_t)$ of algebres on Ω. The unknown process becomes a family of homomorphisms from $\mathcal{A}$ to $\mathcal{B}_t$, $f \longmapsto f \circ X_t$. The initial value is a homomorphism from $\mathcal{A}$ to $\mathcal{B}_0$, $f \longmapsto f(X_0)$ (for example, the mapping $f \longmapsto f(x) \, 1$). Then it appears that the Brownian r.v.'s B_t^α describe something 'flat' while X_t is 'curved'. The coefficients L_α and L are now mappings from $\mathcal{A}$ to itself (note the loss of generality: we wouldn't khow what to do with rough coefficients). Writing that the mappings $f \longmapsto f \circ X_t$ are homomorphisms, we get interesting relations:

1) $L_\alpha(fg) - fL_\alpha g - (L_\alpha f)\, g = 0$ (each L_α is a derivation of $\mathcal{A}$);
2) $L(fg) - fLg - (Lf)\, g = \sum_\alpha L_\alpha f L_\alpha g$.

We now extend the formalism to a non-commutative situation.

We replace the algebra $\mathcal{A}$ by any unital complex algebra with involution (all homomorphisms being assumed to preserve involutions and units). We assume for simplicity that $\mathcal{A}$ is a concrete $*$-algebra (not a C–algebra in general) of bounded operators on some (pre)Hilbert space $\mathcal{J}$, called the *initial space.* In the classical situation $\mathcal{J}$ could be the L^2 space of some 'Lebesgue measure' on the manifold.

The source of 'non-commutative noise' (the flat space) is a Fock space Φ over $\mathcal{H} = L^2(\mathbb{R}_+, \mathcal{K})$; the dimension of $\mathcal{K}$ is called the *multiplicity* and can be infinite. We fix an orthonormal basis (e^α) of $\mathcal{K}$. The upper index corresponds to the usual position of indexes for the components of a Brownian motion, but is troublesome from the point of view of linear algebra. Then Φ is isomorphic to the L^2 space of a family of independent Brownian motions (B_t^α). As before, Φ has for each t a tensor product decomposition $\Phi_t \otimes \Phi^t$. We put $\Psi = \mathcal{J} \otimes \Phi$, $\Psi_t = \mathcal{J} \otimes \Phi_t$. The algebra $\mathcal{B}_t$ consists of all bounded operators on Ψ which operate on the first factor Ψ_t. Then our 'diffusion' consists of a family of homomorphisms $X_t : \mathcal{A} \longmapsto \mathcal{B}_t$, while an initial homomorphism X_0 is given, usually $f \longmapsto f \otimes I$.

The classical Brownien motion is replaced by a huge collection of operators. We have first all *creators and annihilators* on Fock space: creators $a^\alpha(t)$, annihilators $a_\alpha(t)$. Then we also need analogues of the number operator. A bounded operator

M on $\mathcal{K}$ gives rise to an operator $M_t = I_{[0,t]} \otimes M$ on $\mathcal{H} = L^2(\mathbb{R}_+) \otimes \mathcal{K}$, which is extended by differential second quantization as an operator on Fock space. The most important case for us is that of $M = e_\beta \otimes e^\alpha$ of rank 1: then its differential second quantization is denoted by $a_\beta^\alpha(t)$, and this matrix plays the role of N_t above.

All these notations can be unified if we introduce an additional index: we call $a_0^\alpha(t)$ and $a_\alpha^0(t)$ the above creator and annihilator, and $a_0^0(t)$ is multiplication by the scalar function $I_{[0,t]}$.

All these operators together constitute the non-commutative Boson noise. The creation and annihilation operators describe the 'diffusion part', and the huge matrix provides the 'jumps'.

Let us return for an instant to classical Brownian motion dB_t^α. Taking into account the definition $dB_t^0 = dt$, we may rewrite the classical Ito table as follows

$$dB_t^\rho \, dB_t^\sigma = \widehat{\delta}^{\rho\sigma} \, dt \; ,$$

where the 'Evans symbol' $\widehat{\delta}^{\rho\sigma}$ differs from the usual Kronecker symbol by the relation $\widehat{\delta}^{00} = 0$. General stochastic integrals for a family $\boldsymbol{H}$ of operators defined on a suitable domain (generally the so-called *exponential domain* which are *adapted,* i.e. $H_\sigma^\rho(t)$ acts only on the first factor of the decomposition $\Psi = \mathcal{J} \otimes \Phi = (\mathcal{J} \otimes \Phi_t) \otimes \Phi^t$. Such a stochastic integral can be written

$$I_t(\boldsymbol{H}) = \eta + \sum_{\rho\sigma} \int_0^t H_\sigma^\rho(s) \, da_\rho^\sigma(s) \; .$$

Here η acts only on the initial space. The differential elements $da_\rho^\sigma(t)$ are combined according to the *Evans table*

$$da_\lambda^\mu(t) \, da_\rho^\sigma(t) = \widehat{\delta}_\lambda^\sigma \, da_\rho^\mu(t) \; ,$$

which in the case of creators and annihilators reduce to the canonical commutation relations. The rigorous meaning to be given to this equation (as an 'integration by parts' formula for operators was one of the main points of the seminal paper by Hudson and Parthasarathy, and was freed from thorny domain problems by Attal.

We now generalize the Ito theory of (non-linear) s.d.e.'s. We are given an algebra $\mathcal{A}$ of operators on the initial Hilbert space $\mathcal{J}$, which plays the role of a manifold algebra $C^?$??, and a family of mappings $\mathbf{L}_\sigma^\rho$ from $\mathcal{A}$ to itself. An 'initial mapping' X_0 from $\mathcal{A}$ to the algebra of all operators on $\mathcal{J}$ (also considered as operators on Ψ) is also given, and the object to be constructed is an adapted family of mappings X_t from $\mathcal{A}$ to operators on Ψ, adapted meaning that X_t acts only up to time t. Instead of writing $X_t(F)$ for $F \in \mathcal{A}$, we will write $F \circ X_t$, as if X_t were an E–valued r.v. and F were a function on E. This is heretical, but brings closer together que classical and the non-commutative definitions. The equation of the flow then is:

$$F \circ * X_t = F \circ * X_0 + \sum_{\rho\sigma} \int_0^t \mathbf{L}_\sigma^\rho F \circ * X_s \, da_\rho^\sigma(s) \; .$$

More precisely, one assumes most of the time that X_0 is a (norm contracting) homomorphism between algebras, and one expects that the same will hold true for X_t.

This puts some formal restrictions (as in the classical case) on the flow coefficients $\mathbf{L}_\sigma^\rho$. They are called *structure equations,* but of course they bear no relation with the structure equations of martingales!

$$(1) \qquad\qquad \mathbf{L}_\sigma^\rho(I) = 0 \; ,$$

$$(2) \qquad\qquad (\mathbf{L}_\rho^\sigma(F)) = \mathbf{L}_\sigma^\rho(F) \; ,$$

$$(3) \qquad \mathbf{L}_\sigma^\rho(FG) - F\mathbf{L}_\sigma^\rho(G) - \mathbf{L}_\sigma^\rho(F)\,G = \sum_\alpha \mathbf{L}_\sigma^\alpha(F)\,\mathbf{L}_\alpha^\rho(G)| \; .$$

Evans gave the first existence theorem for quantum flows, assuming the coefficients were bounded and the multiplicity finite. Even in this simple looking case, checking that X_t is a homomorphism is highly non-trivial! This is even more true of the extension of Evan' result by Mohari, which relaxes both assumptions.

For some time much effort was dedicated to proving a theorem on quantum flows which would include the global existence theorem for smooth Ito equations on a smooth *compact* manifold, but they stopped a little short of the goal: the best theorems I know (Fagnola, Chebotarev) demand some ellipticity.

I will finish this talk with an introduction to the Bhat–Parthasarathy approch to Markov processes, in the most elementary case: time is reduced to 'before' and 'after'.

We recall the definition of a *(transition) kernel* in classical probability. Given two measurable spaces $(E_0, \mathcal{E}_0)$ and $(E_0, \mathcal{E}_1)$, a kernel P from E_0 to E_1 associates with every point x of E_0 a probability measure $P(x, dy)$ on E_1, depending measurably on x. Thus for every bounded measurable function f on E_1 (note the reverse order), the function $Pf : x \longmapsto \int P(x, dy) f(y)$ is measurable on E_0, and P may be described as a mapping from (measurable) functions on E_1 to (measurable) functions on E_0, which is positive, transforms 1 into 1, and satisfies a countable additivity property.

Every measurable mapping h from E_0 to E_1 defines a kernel H, such that $H(x, dy) = \varepsilon_{h(x)}(dy)$. Such a kernel is deterministic, in the following sense: if one imagines a kernel $P(x, dy)$ as the probability distribution of bullets fired on E_1 from a gun located at $x \in E_0$, then the gun corresponding to the kernel H has no spreading: all bullets from x fall at the same point $h(x)$. This is reflected into the algebraic property that $H(fg) = H(f)\,H(g)$ for any two (bounded measurable) functions f, g on F.

Let P be a kernel from E_0 to E_1, and let μ be a probability law (a 'state' as physicists say) on E_0. Then we may build a probability space $(\Omega, \mathcal{F}, \mathbb{P})$, a pair X_0, X_1 of random variables on Ω, taking values in E_0, E_1, such that the law of X_0 is μ and the conditional law of X_1 given that $X_0 = x$ is $P(x, dy)$. The construction is well known: we take Ω to be $E_0 \times E_1$ with the product σ–field, we define X_0 and X_1 to be the two coordinate mappings on $E_0 \times E_1$, and we define the law $\mathbb{P}$ as the only measure on Ω such that — as usual, the tensor product $a_0 \otimes a_1$ of two bounded measurable functions $a_0(x), a_1(y)$ on E_0, E_1 respectively denotes the function $a_0(x)\,a_1(y)$ on $E_0 \times E_1$

$$\mathbb{E}[a_0 \otimes a_1] = \int_{E_0} \mu(dx)\, a_0(x)\, Pa_1(x) \; .$$

We will now translate this construction into non-commutative language. What we get is a well known construction due to Stinespring, which will lead us to the proper definition of non-commutative kernels. Before reading Bath–Parthasarathy I had never understood that construction properly, and even less that it belonged to probability theory. What B–P have done is to iterate Stinespring's construction, extend it to the continuous case, and then beginning to study the process.

It is a fundamental principle of non-commutative geometry that a 'non-commutative space' E *is* an abstract algebra $\mathcal{A}$, whose elements intuitively represent 'functions' on E — additional structure on $\mathcal{A}$ will tell whether they must be undertood as measurable, continuous, differentiable... functions.

Thus the two spaces E_0, E_1 become now two algebras $\mathcal{A}_0, \mathcal{A}_1$, our 'kernel from E_0 to E_1' will be a linear mapping P from $\mathcal{A}_1$ to $\mathcal{A}_0$, and our measure μ will be a state on $\mathcal{A}_0{}^*$. As a minor technical assumption, we will assume $\mathcal{A}_0, \mathcal{A}_1$ are C-*algebras*. The intuitive idea is that in this case we are dealing with the non–commutative analogue of two *compact* spaces and their algebras of continuous functions, and of *Feller kernels* which preserve continuity. Then we need not worry about the countable additivity of measures and kernels. If the reader does not know yet what a C-algebra is, let him simply omit the corresponding details.

It is clear that we must assume P maps I_1 to I_0, and is a positive mapping. However, and this will be our main point, *positivity is not sufficient.*

We are going to construct a pair or 'random variables X_0, X_1 with values in E_0, E_1' such that the law of X_0 is μ, and P somehow describes the conditioning of X_1 by X_0. The idea is to extract, from the commutative case, the way the Hilbert space $L^2(\mathbb{P})$ is constructed from $L^2(\mu)$ and the transition kernel P. To achieve this, we first construct the analogue of $L^2(\mu)$. That is, we give ourselves a Hilbert space $\mathcal{K}_0$, a representation X_0 from $\mathcal{A}_0$ on $\mathcal{K}_0$, and a unit vector ψ implementing the state μ

$$\mu(a_0) = \langle\, \psi\,,\, (a_0 \circ X_0)\, \psi \,\rangle (a_0 \in \mathcal{A}_0)\ .$$

We recall that $I_0 \circ X_0$ is assumed to be the identity operator on $\mathcal{K}_0$. We may for instance construct $\mathcal{K}_0$ using the so called GNS construction. Still, it is better to keep some freedom here: changing the initial vector will then give different states μ — though a simultaneaous construction for *all* possible initial states (as in the classical case) could not be achieved with a separable $\mathcal{K}_0$.

Our purpose is to construct a Hilbert space $\mathcal{K}_1$ containing $\mathcal{K}_0$, a representation X_1 from $\mathcal{A}_1$ on $\mathcal{K}_1$ ($I_1 \circ X_1$ being the identity of $\mathcal{K}_1$), such that the following property holds (the notation $a_0 0 X_0$ means $X_0(a_0)$, in order to stress the intuitive meaning of a_0 as a function on E_0 ; similarly for $a_1 0 X_1$).

$$\langle\, \psi\,,\, a_1 \circ X_1\, a_0 \circ X_0 \psi \,\rangle = \langle\, \psi\,,\, (Pa_1\, a_0) \circ X_0 \psi \,\rangle\ .$$

Though $a_0 \circ X_0$ is an operator on $\mathcal{K}_0$, not on $\mathcal{K}_1$, still it can be applied to $\psi \in \mathcal{K}_0{}^*$. On the other hand $a_0 \circ X_0\, a_1 \circ X_1 \psi$ would be meaningless. Thus we are constructing something coarser than a 'joint law' for X_0 and X_1 — intuitively speaking, P predicts the value of X_1 while X_0 is under observation.

A necessary condition for the existence of such a Hilbert space $\mathcal{K}_1$ is that, given any finite families of elements a_i^0 of $\mathcal{A}_0$ and a_i^1 of $\mathcal{A}_1$, and putting $A = \sum_i a_i^1 \circ X_1\, a_i^0 \circ X_0$,

---

   * Since E_0 and E_1 do not exist as spaces, one generally says P is a kernel from $\mathcal{A}_1$ to $\mathcal{A}_0$.

   * In fact, nothing depends on the particular choice of $\psi \in \mathcal{K}_0$.

$\langle \psi, AA\psi \rangle$ should be positive, that is,

$$\mu(\sum_{ij} a_j^{0*} P(a_j^{1*} a_i^1) a_i^0) \geq 0 .$$

It should be a property of the mapping P that the preceding assumption is satisfied for an arbitrary state μ on $\mathcal{A}_0$, and thus we are led naturally to the basic definition of *complete positivity*

$$\sum_{ij} a_j^{0*} P(a_j^{1*} a_i^1) a_i^0 \geq 0 \text{ for arbitrary } a_i^1 \in \mathcal{A}_1, \, a_i^0 \in \mathcal{A}_0.$$

A trivial case where this condition is satisfied is that of an algebra homomorphism P from $\mathcal{A}_1$ to $\mathcal{A}_0$ (corresponding to the 'no spreading' case in the commutative situation).

If this condition is satisfied, we can construct the Hilbert space $\mathcal{K}_1$ and the representation X_1, using the following procedure, which generalizes the GNS method:

1) Provide the algebraic tensor product $\mathcal{K}_0 \otimes \mathcal{A}_1$ with the only hermitian form such that, for $k, h \in \mathcal{K}_0$, $b_1, a_1 \in \mathcal{A}_1$

$$\langle k \otimes b_1 , h \otimes a_1 \rangle = \langle k, P(b_1 a_1) h \rangle .$$

It is not difficult to check that we get in this way a positive (possibly degenerate) scalar product. Otherwise stated, we have for finite families $k_i, h_i \in \mathcal{K}_0$, $b_i, a_i \in \mathcal{A}_1$

$$\sum_{ij} \langle k_j, P(b_j a_i) h_i \rangle \geq 0 .$$

This follows from the complete positivity assumption on P if $h_i = (a_i^0 \circ X_0) \psi$, $k_i = (b_i^0 \circ X_0) \psi$ for some vector $\psi \in \mathcal{K}_0$ and elements a_i^0, b_i^0 of $\mathcal{A}_0$. Then we get the result by density for arbitrary $h_i, k_i \in \mathcal{K}_0$ if the representation $\mathcal{K}_0$ is cyclic with generating vector ψ (GNS case). Finally, we decompose $\mathcal{K}_0$ into a direct sum of cyclic representations to reach the general case.

2) Knowing this scalar product is positive, we construct the Hilbert space $\mathcal{K}_1$ by killing the null-space and completing. If we map $h \in \mathcal{K}_0$ to $h \otimes I_1 \in \mathcal{K}_0 \otimes \mathcal{A}_1$, we get an isometry: its image does not intersect the null space, and we may identify h with $h \otimes I_1$, thus imbedding $\mathcal{K}_0$ into $\mathcal{K}_1$. We may identify ψ with $\psi \otimes I_1$.

We define a representation X_1 of $\mathcal{A}_1$ in $\mathcal{K}_1$ as follows. Given $a \in \mathcal{A}_1$, we put

$$(a \circ X_1) (h \otimes b) = h \otimes (ab) .$$

It is clear that this operation behaves well with respect to sums and products, a little less obvious that $(a \circ X) = a \circ X$, and that the action of a is bounded with norm at most $\| a \|$. To prove this last point, assume $\| a \| > 1$ and remark that $aa > I$ may be written as $aa = I - cc$. Then

$$\| \sum_i h_i \otimes ab_i \|^2 = \sum_{ij} \langle h_j \otimes (ab_j), h_i \otimes ab_i \rangle = \sum_{ij} \langle h_j, P(b_j a_j a_i b_i) h_i \rangle$$

$$= \| \sum_i h_i \otimes b_i \|^2 - \sum_{ij} \langle h_j, P(b_j c_j c_i b_i) h_i \rangle$$

and the last term to the right is positive. Then the operators are extended to $\mathcal{K}_1$ by continuity and our representation of $\mathcal{A}_1$ is constructed.

In contrast with this, one generally cannot define a representation of $a \in \mathcal{A}_0$ in $\mathcal{K}_1$, extending its action $a \circ X_0$ on $\mathcal{K}_0$, by the formula

$$(a \circ X_0)\,(h \otimes b) = ((a \circ X_0)\,h) \otimes b \ .$$

This operation behaves well with respect to sums and products, but *not* with respect to adjoints, and generally it is *not* continuous. We may still consider X_0 as a representation of $\mathcal{A}_0$ in $\mathcal{K}_1$, but not quite in the usual sense: $I_0 \circ X_0$ is not the identity operator, but the orthogonal projection on $\mathcal{K}_0$. If we really want to have a representation respecting the unit, then we assume $\mathcal{A}_0$ has a multiplicative state δ and put, for $x = y + z \in \mathcal{K}_0 \oplus \mathcal{K}_0^\perp$

$$(a \circ X_0)\,x = (a \circ X_0)\,y + \delta(a)\,z \ .$$

The GNS construction is unique up to isomorphism, under an assumption of cyclicity. Here the assumption that ensures uniqueness is the following: *the (closed) invariant subspace for the representation X_1 generated by $\mathcal{K}_0$ is the whole of $\mathcal{K}_1$.* Otherwise stated, the vectors $(a_1 \circ X_1)\,h$ $(a_1 \in \mathcal{A}_1, h \in \mathcal{K}_0)$ form a total set in $\mathcal{K}_1$. This property it satisfied here from the construction, and it is easily seen that any two situations for which it is satisfied are unitarily equivalent. This assumption is called *minimality*.

This is probably sufficient to give the basic idea — and then, to work; construct a Markov process in continuous time, starting from a semigroup of completely positive mappings, and see whether you can give a meaning, say, to the strong Markov property! (One can, as shown recently by Attal and Parthasarathy).

A topic we haven't been able to describe is the recent one of *free independence*.

References

[1] ACCARDI, L., FRIGERIO, A. and LEWIS, J. Quantum stochastic processes, *Publ. RIMS Kyoto*, **18**, 1982, p. 94–133.

[2] ATTAL, S. An algebra of non commutative bounded semimartingales. Square and angle quantum brackets, *Journ. Funct. Analysis*, **124** 1994, p. 292–332.

[3] ATTAL, S. and MEYER, P.A. Interprétation probabiliste et extension des intégrales stochastiques non commutatives, *Sem. Prob. XXVII*, Springer Verlag, L.N.M. **1557**, 1993, p. 312–327.

[4] AZÉMA, J. Sur les fermis aliatoires, *Sem. Prob. XIX*, Springer Verlag, L.N.M. **1123**, 1985, p. 397–495.

[5] BARNETT, C., STREATER, R.F. and WILDE, I.F. The Ito-Clifford integral I, *J. Funct. Anal.* 48, 1982, p. 172–212. — II, *J. London Math. Soc.* 27, 1983, p. 373–384. — III, Markov properties of solutions to s.d.e.'s, *Comm. Math. Phys.* 89, 1983, p. 13–17. —IV, a Radon–Nikodym theorem and bracket processes, *J. Oper. Th.* **11**, 1984, p. 255–271.

[6] BHAT, B.V.R. and PARTHASARATHY, K.R. Kolmogorov's existence theorem for Markov processes in C^*-algebras, *Proceedings of the Indian Academy of Sciences*, **104**, 1994, p. 253–262.

[7] EMERY, M. On the Azéma martingales, *Sem. Prob. XXIII*, Springer LN **1372**, 1989, p. 66–87.

[8] EMERY, M. Sur les martingales d'Azima (suite, *Sem. Prob. XXIV*, Springer LN **1426**, 1990, p. 442–447.

[9] EMERY, M. On the chaotic representation property for martingales, *to appear, Proc. of the 1993 Kolmogorov Semester on Prob. and Stat.*, St Petersburgh.

[10] EVANS, M. Existence of quantum diffusions, *Prob. Th. Rel. Fields*, **81**, 1989, p. 473–483.

[11] EVANS, M. and HUDSON, R.L. Multidimensional quantum diffusions, *QP.III*, p. 69–88 ; Perturbation of quantum diffusions, *J. London Math. Soc.*, **41**, 1992, p. 373–384.

[12] HUDSON, R.L., KARANDIKAR, R.L. and PARTHASARATHY, K.R. Towards a theory of non commutative semimartingales adapted to brownian motion and a quantum Ito's formula, *Theory and Applications of Random Fields, Bangalore 1982*, LNCI **49**, 1983, p. 96–110.

[13] HUDSON, R.L. and PARTHASARATHY, K.R. Quantum Ito's formula and stochastic evolutions, *Comm. Math. Phys.*, **93**, 1984, p. 301–303 ; Unification of fermion and boson stochastic calculus, *Comm. Math. Phys.*, **104**, 1986, p. 457–470.

[14] MEYER, P.A. *Quantum Probability for Probabilists*, Springer LNM **1538**, 2nd edition 1995.

[15] PARTHASARATHY, K.R. *An Introduction to Quantum Stochastic Calculus*, Birkhäuser, Basel 1992.

[16] PARTHASARATHY, K.R. and SINHA, K.B. Boson-Fermion transformations in several dimensions, *Pramana J. Phys.*, **27**, 1986, p. 105–116 ; Markov chains as Evans-Hudson diffusions in Fock space. *Sém. Prob. XXIV*, LNM **1426**, 1990, p. 362–369 ; Representation of bounded quantum martingales in Fock space, *Journ. Funct. Anal.*, **67**, 1986, p. 126–151.

[17] SCHÜRMANN, M. The Azéma martingales as components of quantum independent increment processes, *Sem. Prob. XXV*, LNM **1485**, p. 24–30.

[18] SPEICHER, R. A new example of "independence" and "white noise", *Prob. Th. Rel. Fields*, **84**, 1990, p. 141–159. Survey of stochastic integration on the full Fock space, *QP.VI*, p. 421–437.

STOCHASTIC ANTICIPATING CALCULUS

DAVID NUALART,[*] *Universitat de Barcelona*

The purpose of the stochastic anticipating calculus is to develop a differential and integral calculus involving stochastic processes which are not necessarily adapted to the Brownian motion $\{W_t, t \geq 0\}$. This stochastic calculus is mainly used to formulate and solve stochastic differential equations of the form

$$\begin{cases} dX_t = \sigma(t, X_t)dW_t + b(t, X_t)dt, \\ X_0 \in \mathbb{R}^m, \end{cases}$$

where the coefficients $\sigma(t,x)$, $b(t,x)$ or the initial condition X_0 depend on the whole trajectory of the Brownian motion W.

1. Stochastic integrals of nondapted processes

The main problem in giving a rigorous sense to a stochastic differential equation with random coefficients is to define stochastic integrals of the form

$$\int_0^t u_s dW_s,$$

where $u = \{u_t, t \geq 0\}$ is a nonadapted stochastic process. In the following sections we will review different approaches that have been used during the last years to introduce stochastic integrals of nonadapted processes.

1.1. *The Wiener chaos expansion* We will assume that $W = \{W_t, t \in [0,1]\}$ is a standard Wiener process defined on the canonical probability space $(\Omega, \mathcal{F}, P)$. That is, $\Omega = C_0([0,1])$, P is the Wiener measure and $\mathcal{F}$ is the completion of the Borel σ-field of Ω with respect to P. Then $W_t(\omega) = \omega(t)$ is the canonical process.

Suppose that $u = \{u_t, t \in [0,1]\}$ is a measurable process such that $E \int_0^1 u_t^2 dt < \infty$. The process u is not supposed to be adapted to the filtration generated by the Brownian motion W. For each $t \in [0,1]$ a.e. u_t is a square integrable random variable that can be developed into a sum of orthogonal multiple stochastic integrals (cf. Itô [20]):

$$u_t = \sum_{n=0}^{\infty} I_n(f_n(\cdot, t)),$$

where I_n denotes the multiple stochastic integral of order n, and $f_n \in L^2([0,1]^{n+1})$ is symmetric in the first n variables. The *Skorohod integral* of the process u, denoted

---

*Postal address: Departament d'Estadistica, Facultat de Mathematiques, Universitat de Barcelona, Gran Via de les Corts Catalanes 585, 08007 Barcelona, Spain.

by $\delta(u) = \int_0^1 u_t dW_t$, is defined by

$$\delta(u) = \sum_{n=0}^{\infty} I_{n+1}(\tilde{f}_n),$$

provided the above series converges in $L^2(\Omega)$, and where $\tilde{f}_n$ denotes the symmetrization of f_n in all its variables. Namely, the Skorohod integral of a process $u \in L^2([0,1] \times \Omega)$ exists if and only if $\sum_{n=0}^{\infty}(n+1)!\|\tilde{f}_n\|^2 < \infty$. This notion of stochastic integral was introduced by Skorohod in [53] and it is a generalization of the Itô stochastic integral. Indeed, if the process $u \in L^2([0,1] \times \Omega)$ is adapted then it is Skorohod integrable and $\delta(u)$ coincides with the classical Itô integral.

1.2. *Anticipating integrals and Malliavin calculus* In 1982 Gaveau and Trauber proved in [16] that the Skorohod integral coincides with the adjoint of the derivative operator on the Wiener space. That is, $\delta(u) = \int_0^1 u_t dW_t$ is the square integrable random variable characterized by the duality relation

$$E(F\delta(u)) = E(\int_0^1 D_t F u_t dt),$$

for any smooth random variable F of the form $F = f(W(t_1), \ldots, W(t_n))$, where f is an infinitely differentiable function with bounded derivatives. In the above formula, the derivative $D_t F$ is defined by

$$D_t F = \sum_{i=1}^{n} \frac{\partial f}{\partial x_i}(W(t_1), \ldots, W(t_n))\mathbf{1}_{[0,t_i]}(t).$$

The stochastic calculus of variations on the Wiener space (also called Malliavin calculus) is an infinite differential calculus introduced by Malliavin in [29] in order to provide a probabilistic proof of Hörmander's hypoellipticity theorem. This calculus was later developed by several authors, among them let us mention Bismut [4], Stroock [54] and Watanabe [57]. The techniques of the Malliavin calculus have allowed to develop a stochastic calculus for the Skorohod integral (see Sekiguchi and Shiota [52], Nualart and Pardoux [36]), which extends the classical Itô calculus introduced in the early 40's. Here are some of the basic features of the Skorohod stochastic calculus:

(1) *Local property:* The Skorohod integral verifies $\int_0^1 u_t dW_t = 0$ almost surely on the set $\{\int_0^1 u_t^2 dt = 0\}$, provided that u belongs to the space $\mathbb{L}^{1,2} := L^2([0,1]; \mathbb{D}^{1,2})$ which is included into the domain of the Skorohod integral. Here $\mathbb{D}^{1,2}$ denotes the closure of the class of smooth random variables by the Sobolev norm

$$\|F\|_{1,2}^2 := E(F^2) + E(\int_0^1 (D_t F)^2 dt).$$

(2) *Properties of the indefinite Skorohod integral:* Suppose that the process u belongs to the space $\mathbb{L}^{1,2}$. Maximal inequalities are not known for the Skorohod

integral. Nevertheless, applying Meyer's inequalities one can show that the process $X_t = \int_0^t u_s dW_s$, $t \in [0,1]$, has a continuous version provided that for some $p > 2$ one has

$$E\left(\int_0^1 \left(\int_0^1 (D_s u_t)^2 ds \right)^{\frac{p}{2}} dt \right) < \infty.$$

On the other hand, X_t possesses a nontrivial quadratic variation analogous to the quadratic variation of the Itô integral, namely,

$$\sum_{i=0}^{n-1} (X_{t_{i+1}} - X_{t_i})^2 \to \int_0^1 u_s^2 ds,$$

in L^1 as $\max_i(t_{i+1} - t_i)$ tends to zero, where $0 = t_0 < t_1 < \cdots < t_n = 1$.

(3) *Change-of-variables formula:* Suppose that $X_t = \int_0^t u_s dW_s$, the integrand u belongs to the space $\mathbb{L}^{2,4} := L^4([0,1]; \mathbb{D}^{2,4})$ (that is, u has two derivatives that have moments of fourth order), and $F : \mathbb{R} \to \mathbb{R}$ is a twice continuously differentiable function. Then then the following generalized Itô's formula holds (in differential form):

$$dF(X_t) = F'(X_t)u_t dW_t + \frac{1}{2}F''(X_t)u_t^2 dt + F''(X_t)u_t \left(\int_0^t D_t u_s dW_s \right) dt.$$

1.3. *Stochastic integrals defined via a regularization of the Brownian paths* A natural approach to define the stochastic integral of a nonadapted process u with respect to the Brownian motion W consists in approximating the paths of W by regular functions W^n and define the stochastic integral of u as the limit of the ordinary integrals $\int_0^1 u_s dW_s^n$. Different regularization procedures are possible:

I)   We can consider, for instance, a *polygonal interpolation* of W given its values on the points of a partition $\pi = \{0 = t_0 < t_1 < \cdots < t_n = 1\}$ of $[0,1]$. Set

$$\dot{W}_t^\pi = \sum_{i=0}^{n-1} \frac{W_{t_{i+1}} - W_{t_i}}{t_{i+1} - t_i} \mathbf{1}_{[t_i,t_{i+1}]}(t).$$

Given a measurable process $u = \{u_t, t \in [0,1]\}$ such that $\int_0^1 |u_s| ds < \infty$ almost surely, we define

$$\int_0^1 u_t \circ dW_t = \lim_{|\pi|\downarrow 0} \int_0^1 u_t \dot{W}_t^\pi dt,$$

provided the limit exists in probability. This integral is called the *(extended) Stratonovich integral.*

II)   We can also approximate the Brownian path W_t by $(\varphi_\epsilon * W)_t$, where $\{\varphi_\epsilon, \epsilon > 0\}$ is an approximation of the identity. In this way we can introduce the following definition of the stochastic integral

$$\int_0^1 u_t \circ_\varphi dW_t = \lim_{\epsilon\downarrow 0} \int_0^1 u_t(\varphi_\epsilon * \dot{W})_t dt,$$

provided the limit exists in probability. In particular, if we take

$$\varphi_\epsilon(x) = \frac{1}{2\epsilon}\mathbf{1}_{[-\epsilon,\epsilon]}(x),$$

we obtain

$$\int_0^1 u_t \circ_\varphi dW_t = \lim_{\epsilon\downarrow 0} \int_0^1 u_t \frac{W_{(t+\epsilon)\wedge 1} - W_{(t-\epsilon)\vee 0}}{2\epsilon} dt.$$

This type of integral is called *symmetric stochastic integral,* and, among other authors, was studied by Russo and Vallois in [50].

The extended Stratonovich integral and the symmetric integral defined through a symmetric approximation of the identity φ_ϵ coincide under "nice" hypotheses, and have the following properties:

(A) If u_t is a continuous semimartingale the Stratonovich and the symmetric integral coincide with the classical Stratonovich integral.

(B) Suppose that $u \in \mathbb{L}^{1,2}$, and the kernel $D_s u_t$ possesses (possibly different) continuous versions in $L^2(\Omega)$ in the regions $\{s \leq t\}$ and $\{s \geq t\}$. Then the Stratonovich integral equals to the Skorohod integral plus a complementary term:

$$(1) \qquad \int_0^1 u_t \circ dW_t = \int_0^1 u_t dW_t + \frac{1}{2}\int_0^1 (D_t u_{t+} + D_t u_{t-})dt.$$

This equality has been used to obtain L^p estimates for the Stratonovich integral.

(C) The change-of-variable formula for the Stratonovich integral coincides with that of the classical analysis, that means, if $X_t = \int_0^t u_s \circ dW_s$, under some technical hypotheses, we have

$$dF(X_t) = F'(X_t)u_t \circ dW_t.$$

III)   The preceding procedure can be modified by choosing a nonsymmetric approximation of the identity like

$$\varphi_\epsilon(x) = \frac{1}{\epsilon}\mathbf{1}_{[-\epsilon,0]}(x),$$

or by choosing a nonanticipating polygonal approximation of the form

$$\dot{W}_t^\pi = \sum_{i=1}^{n-1} \frac{W_{t_i} - W_{t_{i-1}}}{t_i - t_{i-1}}\mathbf{1}_{[t_i,t_{i+1}]}(t),$$

where $t_i = \frac{i}{n}$. In this way we obtain the *forward stochastic integral,* denoted by $\int_0^1 u_t * dW_t$. This type of stochastic integral has been studied by Berger and Mizel [3], Asch and Potthoff [1] and Russo and Vallois [50], among others. The forward integral is actually an extension of the Itô integral, and it follows the same change-of-variables formula, that is,

$$dF(X_t) = F'(X_t)u_t dW_t + \frac{1}{2}F''(X_t)u_t^2 dt.$$

1.4. *Stochastic integrals based on the Fourier expansion of the integrand* Fix a complete orthonormal system $\{e_i, i \geq 1\}$ in $L^2([0,1])$. A stochastic process $u \in L^2([0,1] \times \Omega)$ can be expanded into the following random sum

$$u_t = \sum_{i=1}^{\infty} \langle u, e_i \rangle e_i(t).$$

Then, we can define the L^2-*stochastic integral* of u as follows

$$L^2 - \int_0^1 u_t dW_t = \sum_{i=1}^{\infty} \langle u, e_i \rangle \int_0^1 e_i(t) dW_t,$$

provided that the series converges in $L^2(\Omega)$ and the limit does not depend on the basis $\{e_i\}$. This approach to define stochastic integrals has been developed in a series of papers by Ogawa [43, 44, 45] and Rosinski [49].

If the process u belongs to the space $\mathbb{L}^{1,2}$ and the kernel Du is nuclear a.s., then one has the following relationship between the different stochastic integrals:

$$L^2 - \int_0^1 u_t dW_t = \int_0^1 u_t \circ dW_t = \int_0^1 u_t dW_t + \mathrm{Tr}\,(Du).$$

1.5. *Stochastic integrals based on the enlargement of the Brownian filtration* Given a continuous stochastic process $u = \{u_t, t \in [0,1]\}$ we can consider the family of σ-fields defined by

$$\mathcal{G}_t = \sigma\{u_s, W_s; 0 \leq s \leq t\}.$$

If the Brownian motion W happens to be a semimartingale with respect to the family of σ-fields $\{\mathcal{G}_t, t \in [0,1]\}$ we could define the stochastic integral $\int_0^1 u_t dW_t$ using the semimartingale theory. In this case, the semimartingale integral $\int_0^1 u_t dW_t$ coincides with the forward integral $\int_0^1 u_t * dW_t$ defined by means of a causal regularization of the Brownian paths. The increasing family of σ-fields $\{\mathcal{G}_t, t \in [0,1]\}$ constitutes a particular example of the general theory of enlargement of the Brownian filtration developed by Jeulin in [22].

Consider the special case where the process u has the form $u_t = v_t(F)$, where F is a d-dimensional random vector and $\{v_t(x), t \in [0,1], x \in \mathbb{R}^d\}$ is an adapted process depending on the parameter $x \in \mathbb{R}^d$. In this situation $\mathcal{G}_t = \mathcal{F}_t \vee \sigma\{F\}$, and a criterion for W to be a $\mathcal{G}_t$-semimartingale (see Jacod [21]) is that for all $t \in [0,1]$ the conditional law of F given $\mathcal{F}_t$ is absolutely continuous with respect to a fixed measure μ on $\mathbb{R}^d$.

2. Anticipating stochastic differential equations

The solution of a stochastic differential equation will not be adapted to the Brownian filtration when the initial condition or the coefficients are random and nonadapted. In this case the anticipating stochastic calculus has allowed to formulate and solve this type of stochastic equations. We will describe in this section some of the recent developments of this theory.

2.1. *Random initial condition and Stratonovich formulation* If the nonadaptability comes only from the initial condition, and the stochastic differential equation is formulated using the Stratonovich integral, there is a simple trick wich allows to find a solution:

Consider the Stratonovich stochastic differential equation

$$(1) \qquad X_t = X_0 + \sum_{i=1}^{d} \int_0^t \sigma_i(X_s) \circ dW_s^i + \int_0^t b(X_s)ds,$$

where $\sigma_i, b : \mathbb{R}^m \to \mathbb{R}^m$ are deterministic smooth functions, the initial condition X_0 is an m-dimensional random vector, and W is a d-dimensional Wiener process. We can define a stochastic flow of transformations $\{\varphi_t(x), t \in [0,1], x \in \mathbb{R}^m\}$ of $\mathbb{R}^m$ associated to Eq. (1):

$$\varphi_t(x) = x + \sum_{i=1}^{d} \int_0^t \sigma_i(\varphi_s(x)) \circ dW_s^i + \int_0^t b(\varphi_s(x))ds.$$

Then if we replace the deterministic initial point x by the random initial condition X_0, the resulting process $X_t = \varphi_t(X_0)$ will be a nonadapted solution of the anticipating Eq. (1).

This is a consequence of the substitution formula for the Stratonovich integral:

$$\int_0^t \sigma_i(\varphi_s(x)) \circ dW_s^i \bigg|_{x=X_0} = \int_0^t \sigma_i(\varphi_s(X_0)) \circ dW_s^i,$$

which can be proved using only the definition of this integral.

To show the *uniqueness* of the solution $\varphi_t(X_0)$ one has to specify a suitable class of processes $\mathcal{C}$ where the solution will be unique. Using the stochastic calculus for the Skorohod integral, in [41] the uniqueness is established in a class of processes denoted by $\mathbb{L}_{C,loc}^{1,4}$. The main idea is as follows. Let Y_t be another solution to Eq. (1) in this class. Then, using an anticipating version of the Itô-Ventzell formula one shows that

$$d(\varphi_t^{-1}(Y_t)) = 0,$$

which implies $\varphi_t^{-1}(Y_t) = X_0$, and, hence, $Y_t = \varphi_t(X_0)$.

A more direct approach is developed in [24], where the uniqueness is established in the class of processes which are limit of the corresponding approximating solutions.

Remarks:

(1) The existence and uniqueness of a solution to Eq. (1) have been established also in the case of a random drift coefficient b in [41], and in the one-dimensional case assuming that σ and b are random in [23]. The case of a random diffusion coefficient in higher dimensions is an open problem. An infinite dimensional extension of the substitution argument and its application to solve Hilbert-valued anticipating stochastic differential equations has been developed in [17].

(2) Probabilistic properties of the solution X_t of Eq. (1) can be deduced from similar properties for the stochastic flow $\varphi_t(x)$, that is, properties for diffusion processes which hold uniformly with respect to the initial condition. Examples of these properties are the principle of large deviations [32], the characterization of the topological support of law of the solution [31], and the existence and regularity of the density of the solution at a given time [30, 11].

2.2. *Linear Skorohod stochastic differential equations* Suppose that we want to solve a stochastic differential equation of the form (1) but replacing the Stratonovich by the Skorohod integral. In this case we cannot use the substitution approach which works very well for the Stratonovich equations. On the other hand, we cannot use a fixed point argument in order to show the existence and uniqueness of a solution as it is done for the Itô stochastic equations. There is no general theory to deal with these equations, although in some particular cases (σ linear, and the one dimensional case) one can show the existence and the uniqueness of a solution. In oder to illustrated the behaviour of the solution to these equations, let us consider a simple linear example.

Let $\sigma : [0,1] \to \mathbb{R}$ be a deterministic and square integrable function, and consider the equation

$$(2) \qquad X_t = X_0 + \int_0^t \sigma_s X_s dW_s, \quad t \in [0,1].$$

Then, if $X_0 \in L^p(\Omega)$, for some $p > 2$, there exists a unique solution of Eq. (2) which is given by

$$(3) \qquad X_t = X_0 \left(\omega. - \int_0^{\cdot} \sigma_s ds \right) e^{\int_0^t \sigma_s dW_s - \frac{1}{2} \int_0^t \sigma_s^2 ds}.$$

One can show that the process $\{X_t\}$ given by (3) satisfies (2), taking into account the duality formula for the operator δ, by proving that for any smooth random variable G one has

$$(4) \qquad E(X_t G) = E(X_0 G) + E\left(\int_0^t \sigma_s X_s D_s G ds \right).$$

The equality (4) can be easily proved applying Girsanov theorem twice, and using the property

$$\frac{d}{ds}((D_s G)(\omega. - \int_0^{\cdot} \sigma_s ds)) = -\sigma_s((D_s G)(\omega. - \int_0^{\cdot} \sigma_s ds)).$$

This is the approach that was introduced by R. Buckdahn in [5] in order to solve quasilinear Skorohod differential equations. In [7] this technique is used in order to deal with equations with linear random coefficients, using an extended version of the Girsanov theorem.

The linear multidimensional case cannot be treated with this change of measure argument, unless $\sigma_i(t, x)$ depends only on the i-th coordinate of x for each i. In order to deal with multidimensional linear equations, it is useful to interpret the process (3) in terms of the Wick product. In this way we have

$$(5) \qquad X_t = X_0 \Diamond e^{\int_0^t \sigma_s dW_s - \frac{1}{2}\int_0^t \sigma_s^2 ds}.$$

The Wick product was introduced by G.C. Wick in 1950 in the context of quantum field theory, and now it plays an important role in mathematical physics and stochastic analysis. The Wick product of is defined by:

$$I_n(f) \Diamond I_m(g) = I_{n+m}(f \tilde\otimes g).$$

This definition can be extended by linearity to finite combinations of multiple integrals, and more generally to Wiener functionals $F = \sum_{n=0}^\infty I_n(f_n)$, and $G = \sum_{n=0}^\infty I_n(g_n)$, such that the following series converge

$$(6) \qquad \sum_{n=0}^\infty n!2^{2n}\|f_n\|^2 < \infty, \quad \sum_{n=0}^\infty n!2^{2n}\|g_n\|^2 < \infty.$$

An open question is to find sufficient conditions weaker that (6) for this series to converge in $L^2(\Omega)$, or in $L^1(\Omega)$. For instance, we do not know if this convergence holds when $F, G \in \mathbb{D}^\infty$.

From Eq. (5) we see that the solution of a linear Skorohod stochastic differential equation is obtained by forming the Wick product of the initial condition X_0 with the solution of the Itt equation whose initial condition is one. This rule also works for multidimensional equations. The main problem is to show that the Wick product of the initial random variable X_0 times a martingale exponential is an ordinary random variable. If σ is a deterministic matrix, then the Wick product $X_0 \Diamond \exp(\sigma W_t - \frac{1}{2}\sigma\sigma^* t^2)$ is an ordinary random variable provided X_0 has bounded derivatives up to some order (see [10]).

Skorohod stochastic differential equations behave as infinite dimensional first order hyperbolic stochastic partial differential equations. Indeed, consider the onedimensional Skorohod stochastic differential equation:

$$(7) \qquad X_t = X_0 + \int_0^t \sigma(X_s)dW_s + \int_0^t b(X_s)ds.$$

Using the relationship between the Skorohod and extended Stratonovich integral (see (1)) we can formally rewrite Eq. (7) in the following form

$$(8) \qquad \frac{\partial X_t}{\partial t} = \sigma(X_t) \circ \frac{dW_t}{dt} + \frac{1}{2}\sigma'(X_t)(D_{t+}X_t + D_{t-}X_t) + b(X_t).$$

Observe that Eq. (8) can be regarded as a first order hyperbolic stochastic differential equation on the Wiener space, and one can try to apply a stochastic version of the characteristic curves method developed by Kunita in [25].

In general, although the coefficients of (7) were smooth and with bounded derivatives, the solution might explode in finite time. This behaviour has also been noticed by Buckdahn in [9]. In this article, by means of the Itt formula for the Skorohod integral, an explicit solution is obtained for the solution of the equation (7) when $k = d = 1$ and the coefficients σ and b do not depend on t and are of class C_b^2.

2.3. *Stochastic boundary value problems* Another type of stochastic differential equations whose solution is not adapted to the Brownian filtration are equations where a functional condition on the values of the process at times $t = 0$ and $t = 1$ is imposed. Consider, in general a stochastic differential equation on $\mathbb{R}^d$ of the form

$$(9) \qquad \begin{cases} dX_t = \sum_{i=1}^d \sigma_i(X_t) \circ dW_t^i + b(X_t)dt, & 0 \le t \le 1, \\ h(X_0, X_1) = 0, \end{cases}$$

that means, instead of giving the value of the process at time zero, we impose a boundary condition of the form $h(X_0, X_1) = 0$. The coefficients σ_i, b and h are deterministic functions. The existence and uniqueness of a solution for an equation of this type has been investigated in some particular cases. On the other hand, one has studied what type of Markov property is satisfied by the solution. The following particular situations have been considered:

(a) The case where the coefficients are affine functions has been studied in [42]. That means, we assume in (9) that $\sigma_i(x) = G_i x + b_i$, $b(x) = Fx + a$, and $h(x, y) = H_0 x + H_1 y - h_0$, where F, G_i, H_0 and H_1 are $d \times d$ matrices, and a, b_i and h_0 are vectors in $\mathbb{R}^d$. We suppose that the matrix $[H_0 : H_1]$ has rank d. Particular examples of this type of boundary conditions: *i)* the case where we fix the first l coordinates of X_0, $1 \le l \le d$, and the last $d - l$ coordinates of X_1, *ii)* the case of a periodic boundary condition $X_0 = X_1$.

Applying the anticipating stochastic calculus one can find the following explicit expression for the solution of Eq. (9):

$$X_t = \Phi_t[(H_0 + H_1\Phi_1)^{-1}(h_0 - H_1\Phi_1 \int_0^1 \Phi_s^{-1} \circ dV_s)] + \Phi_t \int_0^t \Phi_s^{-1} \circ dV_s,$$

where Φ_t denotes the fundamental solution of the linear system

$$d\Phi_t = F\Phi_t dt + \sum_{i=1}^k G_i \Phi_t \circ dW_t^i, \qquad \Phi_0 = I,$$

and $V_t = at + \sum_{i=1}^d b_i W_t^i$. One has to assume that $\det(H_0 + H_1\Phi_1) \neq 0$ a.s.

Concerning the Markov property of the solution, one can show that $\{X_t\}$ is a Markov field (or a reciprocal process) in the sense that the values of X inside any interval $[s, t]$ are conditionally independent of the values outside this interval, given X_t and X_s, in the following cases:

(i) $G_1 = \cdots = G_k = 0$ (Gaussian case)

(ii) $a = b_1 = \cdots = b_k = 0$, and Φ_t is a diagonal matrix for all $t \in [0, 1]$ (for instance, this is true in dimension one).

(b) The case $k = d$ and $\sigma = \mathrm{Id}$ has been discussed in [37]. In this case, the existence and uniqueness of a solution is a pathwise problem which can be treated assuming some monotonicity properties on the function b. On the other hand, the Markov field property only holds in the linear case when $d = 1$. This negative result has been proved using an anticipating version of the Girsanov theorem due to Kusuoka. Roughly speaking, in the nonadapted case, the Girsanov density is not a multiplicative functional of the solution and this prevents the Markov property to be preserved by the corresponding nonlinear transformation.

(c) The existence and uniqueness of a solution to (9) in the one-dimensional case has been obtained in [13] assuming a linear boundary condition of the form $F_0 X_0 + F_1 X_1 = h_0$. The coefficients b and σ are of class $\mathcal{C}^4$ with bounded derivatives, and $F_0 F_1 \neq 0$. On the other hand, if σ is linear ($\sigma(x) = \alpha x$), $h_0 \neq 0$, and assuming that b is of class $\mathcal{C}^2$, then one can show that the solution $\{X_t\}$ is a Markov field only if the drift is of the form $b(x) = Ax + Bx \log |x|$, where $|B| < 1$.

(d) In [15] a stochastic version of the invariant imbedding method has been succesfully applied to establish the existence and uniqueness of a solution to (9) in a wide class of situations.

3. Further applications of the anticipating stochastic calculus

Let us conclude this review by discussing some applications and recent extensions of the stochastic calculus for nonadapted processes.

(I) As a successful application of the anticipating calculus, we have to mention the extension of the Girsanov theorem to anticipating shifts on the Wiener space and its application to the analysis of the Markov property of stochastic differential equations with boundary conditions.

The Girsanov theorem for transformations of the Wiener measure under shifts of the form $\omega_t \mapsto \omega_t + \int_0^t u_s ds$, where u is a non necessarily adapted square integrable process, goes back to the work by Cameron and Martin [12]. Recent extensions are due to Ramer [48], Kusuoka [27], Buckdahn [6], and Ustunel and Zakai [56].

We have already discussed the application of the anticipating Girsanov theorem to study the Markov property of stochastic boundary value problems of the form (9). This method has also been used to investigate the Markov field property for solutions to stochastic partial differential equations driven by a white noise. As an illustration of this type of application let us mention the following result obtained in [14]:

Consider the elliptic stochastic partial differential equation:

$$(1) \qquad \begin{cases} \Delta u(x) = \dot{W} + f(u(x)), & x \in D \\ u|_{\partial D} = 0, \end{cases}$$

where D is a bounded domain in $\mathbb{R}^k$ ($k = 1, 2, 3$) with a smooth boundary, f is a continuously differentiable function such that $f' \geq 0$, and $\dot{W}$ is a white noise on D. Then the solution u is a germ Markov field if and only if $f(x) = ax + b$.

If $k = 1$ and $D = (0, 1)$, Eq. (1) is a one-dimensional second order stochastic differential equation with Dirichlet boundary condition that has been analyzed in [38]. Again the solution has the Markov field property iff the function f is affine.

(II) A martingale $M = \{M_t, t \geq 0\}$ is called *normal* if $\langle M, M \rangle_t = t$. Multiple stochastic integrals can be defined with respect to normal martingales. Recently, a formula for the product of two multiple stochastic integrals has been deduced in [51]. Using the Wiener chaos expansion and this formula one can develop an anticipating stochastic calculus for normal martingales [28]. However some of the nice properties of the Skorohod integral (like the local property or the change-of-variables formula) have not yet been established.

References

[1] J. Asch and J. Potthoff: Itô's lemma without nonanticipatory conditions. *Probab. Theory Rel. Fields* **88** (1991) 17–46.

[2] M. A. Berger: A Malliavin–type anticipative stochastic calculus. *Ann. Probab.* **16** (1988) 231–245.

[3] M. A. Berger and V. J. Mizel: An extension of the stochastic integral. *Ann. Probab.* **10** (1982) 435–450.

[4] J. M. Bismut: Martingales, the Malliavin Calculus and hypoellipticity under general Hörmander's condition. *Z. für Wahrscheinlichkeitstheorie verw. Gebiete* **56** (1981) 469–505.

[5] R. Buckdahn: Quasilinear partial stochastic differential equations without nonanticipation requirement. Preprint 176, Humboldt Universität, Berlin, 1988.

[6] R. Buckdahn: Anticipative Girsanov transformations. *Probab. Theory Rel. Fields* **89** (1991) 211–238.

[7] R. Buckdahn: Linear Skorohod stochastic differential equations. *Probab. Theory Rel. Fields* **90** (1991) 223–240.

[8] R. Buckdahn: Anticipative Girsanov transformations and Skorohod stochastic differential equations. Memoires of the AMS **533**, 1994.

[9] R. Buckdahn: Skorohod stochastic differential equations of diffusion type. *Probab. Theory Rel. Fields* **92** (1993) 297–324.

[10] R. Buckdahn and D. Nualart: Linear stochastic differential equations and Wick products. *Probab. Theory Rel. Fields.* **99** (1994) 501–526.

[11] M.E. Caballero, B. Fernandez, and D. Nualart: Smoothness of distributions for solutions of anticipating stochastic differential equations. *Stochastics and Stochastics Reports* **52** (1995) 303-322.

[12] R. H. Cameron and W. T. Martin: Transformation of Wiener integrals by nonlinear transformations. *Trans. Amer. Math. Soc.* **66** (1949) 253–283.

[13] C. Donati-Martin: Equations différentielles stochastiques dans $\mathbb{R}$ avec conditions au bord. *Stochastics and Stochastics Reports* **35** (1991) 143–173.

[14] C. Donati-Martin and D. Nualart: Markov property for elliptic stochastic partial differential equations. *Stochastics and Stochastics Reports* **46** (1994) 107–115.

[15] J. Garnier: Stochastic invariant imbedding: Application to stochastic differential equations with boundary conditions. *Probab. Theory Rel. Fields* **102** (1995) 249–271.

[16] B. Gaveau and P. Trauber: L'intégrale stochastique comme opérateur de divergence dans l'espace fonctionnel. *J. Functional Anal.* **46** (1982) 230–238.

[17] A. Grorud, D. Nualart, and M. Sanz: Hilbert-valued anticipating stochastic differential equations. *Ann. Inst. Henri Poincaré* **30** (1994) 133–161.

[18] T. Hida, H. H. Kuo, J. Potthoff and L. Streit: *White Noise: An Infinite Dimensional Calculus.* Kluwer, 1993.

[19] P. Imkeller and D. Nualart: Integration by parts on Wiener space and the existence of occupation densities. *Ann. Probab.* **22** (1994) 469–493.

[20] K. Itô: Multiple Wiener integral. *J. Math. Soc. Japan* **3** (1951) 157–169.

[21] J. Jacod: Grossissement initial, hypothèse (H'), et théorème de Girsanov. *Lecture Notes in Math.* **1118** (1985) 15–35.

[22] T. Jeulin: *Semimartingales et grossissement d'une filtration.* Lecture Notes in Math. **833**, Springer–Verlag, 1980.

[23] A. Kohatsu-Higa and J. León: Anticipating stochastic differential equations of Stratonovich type. Preprint.

[24] A. Kohatsu-Higa, J. Lesn and D. Nualart: Stochastic differential equations with random coefficients. Preprint.

[25] H. Kunita: First order stochastic partial differential equations. In: Stochastic Analysis, Proc. Taniguchi Inter. Symp., Katata and Kyoto 1982, North–Holland, 1984, 249-270.

[26] H. H. Kuo and A. Russek: White noise approach to stochastic integration. *Journal Multivariate Analysis* **24** (1988) 218–236.

[27] S. Kusuoka: The nonlinear transformation of Gaussian measure on Banach space and its absolute continuity (I). *J. Fac. Sci. Univ. Tokyo IA* **29** (1982) 567–597.

[28] J. Ma, Ph. Protter and M. San Martin: Anticipating integrals for a class of martingales. Preprint.

[29] P. Malliavin: Stochastic calculus of variations and hypoelliptic operators. In: *Proc. Inter. Symp. on Stoch. Diff. Equations, Kyoto* 1976, Wiley 1978, 195–263.

[30] T. Masuda: Absolute continuity of distributions of solutions of anticipating stochastic differential equations. *J. Functional Anal.* **95** (1991) 414–432.

[31] A. Millet and D. Nualart: Support theorems for a class of anticipating stochastic differential equations. *Stochastics and Stochastics Reports* **39** (1992) 1–24.

[32] A. Millet, D. Nualart and M. Sanz: Large deviations for a class of anticipating stochastic differential equations. *Annals of Probability* **20** (1992) 1902–1931.

[33] D. Nualart: Noncausal stochastic integrals and calculus. In: *Stochastic Analysis and Related Topics*, eds.: H. Korezlioglu and A. S. Üstünel, Lecture Notes in Math. **1316** (1988) 80–129.

[34] D. Nualart: *The Malliavin Calculus and Related Topics.* Springer-Verlag, 1995.

[35] D. Nualart: Analysis on Wiener space and anticipating stochastic calculus. In: École d'Été de Probabilités de Saint Flour 1995. *Lecture Notes in Math.* To appear.

[36] D. Nualart and E. Pardoux: Stochastic calculus with anticipating integrands. *Probab. Theory Rel. Fields* **78** (1988) 535–581.

[37] D. Nualart and E. Pardoux: Boundary value problems for stochastic differential equations. *Ann. Probab.* **19** (1991) 1118–1144 .

[38] D. Nualart and E. Pardoux: Second order stochastic differential equations with Dirichlet boundary conditions. *Stochastic Processes and Their Applications* **39** (1991) 1–24 .

[39] D. Nualart and M. Zakai: Generalized stochastic integrals and the Malliavin calculus. *Probab. Theory Rel. Fields* **73** (1986) 255–280.

[40] D. Nualart and M. Zakai: Generalized multiple stochastic integrals and the representation of Wiener functionals. *Stochastics* **23** (1988) 311–330.

[41] D. Ocone and E. Pardoux: A generalized Itô–Ventzell formula. Application to a class of anticipating stochastic differential equations. *Ann. Inst. Henri Poincaré* **25** (1989) 39–71.

[42] D. Ocone and E. Pardoux: Linear stochastic differential equations with boundary conditions. *Probab. Theory Rel. Fields* **82** (1989) 489–526.

[43] S. Ogawa: Quelques propriétés de l'intégrale stochastique de type noncausal. *Japan J. Appl. Math.* **1** (1984) 405–416.

[44] S. Ogawa: The stochastic integral of noncausal type as an extension of the symmetric integrals. *Japan J. Appl. Math.* **2** (1984) 229–240.

[45] S. Ogawa: Une remarque sur l'approximation de l'intégrale stochastique du type noncausal par une suite des intégrales de Stieltjes. *Tohoku Math. J.* **36** (1984) 41–48.

[46] E. Pardoux: Applications of anticipating stochastic calculus to stochastic differential equations. In: *Stochastic Analysis and Related Topics II*, eds.: H. Korezlioglu and A. S. Üstünel, Lecture Notes in Math. **1444** (1990) 63–105.

[47] E. Pardoux and P. Protter: Two-sided stochastic integrals and calculus. *Probab. Theory Rel. Fields* **76** (1987) 15–50.

[48] R. Ramer: On nonlinear transformations of Gaussian measures. *J. Functional Anal.* **15** (1974) 166–187.

[49] J. Rosinski: On stochastic integration by series of Wiener integrals. *Appl. Math. Optimization* **19** (1989) 137–155.

[50] F. Russo and P. Vallois: Forward, backward and symmetric stochastic integration. *Probab. Theory Rel. Fields* **97** (1993) 403–421.

[51] F. Russo and P. Vallois: Product of two multiple stochastic integrals. Preprint.

[52] T. Sekiguchi and Y. Shiota: L^2-theory of noncausal stochastic integrals. *Math. Rep. Toyama Univ.* **8** (1985) 119–195.

[53] A. V. Skorohod: On a generalization of a stochastic integral. *Theory Probab. Appl.* **20** (1975) 219–233.

[54] D. W. Stroock: Some applications of stochastic calculus to partial differential equations. In: *École d'Été de Probabilités de Saint Flour*, Lecture Notes in Math. **976** (1983) 267–382.

[55] A. S. Üstünel: The Itô formula for anticipative processes with nonmonotonous time via the Malliavin calculus. *Probab. Theory Rel. Fields* **79** (1988) 249–269.

[56] A. S. Üstünel and M. Zakai: Transformations of the Wiener measure under non–invertible shifts. *Probab. Theory Rel. Fields* **99** (1994) 485–500.

[57] S. Watanabe: *Lectures on Stochastic Differential Equations and Malliavin Calculus*, Tata Institute of Fundamental Research, Springer-Verlag, 1984.

FOUNDATION OF ENTROPY, COMPLEXITY AND FRACTALS IN QUANTUM SYSTEMS

MASANORI OHYA,* *Science University of Tokyo*

Abstract

Fundamentals of quantum entropy are totally reviewed from von Neumann to recent works including new formulation of Kolmogorov-Sinai type dynamical entropy. Entropy is one of the most important quantities to describe a chaotic aspect of several different phenomena. There exist many trials to express the complexity of a dynamical system. I proposed Information Dynamics to synthesize the dynamics of state change and the complexity of system in 1991. In this paper, it is shown that the complexity of entropy type becomes a useful tool in the following two points: It generalizes the usual formulation of dynamical entropy so that it can be applied to concrete quantum processes like optical communication. New concept of fractal dimension, so-called the fractal dimension of a state, can be introduced by this entropic complexity.

1. Mathematical Descriptions

Let us fix the notations used throughout this paper. Let μ be a measure on a measurable space $(\Omega, \mathcal{F})$, $P(\Omega)$ be the set of all probability measures on Ω and $M(\Omega)$ be the set of all measurable functions on Ω. We denote the set of all bounded linear operators on a Hilbert space $\mathcal{H}$ by $B(\mathcal{H})$, and the set of all density operators on $\mathcal{H}$ by $\mathfrak{S}(\mathcal{H})$. Moreover, let $\mathbb{S}(\mathcal{A})$ be the set of all states on C*-algebra or von Neumann algebra $\mathcal{A}$. The descriptions of a classical dynamical system(CDS), a quantum dynamical system(QDS) and a general quantum dynamical system(GQDS) are given in the following Table:

	CDS	QDS	GQDS
observable	real r.v. in $M(\Omega)$	Hermitian operator A on $\mathcal{H}$ (self adjoint operator in $B(\mathcal{H})$)	self-adjoint element A in C*-algebra $\mathcal{A}$
state	probability measure $\mu \in P(\Omega)$	density operator ρ on $\mathcal{H}$	p.l.fnal $\varphi \in \mathbb{S}$ with $\varphi(I) = 1$
expectation	$\int_\Omega f d\mu$	$\mathrm{tr}\rho A$	$\varphi(A)$

Table. 1.1 Descriptions of CDS, QDS and GQDS

* Postal address: Department of Information Sciences, Science University of Tokyo, Noda City, Chiba 278, Japan.

2. Classical Entropy

2.1. *Discrete case (Shannon's theory [43])* The concept of entropy was introduced and developed to study the following topics: irreversible behavior, symmetry breaking, amount of information transmission, characterization of states, chaotic property of states, etc. First we start to review classical discrete entropies in this section.

A state in a discrete classical system is described by a probability distribution such that

$$\Delta_n = \left\{ p = \{p_i\}_{i=1}^n; \sum_i p_i = 1, p_i \geq 0 \right\}.$$

The entropy of a state $p = \{p_i\} \in \Delta_n$ is given by

$$S(p) = -\sum_i p_i \log p_i.$$

The relative information (relative entropy) is a quantity to express a sort of difference between two states and is defined by Kullback and Leibler [23] as

$$S(p, q) = \begin{cases} \sum_i p_i \log \frac{p_i}{q_i} & (p \ll q) \\ \infty & \text{otherwise} \end{cases}$$

for any $p, q \in \Delta_n$, where $p \ll q$ means that p is absolutely continuous w.r.t. q, that is, if $p_i = 0$, then $q_i = 0$.

When a system is changed under some internal or external effects, the change of a system is indicated by that of a state. The effect causing the state change is exhibited by a so-called channel, a mapping from a state to a state. Once a state p is changed through such a channel Λ^*, the information transmitted from an initial state p to a final state $q \equiv \Lambda^* p$ is described by the mutual entropy, which is the most important quantity in information theory discovered by Shannon. This entropy is defined by

$$I(p; \Lambda^*) = S(r, p \otimes q) = \sum_{i,j} r_{ij} \log \frac{r_{ij}}{p_i q_j},$$

where the channel $\Lambda^* : \Delta_n \to \Delta_m$ is given by a transition matrix $(p(j|i))$, and $q = \Lambda^* p = \left(\sum_j p(j|i) p_i \right)$, $r_{ij} = p(j|i) p_i$, $p \otimes q = \{p_i q_j\}$.

The fundamental inequality of Shannon is

$$0 \leq I(p; \Lambda^*) \leq \min\{S(p), S(q)\}$$

According to this inequality, the ratio

$$r(p; \Lambda^*) = \frac{I(p; \Lambda^*)}{S(p)}$$

exhibits an efficiency of the channel transformation Λ^* in a communication process.

2.2. *Continuous case* In classical continuous systems, a state is described by a probability measure μ. For an input system $(\Omega, \mathcal{F}, P(\Omega))$ and an output system $(\overline{\Omega}, \overline{\mathcal{F}}, P(\overline{\Omega}))$, a channel is a map Λ^* from $P(\Omega)$ to $P(\overline{\Omega})$, in particular, Λ^* is called a Markov type if it satisfies

$$\Lambda^* \mu(Q) = \int_\Omega \lambda(x, Q)\varphi(dx), \ \varphi \in P(\Omega), \ Q \in \overline{\mathcal{F}},$$

where $\lambda : \Omega \times \overline{\mathcal{F}} \to [0, 1]$ with (i) $\lambda(x, \cdot) \in P(\overline{\Omega})$ for each $x \in \Omega$, (ii) $\lambda(\cdot, Q) \in M(\Omega)$ for each $Q \in \overline{\mathcal{F}}$. In a continuous system, the entropies are defined as follows: Let $F(\Omega)$ be the set of all finite partitions $\{A_k\}$ of Ω. For any $\mu \in P(\Omega)$, the entropy of μ is defined by

$$S(\mu) = \sup\left\{-\sum_k \mu(A_k) \log \mu(A_k); \{A_k\} \in F(\Omega)\right\},$$

which often becomes infinite in a continuous system. For any $\mu, \upsilon \in P(\Omega)$, the relative entropy is given by

$$S(\mu, \upsilon) = \sup\left\{\sum_k \mu(A_k) \log \frac{\mu(A_k)}{\upsilon(A_k)}; \{A_k\} \in F(\Omega)\right\}$$

$$= \begin{cases} \int_\Omega \log\left(\frac{d\mu}{d\upsilon}\right) d\mu & (\mu \ll \upsilon) \\ \infty & \text{otherwise} \end{cases}$$

Let Φ, Φ_0 be two compound states (measures) defined as

$$\Phi(Q_1, Q_2) = \int_{Q_1} \lambda(x, Q_2)\mu(dx), \ Q_1 \in \mathcal{F}, \ Q_2 \in \overline{\mathcal{F}},$$

$$\Phi_0(Q_1, Q_2) = \mu \otimes \Lambda^* \mu(Q_1, Q_2) = \mu(Q_1)\Lambda^* \mu(Q_2).$$

For $\mu \in P(\Omega)$ and a channel Λ^*, the mutual entropy is given by

$$I(\mu; \Lambda^*) = S(\Phi, \Phi_0).$$

3. Quantum Entropy

3.1. *Entropies for density operators* A state in quantum systems is described by a density operator on a Hilbert space $\mathcal{H}$. The entropy of a state ρ was introduced by von Neumann [30] as

$$S(\rho) = -\mathrm{tr}\rho \log \rho.$$

If $\rho = \sum_k p_k E_k$ is the Schatten decomposition (the spectral decomposition of ρ with $\dim E_k = 1$), then

$$S(\rho) = -\sum_k p_k \log p_k.$$

Therefore the von Neumann entropy contains the Shannon entropy as a special case.

Let us summarize the fundamental properties of the entropy $S(\rho)$ [40].

Theorem 3.1: For any density operator $\rho \in \mathbb{S}(\mathcal{H})$, the followings hold:

(1) Positivity : $S(\rho) \geq 0$.

(2) Symmetry : Let $\rho' = U\rho U^*$ for an unitary operator U. Then

$$S(\rho') = S(\rho).$$

(3) Concavity : $S(\lambda \rho_1 + (1 - \lambda)\rho_2) \geq \lambda S(\rho_1) + (1 - \lambda)S(\rho_2)$ for any $\rho_1, \rho_2 \in \mathbb{S}(\mathcal{H})$ and $\lambda \in [0, 1]$.

(4) Additivity : $S(\rho_1 \otimes \rho_2) = S(\rho_1) + S(\rho_2)$ for any $\rho_k \in \mathbb{S}(\mathcal{H}_k)$.

(5) Subadditivity : For the reduced states ρ_1, ρ_2 of $\rho \in \mathbb{S}(\mathcal{H}_1 \otimes \mathcal{H}_2)$,

$$S(\rho) \leq S(\rho_1) + S(\rho_2).$$

(6) Lower Semicontinuity : If $\|\rho_n - \rho\|_1 \to 0 (\equiv \mathrm{tr}|\rho_n - \rho| \to 0)$ as $(n \to \infty)$, then

$$S(\rho) \leq \lim_{n \to \infty} \inf S(\rho_n).$$

(7) Continuity : Let ρ_n, ρ be elements in $\mathbb{S}(\mathcal{H})$ satisfying the following conditions : (i) $\rho_n \to \rho$ weakly as $n \to \infty$, (ii) $\rho_n \leq A \ (\forall n)$ for some compact operator A, and (iii) $-\sum_k a_k \log a_k < +\infty$ for the eigenvalues $\{a_k\}$ of A. Then $S(\rho_n) \to S(\rho)$.

(8) Strong Subadditivity : Let $\mathcal{H} = \mathcal{H}_1 \otimes \mathcal{H}_2 \otimes \mathcal{H}_3$ and denote the reduced states $\mathrm{tr}_{\mathcal{H}_i \otimes \mathcal{H}_j}\rho$ by ρ_k and $\mathrm{tr}_{\mathcal{H}_k}\rho$ by ρ_{ij}. Then

$$S(\rho) + S(\rho_2) \leq S(\rho_{12}) + S(\rho_{23}) \text{ and } S(\rho_1) + S(\rho_2) \leq S(\rho_{13}) + S(\rho_{23}).$$

For two states $\rho, \sigma \in \mathbb{S}(\mathcal{H})$, the relative entropy [24, 45] is defined by

$$S(\rho, \sigma) = \begin{cases} tr\rho \left(\log \rho - \log \sigma \right) & (\rho \ll \sigma) \\ \infty & \text{otherwise} \end{cases}$$

where $\rho \ll \sigma$ means that $\mathrm{tr}\sigma A = 0 \Rightarrow \mathrm{tr}\rho A = 0$ for $A \geq 0$.

Let Λ^* be a channel, that is, a mapping from $\mathbb{S}(\mathcal{H})$ to $\mathbb{S}(\overline{\mathcal{H}})$ and take the compound states[32, 33] (corresponding to joint distributions in CDS)

$$\theta_E = \sum_k p_k E_k \otimes \Lambda^* E_k, \quad \theta_0 = \rho \otimes \Lambda^* \rho.$$

The quantum mutual entropy for a state $\rho \in \mathbb{S}(\mathcal{H})$ and a channel Λ^* should satisfy the following conditions: (1) If the channel is trivial, i.e., Λ^* is an identity map, then the mutual entropy $I(\rho; \Lambda^*) = S(\rho)$. (2) If the input and output systems are claasical, then $I(\rho; \Lambda^*) =$ the classical mutual entropy. (3) The Scannon's fundamental inequality $0 \leq I(\rho; \Lambda^*) \leq \min\{S(\rho), S(\Lambda^* \rho)\}$ should be held. Such a quantum mutual entropy is given by[32]

$$I(\rho; \Lambda^*) = \sup \left\{ S(\theta_E, \theta_0); E = \{E_k\} \right\}.$$

where the supremum is taken over all Schatten decompositions. If the channel is linear, then it becomes

$$I(\rho; \Lambda^*) = \sup \left\{ \sum_k p_k S(\Lambda^* E_k, \Lambda^* \rho); E = \{E_k\} \right\}.$$

When ρ is a classical state (p_k), the Schatten decomposition is unique, $\rho = \sum_k p_k \delta_k$ with $\delta_k(j) \equiv \delta_{kj}$ (Kronecker's delta) and the mutual entropy is written as

$$I(\rho; \Lambda^*) = \sum_k p_k S(\Lambda^* \delta_k, \Lambda^* \rho).$$

The above quantum mutual entropy satisfies all three conditions above and contains other definitions of the mutual entropy as special cases[32, 39]. The following fundamental inequality of Shannon's type (the condition (3)) was proved in [32]:

Theorem 3.2:

$$0 \leq I(\rho; \Lambda^*) \leq \min\{S(\rho), S(\Lambda^* \rho)\}.$$

3.2. *Channeling transformations* A general quantum system containing continuous cases is described by a C*-algebra or a von Neumann algebra [10, 11, 20]. Let $\mathcal{A}$ be a C*-algebra (i.e., complex normed algebra with involution $*$ such that $\|A\| = \|A^*\|$, $\|A^* A\| = \|A\|^2$ and complete w.r.t. $\|\cdot\|$) and $\mathbb{S}(\mathcal{A})$ be the set of all states on $\mathcal{A}$ (i.e., positive continuous linear functionals φ on $\mathcal{A}$ such that $\varphi(I) = 1$ if $I \in \mathcal{A}$).

A channel is a mapping from $\mathbb{S}(\mathcal{A})$ to $\mathbb{S}(\overline{\mathcal{A}})$. Almost all physical transformations are described by this mapping [31, 35]. First give the mathematical definition of channels.

Definition 3.1: Let $(\mathcal{A}, \mathbb{S}(\mathcal{A}), \alpha(\mathbf{R}))$ be an input system and $(\overline{\mathcal{A}}, \mathbb{S}(\overline{\mathcal{A}}), \overline{\alpha}(\mathbf{R}))$ be an output system. Take any $\varphi, \psi \in \mathbb{S}(\overline{\mathcal{A}})$.

(1) Λ^* is linear if $\Lambda^*(\lambda\varphi + (1-\lambda)\psi) = \lambda\Lambda^*\varphi + (1-\lambda)\Lambda^*\psi$ holds for any $\lambda \in [0, 1]$.

(2) Λ^* is completely positive (CP) if Λ^* is linear and its dual $\Lambda : \overline{\mathcal{A}} \to \mathcal{A}$ satisfies

$$\sum_{i,j=1}^{n} A_i^* \Lambda(\overline{A}_i^* \overline{A}_j) A_j \geq 0$$

for any $n \in \mathbf{N}$ and any $\{\overline{A}_i\} \subset \overline{\mathcal{A}}$, $\{A_i\} \subset \mathcal{A}$.

(3) Λ^* is Schwarz type if $\Lambda(\overline{A}^*) = \Lambda(\overline{A})^*$ and $\Lambda(\overline{A})^*\Lambda(\overline{A}) \leq \Lambda(\overline{A}^*\overline{A})$.

(4) Λ^* is stationary if $\Lambda \circ \alpha_t = \overline{\alpha}_t \circ \Lambda$ for any $t \in \mathbf{R}$.

(5) Λ^* is ergodic if Λ^* is stationary and $\Lambda^*(exI(\alpha)) \subset exI(\overline{\alpha})$.

(6) Λ^* is orthogonal if any two orthogonal states $\varphi_1, \varphi_2 \in \mathbb{S}(\mathcal{A})$ (denoted by $\varphi_1 \perp \varphi_2$) implies $\Lambda^*\varphi_1 \perp \Lambda^*\varphi_2$.

(7) Λ^* is deterministic if Λ^* is orthogonal and bijective.

(8) For a subset $\mathcal{S}$ of $\mathbb{S}(\mathcal{A})$, Λ^* is chaotic for $\mathcal{S}$ if $\Lambda^*\varphi_1 = \Lambda^*\varphi_2$ for any $\varphi_1, \varphi_2 \in \mathcal{S}$.

(9) Λ^* is chaotic if Λ^* is chaotic for $\mathbb{S}(\mathcal{A})$.

Most of channels appeared in physical processes are CP channels. Examples of such channels are the followings [35]: Take a density operator ρ as an input state.
(1)Unitary evolution: Let H be the Hamiltonian of a system.

$$\rho \to \Lambda_t^*\rho = AdU_t(\rho) \equiv U_t\rho U_t^*,$$

where $t \in \mathbf{R}$, $U_t = \exp(-itH)$.

(2)Semigroup evolution: Let $V_t(t \in \mathbf{R}^+)$ be an one parameter semigroup on $\mathcal{H}$.

$$\rho \to \Lambda_t^*\rho = V_t\rho V_t^*, \quad t \in \mathbf{R}^+.$$

(3)Quantum measurement: When a measuring apparatus is prepared by an positive operator valued measure $\{Q_n\}$ and the state ρ changes to a state $\Lambda^*\rho$ after this measurement,

$$\rho \to \Lambda^*\rho = \sum_n Q_n\rho Q_n.$$

*(4) Reduction:*If a system Σ_1 interacts with an external system Σ_2 described by another Hilbert space $\mathcal{K}$ and the initial states of Σ_1 and Σ_2 are ρ and σ, respectively, then the combined state θ_t of Σ_1 and Σ_2 at time t after the interaction between two systems is given by

$$\theta_t \equiv U_t(\rho \otimes \sigma)U_t^*,$$

where $U_t = \exp(-itH)$ with the total Hamiltonian H of Σ_1 and Σ_2. A channel is obtained by taking the partial trace w.r.t. $\mathcal{K}$ such as

$$\rho \to \Lambda_t^*\rho \equiv \mathrm{tr}_\mathcal{K}\theta_t.$$

*(5)Optical communication processes:*Quantum communication process is described by the following scheme.

$$
\begin{array}{ccc}
& \nu \in \mathbb{S}(\mathcal{K}) & \\
& \downarrow & \\
\mathbb{S}(\mathcal{H}) \ni \rho & \longrightarrow & \bar{\rho} = \Lambda^*\rho \in \mathbb{S}(\mathcal{H}) \\
& \downarrow & \\
& \mathrm{Loss} &
\end{array}
$$

$$
\begin{array}{ccc}
\mathbb{S}(\mathcal{H}) & \xrightarrow{\ \Lambda^*\ } & \mathbb{S}(\mathcal{H}) \\
\gamma^* \downarrow & & \uparrow a^* \\
\mathbb{S}(\mathcal{H} \otimes \mathcal{K}) & \xrightarrow{\ \pi^*\ } & \mathbb{S}(\mathcal{H} \otimes \mathcal{K})
\end{array}
$$

The above maps γ^*, a^* are given as

$$
\begin{aligned}
\gamma^*(\rho) &= \rho \otimes \nu, & \rho \in \mathbb{S}(\mathcal{H}), \\
a^*(\theta) &= \mathrm{tr}_\mathcal{K}\theta, & \theta \in \mathbb{S}(\mathcal{H} \otimes \mathcal{K}),
\end{aligned}
$$

where ν is a noise coming from the outside of the system. The map π^* is a certain channel determined by physical properties of the device transmitting information. Hence the channel for the above process is given as

$$\Lambda^*\rho \equiv \mathrm{tr}_\mathcal{K}\pi^*(\rho \otimes \nu) = (a^* \circ \pi^* \circ \gamma^*)(\rho)$$

An example of this channel is the attenuation channel defined as follows: Take $\nu = |0\rangle \langle 0| = $ vacuum state. Then

$$\Lambda^*\rho = |\alpha\theta\rangle \langle\alpha\theta|, \ 0 \le |\alpha|^2 \le 1$$

for a coherent state $\rho = |\theta\rangle \langle\theta|$.

Note that there exists a special channel named "lifting"[3], and it is a useful concept to characterize quantum communication or stochastic proccsess [2, 17].

3.3. *Entropies for general quantum states* The entropy (uncertainty) of a state $\varphi \in S$ seen from the reference system S, a weak $*$-compact convex subset of $\mathbb{S}$, was introduced in [34]. This entropy contains von Neumann's entropy and classical entropy as special cases.

Every state $\varphi \in S$ has a maximal measure μ pseudosupported on exS (extreme points in S) such that

$$\varphi = \int_S \omega d\mu.$$

The measure μ giving the above decomposition is not unique unless S is a Choquet simplex [12], so that we denote the set of all such measures by $M_\varphi(S)$. Take

$$D_\varphi(S) \equiv$$

$$\left\{ \mu \in M_\varphi(S); \exists \{\mu_k\} \subset \mathbf{R}^+ \text{and } \{\varphi_k\} \subset exS \text{ s.t. } \sum_k \mu_k = 1, \mu = \sum_k \mu_k \delta(\varphi_k) \right\},$$

where $\delta(\varphi)$ is the delta measure concentrated on $\{\varphi\}$. Put

$$H(\mu) = -\sum_k \mu_k \log \mu_k$$

for a measure $\mu \in D_\varphi(S)$. Then the entropy of a general state $\varphi \in S$ w.r.t. S is defined by

$$S^S(\varphi) = \begin{cases} \inf\{H(\mu); \mu \in D_\varphi(S)\} & (D_\varphi(S) \neq \emptyset) \\ \infty & (D_\varphi(S) = \emptyset) \end{cases}$$

This entropy (mixing S-entropy) of a general state φ satisfies the following properties [34, 35]. When S is the total space $\mathbb{S}$, we simply denote $S^S(\varphi)$ by $S(\varphi)$.

Theorem 3.3: When $A = B(\mathcal{H})$ and $\alpha_t = Ad(U_t)$ (i.e., $\alpha_t(A) = U_t^* A U_t$ for any $A \in A$) with a unitary operator U_t, for any state φ given by $\varphi(\cdot) = \mathrm{tr}\rho\cdot$ with a density operator ρ, the following facts hold:

(1) $S(\varphi) = -\mathrm{tr}\rho \log \rho$.

(2) If φ is an α-invariant faithful state and every eigenvalue of ρ is non-degenerate, then $S^I(\varphi) = S(\varphi)$.

(3) If $\varphi \in K(\alpha)$, then $S^K(\varphi) = 0$.

Theorem 3.4: For any $\varphi \in K(\alpha)$, we have

(1) $S^K(\varphi) \leq S^I(\varphi)$.

(2) $S^K(\varphi) \leq S(\varphi)$.

This $\mathcal{S}$ (or mixing) entropy gives a measure of the uncertainty observed from the reference system $\mathcal{S}$ so that it has the following merits : Even if the total entropy $S(\varphi)$ is infinite, $S^{\mathcal{S}}(\varphi)$ is finite for some $\mathcal{S}$, hence it explains a sort of symmetry breaking in $\mathcal{S}$. Other similar properties as $S(\rho)$ hold for $S^{\mathcal{S}}(\varphi)$ [34, 35]. This entropy can be appllied to characterize normal states and quantum Markov chains in von Neumann algebras[5].

The relative entropy for two general states φ and ψ was introduced by Araki [6, 7] and Uhlmann [44] and their relation is considered in [15, 18, 19].

$\langle$Araki's definition$\rangle$

Let $\mathcal{N}$ be σ-finite von Neumann algebra acting on a Hilbert space $\mathcal{H}$ and φ, ψ be normal states on $\mathcal{N}$ given by $\varphi(\cdot) = \langle x, \cdot x \rangle$ and $\psi(\cdot) = \langle y, \cdot y \rangle$ with $x, y \in \mathcal{K}$ (a positive natural cone). The operator $S_{x,y}$ is defined by

$$S_{x,y}(Ay + z) = s^{\mathcal{N}}(y)A^*x, \ \ A \in \mathcal{N}, \ s^{\mathcal{N}'}(y)z = 0,$$

on the domain $\mathcal{N} y + (I - s^{\mathcal{N}'}(y))\mathcal{H}$, where $s^{\mathcal{N}}(y)$ is the projection from $\mathcal{H}$ to $\{\mathcal{N}'y\}^-$, the $\mathcal{N}$-support of y. Using this $S_{x,y}$, the relative modular operator $\Delta_{x,y}$ is defined as $\Delta_{x,y} = (S_{x,y})^* \overline{S_{x,y}}$, whose spectral decomposition is denoted by $\int_0^\infty \lambda de_{x,y}(\lambda)$ ($\overline{S_{x,y}}$ is the closure of $S_{x,y}$). Then the Araki relative entropy is given by

$$S(\psi, \varphi) = \begin{cases} \int_0^\infty \log \lambda d \langle y, e_{x,y}(\lambda) y \rangle & (\psi \ll \varphi) \\ \infty & \text{otherwise} \end{cases},$$

where $\psi \ll \varphi$ means that $\varphi(A^*A) = 0$ implies $\psi(A^*A) = 0$ for $A \in \mathcal{N}$.

$\langle$Uhlmann's definition$\rangle$

Let $\mathcal{L}$ be a complex linear space and p, q be two seminorms on $\mathcal{L}$. Moreover, let $H(\mathcal{L})$ be the set of all positive hermitian forms α on $\mathcal{L}$ satisfying $|\alpha(x, y)| \leq p(x)q(y)$ for all $x, y \in \mathcal{L}$. Then the quadratical mean $QM(p, q)$ of p and q is defined by

$$QM(p, q)(x) = \sup\{\alpha(x, x)^{1/2}; \ \alpha \in H(\mathcal{L})\}, \ \ x \in \mathcal{L},$$

and there exists a function $p_t(x)$ of $t \in [0, 1]$ for each $x \in \mathcal{L}$ satisfying the following conditions:

(1) For any $x \in \mathcal{L}$, $p_t(x)$ is continuous in t,

(2) $p_{1/2} = QM(p, q)$,

(3) $p_{t/2} = QM(p, p_t)$,

(4) $p_{(t+1)/2} = QM(p_t, q)$.

This seminorm p_t is denoted by $QI_t(p, q)$ and is called the quadratical interpolation from p to q. It is shown [44] that for any positive hermitian forms α, β, there exists a unique function $QF_t(\alpha, \beta)$ of $t \in [0, 1]$ with values in the set $H(\mathcal{L})$ such that $QF_t(\alpha, \beta)(x, x)^{1/2}$ is the quadratical interpolation from $\alpha(x, x)^{1/2}$ to $\beta(x, x)^{1/2}$. The relative entropy functional $S(\alpha, \beta)(x)$ of α and β is defined as

$$S(\alpha, \beta)(x) = -\lim_{t \to 0} \inf \frac{1}{t} \left\{ QF_t(\alpha, \beta)(x, x) - \alpha(x, x) \right\}$$

for $x \in \mathcal{L}$. Let $\mathcal{L}$ be a *-algebra $\mathcal{A}$ and φ, ψ be positive linear functionals on $\mathcal{A}$ defining two hermitian forms φ^L, ψ^R such as $\varphi^L(A, B) = \varphi(A^* B)$ and $\psi^R(A, B) = \psi(BA^*)$. Then the relative entropy of φ and ψ is defined by

$$S(\psi, \varphi) = S(\psi^R, \varphi^L)(I).$$

Next we discuss the mutual entropy in GQDS.

For $\varphi \in \mathcal{S}(\mathcal{A}) \subset \mathbb{S}(\mathcal{A})$ and $\Lambda^* : \mathbb{S}(\mathcal{A}) \to \mathbb{S}(\overline{\mathcal{A}})$, define the compound states by

$$\Phi_\mu^{\mathcal{S}} = \int_{\mathcal{S}} \omega \otimes \Lambda^* \omega d\mu$$

and

$$\Phi_0 = \varphi \otimes \Lambda^* \varphi$$

The first compound state generalizes the joint probability in CDS and it exhibits the correlation between the initial state φ and the final state $\Lambda^* \varphi$ [34].

The mutual entropy w.r.t. $\mathcal{S}$ and μ is

$$I_\mu^{\mathcal{S}}(\varphi; \Lambda^*) = S(\Phi_\mu^{\mathcal{S}}, \Phi_0)$$

and the mutual entropy w.r.t. $\mathcal{S}$ is defined as [35]

$$I^{\mathcal{S}}(\varphi; \Lambda^*) = \lim_{\varepsilon \to 0} \sup \left\{ I_\mu^{\mathcal{S}}(\varphi; \Lambda^*); \mu \in F_\varphi^\varepsilon(\mathcal{S}) \right\},$$

where

$$F_\varphi^\varepsilon(\mathcal{S}) = \begin{cases} \{\mu \in D_\varphi(\mathcal{S}); S^{\mathcal{S}}(\varphi) \leq H(\mu) \leq S^{\mathcal{S}}(\varphi) + \varepsilon < +\infty\} \\ M_\varphi(\mathcal{S}) \quad \text{if} \quad S^{\mathcal{S}}(\varphi) = +\infty \end{cases}$$

$$D_\varphi(\mathcal{S}) = \left\{ \mu \in M_\varphi(\mathcal{S}); \exists \{\mu_k\} \subset \mathbf{R}^+ \text{ s.t } \mu = \sum_k \mu_k \delta(\varphi_k), \varphi_k \in exS, \sum_k \mu_k = 1 \right\}.$$

The following fundamental inequality is satisfied for almost all physical cases.

$$0 \leq I^{\mathcal{S}}(\varphi; \Lambda^*) \leq S^{\mathcal{S}}(\varphi)$$

The main properties of the relative entropy and the mutual entropy are shown in the following theorems [6, 7, 15, 18, 19, 42, 40, 44].

Theorem 3.5

(1) Positivity : $S(\varphi, \psi) \geq 0$ and $S(\varphi, \psi) = 0$ iff $\varphi = \psi$.

(2) Joint Convexity : $S(\lambda\psi_1 + (1 - \lambda)\psi_2, \lambda\varphi_1 + (1 - \lambda)\varphi_2) \leq \lambda S(\psi_1, \varphi_1) + (1 - \lambda)S(\psi_2, \varphi_2)$ for any $\lambda \in [0, 1]$.

(3) Additivity : $S(\psi_1 \otimes \psi_2, \varphi_1 \otimes \varphi_2) = S(\psi_1, \varphi_1) + S(\psi_2, \varphi_2)$.

(4) Lower Semicontinuity : If $\lim_{n\to\infty} \|\psi_n - \psi\| = 0$ and $\lim_{n\to\infty} \|\varphi_n \to \varphi\| = 0$, then $S(\psi, \varphi) \leq \liminf_{n\to\infty} S(\psi_n, \varphi_n)$. Moreover, if there exists a positive number λ satisfying $\psi_n \leq \lambda\varphi_n$, then $\lim_{n\to\infty} S(\psi_n, \varphi_n) = S(\psi, \varphi)$.

(5) Monotonicity : For a channel Λ^* from $\mathbb{S}$ to $\overline{\mathbb{S}}$,

$$S(\Lambda^*\psi, \Lambda^*\varphi) \leq S(\psi, \varphi).$$

(6) Lower Bound : $\|\psi - \varphi\|^2/4 \leq S(\psi, \varphi)$.

Theorem 3.6: For a state $\varphi(\cdot) = \mathrm{tr}\rho\cdot$ and a channel Λ^*, we have

(1) if Λ^* is deterministic, then $I(\varphi; \Lambda^*) = S(\varphi)$;

(2) if Λ^* is chaotic, then $I(\varphi; \Lambda^*) = 0$;

(3) if Λ^* is ergodic and φ is stationary for a time evolution $\alpha_t = AdU_t$, and if every eigenvalue of ρ is nonzero and nondegenerate, then $I(\varphi; \Lambda^*) = S(\Lambda^*\varphi)$.

Before closing this section, we mention the dynamical entropy introduced by Connes, Narnhofer and Thirring [13].

The CNT entropy $H_\varphi(\mathcal{M})$ of C*-subalgebra $\mathcal{M} \subset \mathcal{A}$ is defined by

$$H_\varphi(\mathcal{M}) \equiv \sup_{\varphi=\sum_j \mu_j \varphi_j} \sum_j \mu_j S(\varphi_j \mid_\mathcal{M}, \varphi \mid_\mathcal{M})$$

where the supremum is taken over all finite decompositions $\varphi = \sum_j \mu_j\varphi_j$ of φ and $\varphi \mid_\mathcal{M}$ is the restriction of φ to $\mathcal{M}$. subalgebra There are some relations between the mixing entropy $S^S(\varphi)$ and the CNT entropy [28, 40].

Theorem 3.7:

(1) For any state φ on a unital C*-algebra $\mathcal{A}$,

$$S(\varphi) = H_\varphi(\mathcal{A}).$$

(2) Let $(\mathcal{A}, G, \alpha)$ be a G-finite W*-dynamical system, φ be a G-invariant normal state of $\mathcal{A}$, then

$$S^{I(\alpha)}(\varphi) = H_\varphi(\mathcal{A}^\alpha),$$

where $\mathcal{A}^\alpha$ is the fixed points algebra of $\mathcal{A}$ w.r.t. α.

(3) Let $\mathcal{A}$ be the C*-algebra $C(\mathcal{H})$ of all compact operators on a Hilbert space $\mathcal{H}$, and G be a group, α be a *-automorphic action of G-invariant density operator. Then

$$S^{I(\alpha)}(\rho) = H_\rho(\mathcal{A}^\alpha).$$

(4) There exists a model [28] such that

$$S^{I(\alpha)}(\varphi) > H_\varphi(\mathcal{A}^\alpha) = 0.$$

4. Information Dynamics

Information dynamics is a synthesis of the dynamics of state change and the complexity of states [20, 36, 38]. It is a trial to provide a new view for the study of chaotic behavior of systems.

Let $(\mathcal{A}, \mathbb{S}, \alpha(G))$ be an input (or initial) system and $(\overline{\mathcal{A}}, \overline{\mathbb{S}}, \overline{\alpha}(\overline{G}))$ be an output (or a final) system. Here $\mathcal{A}$ is the set of all objects to be observed and $\mathbb{S}$ is the set of all means getting the observed value, $\alpha(G)$ is a certain evolution of the system. Often we have $\mathcal{A} = \overline{\mathcal{A}}$, $\mathbb{S} = \overline{\mathbb{S}}$, $\alpha = \overline{\alpha}$. Therefore we claim

[Giving a mathematical structure to input and output triples
$\equiv$ Having a theory]

The dynamics of state change is described by a channel $\Lambda^* : \mathbb{S} \to \overline{\mathbb{S}}$ (sometimes $\mathbb{S} \to \mathbb{S}$). The fundamental point of ID is that ID contains two complexities in itself.

Let $(\mathcal{A}_t, \mathbb{S}_t, \alpha^t(G^t))$ be the total system of $(\mathcal{A}, \mathbb{S}, \alpha)$ and $(\overline{\mathcal{A}}, \overline{\mathbb{S}}, \overline{\alpha})$, and $\mathcal{S}$ be a subset of $\mathbb{S}$ in which we are measuring observables(e.g., $\mathcal{S} = I(\alpha), K(\alpha)$ in C*-system).

C is the complexity of a state φ measured from $\mathcal{S}$ and T is the transmitted complexity associated with the state change $\varphi \to \Lambda^*\varphi$, both of which should satisfy the following properties :

(i) For any $\varphi \in \mathcal{S} \subset \mathbb{S}$,

$$C^{\mathcal{S}}(\varphi) \geq 0, \ T^{\mathcal{S}}(\varphi; \Lambda^*) \geq 0.$$

(ii) For any orthogonal bijection $j : ex\mathbb{S} \to ex\mathbb{S}$ (the set of all extreme points in $\mathbb{S}$),

$$C^{j(\mathcal{S})}(j(\varphi)) = C^{\mathcal{S}}(\varphi),$$

$$T^{j(\mathcal{S})}(j(\varphi); \Lambda^*) = T^{\mathcal{S}}(\varphi; \Lambda^*).$$

(iii) For $\Phi \equiv \varphi \otimes \psi \in \mathcal{S}_t \subset \mathbb{S}_t,$

$$C^{\mathcal{S}_t}(\Phi) = C^{\mathcal{S}}(\varphi) + C^{\overline{\mathcal{S}}}(\psi).$$

(iv) $0 \leq T^{\mathcal{S}}(\varphi; \Lambda^*) \leq C^{\mathcal{S}}(\varphi).$

(v) $T^{\mathcal{S}}(\varphi; id) = C^{\mathcal{S}}(\varphi)$, where "id" is an identity map from $\mathbb{S}$ to $\mathbb{S}$.

Instead of (iii), when "(iii') $\Phi \in \mathcal{S}_t \subset \mathbb{S}_t$, put $\varphi \equiv \Phi \mid_{\mathcal{A}}$ (i.e., the restriction of Φ to $\mathcal{A}$), $\psi \equiv \Phi \mid_{\overline{\mathcal{A}}}$, $C^{\mathcal{S}_t}(\Phi) \leq C^{\mathcal{S}}(\varphi) + C^{\overline{\mathcal{S}}}(\psi)$ " is satisfied, C and T is called the pair of strong complexity. Therefore ID can be considered as the following definition 4.1.

*Definition 4.1:*Information Dynamics (ID) is defined by

$$\left(\mathcal{A}, \mathbb{S}, \alpha(G); \overline{\mathcal{A}}, \overline{\mathbb{S}}, \overline{\alpha}(\overline{G}); \; \Lambda^*; \; C^{\mathcal{S}}(\varphi), T^{\mathcal{S}}(\varphi; \; \Lambda^*)\right)$$
$$\text{and some relations } R \text{ among them.}$$

Thus, in the framework of ID, we have to

(i) determine mathematically

$$\mathcal{A}, \mathbb{S}, \alpha(G); \overline{\mathcal{A}}, \; \overline{\mathbb{S}}, \overline{\alpha}(\overline{G}),$$

(ii) choose Λ^* and R, and

(iii) define $C^{\mathcal{S}}(\varphi), T^{\mathcal{S}}(\varphi; \Lambda^*).$

Information Dynamics can be applied to the study of chaos in the following ways:

(a) ψ is more chaotic than φ as seen from the reference system $\mathcal{S}$ if $C^{\mathcal{S}}(\psi) \geq C^{\mathcal{S}}(\varphi).$

(b) When φ changes to $\Lambda^*\varphi$, a degree of chaos associated to this state change is given by

$$D^{\mathcal{S}}(\varphi; \Lambda^*) = C^{\overline{\mathcal{S}}}(\Lambda^*\varphi) - T^{\mathcal{S}}(\varphi; \Lambda^*).$$

In ID, several different topics can be treated from the same standpoint, so that we can find a new clue bridging several different fields. For examples, we may have the following applications:

(1) The study of optical communication processes.

(2) Formulation of fractal dimensions of states, and the study of complexity for some systems.

(3) Definition of genetic matrix for genome sequences and construction of phylogenetic tree for evolution of species.

(4) Entropic complexities $\Longrightarrow$ KS type complexities (entropy) $\Longrightarrow$ classification of dynamical systems.

(5) The study of optical illusion (psychology).

(6) The study of some economic models.

In this paper, we discuss (2) and (4) above.

5. Entropic Complexity in GQS

We introduce three types of entropic complexity here. As before, let $(\mathcal{A}, \mathbb{S}(\mathcal{A}), \alpha(G))$, and $(\overline{\mathcal{A}}, \overline{\mathbb{S}}(\overline{\mathcal{A}}), \overline{\alpha}(\overline{G}))$ be C^*-systems. Let $\mathcal{S}$ be a weak $*$-compact convex subset of $\mathbb{S}(\mathcal{A})$ and $M_\varphi(\mathcal{S})$ be the set of all maximal measure μ on $\mathcal{S}$ with the fixed barycenter φ

$$\varphi = \int_\mathcal{S} \omega d\mu.$$

Moreover let $F_\varphi(\mathcal{S})$ be the set of all measures of finite support with the fixed barycenter φ. Three pairs of complexity are :

$$T^\mathcal{S}(\varphi; \Lambda^*) \equiv \sup \left\{ \int_\mathcal{S} S(\Lambda^*\omega, \Lambda^*\varphi) d\mu; \mu \in M_\varphi(\mathcal{S}) \right\}$$

$$C_T^\mathcal{S}(\varphi) \equiv T^\mathcal{S}(\varphi; id)$$

$$I^\mathcal{S}(\varphi; \Lambda^*) \equiv \sup \left\{ S\left(\int_\mathcal{S} \omega \otimes \Lambda^*\omega d\mu, \varphi \otimes \Lambda^*\varphi \right) \mu \in M_\varphi(\mathcal{S}) \right\}$$

$$C_I^\mathcal{S}(\varphi) \equiv I^\mathcal{S}(\varphi; id)$$

$$J^\mathcal{S}(\varphi; \Lambda^*) \equiv \sup \left\{ \int_\mathcal{S} S(\Lambda^*\omega, \Lambda^*\varphi) d\mu; \mu \in F_\varphi(\mathcal{S}) \right\}$$

$$C_j^\mathcal{S} \equiv J^\mathcal{S}(\varphi; id)$$

These complexities with the mixing $\mathcal{S}$-entropy $S^\mathcal{S}(\varphi)$ and the CNT entropy $H_\varphi(\mathcal{A})$ satisfy the following relations [29, 38].

Theorem 5.1:

(1) $0 \leq I^\mathcal{S}(\varphi; \Lambda^*) \leq T^\mathcal{S}(\varphi; \Lambda^*) \leq J^\mathcal{S}(\varphi; \Lambda^*)$.

(2) $C_I(\varphi) = C_T(\varphi) = C_J(\varphi) = S(\varphi) = H_\varphi(\mathcal{A})$ (for the case $\mathbb{S} = \mathcal{S}$).

(3) When $\mathcal{A} = \overline{\mathcal{A}} = B(\mathcal{H})$, for any density operator ρ

$$0 \leq I^\mathcal{S}(\rho; \Lambda^*) = T^\mathcal{S}(\rho; \Lambda^*) \leq J^\mathcal{S}(\rho; \Lambda^*).$$

6. KS Type Complexities

The Kolmogorov-Sinai (KS) dynamical entropy is used in CDS to compute the mean information and the degree of chaos for dynamical systems[9].

Let θ (resp. $\overline{\theta}$) be an automorphism of $\mathcal{A}$ (resp. $\overline{\mathcal{A}}$) such that

$$\varphi \circ \theta = \varphi$$

and Λ be a convariant (i.e., $\Lambda \circ \theta = \bar{\theta} \circ \Lambda$) CP map from $\bar{A}$ to A. Take a finite subalgebra A_k (resp.$\bar{A}_k$) of A (resp.$\bar{A}$) and a unital map α_k (resp.$\bar{\alpha}_k$) from A_k (resp.$\bar{A}_k$) to A (resp.$\bar{A}$). Put $\alpha^M \equiv (\alpha_1, \alpha_2, \cdots, \alpha_M)$, $\bar{\alpha}_\Lambda^N \equiv (\Lambda \circ \bar{\alpha}_1, \Lambda \circ \bar{\alpha}_2, \cdots, \Lambda \circ \bar{\alpha}_N)$. Then two compound states for α^M and $\bar{\alpha}_\Lambda^N$ w.r.t $\mu \in M_\varphi(\mathcal{S})$ are

$$\Phi_\mu^{\mathcal{S}}\left(\alpha^M\right) = \int_{\mathcal{S}} \overset{M}{\underset{m=1}{\otimes}} \alpha_m^* \omega d\mu,$$

$$\Phi_\mu^{\mathcal{S}}\left(\alpha^M \cup \bar{\alpha}_\Lambda^N\right) = \int_{\mathcal{S}} \overset{M}{\underset{m=1}{\otimes}} \alpha_m^* \omega \overset{N}{\underset{n=1}{\otimes}} \bar{\alpha}_n^* \Lambda^* \omega d\mu.$$

Define the transmitted complexities as follows:

$$T^{\mathcal{S}}(\varphi; \ \alpha^M, \bar{\alpha}_\Lambda^N)$$
$$\equiv \ \sup \left\{ \int_{\mathcal{S}} S\left(\overset{M}{\underset{m=1}{\otimes}} \alpha_m^* \omega \overset{N}{\underset{n=1}{\otimes}} \bar{\alpha}_n^* \Lambda^* \omega, \Phi_\mu^{\mathcal{S}}(\alpha^M) \otimes \Phi_\mu^{\mathcal{S}}(\bar{\alpha}_\Lambda^N) \right) d\mu \ ; \right.$$
$$\left. \mu \in M_\varphi(S) \right\}$$

$$I^{\mathcal{S}}(\varphi; \ \alpha^M, \bar{\alpha}_\Lambda^N)$$
$$\equiv \ \sup \left\{ S\left(\Phi_\mu^{\mathcal{S}}(\alpha^M \cup \bar{\alpha}_\Lambda^N), \Phi_\mu^{\mathcal{S}}(\alpha^M) \otimes \Phi_\mu^{\mathcal{S}}(\bar{\alpha}_\Lambda^N) \right) \ ; \mu \in M_\varphi(S) \right\} \ ;$$

$$J^{\mathcal{S}}(\varphi; \ \alpha^M, \bar{\alpha}_\Lambda^N)$$
$$\equiv \ \sup \left\{ \int_{\mathcal{S}} S\left(\overset{M}{\underset{m=1}{\otimes}} \alpha_m^* \omega \overset{N}{\underset{n=1}{\otimes}} \bar{\alpha}_n^* \Lambda^* \omega, \Phi_\mu^{\mathcal{S}}(\alpha^M) \otimes \Phi_\mu^{\mathcal{S}}(\bar{\alpha}_\Lambda^N) \right) d\mu_f \ ; \right.$$
$$\left. \mu_f \in F_\varphi(\mathcal{S}) \right\}.$$

In the case of $\mathbb{S} = \mathcal{S}$, we denote $T^{\mathcal{S}}$ by T for simplicity, and so do I, J. When $A_k = A_0 = \bar{A}_k$, $A = \bar{A}$, $\theta = \bar{\theta}$, $\alpha_k = \theta^{k-1} \circ \alpha = \bar{\alpha}_k$ ($\alpha : A_0 \to A$; unital CP),

$$\tilde{T}_\varphi^{\mathcal{S}}(\theta, \alpha, \Lambda^*) \equiv \limsup_{N \to \infty} \frac{1}{N} T^{\mathcal{S}}(\varphi; \ \alpha^N, \bar{\alpha}_\Lambda^N)$$

$$\tilde{T}_\varphi^{\mathcal{S}}(\theta, \Lambda^*) \equiv \sup_\alpha \tilde{T}_\varphi^{\mathcal{S}}(\theta, \alpha, \Lambda^*)$$

$\tilde{T}_\varphi^{\mathcal{S}}(\theta, \Lambda^*)$ is the mean transmitted complexity w.r.t θ and Λ^*. We similarly define $\tilde{I}_\varphi^{\mathcal{S}}, \tilde{J}_\varphi^{\mathcal{S}}$. Then the CNT type theorem [29] holds for these complexties.

Theorem 6.1: If there exist $\alpha_m' : A \to A_m$ such that $\alpha_m \circ \alpha_m' = id$, $\bar{\alpha}_m \circ \bar{\alpha}_m' = id$ as $m \to \infty$, then

$$\tilde{T}_\varphi^{\mathcal{S}}(\theta, \Lambda^*) = \lim_{m \to \infty} \tilde{T}_\varphi^{\mathcal{S}}(\theta, \alpha_m, \Lambda^*)$$

Same for $\tilde{I}_\varphi^\mathcal{S}$, $\tilde{J}_\varphi^\mathcal{S}$.

Our complexties generalize usual dynamical entropy in the following senses.

(1) $T^\mathcal{S}(\varphi;\ id, \Lambda^*) = T^\mathcal{S}(\varphi, \Lambda^*)$, where $id : \mathcal{A} \to \mathcal{A}$.

(2) When $\mathcal{A}_n, \mathcal{A}$ are abelian C*-algebras and α_k is an embedding,

$$T(\mu; \alpha^M) = S_\mu^{\text{classical}} \left(\bigvee_{m=1}^{M} \tilde{A}_m \right)$$

$$I(\mu; \alpha^M, \bar{\alpha}^N) = I_\mu^{\text{classical}} \left(\bigvee_{n=1}^{M} \tilde{A}_n, \bigvee_{n=1}^{N} \tilde{B}_n \right)$$

for any finite partitions $\tilde{A}_n, \tilde{B}_n$ on the probability space $(\Omega = \text{spec}(\mathcal{A}), \mathcal{F}, \mu)$

(3) When Λ is the restriction of $\mathcal{A}$ to a subalgebra $\mathcal{M}$ of $\mathcal{A}$; $\Lambda = |_\mathcal{M}$,

$$J(\varphi; |_\mathcal{M}) = J(\varphi;\ id; |_\mathcal{M}) = H_\varphi(\mathcal{M}) = \text{CNT entropy.}$$

Moreover, the dynamical entropy $\tilde{H}_\varphi(\theta)$ for θ is given by [13, 8]

$$\tilde{H}_\varphi(\theta; \mathcal{M}) = \limsup_{N \to \infty} \frac{1}{N} H_\varphi(\mathcal{M} \vee \theta \mathcal{M} \vee \cdots \vee \theta^{N-1} \mathcal{M})$$

$$\implies \tilde{H}_\varphi(\theta) = \sup_{\mathcal{M}} \{ \tilde{H}_\varphi(\theta; \mathcal{M});\ \mathcal{M} \subset \mathcal{A} \}.$$

This $\tilde{H}_\varphi$ is equivalent to our complexity J: Let $\mathcal{M} \subset \mathcal{A}_0, \mathcal{A} = \otimes^\mathbf{N} \mathcal{A}_0, \theta \in Aut(\mathcal{A})$, $\alpha^N \equiv (\alpha, \theta \circ \alpha, \cdots, \theta^{N-1} \circ \alpha)$, $\alpha = \bar{\alpha}; \mathcal{A}_0 \to \mathcal{A}$ (embedding) and $\mathcal{M}_N \equiv \otimes_1^N \mathcal{M}$. Then we have

$$\tilde{J}_\varphi(\theta; \mathcal{M}) = \limsup_{N \to \infty} \frac{1}{N} J(\varphi;\ \alpha^N; |_{\mathcal{M}_N})$$

and

$$\tilde{J}_\varphi(\theta) \equiv \sup_{\mathcal{M}} \{ \tilde{J}_\varphi(\theta;\ \mathcal{M});\ \mathcal{M} \subset \mathcal{A}_0.$$

Similarly, $\tilde{T}_\varphi^\mathcal{S}(\theta), \tilde{I}_\varphi^\mathcal{S}(\theta)$ are computed.

Finally we note that the quantum KS dynamical entropy can be formulated [4] through quantum Markov chain of Accardi[1].

7. Model Computation

Numerical computation of the dynamical entropy can be used to see which state (or modulated state) is most effective for optical fiber communication [41]. Let $\mathcal{H}_0$ and $\overline{\mathcal{H}}_0$ be input and output Hilbert spaces, respectively. In order to send a state carrying information to the output system, we might need to modulate the state in proper way. A modulation $\mathcal{M}$ is a channel, denoted by $\Gamma^*_{(\mathcal{M})}$, from $\mathbb{S}(\mathcal{H}_0)$ to a certain state space $\mathbb{S}(\mathcal{H}_{(\mathcal{M})})$ on a proper Hilbert space $\mathcal{H}_{(\mathcal{M})}$. Take

$$\mathcal{A} \equiv \overset{\infty}{\underset{i=-\infty}{\otimes}} B(\mathcal{H}_0), \quad \overline{\mathcal{A}} \equiv \overset{\infty}{\underset{i=-\infty}{\otimes}} B(\overline{\mathcal{H}}_0).$$

Let $\mathbb{S}$ (resp. $\overline{\mathbb{S}}$) be the set of all density operators in $\mathcal{A}$ (resp. $\overline{\mathcal{A}}$). Let θ (resp. $\overline{\theta}$) be a shift on $\mathcal{A}$ (resp. $\overline{\mathcal{A}}$) and α (resp. $\overline{\alpha}$) be an embedding map from $B(\mathcal{H}_0)$ to $\mathcal{A}$, (resp. $B(\overline{\mathcal{H}}_0)$ to $\overline{\mathcal{A}}$) as before. Let Λ^* be an attenuation channel (i.e., $\Lambda^*_{\sqrt{\eta},\sqrt{1-\eta}}$ with the transmission rate η discussed in Sec.3.2). Put $\tilde{\Lambda} \equiv \overset{\infty}{\underset{i=-\infty}{\otimes}} \Lambda$ and $\tilde{\Gamma}_{(\mathcal{M})} \equiv \overset{\infty}{\underset{i=-\infty}{\otimes}} \Gamma_{(\mathcal{M})}$ and define

$$\alpha^N_{(\mathcal{M})} \equiv (\alpha \circ \tilde{\Gamma}_{(\mathcal{M})}, \cdots, \theta^{N-1} \circ \alpha \circ \tilde{\Gamma}_{(\mathcal{M})}),$$

$$\overline{\alpha}^N_{\tilde{\Lambda}(\mathcal{M})} \equiv (\tilde{\Gamma}_{(\mathcal{M})} \circ \tilde{\Lambda} \circ \overline{\alpha}, \cdots, \tilde{\Gamma}_{(\mathcal{M})} \circ \tilde{\Lambda} \circ \overline{\theta}^{N-1} \circ \overline{\alpha}).$$

We here only consider two modulations PAM (pulse amplitude modulation) and PPM (pulse position modulation). The modulations $\Gamma^*_{(PAM)}$ for PAM and $\Gamma^*_{(PPM)}$ for PPM are written as

$$\Gamma^*_{(PAM)}(E_n) = |n\rangle\langle n|$$

$$\Gamma^*_{(PPM)}(E_n) = \underbrace{|0\rangle\langle 0| \otimes \cdots \otimes |0\rangle\langle 0| \otimes \overbrace{|d\rangle\langle d|}^{n-th} \otimes |0\rangle\langle 0| \otimes \cdots \otimes |0\rangle\langle 0|}_{M},$$

where E_n is a pure state in $S(\mathcal{H}_0)$ coding the n-th symbol and $|k\rangle\langle k|$ is a k-photon number state. For a stationary initial state $\rho = \sum_m \mu_m \overset{\infty}{\underset{i=-\infty}{\otimes}} \rho_m^{(i)} \in \mathbb{S}$ with $\rho_m^{(i)} = \sum_{n_i} \lambda_{n_i}^{(m)} E_{n_i}$, $E_{n_i} \in \text{ex } S(\mathcal{H}_0)$ and the attenuation channel Λ^*, the transmitted complexities of mutual entropy type for two modulations PAM and PPM are calculated as

$$I(\rho;\ \alpha^N_{(PAM)}, \overline{\alpha}^N_{\Lambda(PAM)})$$

$$= \sum_{j_0=0}^{M} \cdots \sum_{j_{N-1}=0}^{M} \sum_{n_0=J_0}^{M} \cdots \sum_{n_{N-1}=J_{N-1}}^{M} \left(\sum_m \mu_m \prod_{k=0}^{N-1} \lambda^{(m)}_{n_k}\right)\left(\prod_{k'=0}^{N-1} |C^{n_{k'}}_{j_{k'}}|^2\right)$$

$$\times\ \left\{ \log \prod_{k=0}^{N-1} |C^{n_k}_{j_k}|^2 - \log\left(\sum_{n'_0=J_0}^{M} \cdots \sum_{n'_{N-1}=J_{N-1}}^{M} \left(\sum_{m'} \mu_{m'} \prod_{k'=0}^{N-1} \lambda^{(m')}_{n'_{k'}}\right)\right.\right.$$

$$\times\ \left.\left.\left(\prod_{k''=0}^{N-1} |C^{n'_{k''}}_{j_{k''}}|^2\right)\right)\right\}.$$

$$I(\rho;\ \alpha^N_{(PPM)}, \overline{\alpha}^N_{\Lambda(PPM)})$$

$$= -\sum_{n_0=1}^{M} \cdots \sum_{n_{N-1}=1}^{M} \left(\sum_m \mu_m \prod_{k=0}^{N-1} \lambda^{(m)}_{n_k}\right)\left\{\sum_{p=1}^{N} \sum_{\{q_1,\cdots,q_p\}\subset\{1,2,\cdots,N\}}\right.$$

$$\times\ \left.\left(\sum_{\ell_1=1}^{d} \cdots \sum_{\ell_p=1}^{d} |C^d_{\ell_1}|^2 \cdots |C^d_{\ell_p}|^2 (1-\eta)^{N-p} \eta^p \log\left(\sum_{m'} \mu_{m'} \prod_{k'=0}^{p} \lambda^{(m')}_{n_{q_{k'}}}\right)\right)\right\}.$$

where

$$|C^{n_i}_{j_i}|^2 = \frac{n_i!}{j_i!(n_i-j_i)!} \eta^{j_i} (1-\eta)^{(n_i-j_i)}.$$

Then we have[41]

Theorem 7.1: If $\mathcal{A} = \overline{\mathcal{A}}$, $\theta = \overline{\theta}$, $\alpha = \overline{\alpha}$ and $d \geq N$, then

$$\widetilde{I}_\rho\left(\theta, \alpha_{(PPM)}, \Lambda^*\right) \geq \widetilde{I}_\rho\left(\theta, \alpha_{(PAM)}, \Lambda^*\right).$$

8. Fractal Dimension of State

Usual fractal theory mostly treats geometrical sets [25]. It is desirable to extend the fractal theory so as to be applicable to some other objects. For this purpose, we introduced the fractal dimensions for general states in[37]. First we recall two usual fractal dimensions of geometrical sets.

⟨Scaling dimension ⟩

We observe a complex set F built from a fundamental pattern. If the number of the patterns observed is $N(1)$ when the scale is very rough, say 1,and the number is $N(r)$ when the scale is r, then we call the dimension defined through

$$d_s(F) = \frac{\log(N(r)/N(1))}{\log(1/r)}$$

the scaling dimension of the set F.

⟨ Capacity dimension ⟩

Let us cover a set F in the n-dimension Euclidean space R^n by a certain convex set with the diameter ε. If the smallest number of the convex sets needed to cover the set F is $N(\varepsilon)$, then we call the dimension given by

$$d_c(F) = \lim_{\varepsilon \to 0} \frac{\log N(\varepsilon)}{\log (1/\varepsilon)}$$

the capacity dimension (or the ε-entropy dimension) of the set F.

These two fractal dimensions become equal for almost all sets in which we can compute these dimensions.

The ε-entropy is extensively studied by Kolmogorov [22] and his ε-entropy is defined for a probability measure, which gives us an idea to define the ε-entropy for a general quantum state.

Kolmogorov introduced the notion of ε-entropy in probability space $(\Omega, \mathcal{F}, \mu)$. His formulation is as follows: For two random variables $f, g \in M(\Omega)$, the mutual entropy $I(f, g)$ is defined by the joint probability measure $\mu_{f,g}$ and the direct product measure $\mu_f \otimes \mu_g$ such that

$$I(f, g) = S(\mu_{f,g}, \mu_f \otimes \mu_g),$$

where $S(\cdot, \cdot)$ is the relative entropy [23].

The ε-entropy of Kolmogorov for a random variable f valued on a metric space (X, d) is given by

$$S_K(f, \varepsilon) \equiv \inf \{I(f, g) ; g \in M_d(f, \varepsilon)\},$$

where

$$M_d(f, \varepsilon) \equiv \{g \in M(\Omega) ; \|f - g\| \le \varepsilon\}$$

with

$$\|f - g\| \equiv \sqrt{\int_{X \times X} d(x, y)^2 \, d\mu_{fg}(x, y)}.$$

For a general probability measure μ on $(\Omega, \mathcal{F})$, the Kolmogorov ε-entropy $S_K(\mu; \varepsilon)$ is given by

$$S_K(\mu; \varepsilon) = \inf \{S(\mu_{co}, \mu \otimes \bar{\mu}) ; \bar{\mu} \in P_0(\Omega)\},$$

where μ_{co} is the joint (compound) probability measure of μ and $\bar{\mu}$ and $P_0(\Omega)$ is the set of all probability measures $\bar{\mu}$ satisfying $\|\mu - \bar{\mu}\| \le \varepsilon$.

We introduced the ε-entropy of a general quantum state φ and the fractal dimensions of the state φ [35, 37, 38]. Let $\mathcal{C}$ be the set of all channels physically interested and define two sets

$$\mathcal{C}_1(\Lambda^*; \varphi) = \{\Gamma^* \in \mathcal{C}; \Gamma^* \varphi = \Lambda^* \varphi\},$$

$$C_2\left(\varphi;\varepsilon\right) = \left\{\Gamma^* \in \mathcal{C}\,;\, \|\varphi - \Gamma^*\varphi\| \le \varepsilon\right\}.$$

Then the ε-entropy of a state of φ w.r.t. $\mathcal{S}$ is defined by means of the transmitted complexity $T^{\mathcal{S}}\left(\varphi;\Lambda^*\right)$ as

$$S^{\mathcal{S}}_{\mathcal{C},T}\left(\varphi;\varepsilon\right) = \inf\left\{T^{\mathcal{S}}_{\max,\mathcal{C}}\left(\varphi;\Lambda^*\right)\,;\,\Lambda^* \in C_2\left(\varphi;\varepsilon\right)\right\},$$

where

$$T^{\mathcal{S}}_{\max,\mathcal{C}}\left(\varphi;\Lambda^*\right) = \sup\left\{T^{\mathcal{S}}\left(\varphi;\Gamma^*\right)\,;\,\Gamma^* \in C_1\left(\Lambda^*;\varphi\right)\right\}.$$

When $\mathcal{S} = \mathbb{S}$ and $\mathcal{C}$ is the set of all channels on $\mathbb{S}$, our ε-entropy is simplely denoted by $S_{\mathrm{O,T}}\left(\varphi;\varepsilon\right)$.

The capacity dimension of a state φ w.r.t. $\mathcal{S}$ and $\mathcal{C}$ is defined by

$$d^{\mathcal{S}}_{\mathcal{C},T}\left(\varphi\right) \equiv \lim_{\varepsilon \to 0} d^{\mathcal{S}}_{\mathcal{C},T}\left(\varphi;\varepsilon\right),$$

where

$$d^{\mathcal{S}}_{\mathcal{C},T}\left(\varphi;\varepsilon\right) = \frac{S^{\mathcal{S}}_{\mathcal{C},T}\left(\varphi;\varepsilon\right)}{\log\left(1/\varepsilon\right)}.$$

The above $d^{\mathcal{S}}_{\mathcal{C},T}\left(\varphi;\varepsilon\right)$ is called the capacity dimension of ε-order. The information dimension of a state φ for $\mathcal{S}$ and $\mathcal{C}$ of ε order is defined by

$$d^{\mathcal{S}}_{I,\mathcal{C},T}\left(\varphi;\varepsilon\right) = \frac{S^{\mathcal{S}}_{\mathcal{C},T}\left(\varphi;\varepsilon\right)}{S^{\mathcal{S}}\left(\varphi\right)}.$$

These ε-entropy and fractal dimensions are applied to several physical phenomena and mathematical objects. For instance, we can classify the shapes of the seas of the moon and rivers [27] and consider a symmetry breaking in Ising system [26]. Here we state the main results concerning the Gaussian measures [21]. For a random variable $f = (f_1, \cdots, f_n)$ from Ω to R^n , the random variable norm $\|\cdot\|_{\mathrm{R.V}}$ of the measure μ_f associated with f is defined by

$$\|\mu_f\|_{\mathrm{R.V}} = \sqrt{\frac{1}{n}\sum_{i=1}^{n}\int_{\Omega}|f_i|^2 d\mu}.$$

*Theorem 8.1:*If the distance of two states is defined through the above random variable norm on $\Omega = R^n$ and the transmitted complexity T is the mutual entropy in CDS, then

$$(1)\quad S_{\mathrm{O,I}}(\mu_f;\varepsilon) = S_{\mathrm{K}}(f;\varepsilon) = \frac{1}{2}\sum_{i=1}^{n}\log\max\left(\frac{\lambda_i}{\theta^2},1\right)$$

where $\lambda_1,\dots,\lambda_n$ are the eigenvalues of the covariance operator R for μ_f and θ^2 is a constant uniquely determined by the equation $\sum_{i=1}^{n}\min(\lambda_i,\theta^2) = \varepsilon^2$

(2) $d_{\mathrm{O,I}}(\mu_f) = d_{\mathrm{K}}(\mu_f) = n$

According to this theorem the Kolmogorov ε-entropy coincides with our ε-entropy when the norm of states is given by the random variable norm on R^n. The difference between $S_{\mathrm{O,T}}$ and S_{K} come from the norm for states taken. When we take the norm of states by the total variation, namely,

$$\|\mu\| = |\mu|(\Omega).$$

Let $\Omega = R$ for simplicity. An input state μ is described by the mean 0 and the covariance σ^2 and we take the set $\mathcal{C}$ of the channels Λ^* sending a Gaussian measure to a Gaussian measure with a noise expressed by one-dimensional Gaussian measure $\mu_0 = [0, \sigma_0^2]$ so that the output state $\Lambda^*\mu$ is represented by $[0, \alpha^2\sigma^2 + \sigma_0^2]$ with a certain constant α. Since the channel Λ^* depends on α and σ_0^2, we put $\Lambda^* = \Lambda^*_{(\bar{\alpha},\sigma_0^2)}$.

Theorem 8.2: When $\|\mu - \Lambda^*_{(\bar{\alpha},\sigma_0^2)}\mu\| = \delta$, we denote $\alpha^2\sigma^2 + \sigma_0^2$ by C_δ. If α^2 satisfies $\alpha^2 \leq \frac{C_\delta - \delta}{\sigma^2}$, then we have

(1) $S_{\mathcal{C},\mathrm{O,I}}(\mu;\varepsilon) = \frac{1}{2}\log\frac{1}{\varepsilon} + \frac{1}{2}\log\dfrac{\sigma^2}{\left(1 + \frac{\sqrt{2\pi}}{4}(\varepsilon + o(\varepsilon))\right)} > S_{\mathrm{K}}(\mu;\varepsilon) = 0,$

(2) $d_{\mathcal{C},\mathrm{O,I}}(\mu) = \dfrac{1}{2},$

where $o(\varepsilon)$ is the order of ε : $\lim_{\varepsilon \to 0} o(\varepsilon) = 0$.

This theorem tells the difference between the Kolmogorov ε-entropy and our ε-entropy. It concludes that our fractal dimension enables to classify the Gaussian measures.

References

[1] L. Accardi, Noncommutative Markov chains, International School of Mathematical Physics, Camerino, pp. 268–295, 1974.

[2] L. Accardi, M. Ohya and H. Suyari, Computation of mutual entropy in quantum Markov chains, Open Sys. Information Dyn., **2**, pp337-354, 1994.

[3] L. Accardi and M. Ohya, Compound channels, transition expectations and liftings, to appear in J. Multivariate Analysis.

[4] L. Accardi, M. Ohya and N. Watanabe, Dynamical entropy through quantum Markov chain, to appear in Open System and Information Dynamics.

[5] S.Akashi, Superposition representability problems of quantum information channels, to appear in Open Systems and Information Dynamics.

[6] H. Araki, Relative entropy of states of von Neumann algebras, Publ. RIMS, Kyoto Univ., **11**, pp. 809–833, 1976.

[7] H. Araki, Relative entropy for states of von Neumann algebras II, Publ. RIMS, Kyoto Univ., **13**, pp. 173–192, 1977.

[8] F. Benatti, Deterministic Chaos in Infinite Quantum Systems, Trieste Notes in Physics, Springer-Verlag, 1993.

[9] P. Billingsley, Ergodic Theory and Information, Wiley, New York, 1965.

[10] O. Bratteli and D.W. Robinson, Operator Algebras and Quantum Statistical Mechanics I, Springer, New York, Berlin, Heidelberg, 1979.

[11] O. Bratteli and D.W. Robinson, Operator Algebras and Quantum Statistical Mechanics II, Springer, New York, Berlin, Heidelberg, 1981.

[12] G. Choquet, Lecture Analysis I, II, III, Bengamin, New York, 1969.

[13] A. Connes, H. Narnhofer and W. Thirring, Dynamical entropy of C*-algebras and von Neumann algebras, Commun. Math. Phys., **112**, pp. 691-719, 1987.

[14] A. Connes and E. Størmer, Entropy for automorphisms of II_1 von Neumann algebras, Acta Math., **134**, pp. 289–306, 1975.

[15] M.J. Donald, On the relative entropy, Commun. Math. Phys., **105**, pp. 13–34, 1985.

[16] G.G. Emch, Positivity of the K-entropy on non-abelian K -flows, Z. Wahrscheinlichkeitstheorie verw. Gebiete, **29**, pp. 241–252, 1974.

[17] K.H.Fichtner, W.Freudenberg and V.Liebscher, Beam splitting and time evolutions of Boson systems, preprint.

[18] F. Hiai, M. Ohya and M. Tsukada, Sufficiency, KMS condition and relative entropy in von Neumann algebras, Pacific J. Math., **96**, pp. 99–109, 1981.

[19] F. Hiai, M. Ohya and M. Tsukada, Sufficiency and relative entropy in *-algebras with applications to quantum systems, Pacific J. Math., **107**, pp. 117–140, 1983.

[20] R. S. Ingarden, A. Kossakowski and M. Ohya, Information Dynamics and Open Systems, Kluwer, 1997.

[21] K. Inoue, T. Matsuoka and M. Ohya, New approach to ε-entropy and its comparison with Kolmogolov's ε-entropy, SUT preprint.

[22] A. N. Kolmogorov, Theory of transmission of information, Amer. Math. Soc. Translation, Ser.2, **33**, pp. 291–321, 1963.

[23] S. Kullback and R. Leibler, On information and sufficiency, Ann. Math. Stat., **22**, pp. 79–86, 1951.

[24] G. Lindblad, Completely positive maps and entropy inequalities, Commun. Math. Phys., **40**, pp. 147–151, 1975.

[25] B. B. Mandelbrot, The Fractal Geometry of Nature, W. H. Freemann and company, San Francisco, 1982.

[26] T. Matsuoka and M. Ohya, Fractal dimensions of states and its application to Ising model, Rep. Math. Phys., **36**, pp. 365–379, 1995.

[27] T. Matsuoka and M. Ohya, Fractal dimension of states and its application to shape analysis problem, to appear

[28] N. Muraki, M. Ohya and D. Petz, Note on entropy of general quantum systems, Open Systems and Information Dynamics, **1**, No.1, pp. 43–56, 1992.

[29] N. Muraki and M. Ohya, Entropy functionals of Kolmogorov-Sinai type and their limit theorems, Letters in Math. Phys., **36**, pp. 327-335, 1996.

[30] J. von Neumann, Die Mathematischen Grundlagen der Quantenmechanik, Springer- Berlin, 1932.

[31] M. Ohya, Quantum ergodic channels in operator algebras, J. Math. Anal. Appl., **84**, pp. 318–327, 1981.

[32] M. Ohya, On compound state and mutual information in quantum information theory, IEEE Trans. Information Theory, **29**, pp. 770-777, 1983.

[33] M. Ohya, Note on quantum probability, L. Nuovo Cimento, **38**, pp. 402–406, 1983.

[34] M. Ohya, Entropy transmission in C*-dynamical systems, J. Math. Anal. Appl., **100**, pp. 222–235, 1984.

[35] M. Ohya, Some aspects of quantum information theory and their applications to irreversible processes, Rep. Math. Phys., **27**, pp. 19–47, 1989.

[36] M. Ohya, Information dynamics and its application to optical communication processes, Lecture Notes in Physics, **378**, Springer, pp. 81–92, 1991.

[37] M. Ohya, Fractal dimensions of states, Quantum Probability and Related Topics, **6**, pp. 359-369, World Scientific, Singapore, 1991.

[38] M. Ohya, State change, complexity and fractal in quantum systems, Quantum Communications and Measurement, Plenum, pp. 309 – 320, 1995.

[39] M. Ohya, Fundamentals of quantum mutual entropy and capacity, SUT preprint.

[40] M. Ohya and D. Petz, Quantum Entropy and its Use, Springer-Verlag, 1993.

[41] N. Ohya, and N. Watanabe, Note on irreversible dynamics and quantum information, to appear in the Alberto Frigerio conference proceedings.

[42] D.Petz, Sufficient subalgebras and the relative entropy of states on a von Neumann algebra, Commun. Math. Phys., **105**, pp123-131, 1986.

[43] C.E. Shannon, Mathematical theory of communication, Bell System Tech. J., **27**, pp. 379–423, 1948.

[44] A. Uhlmann, Relative entropy and the Wigner-Yanase-Dyson-Lieb concavity in interpolation theory, Commun. Math. Phys., **54**, pp. 21–32, 1977.

[45] H. Umegaki, Conditional expectation in an operator algebra IV, (entropy and information), Kodai Math. Sem. Rep., **14**, pp. 59–85, 1962.

[46] A.Wehrl, General properties of entropy, Rev. Mod. Phys., **50**, pp221-260, 1978.

DIRICHLET FORMS ON INFINITE-DIMENSIONAL 'MANIFOLD-LIKE' STATE SPACES: A SURVEY OF RECENT RESULTS AND SOME PROSPECTS FOR THE FUTURE

MICHAEL RÖCKNER,* *Universität Bielefeld*

Dedicated to our admired colleague, teacher, and friend Professor Masatoshi Fukushima on the occasion of his 60th birthday.

Abstract

We give a (to some extent pedagogical) survey on recent results about Dirichlet forms on infinite-dimensional 'manifold-like' state spaces including path and loop spaces as well as spaces of measures. The latter are associated with interacting Fleming–Viot processes resp. infinite particle systems. Also some new results, further developing the Dirichlet form approach to infinite particle systems, are enclosed. Finally, a brief summary of other research activities in the theory of Dirichlet forms is given and some prospects for the future are indicated.

DIRICHLET FORMS; INFINITE-DIMENSIONAL MANIFOLDS; PATH AND LOOP SPACES; FLEMING–VIOT PROCESSES; INFINITE PARTICLE SYSTEMS; EUCLIDEAN QUANTUM FIELDS; LATTICE GIBBS STATES; ERGODICITY

AMS 1991 SUBJECT CLASSIFICATION: PRIMARY 31C25
SECONDARY 58B99; 58G32; 60G57; 60J45; 60J60; 60K35; 81S20; 81T08; 82B26; 82B31

1. Introduction

The purpose of this paper is to give a survey of recent results on Dirichlet forms and their associated Markov processes on infinite-dimensional 'manifold-like' state spaces. At the same time we try to indicate directions of future research in this area. The selection of topics presented here is very much biased and concentrates on results mostly achieved by the author in collaboration with a number of colleagues. The exposition is to some extent pedagogical, adressing the non-expert reader, supressing technicalities as much as possible, and giving more emphasis to ideas and prospects rather than proofs. We start with the classical case of gradient-type Dirchlet forms on finite-dimensional manifolds in Section 2, summarizing the necessary modifications to go to infinite-dimensional situations. The latter are presented in the subsequent Sections 3 to 6 in the shape of three *model-case studies*. The first concerns gradient-type Dirichlet forms on Banach (or topological vector) spaces. Special emphasis is given here to the classical Wiener space (cf. Section 3). In the second and third model-case studies the underlying infinite-dimensional 'manifolds' are *no longer flat*. More

---

* Postal address: Fakultät für Mathematik, Universität Bielefeld, Postfach 100131, 33501 Bielefeld, Germany.

precisely, we consider Dirichlet forms on loop spaces over compact finite-dimensional Riemannian manifolds (cf. Section 4) resp. Dirichlet forms on state spaces of measures over a polish space (cf. Sections 5 and 6). The latter are related to Fleming–Viot processes resp.infinite particle systems. We emphasize that in Section 6 all underlying hard results are borrowed from [Y95], [Os95]. Our contribution here is only to provide the suitable geometric framework, i.e., we identify a natural underlying tangent bundle structure which, in particular, enables us to introduce more general Dirichlet forms than the ones studied in [Y95], [Os95] (e.g. we can allow non-constant diffusion coefficients). Thus, Section 6 contains really new results and should hence be of interest also for experts. Subsequently, in Section 7 we describe a recent application of gradient-type Dirichlet forms on Banach spaces (cf. Section 3) to the study of Gibbs states in statistical mechanics. Finally, we include a list of other topics presently being studied intensively within Dirichlet space theory. However, we do not claim any completeness of this list, neither w.r.t. the topics nor the scientists mentioned. On the contrary we would like to apologize at this point towards all people who should have been also mentioned or whose work should have been quoted. To keep the cited literature within bearable limits we mostly refer to the recent monographs [BH91], [MR92], [FOT94] and the references therein as well as the proceedings [FFGKRS93], [MRYa95].

Concerning the prospects for the future I do not wish to single out any special sub-branches in Dirichlet form theory that I personally think are of particular importance. I only want to express my sincere conviction that this theory and its applications bear a substantial amount of deep mathematics that only partially has been discovered. A lot of progress has been made, but we are far from being even only remotely close to the frontiers. I hope that this paper will contribute to attract more attention of both the working probabilist and analyst to the theory of Dirichlet forms, an area of mathematics which, as I am convinced, will be extremely active also in the next decade and, most probably, much further into the next century.

2. Review of the finite-dimensional case and motivations

Let (M, g) be an oriented complete Riemannian manifold with volume element dx. Consider the corresponding Laplace-Beltrami operator $L := \frac{1}{2}\Delta$ on (real) $L^2(M; dx)$ defined as the closure of the restriction of L to $C_0^\infty(M)$ ($:=$ the set of all smooth compactly supported functions on M). Hence L is a self-adjoint operator on $L^2(M; dx)$ and let us denote its corresponding operator semigroup by $(T_t)_{t\geq0}$, i.e., $T_t := e^{\frac{1}{2}\Delta t}$, $t \geq 0$, on $L^2(M; dx)$. As is well-known, L generates the Brownian motion $\mathbb{M} := (P_x)_{x\in M}$ on M, i.e., each P_x is a probability measure on $\Omega :=$ up to the life time ζ continuous paths $\omega : [0, \zeta(\omega)[\longrightarrow M$ with $\omega(0) = x$, such that for all $u \in L^2(M; dx)$, $t \geq 0$,

$$x \mapsto \int_\Omega u(\omega(t))\, P_x(d\omega) \text{ is a } dx\text{-version}$$

(1)   of (the $L^2(dx)$-class) $T_t u$.

Integrating by parts for $u, v \in C_0^\infty(M)$ we obtain

$$(2) \qquad -\int_M Lu\, v\, dx \;=\; \frac{1}{2}\int_M g(\nabla u, \nabla v)\, dx$$
$$=:\; \mathcal{E}(u,v)$$

and taking the completion $D(\mathcal{E}) := \overline{C_0^\infty(M)}$ of $C_0^\infty(M)$ w.r.t. the norm $(\mathcal{E} + (\ ,\ )_{L^2(M;dx)})^{1/2}$ we get the associated *Dirichlet form* $(\mathcal{E}, D(\mathcal{E}))$. Note that $D(\mathcal{E}) \subset L^2(M;dx)$ and that $D(\mathcal{E})$ is nothing but $H_0^{1,2}(M;dx)$, the classical Sobolev space on M of order 1 in $L^2(M;dx)$. $(\mathcal{E}, D(\mathcal{E}))$ is also sometimes called *Dirchlet space*. $(\mathcal{E}, D(\mathcal{E}))$ contains the entire information about L, hence about $(T_t)_{t \geq 0}$ and thus about $\mathbb{M}$. In fact, they are all 'equivalent objects':

$$(3) \qquad \mathcal{E} \longleftrightarrow L \longleftrightarrow (T_t)_{t \geq 0} \longleftrightarrow \mathbb{M}$$

The advantage of $\mathcal{E}$ over $(T_t)_{t \geq 0}$ is that it is more explicit, and w.r.t. L it is a 'first order' object and hence (in particular, in regard to domain questions) easier to analyse resp. to construct in much more general situations, as we shall see below. The heart of Dirichlet space theory is to use this simpler object $(\mathcal{E}, D(\mathcal{E}))$ to study the corresponding Markov process, that is, the Brownian motion $\mathbb{M}$ directly, and vice versa, to use $\mathbb{M}$ to deduce information about the two analytic objects L and $(T_t)_{t \geq 0}$ via the Dirichlet form $(\mathcal{E}, D(\mathcal{E}))$. In other words the theory of Dirichlet forms can be described as classical potential theory via a Hilbert space approach (i.e., within an L^2-framework).

As mentioned in the introduction the main purpose of this paper is to explain how to 'lift' this scheme resp. machinery resp. theory to cases where the manifold M is infinite-dimensional. The first problem one meets is that there is no reasonable exact analogue of the volume element dx (resp. the Lebesgue measure in the flat case) and that, nevertheless, one *needs* an underlying *reference measure*. So, let us first modify the above finite-dimensional situation by inserting a density $\rho : M \to [0, \infty)$ in front of dx. The Dirichlet form $(\mathcal{E}, D(\mathcal{E}))$ can still be written down explicitly:

$$\mathcal{E}(u,v) := \frac{1}{2}\int g(\nabla u, \nabla v)\, \rho\, dx\,,$$
$$u, v \in D(\mathcal{E}) := H_0^{1,2}(M; \rho dx) :=$$
$$(4) \qquad \text{completion of } C_0^\infty(M) \text{ w.r.t. } (\mathcal{E} + (\ ,\ )_{L^2(M;\rho dx)})^{1/2}$$

However, one has to make sure that this completion $H_0^{1,2}(M; \rho dx)$ still embeds (*one-to-one*) into $L^2(M; \rho dx)$. This problem is known as the *closability problem* and is a key problem of the theory. Above it can easily be overcome by imposing very weak conditions on ρ (as e.g. lower-semi-continuity, see the references in [MR92]). Again, as comes out of the general theory (cf. below) we have the correspondence (3). E.g., the operator L has the heuristic representation

$$(5) \qquad L = \frac{1}{2}\Delta + \frac{1}{2}\nabla \log\rho \,\cdot\, \nabla\,.$$

Though we have quite limited information about the corresponding semigroup $(T_t)_{t \geq 0}$, the associated process $\mathbb{M}$, the so-called *distorted Brownian motion*, has been completely analysed through the Dirichlet form (4) by M. Fukushima (see the references in [FOT94]).

The advantage of this modification is that infinite-dimensional analogues of ρdx *do* exist. Before we make the transition to infinite dimensions let us first briefly summarize the underlying abstract scheme resp. theory.

We confine ourselves to *symmetric Dirichlet forms* since essentially only these will appear in the concrete model-cases presented in subsequent sections (see [MR92] for the general case).

We arrive at the abstract scheme of Dirichlet space theory as follows: replace the manifold M by a general Hausdorff topological space E and dx (resp. ρdx) by a σ-finite measure m on its Borel σ-algebra $\mathcal{B}(E)$. An abstract (symmetric) Dirichlet form $(\mathcal{E}, D(\mathcal{E}))$ on $L^2(E; m)$ is a positive definite symmetric bilinear form $\mathcal{E} : D(\mathcal{E}) \times D(\mathcal{E}) \to \mathbb{R}$ on a dense linear domain $D(\mathcal{E}) \subset L^2(E; m)$ which is *closed* (i.e., $D(\mathcal{E})$ is complete w.r.t. the norm $(\mathcal{E} + (\ , \ )_{L^2(E;m)})^{1/2})$ such that the following property (called *contraction* or *Dirichlet property*) holds:

$$(6) \qquad u^{\#} := \min(\max(u, 0), 1) \in D(\mathcal{E}) \text{ and } \mathcal{E}(u^{\#}, u^{\#}) \leq \mathcal{E}(u, u) \text{ for all } u \in D(\mathcal{E}).$$

In order to have the complete correspondence (3) including the process $\mathbb{M}$ as well as all the theory of Dirichlet forms available in this general case, we need, however, an additional assumption on $(\mathcal{E}, D(\mathcal{E}))$. It is clear from (1) that the existence of $\mathbb{M}$ implies some sort of regularity of $(T_t)_{t \geq 0}$ hence of $(\mathcal{E}, D(\mathcal{E}))$. This additional property of $(\mathcal{E}, D(\mathcal{E}))$ is called *quasi-regularity*. It is an analytic property that characterizes all Dirichlet forms which are associated with right continuous strong Markov processes. Its discovery was the final step of a long development within the theory which originates from early purely analytic work by A. Beurling and J. Deny which was given its probabilistic counterpart through fundamental work by M. Fukushima and M.L. Silverstein under some additional regularity hypothesis. The above complete characterization was finally achieved in [AMR93] (see also [MR92]). But in the meantime numerous people (as e.g. A. Ancona, N. Bouleau, S. Carrillo–Menendez, E.B. Dynkin, P. Fitzsimmons, F. Hirsch, R. Høegh–Krohn, S. Kusuoka, Y. LeJan, Y. Oshima, M. Takeda, B. Schmuland, ...) made important contributions to the development (cf. the 'Notes' in [BH91], [MR92], and [FOT94]). So, if the Dirichlet form $(\mathcal{E}, D(\mathcal{E}))$ is quasi-regular, the full correspondence (3) (including the theory that goes with it) is at our disposal, in particular, there exists an associated process $\mathbb{M}$. We refer to [MR92, Diagrams 1,2,3 on pp. 14, 27, 39 respectively] for the precise description of the connecting arrows '$\longleftrightarrow$' in (3) in this general case.

Though not complicated, we shall not give the precise definition of quasi-regularity, but instead look into the following more concrete situation, sufficient to understand the examples later. Let us assume that in (3) $\mathbb{M}$ has continuous sample paths and no path with a finite life-time terminates inside E. Then we have the following representation (cf. [FOT94, Chap. 5] and [MR92, Chap. VI]) for our abstract

Dirichlet form $(\mathcal{E}, D(\mathcal{E}))$:

$$(7) \qquad \mathcal{E}(u, v) = \frac{1}{2} \int d\mu_{\langle u,v \rangle} \,, \quad u, v \in D(\mathcal{E}) \,,$$

where $\mu_{\langle u,v \rangle}$ are signed measures (so-called *energy-measures*) on $\mathcal{B}(E)$ satisfying the chain rule

$$(8) \qquad \mu_{\langle uw,v \rangle} = u\mu_{\langle w,v \rangle} + w\mu_{\langle u,v \rangle} \,,$$

$u, v, w \in D(\mathcal{E})$, bounded. In the classical situation above and all cases in the subsequent sections we have $\mu_{\langle u,v \rangle} \ll m$, i.e.,

$$(9) \qquad \mu_{\langle u,v \rangle} = \Gamma(u, v) \cdot m$$

(where $\Gamma(\cdot, \cdot)$ is called *square-field operator*), and more precisely even

$$(10) \qquad \Gamma(u, v) = g(\nabla u, \nabla v) \,.$$

Dirichlet forms satisfying (7) – (10) are called *of gradient-type.*

Our aim below is to present a 'recepy' how to identify resp. construct the Dirichlet form (of gradient-type for simplicity) which is appropriate for the respective application one is interested in, and make the above scheme (i.e., correspondence (3)) available and the theory behind it work. Recall we still fix E, m as above.

The following ingredients are sufficient:

(G) We need to find a suitable *gradient* (satisfying the chain rule) on a space D of sufficiently many continuous test functions in $L^2(E; m)$. In particular, we need to find a corresponding ('tangent'-) bundle structure. This will give us what is called a *pre-Dirichlet form*, namely $\mathcal{E}(u, v)$ for $u, v \in D$ on $L^2(E; m)$.

(C) We need to check whether $(\mathcal{E}, D)$ is *closable* on $L^2(E; m)$, i.e., whether the continuous extension of the inclusion map $D \subset L^2(E; m)$ to the completion $D(\mathcal{E}) := \overline{D}$ of D w.r.t. $(\mathcal{E} + (\ ,\ )_{L^2(E;m)})^{1/2}$ is one-to-one. $(\mathcal{E}, D(\mathcal{E}))$, called the *closure* of $(\mathcal{E}, D)$ on $L^2(E; m)$, is then automatically a Dirichlet form, since the contraction property (6) immediately follows from the chain rule. In analogy to (4), $D(\mathcal{E})$ can be considered as a generalized Sobolev space $H_0^{1,2}(E; m)$ on E.

(T) We have to show whether the following *tightness* condition holds: there exist compact $K_n \subset E$, $n \in \mathbb{N}$, such that $\{u \in D(\mathcal{E}) | u = 0 \text{ on } E \backslash K_n \text{ for some } n \in \mathbb{N}\}$ is dense in $D(\mathcal{E})$ w.r.t. $(\mathcal{E} + (\ ,\ )_{L^2(E;m)})^{1/2}$. This implies that $(\mathcal{E}, D(\mathcal{E}))$ is quasi-regular.

We shall not formally prove that (G), (C), (T) imply that $(\mathcal{E}, D(\mathcal{E}))$ is in fact a quasi-regular Dirichlet form, though we already gave some indication concerning the rôle of (G) and (C). We rather ask the reader to simply believe us at this point or to consult [MR92, Chap. I and Chap. IV, Sections 1–3]. Instead, we present a number of examples where (G), (C), and (T) have been verified in detail and where the underlying geometry and the resulting processes are of particular interest. The latter will be strong Markov processes which (because of (G)) automatically have continuous sample paths, i.e., they are *diffusions*. Furthermore, in all examples below (except for the last in Section 6 if b or d or c is not identically equal to zero) these diffusions have m as *symmetrizing* and *invariant measure*.

Remark 1 Summarizing the above we know that if we can check (G), (C), (T), then the correspondence (3) holds (in particular there exists an associated diffusion) and the entire theory of Dirichlet forms is available. We emphasize that this also means that (essentially) all results in [FOT94] are applicable though they are proved there only for *regular* Dirichlet forms on *locally compact* separable metric *state spaces*. These results, however, generalize immediately to quasi-regular Dirichlet forms on arbitrary state spaces by the so-called *transfer method* developed in [MR92, Chap. VI]. Thus, we can e.g. apply spectral synthesis, capacity-methods, quasi-sure analysis, Beurling-Deny-LeJan formulae, Revuz correspondence, stochastic analysis by additive functionals, Itô-Fukushima decomposition, forward/backward martingale decomposition, perturbation theory, traces, general criteria for conservativity, transience, recurrence, and for tightness of path space measures, etc.

We conclude this section with some comments on (G): one might ask the question whether we are interested in rather having a gradient than just a square field operator Γ (cf. (9)). It is well-known and, in fact, quite trivial (cf. [BH91, Chap. V, Exercise 5.9]) that for each such Γ one can construct various (of course, non-canonical) abstract bundle structures and gradients so that (10) holds. But this is almost useless for practical purposes (e.g., it does not help at all in solving the key problem (C) in concrete applications). A gradient-type Dirichlet form on a natural tangent bundle adapted to the respective situation, however, is very well suited for a deep analysis and, in particular, for constructing new Dirichlet forms from it. Thus, one can fully exploit the robustness of the theory w.r.t. all kinds of perturbations and apply the respective very well-developed techniques. We shall see this in some detail in Section 6 below.

3. Infinite-dimensional state space — flat case: Banach spaces

Let us restrict to a well-studied special case. We assume that

$$E := C_0([0,1] \to \mathbb{R}^d) := \{z : [0,1] \to \mathbb{R}^d \,|\, z \text{ is continuous and } z(0) = 0\}$$

equipped with the uniform norm, i.e., we replace our finite-dimensional manifold M in Section 2 by the *classical Wiener space*. Instead of the volume element dx we take

$$m := Wiener\ measure \text{ on } E \ .$$

To realize condition (G) we take as test function space D the space $\mathcal{F}C_b^\infty$ of bounded smooth cylinder functions, i.e.,

$$\mathcal{F}C_b^\infty := \{f(l_1,\ldots,l_n) \,|\, n \in \mathbb{N},\ f \in C_b^\infty(\mathbb{R}^n),\ l_1,\ldots,l_n \in E'\}$$

where E' denotes the dual of E and $C_b^\infty(\mathbb{R}^n)$ the space of all infinitely differentiable functions on $\mathbb{R}^n$ with all partial derivatives bounded. The bundle structure is in this case trivial, namely we take as 'tangent space' to E at every $z \in E$ the same space, namely the classical *Cameron-Martin space*

$$H := \{h \in E \mid h \text{ is absolutely continuous and}$$

$$\|h\|_H^2 := \int_0^1 |h'|_{\mathbb{R}^d}^2 dt < \infty\}.$$

Then $E' \subset H' \equiv H \subset E$ densely and continuously. Let us denote the embedding $E' \subset H$ by j_H. Then we can define the gradient for $u \in \mathcal{F}C_b^\infty$ by

$$\nabla u(z) := j_H(u'(z)) \ (\in H) \text{ for } z \in E \,,$$

where $u'(z) \ (\in E')$ is the Fréchet derivative of u in z, and we obtain the pre-Dirichlet form

$$\mathcal{E}(u, v) := \frac{1}{2} \int_E \langle \nabla u(z), \nabla v(z) \rangle_H \ m(dz) \,,$$

(11) $$u, v \in \mathcal{F}C_b^\infty \,.$$

Thus, (G) is verified. It is also well-known that by the choice of H $(\mathcal{E}, \mathcal{F}C_b^\infty)$ is closable on $L^2(E; m)$ (cf. e.g. [MR92, Chap. II, Corollary 3.13]), i.e., (C) holds. The corresponding closure $(\mathcal{E}, D(\mathcal{E}))$ is of fundamental importance in the *Malliavin calculus* (cf. [MR92, Chap. II, Remark 3.14], [BH91, Chap. III] and the references therein). Also (T) holds for the Dirichlet form $(\mathcal{E}, D(\mathcal{E}))$ (e.g. by the general result in [RS92] or [MR92, Chap. IV, Subsection 4b)]). Hence the above scheme (cf. (3)) applies. The corresponding diffusion is nothing but the Ornstein–Uhlenbeck process on Wiener space. It solves the stochastic differential equation

$$dX_t = dW_t^H - \frac{1}{2} X_t \, dt \,,$$

where W^H is the Brownian motion on E with covariance $\langle \ , \ \rangle_H$ (cf. e.g. [AR91, Section 7.I]).

A more complicated situation arises if above in case $d = 2, 3$ instead of Wiener measure one takes m to be equal to the *polymer measure*. Since m is no longer Gaussian and not absolutely continuous w.r.t. Wiener measure if $d = 3$, it has only been discovered recently that (C) (and also (T)) holds so that everything above extends to this case except that the stochastic differential equation for the associated process cannot be written down. (cf. [AHRZho95] and [ARZho95] for details).

Also topological vector spaces E other than $C_0([0,1], \mathbb{R}^d)$ replacing our manifold M of Section 2 have been studied, as e.g. $E := \mathcal{S}'(\mathbb{R}^2)$ (i.e., the space of tempered distributions on $\mathbb{R}^2$). In this case again a fixed tangent space at each point of $\mathcal{S}'(\mathbb{R}^2)$ has been chosen, namely $H := L^2(\mathbb{R}^2; dx)$. Thus,

$$\mathcal{S}(\mathbb{R}^2) \subset L^2(\mathbb{R}^2; dx) \subset \mathcal{S}'(\mathbb{R}^2) \,.$$

m was taken to be, for example, a *Euclidean (infinite volume) Φ_2^4-quantum field*. Defining the gradient similarly as above one thus realizes (G) and obtains the pre-Dirichlet form

$$\mathcal{E}(u,v) \;=\; \int_{\mathcal{S}'(\mathbb{R}^2)} \langle \nabla u(z), \nabla v(z) \rangle_{L^2(\mathbb{R}^2;dx)} \, m(dz) \,,$$

$$(12) \qquad\qquad u,v \in \mathcal{F}C_b^\infty \,,$$

on $L^2(\mathcal{S}'(\mathbb{R}^2);m)$. Both (C) and (T) have also been verified (cf. [AR90], [AR89] resp.). In this case the corresponding process is a weak solution of the non-linear stochastic differential equation

$$(13) \qquad dX_t = dW_t^H + \frac{1}{2}((\Delta - 1)(X_t) - 4 : X_t^3 :)dt \,,$$

where $: z^3 :$ is the usual renormalized power of $z \in \mathcal{S}'(\mathbb{R}^2)$. We refer to [AR91] for details.

One main remaining open problem here is, whether the above process is the unique weak solution of (13). There has been a lot of progress also in this direction (see e.g. [RZ92, RZ94], [AKR95a]), but in cases where the drift is as singular as in (13) all results achieved so far do not apply. It seems that a completely new approach is required.

4. Infinite-dimensional state space — non-flat case: loop spaces over manifolds

Let (M,g) be an oriented compact (finite-dimensional) Riemannian manifold without boundary. Let $x_0 \in M$ be fixed and

$$E := \{\sigma : [0,1] \to M \mid \sigma \text{ continuous}, \ \sigma(0) = \sigma(1) = x_0\}$$

equipped with the topology of uniform convergence. Let m be *pinned Wiener measure* on E. So, E is our 'manifold-like' state space, but to explain the tangent bundle in this case we first have to look at the 'pinned' Cameron-Martin space 'at x_0', i.e.,

$$H_0 := \{h : [0,1] \to T_{x_0}M \ \mid \ h \text{ absolutely continuous}, \ h(0) = h(1) = 0$$

$$\text{and } \|h\|_{H_0}^2 := \int_0^1 g_{x_0}(h'(s), h'(s))ds < \infty\} \,.$$

Let $\tau_t(\sigma) : T_{x_0}M \to T_{\sigma(t)}M$, $t \in [0,1]$, be the stochastic parallel transport along σ (w.r.t. the Levy-Cevita connection on M). Now, we define the tangent space to E at $\sigma \in E$ by

$$H_\sigma := T_\sigma E := \{(\tau_t(\sigma)h(t))_{t \in [0,1]} \mid h \in H_0\} \,,$$

that is, H_σ consists of all vector fields along σ obtained by stochastic parallel translation from any $h \in H_0$. Our test function space in this case is

$$\mathcal{F}C^\infty := \{\sigma \mapsto f(\sigma(s_1), \dots, \sigma(s_n)) \mid n \in \mathbb{N}, \ f \in C^\infty(M^n), \ s_1, \dots, s_n \in [0,1]\} \,.$$

In order to obtain a gradient we first define the *directional derivative* of $\sigma \mapsto u(\sigma) :=$
$f(\sigma(s_1), \ldots, \sigma(s_n))$ w.r.t. $(\tau_t(\sigma)h(t))_{t\in[0,1]}$ by

$$\partial_h u(\sigma) := \sum_{i=1}^n g_{\sigma(s_i)} \left(\nabla_i f(\sigma(s_1), \ldots, \sigma(s_n)), \tau_{s_i}(\sigma)h(s_i) \right), \quad \sigma \in E ,$$

where $\nabla_i f$ is the gradient of f on (M,g) w.r.t. to its i-th coordinate. Clearly, for
$\sigma \in E$,

$$h \mapsto \partial_h u(\sigma)$$

is continuous on H_0 w.r.t. $\| \ \|_{H_0}$. Hence we can define the gradient $\nabla u(\sigma)$ as the
unique element in H_0 such that

$$\langle \nabla u(\sigma), h \rangle_{H_0} = \partial_h u(\sigma) \quad \text{for all } h \in H_0 ,$$

and condition (G) is realized. We obtain a (w.r.t. (M,g) intrinsic) pre-Dirichlet form

$$\mathcal{E}(u,v) = \frac{1}{2} \int_E \langle \nabla u(\sigma), \nabla v(\sigma) \rangle_{H_0} \, m(d\sigma) ,$$
$$u, v \in \mathcal{F}C_b^\infty ,$$

on $L^2(E; m)$ which (by construction of ∇ or more precisely the choice of the tangent
space H_σ, $\sigma \in E$) has been shown to be closable (see e.g. [D92]), so (C) is fulfilled.

(T) has been proved in [DR92]. So, the scheme of Dirichlet space theory applies.
The corresponding diffusion is the analogue of an Ornstein-Uhlenbeck process, but
with values in a loop space (cf. [DR92]). So far, the Dirichlet form approach gives
the only construction of this process.

Remark 2 A general method how to prove (T) in (not necessarily flat) infinite-
dimensional situations including the above example, as well as the free loop space (cf.
[ALR93]) was presented in [RS95].

To describe all open problems in this model-case would be beyond the scope of this
paper. Let us only mention that the study of the geometry and, in particular, the
topology of the underlying manifold through the above Dirichlet form or its associated
process is more than fascinating.

For example, by a recent result of S. Aida and S. Kusuoka it follows that the
Dirichlet form (or equivalently the process) is ergodic if and only if the manifold
is simply connected. However, it is not known so far whether in this case the
corresponding operator L (cf. (3)), which can be interpreted as a kind of Laplace-
Beltrami operator on the loop space E, has a spectral gap. There is a a lot of activity
going on in this direction (S. Aida, B. Driver, K.D. Elworthy, E. Hsu, Z.M. Ma,...)
and also in developing a natural geometry on the infinite-dimensional manifold E
itself (A.B. Cruzeiro, O. Enchev, S. Fang, P. Malliavin, D.W. Stroock,...).

5. Infinite-dimensional state space — non flat case: spaces of probability measures on a polish space

Let S be a polish space (i.e., a complete separable metric space) with Borel σ-algebra $\mathcal{B}(S)$, and replace our manifold M of Section 2 by

$$E := \mathcal{M}_1(S) := \text{all probability measures on } \mathcal{B}(S)$$

equipped with the weak topology. As a substitute for the volume element dx we take a probability measure $m := m_{\theta,\nu_0}$ on $\mathcal{B}(E)$ defined as follows:

$$m(A) := P\left[\sum_{i=1}^{\infty} \rho_i \epsilon_{\xi_i} \in A\right], \quad A \in \mathcal{B}(E),$$

where ϵ_x, $x \in S$, denotes Dirac measure in x, $\xi_i : \Omega \to S$, $i \in \mathbb{N}$, are i.i.d. with distribution $\nu_0 \in \mathcal{M}_1(S)$ and $(\rho_1, \rho_2, \dots) : \Omega \to [0, \infty)^{\mathbb{N}}$ is Poisson-Dirichlet distributed with θ (cf. [EK93, Theorem 8.1]). As the 'tangent space' at $\mu \in E$ we take

$$T_\mu E := L^2(S; \mu)$$

and as test functions

$$\mathcal{F}C^\infty := \left\{\mu \ \mapsto \ f\left(\int \varphi_1 d\mu, \dots, \int \varphi_n d\mu\right)\middle| \ n \in \mathbb{N}, \right.$$
$$\left. f \in C^\infty(\mathbb{R}^n), \ \varphi_1, \dots, \varphi_n \in \mathcal{B}_b(S)\right\},$$

where $\mathcal{B}_b(S)$ denotes the set of all bounded $\mathcal{B}(S)$-measurable functions on S. Now we define the gradient for $u \in \mathcal{F}C^\infty$ and $\mu \in E$ by

$$\nabla u(\mu) := (x \mapsto D_x u(\mu), \ x \in S),$$

where

$$D_x u(\mu) := \frac{d}{ds} u((1-s)\mu + s\,\epsilon_x)\bigg|_{s=0}.$$

If $u(\mu) = f(\int \varphi_1 d\mu, \dots, \int \varphi_n d\mu)$, $\mu \in E$, then by the chain rule for all $x \in S$, $\mu \in E$,

$$D_x u(\mu) = \sum_{i=1}^{n} \partial_i f\left(\int \varphi_1 d\mu, \dots, \int \varphi_n d\mu\right)\left(\varphi_i(x) - \int \varphi_i d\mu\right),$$

where ∂_i denotes the derivative w.r.t. the i-th coordinate. This shows that $x \mapsto D_x u(\mu)$ is in $\mathcal{B}_b(S)$, hence in $L^2(S; \mu)$. In other words ∇u is a vector field on E (w.r.t. the tangent bundle $(L^2(S; \mu))_{\mu \in E}$). Now we have the pre-Dirichlet form

$$\mathcal{E}(u, v) \ = \ \frac{1}{2} \int_E \langle \nabla u(\mu), \nabla v(\mu)\rangle_{L^2(S;\mu)} \, m(d\mu)$$
$$u, v \in \mathcal{F}C^\infty$$

on $L^2(E; m)$ and (G) is realized. Also (C) holds, which immediately follows from the identity

$$(14) \qquad \mathcal{E}(u, v) = \int_E (-Lu)(\mu)\, v(\mu)\, m(d\mu)$$

for all $u, v \in \mathcal{F}C^\infty$, where

$$\begin{aligned} Lu(\mu) \;=\;& \frac{1}{2} \int_S \int_S \frac{\partial}{\partial \epsilon_x}\left(\frac{\partial u}{\partial \epsilon_y}\right)(\mu)(\epsilon_x(dy) - \mu(dy))\, \mu(dx) \\ &+ \frac{\theta}{2} \int_S \int_S \left(\frac{\partial u}{\partial \epsilon_y}(\mu) - \frac{\partial u}{\partial \epsilon_x}(\mu)\right) \nu_0(dy)\mu(dx) \end{aligned}$$

with $\frac{\partial u}{\partial \epsilon_x}(\mu) := \left.\frac{d}{ds} u(\mu + s\epsilon_x)\right|_{s=0}$. Hence the closure $(\mathcal{E}, D(\mathcal{E}))$ of $(\mathcal{E}, \mathcal{F}C^\infty)$ on $L^2(E; m)$ is a Dirichlet form, and again $D(\mathcal{E})$ is a kind of first order Sobolev space $H_0^{1,2}(E; m)$ on $E = \mathcal{M}_1(S)$. The general method from [RS95] to show (T) also applies here. So, the entire theory of Dirichlet forms described in Section 2 applies here as well (cf. [ORS95] for details). The corresponding diffusion is just the well-studied *Fleming–Viot process (with parent independent mutation)* appearing in population genetics (cf. [EK93]).

Remark 3 We would like to emphasize that by definition $(\mathcal{E}, D(\mathcal{E}))$ above depends in fact only on the σ-algebra $\mathcal{B}(S)$ and not on the topology of S.

One challenging problem in this model-case is to study the geometry of $(\mathcal{E}, D(\mathcal{E}))$ respectively of the Fleming–Viot process on $\mathcal{M}_1(S)$ in more detail. In [OR94] a few steps in this direction have been done by analysing the corresponding intrinsic metric. It turns out that the latter is equivalent to the variation norm. Furthermore, it is shown in [OR94], that in some sense the Fleming–Viot process is a (time-changed) Brownian motion when considered w.r.t. the 'correct' geometry on $\mathcal{M}_1(S)$. In finite dimensions, i.e., the case where $S = \{1, \dots, d\}$ this can be proved explicitly, and if ν_0 is the uniform distribution on $\{1, \dots, d\}$ the Fleming–Viot is just the Brownian motion on the $(d-1)$-dimensional sphere (cf. [OR94, Remark 3.3]).

6. Infinite-dimensional state space — non flat case: spaces of integer-valued Radon measures on $\mathbb{R}^d$

In this section we replace our manifold M of Section 2 by the space E of all $\mathbb{Z}^+$-valued Radon measures on $\mathbb{R}^d$ equipped with the vague topology (or a closed subset thereof). As the 'tangent space' at $\mu \in E$ in this case we take

$$T_\mu E := L^2(\mathbb{R}^d \to \mathbb{R}^d; \mu),$$

i.e., the space of all μ-square integrable vector fields on $\mathbb{R}^d$. As test functions on E we choose

$$\begin{aligned} \mathcal{F}C_b^\infty \;:=\; \Big\{ \mu \mapsto f\Big(&\int \varphi_1 d\mu, \dots, \int \varphi_n d\mu \Big) \mid n \in \mathbb{N}, \\ &f \in C_b^\infty(\mathbb{R}^n),\ \varphi_1, \dots, \varphi_n \in C_0^1(\mathbb{R}^d) \Big\}, \end{aligned}$$

where $C_0^1(\mathbb{R}^d)$ denotes the set of all continuously differentiable functions on $\mathbb{R}^d$ with compact support.

Now we define our gradient for $u \in \mathcal{F}C_b^\infty$, $\mu \in E$, by

$$\nabla u(\mu) := (x \mapsto \vec{D}_x u(\mu), \ x \in \sup[\mu]) \tag{15}$$

with

$$\vec{D}_x u(\mu) := \left(\frac{d}{ds} u(\mu + \epsilon_{x_i(s)} - \epsilon_x)|_{s=0} \right)_{1 \le i \le d}, \tag{16}$$

where for $1 \le i \le d$

$$x_i(s) = x + s\, e_i$$

with $\{e_i \mid 1 \le i \le d\}$ the canonical basis of $\mathbb{R}^d$. From (18) below we see that for given $\mu \in E$ it is really enough to define $\vec{D}_x u(\mu)$ for $x \in \mathbb{R}^d$ such that μ has mass in x. (Note that (16) is really an intrinsic definition.)

By the chain rule we obtain that for all $x \in \mathbb{R}^d$, $\mu \in E$,

$$\vec{D}_x u(\mu) = \left(\sum_{j=1}^n \partial_i f\left(\int \varphi_1 d\mu, \dots, \int \varphi_n d\mu \right) \partial_i \varphi_j(x) \right)_{1 \le i \le d}, \tag{17}$$

which shows that $x \mapsto \vec{D}_x u(\mu)$ is in $C_0^1(\mathbb{R}^d \to \mathbb{R}^d)$, hence in $L^2(\mathbb{R}^d \to \mathbb{R}^d; \mu)_{\mu \in E}$. This implies that ∇u is a vector field on E (w.r.t. the tangent bundle $(L^2(\mathbb{R}^d \to \mathbb{R}^d; \mu))_{\mu \in E})$. For any finite measure m on the Borel σ-algebra $\mathcal{B}(E)$ we thus have a pre-Dirichlet form

$$\mathcal{E}(u,v) := \frac{1}{2} \int_E \langle \nabla u(\mu), \nabla v(\mu) \rangle_{L^2(\mathbb{R}^d \to \mathbb{R}^d; \mu)} \, m(d\mu)$$
$$u, v \in \mathcal{F}C_b^\infty, \tag{18}$$

on $L^2(E; m)$ and (G) is realized. In [Y95] (see also [Os95]) a whole class of measures m have been analysed which come from *superstable lower regular pair potentials* Φ (see [Y95] for details) so that (C) holds. (More precisely, an operator L has been identified such that

$$\mathcal{E}(u,v) = - \int Lu\, v\, dm \quad \text{for all } u, v \in \mathcal{F}C_b^\infty.)$$

And also (T) has been verified for these type of measures (cf. [Y95], [Os95]).

Remark 4 The reader should note that (18) is a priori different from the definition of $\mathcal{E}$ in [Y95]. A more 'coordinate based' definition has been given there without identifying an appropriate underlying tangent bundle. But it is easily verified that both definitions are in fact the same.

Now, as explained in Section 2 correspondence (3) is also valid in this case (more precisely for the closure $(\mathcal{E}, D(\mathcal{E}))$ of (18) on $L^2(E; m)$) and the complete theory applies. The corresponding diffusion (at least heuristically) weakly solves a stochastic differential equation of type

$$(19) \qquad dX_t^i = -\frac{1}{2} \sum_{j \neq i} \nabla \Phi(X_t^i - X_t^j)dt + dW_t^i \ , \ i \in \mathbb{N} \ ,$$

where $(W^i)_{i \in \mathbb{N}}$ is a family of independent standard Brownian motions (see [Y95] for the precise meaning of (19) in this case.). Hence it models an interacting infinite particle system.

We conclude this section with briefly explaining how to construct new (even non-symmetric) Dirichlet forms from gradient-type Dirichlet forms. This is completely standard and as easy as in the case of finite-dimensional state spaces. We shall do so starting from the closure $(\mathcal{E}, D(\mathcal{E}))$ of (18) for a fixed m as in [Y95]. (But in all examples from the previous sections this can be carried out in exactly the same way; cf. [MR92, Chap. II, Exercises 3.6, 3.9 and Subsection 3e)] and [ORS95, Sect. 2]). In particular, non-constant diffusion coefficients can occur.

For $\mu \in E$ let $A(\mu)$ be a self-adjoint bounded operator on $L^2(\mathbb{R}^d \to \mathbb{R}^d; \mu)$ such that:

(i) There exists $c \in (0, \infty)$ such that $A(\mu) \geq c\, 1$ for all $\mu \in E$.

(ii) $\mu \mapsto \langle A(\mu)f(\mu, \cdot), g(\mu, \cdot)\rangle_{L^2(\mathbb{R}^d \to \mathbb{R}^d; \mu)}$ is $\mathcal{B}(E)$-measurable for all bounded continuous functions $f, g : E \times \mathbb{R}^d \to \mathbb{R}^d$.

Remark 5 Condition (i) is used to prove closability for the form $(\mathcal{E}^{(1)}, \mathcal{F}C_b^\infty)$ defined in (21) below. However, it can be weakened considerably. By the same arguments as e.g. those in [AR91, Sect. 3] or [MR92, Chap. II, Subsection 2b)] one can allow degenerate cases. The respective comparison function ρ in these references can e.g. be taken to be equal to φ^2, where according to [Eb94, Theorem 1.1] φ can be an arbitrary function in $(\mathcal{E}, D(\mathcal{E}))$ such that $\varphi \neq 0$ m-a.e.

Furthermore, let $c \in L^\infty(E; m)$ and $b : \mathbb{R}^d \times E \to \mathbb{R}^d$ be $\mathcal{B}(\mathbb{R}^d) \otimes \mathcal{B}(E)$-measurable such that

$$(20) \qquad \sup_{\mu \in E} \|b(\mu)\|_{L^2(\mathbb{R}^d \to \mathbb{R}^d; \mu)} < \infty \ ,$$

where $b(\mu)(x) := b(\mu, x)$. Let d be another such function, but c and d should be such that

$$\int \left(\langle \nabla u, d \rangle_{L^2(\mathbb{R}^d \to \mathbb{R}^d; \mu)} + u\, c \right) d\mu \geq 0 \quad \text{for all } u \in \mathcal{F}C_b^\infty \text{ with } u \geq 0 \ .$$

Then for some $\alpha \in (0, \infty)$

$$
\begin{aligned}
\mathcal{E}^{(1)}(u, v) \ :=\ & \int \langle A(\mu)\nabla u(\mu), \nabla v(\mu)\rangle_{L^2(\mathbb{R}^d \to \mathbb{R}^d; \mu)} \ m(d\mu) \\
& + \int \langle \nabla u(\mu), b(\mu)\rangle_{L^2(\mathbb{R}^d \to \mathbb{R}^d; \mu)} v(\mu) \ m(d\mu) \\
& + \int \langle \nabla v(\mu), d(\mu)\rangle_{L^2(\mathbb{R}^d \to \mathbb{R}^d; \mu)} u(\mu) \ m(d\mu) \ , \\
& + \int u(\mu) \, v(\mu)(c(\mu) + \alpha) \ m(d\mu)
\end{aligned}
$$

$$
(21) \qquad\qquad u, v \in \mathcal{F}C_b^\infty \ ,
$$

is closable and the closure is a (non-symmetric) *semi-Dirichlet form* in the sense of [MOR95, Definition 2.1]. The proof is exactly analogous to [MR92, Chap II, Subsection 3e)] (see also [MOR95, Section 3.4]).

Also for (non-symmetric) semi-Dirichlet forms a correspondence like (3) holds and the corresponding theory is equally well-developed (cf. [MOR95] and also [MR92]). Hence this entire theory can be applied in this case. In particular, there exists a corresponding diffusion, which solves a martingale problem given by the corresponding generator. If $b \equiv d \equiv c \equiv 0$, this diffusion has m as symmetrizing and invariant measure.

The main problem for future research is to analyse these processes more precisely and to understand the kind of particle systems they describe. Also questions in how far and what sense this approach gives a unique solution to (19) resp., in the last more general case, to the corresponding martingale problem, are of interest. Also the very difficult problem concerning the ergodicity of the resulting processes should be considered. In the case of lattice models this will be discussed in the next section.

7. One application to lattice Gibbs states

In this section we want to describe one (as I think) particularly striking application of Dirichlet space theory to lattice Gibbs states (which is contained in [AKR95b]). For simplicity, we shall, however, introduce the Gibbs states in question not, as usual, via a local specification, but as symmetrizing measures of the solution for the corresponding infinite-dimensional stochastic differential equation.

Consider the following weighted ℓ^2-space of sequences over the lattice $\mathbb{Z}^d$, $d \in \mathbb{N}$:

$$
E := H_{-p} := \left\{ z \in \mathbb{R}^{\mathbb{Z}^d} \ \middle|\ \|z\|_{-p}^2 := \sum z_k^2(1 + |k|)^{-p} < \infty \right\}
$$

where $p > d$ is fixed. Here $|k|$ denotes the Euclidean norm of $k \in \mathbb{Z}^d \subset \mathbb{R}^d$. Define for $k \in \mathbb{Z}^d$ and $z = (z_j)_{j \in \mathbb{Z}^d} \in E$

$$
b_k(z) := -4z_k^3 - 2 \sum_{j \in \mathbb{Z}^d : |j-k|=1} (z_k - z_j)
$$

and $b := (b_k)_{k \in \mathbb{Z}^d}$. (However, all below applies for much more general b, cf. [AKR95b]; we made this choice of b for simplicity.) Let $X^z := (X_k^z)_{k \in \mathbb{Z}^d}$ be the strong solution of

$$
\begin{aligned}
dX^z(t) &= \frac{1}{2}b(X^z(t))dt + dW_t \\
X^z(0) &= z \in E \ ,
\end{aligned}
$$

where $W = (W_k)_{k \in \mathbb{Z}^d}$ is a family of independent real-valued Wiener processes. (For existence and uniqueness see [DoRo79]). Define for $f : E \to \mathbb{R}_+$, $\mathcal{B}(E)$-measurable, and $t \geq 0$

$$
p_t f(z) := E[f(X^z(t))] \ , \ z \in E \ ,
$$

and let $\mathcal{M}$ denote the set of all *tempered* probability measures m on $\mathcal{B}(E)$ w.r.t. which $(p_t)_{t \geq 0}$ is *symmetric*, i.e., for all $t > 0$

$$
\int_E p_t f \ g \ dm = \int_E f \ p_t g \ dm
$$

for all $f, g : E \to \mathbb{R}_+$, $\mathcal{B}(E)$-measurable. 'm is tempered' means that

$$
\left(\int_E |z_k| \ m(dz) \right)_{k \in \mathbb{Z}^d} \in H_{-p'}
$$

for some $p' \in \mathbb{N}$.

Remark 6 It can be shown (cf. [AKR95b], in particular Proposition 5.9) that $\mathcal{M}$ consists of exactly the tempered Gibbs states of a local specification which are nothing but the Φ_d^4-*Gibbs states on the lattice* $\mathbb{Z}^d$ (cf. also the end of Section 3 above for the continuous Φ_d^4-model in case $d = 2$).

Now the question arises whether, after having defined the convex set $\mathcal{M}$ in terms of $(p_t)_{t>0}$, we can also characterize its extreme points $\mathcal{M}_{\mathrm{ex}}$ using $(p_t)_{t>0}$. This question is answered by the following theorem. First we recall that for each $m \in \mathcal{M}$, $(p_t)_{t>0}$ extends to a strongly continuous semigroup $(T_t^m)_{t \geq 0}$ of symmetric contractions on $L^2(E; m)$ (cf. [MR92, Chap. II, Subsection 4a)]).

Theorem 1 Let $m \in \mathcal{M}$. Then $m \in \mathcal{M}_{\mathrm{ex}}$ if and only if $(T_t^m)_{t \geq 0}$ is ergodic (i.e.,

$$
T_t^m u \underset{t \to \infty}{\longrightarrow} \int u \ dm \ \text{in} \ L^2(E; m)).
$$

This was first proved in [AKR95b] (in much more generality; see Theorem 5.15 therein which covers the above special case) using one of the main results in [BoR95] (namely, Theorem 6.15) in an essential way. It extends a classical result by R.A. Holley and D.W. Stroock ([HoSt76]) for the Ising model to this case with non-compact continuous single spin spaces.

So far, no Dirichlet form has appeared in this section, but it plays the central rôle in the method of proof for Theorem 7.2. We remind the reader of the correspondence (3), and this time we enter it through the semigroup $(T_t^m)_{t \geq 0}$ (where $m \in \mathcal{M}$ is given as in Theorem 7.2) which comes by definition from the transition semigroup of a nice process. Hence we know by (3) that there exists an associated quasi-regular (symmetric) Dirichlet form $(\mathcal{E}_m, D(\mathcal{E}_m))$ on $L^2(E; m)$. This knowledge alone is of not much use, but here $(\mathcal{E}_m, D(\mathcal{E}_m))$ can be identified explicitly as the closure of

$$\mathcal{E}_m(u, v) = \int_E \langle \nabla u(z), \nabla v(z) \rangle_{\ell^2(\mathbb{Z}^d)} \, m(dz) ,$$

(22)
$$u, v \in \mathcal{F}C_b^\infty ,$$

i.e., we have a gradient-type Dirichlet form as in Section 3. (Note that here $\ell^2(\mathbb{Z}^d)$ is the tangent space at each point $z \in E$ and that $\ell^2(\mathbb{Z}^d) \subset E$ continuously and densely, hence $\nabla u(z)$ can be defined exactly as in Section 3). The proof that $\mathcal{E}_m$ is given by (22) on $\mathcal{F}C_b^\infty$ is trivial, but it is extremely hard to show that $D(\mathcal{E}_m)$ is really the completion of $\mathcal{F}C_b^\infty$ w.r.t. $(\mathcal{E}_m + (\, , \,)_{L^2(E;m)})^{1/2}$. Recent results on the Markov uniqueness problem mentioned at the end of Section 3 had to be used (cf. [AKR95a,b]).

The ergodicity of $(T_t^m)_{t \geq 0}$ is equivalent to the more tractable *irreducibility* of $(\mathcal{E}^m, D(\mathcal{E}^m))$ (i.e., $u = $ const. whenever $u \in D(\mathcal{E}^m)$ with $\mathcal{E}^m(u, u) = 0$; by the explicit form of $(\mathcal{E}^m, D(\mathcal{E}^m))$ this is equivalent to $u = $ const. whenever $u \in D(\mathcal{E}^m)$ with $\nabla u = 0$, where we also denote the closure of the gradient by ∇). Thus, it remains to show that this property of $(\mathcal{E}^m, D(\mathcal{E}^m))$ holds if and only if $m \in \mathcal{M}_{\mathrm{ex}}$. Also this second step of the proof is quite tough. The key part in it was already done in [BoR95, Theorem 6.15]. We refer to [AKR95b, Sect. 2] for details.

Remark 7 Since a log-Sobolev inequality for $(\mathcal{E}^m, D(\mathcal{E}^m))$ implies a mass gap, hence ergodicity of $(T_t^m)_{t \geq 0}$, because of Theorem 7.2 one can thus only hope to prove a log-Sobolev inequality (resp. a mass gap) if m is extreme.

Future research in this direction will focus on extending Theorem 7.2 within the framework of (mathematical) statistical mechanics to more general models than the ones studied in [AKR95b], in particular, cases where the single spin spaces are themselves (finite or) infinite-dimensional manifolds. Also an extension to particle systems as in the previous section seems feasible and would certainly be of substantial interest.

8. Other recent developments

So far, we have tried to show that in many concrete infinite-dimensional situations there are natural underlying 'geometric structures' and (intrinsically) associated gradient-type Dirichlet forms to be discovered which determine and help to analyse stochastic processes of substantial interest resp. give means to construct them. With such a wide range of possible applications in prospect there are, of course, also enormous research activities within Dirichlet space theory itself. In addition, areas of applications other than those described above are being explored. It is impossible to describe them here in a way that meets the importance of all these developments. We just want to give a (as already emphasized in the introduction highly incomplete) list of corresponding topics (in the shape of key words) to give an impression of the variety of ongoing research in this area of mathematics. We also mention some names of colleagues working specifically with emphasis on Dirichlet forms in the respective directions. However, we do not include topics that have already been discussed in previous sections.

- boundary theory (Chen, Silverstein, Williams, Zheng,...)

- convergence/approximations (Dal Maso, Dell' Antonio, Kuwae, Lyons, Mosco, Posilicano, Uemura, Zhang,...)

- Dirichlet operators (Albeverio, Eberle, Kondratiev, Liskevich, Semenov, Takeda,...)

- fractals (Fukushima, Kusuoka, LeJan, Metz, Sabot, Shima,...)

- geometry induced by Dirichlet forms (Ma, Jost, Sturm,...)

- large deviations (Fang, Fukushima, Mück, Takeda,...)

- Markov uniqueness (Albeverio, Eberle, Fitzsimmons, Kondratiev, Shigekawa, Song,...)

- non-commutative case (Albeverio, Lindsay,...)

- parabolic case (Oshima, Stannat,...)

- partial differential equations (Biroli, Mosco,...)

- pseudo-differential operators (Hoh, Jacob,...)

- statistical mechanics (Albeverio, Kondratiev, Park, Yoo,...)

- stochastic partial differential equations (Albeverio, Mück, Stannat,...)

- theory of Markov processes (Fitzsimmons, Getoor, Ma,...)

Finally, we would like to draw attention to an entirely new framework of so-called *generalized Dirichlet forms* presently being developed (see [Sta95a, b, 96]). It includes both the elliptic (semi-) Dirichlet forms in [MR92] (resp. [MOR95]) and the parabolic ones. In particular, the fairly restrictive *sector condition* (cf. [MR92]) in non-symmetric situations is dropped. This theory covers all cases known so far, and also includes examples which may even have no dual theory at all, but are still related to some (possibly highly non-symmetric) *notion of energy*. This new framework widens the scope of applications enormously and (e.g., in particular, w.r.t. stochastic partial differential equations) a lot of progress is expected.

Acknowledgements

We would like to thank the organizers for an excellent and very inspiring conference in New York as well as Columbia University and the Italian Academy of Sciences for the hospitality. We also thank Sergio Albeverio, Andreas Eberle, Yu.G. Kondratiev and Wilhelm Stannat for fruitful discussions about this work. Financial support of the German Science Foundation through SFB–343 (Bielefeld) and, in particular, the Faculty of Mathematics at Bielefeld University, that made the participation of the author in the above conference possible, is gratefully acknowledged.

References

[AHRZho95] S. Albeverio, Y.Z. Hu, M. Röckner, X.Y. Zhou: Stochastic quantization of the two-dimensional polymer measure. SFB-343 Preprint-1995.

[AKR95a] S. Albeverio, Y.G. Kondratiev, M. Röckner: Dirichlet operators via stochastic analysis. J. Funct. Anal. 128, 102–138 (1995)

[AKR95b] S. Albeverio, Y.G. Kondratiev, M. Röckner: Ergodicity of L^2-semigroups and extremality of Gibbs states. SFB-343-Preprint 1995.

[ALR93] S. Albeverio, R. Léandre, M. Röckner: Construction of a rotational invariant diffusion on the free loop space. C.R. Acad. Sci. Paris, t. 316, Série I, 287–292 (1993).

[AMR93] S. Albeverio, Z.M. Ma, M. Röckner: Quasi-regular Dirichlet forms and Markov processes. J.Funct. Anal. 111, 118–154 (1993).

[AR89] S. Albeverio, M. Röckner: Dirichlet forms on topological vector spaces – construction of an associated diffusion process. Probab. Th. Rel. Fields 83, 405–434 (1989).

[AR90] S.Albeverio, M.Röckner: Dirichlet forms on topological vector spaces - closability and a Cameron-Martin formula. J. Funct. Anal. 88, 395-436 (1990).

[AR91] S.Albeverio, M.Röckner: Stochastic differential equations in infinite dimensions: solutions via Dirichlet forms. Probab. Th. Rel. Fields 89, 347-386 (1991).

[ARZho95] S.Albeverio, M.Röckner, X.Y.Zhou: Stochastic quantization of the three-dimensional polymer measure. SFB-343-Preprint (1995).

[BoR95] V.I. Bogachev, M.Röckner: Regularity of invariant measures on finite and infinite-dimensional spaces and applications. J. Funct. Anal. 133, 168–223 (1995).

[BH91] N. Bouleau, F. Hirsch: Dirichlet forms and analysis on Wiener space. Berlin - New York: de Gruyter 1991.

[DoRo79] H. Doss, G. Royer: Processus de diffusion associés aux mesures de Gibbs sur $\mathbb{R}^{\mathbb{Z}^d}$. Z. Wahrsch. u. verw. Gebiete 41, 125–158 (1979).

[D92] B. Driver: A Cameron-Martin type quasi-invariance theorem for pinned Brownian motion on a compact Riemannian manifold. Trans. Amer. Soc. 342, 375–395 (1994).

[DR92] B.K. Driver, M.Röckner: Construction of diffusions on path and loop spaces of compact Riemannian manifolds. C.R. Acad. Sci Paris, t 315, Série I, 603–608 (1992).

[Eb94]   A. Eberle: Girsanov-type transformations of local Dirichlet forms: an analytic approach. SFB-343 Preprint (1994). To appear in Osaka J. Math.

[EK93]   S.N. Ethier and T.G. Kurtz: Fleming–Viot processes in population genetics. SIAM J. Control Opt. 31, 345-386, 1993.

[FFGKRS93] E. Fabes, M. Fukushima, L. Gross, C. Kenig, M. Röckner, and D.W. Stroock: Dirichlet Forms, Varenna 1992. Editors: G. Dell' Antonio, V. Mosco. Berlin: Springer 1993.

[FOT94]   M. Fukushima, Y. Oshima and M. Takeda: Dirichlet Forms and Symmetric Markov Processes. Walter de Gruyter, Berlin 1994.

[HoSt76]   R.A. Holley, D.W. Stroock: L_2-theory for the stochastic Ising model. Z. Wahrsch. u. verw. Gebiete 35, 87–101 (1976).

[MR92]   Z.M. Ma and M. Röckner: Introduction to the Theory of (Non-Symmetric) Dirichlet Forms. Springer, Berlin 1992.

[MOR95]   Z.M. Ma, L. Overbeck, M. Röckner: Markov processes associated with Semi-Dirichlet forms. Osaka J. Math. 32, 97–119 (1995).

[MRYa95]   Z.M. Ma, M. Röckner, J.A. Yan: Dirichlet forms and stochastic processes. Berlin: de Gruyter 1995.

[Os95]   H. Osada: Dirichlet form approach to infinite-dimensional Wiener processes with singular interactions. Preprint (1995).

[OR94]   L. Overbeck, M. Röckner: Geometric aspects of finite and infinite-dimensional Fleming–Viot processes. SFB-256-Preprint 1994. Revised and extended version: SFB-343-Preprint (1996).

[ORS95]   L. Overbeck, M. Röckner, B. Schmuland: An analytic approach to Fleming–Viot processes with interactive selection. Ann. Prob. 23, 1-36. 1995.

[RS92]   M. Röckner, B. Schmuland: Tightness of general $C_{1,p}$-capacities on Banach space. J. Funct. Anal. 108, 1-12 (1992).

[RS95]   M. Röckner, B. Schmuland: Quasi-regular Dirichlet forms: examples and counterexamples. Can. J. Math 47, 165–200 (1995).

[RZ92]   M. Röckner, T.S. Zhang: Uniqueness of generalized Schrödinger operators and applications. J. Funct. Anal. 105, 187–231 (1992).

[RZ94]   M. Röckner, T.S. Zhang: Uniqueness of generalized Schrödinger operators – Part II. J. Funct. Anal. 119, 455–467 (1994).

[Sta95a]   W. Stannat: Generalized Dirichlet forms and associated Markov processes. C.R. Acad. Sci Paris, t. 319, 1063–1068 (1994).

[Sta95b]   W. Stannat: Dirichlet forms and Markov processes: a generalized framework including both elliptic and parabolic cases. SFB-343-Preprint 1995. To appear in: Potential Analysis.

[Sta96]   W. Stannat: The theory of generalized Dirichlet forms and its applications in analysis and stochastics. Doktorarbeit Universität Bielefeld 1996.

[Y95]   M.W. Yoshida: Construction of infinite-dimensional interacting diffusion processes through Dirichlet forms. Preprint (1995).

QUANTUM STOCHASTIC CALCULUS AND APPLICATIONS – A REVIEW

KALYAN B. SINHA,* *Indian Statistical Institute*

1. Introduction

It has been a little more than a decade since this subject, as it is understood today, came into being with the seminal paper of Hudson and Parthasarathy [1]. Since then the subject has seen rapid development and many of these can be found in the monographs of Parthasarathy [2] and Meyer [3]. Here I want to discuss some of the more recent developments.

The first section contains notations and a collection of some basic results, the proofs of which can be found in the two monographs mentioned above. The second section deals with quantum stochastic differential equations (q.s.d.e.) with unbounded operator coefficients and Feller condition while the third deals with an application.

2. Notations and Preliminaries

We shall work exclusively in the bosonic (symmetric) Fock space and shall give a few background results, referring the reader to [2] and [3] for the details.

Let h be a complex separable Hilbert space, $\Gamma(h)$ be the symmetric Fock space over h, spanned by the total set of 'exponential vectors'

$$(1) \qquad e(f) = 1 \oplus f \oplus \cdots \oplus \frac{f^{(n)}}{\sqrt{n!}} \oplus \cdots ,$$

where $f \in h$ and $f^{(n)}$ is the n-fold tensor product of f. In most of our discussions, h will be $L^2(\mathbb{R}_+, \mathcal{K})$ with $\mathcal{K}$ another separable Hilbert space, the dimension of which will signify the noise degree of freedom. Let $\mathcal{H}_0$ (a separable Hilbert space) be the initial space or the system space. The $\mathcal{H} = \mathcal{H}_0 \otimes \Gamma(h)$ is the Hilbert space in which we shall work.

Let $\{e_n\}_{n=1}^{\dim \mathcal{K}}$ denote a complete orthonormal system of $\mathcal{K}$. Then the *basic quantum processes* in $\mathcal{H}$ are

$$(2) \qquad
\begin{aligned}
\Lambda_j^k(t)e(f) &= -i \tfrac{d}{d\varepsilon} e\left(e^{i\varepsilon \chi_{[0,t]} \otimes |e_j\rangle\langle e_k|} f \right)\Big|_{\varepsilon=0} , \quad (1 \le j,k \le \dim \mathcal{K}) \\
A_j(t)e(f) &= \int_0^t ds\, f_j(s)e(f) \equiv \int_0^t ds \langle e_j, f(s)\rangle e(f), \\
A_j^+(t)e(f) &= \tfrac{d}{d\varepsilon} e(f + \varepsilon \chi_{[0,t]} e_j)\big|_{\varepsilon=0},
\end{aligned}
\right\}$$

* Postal address: Indian Statistical Institute, Delhi Centre, 7, S.J.S. Sansanwal Marg, New Delhi - 110016, India and Jawaharlal Nehru Centre for Advanced, Scientific Research, Bangalore 560064, India.

where $\chi_{[0,t]}$ is looked upon in the first expression as the multiplication operator (projection) by the indicator function of $[0,t]$ while in the last it is the indicator function itself. If we set

$$Q_j(t) = A_j(t) + A_j^+(t) \text{ and } P_j(t) = -i[A_j(t) - A_j^+(t)],$$

then these define two sets of independent countable families of Brownian motions, but $[P_j(t), Q_k(s)] = -2i(t \wedge s)$. Thus there are two sets of non-commuting Brownian motions in Fock space. Similarly, the selfadjoint operators: $\Lambda_j^j(t) + \sqrt{\ell}\, Q_j(t) + \ell t$ define a family of quantum Poisson processes.

Let us write $h_{[a,b]} = L^2([a,b], \mathcal{K})$, $h_{t]} = h_{[0,t]}$ and $h_{[t} = h_{[t,\infty)}$ and note that the natural continuous tensor product structure of $\Gamma(h)$, viz. $\Gamma(h) = \Gamma(h_{t]}) \otimes \Gamma(h_{[t})$ allows one to define an operator family $\{L(t)\}$ to be *adapted* if $L(t)$ is of the form $L_0(t) \otimes I_{[t}$ for every $t \geq 0$, where $L_0(t)$ is a linear operator in $\mathcal{H}_{t]} \equiv \mathcal{H}_0 \otimes \Gamma(h_{t]})$ and $I_{[t}$ is the identity on $\Gamma(h_{[t})$. The *quantum Ito formulae* are given as :

$$(3) \qquad \left. \begin{aligned} dA_j dA_k^+ &= \delta_{jk}dt,\ dA_j dA_k = dA_j^+ dA_k^+ = 0, \\ d\Lambda_j^k dA_\ell^+ &= \delta_{\ell k}dA_j^+,\ dA_\ell d\Lambda_k^j = \delta_{\ell k}dA_j, \\ d\Lambda_j^k d\Lambda_m^\ell &= \delta_{km}d\Lambda_j^\ell,\ d\Lambda_j^k dA_\ell = dA_\ell^+ d\Lambda_k^j = 0. \end{aligned} \right\}$$

An adapted process $\{L(t)\}$ is integrable w.r.t. the basic quantum processes if it is square-integrable $(\mathcal{D}, \mathcal{E}(\mathcal{M}))$ i.e. if $\int_0^t \|L(s)ue(f)\|^2 ds < \infty$ for every $t \geq 0$, $u \in \mathcal{D}$, a suitable dense subset of $\mathcal{H}_0$ and $f \in \mathcal{M}$, a dense subset of h such that $\mathcal{D} \otimes \mathcal{E}(\mathcal{M}) \subseteq \mathrm{Dom}(L_0(t))$. Here by $\mathcal{E}(\mathcal{M})$ we mean the linear span of $e(f)$, $f \in \mathcal{M}$. One has the estimates,

$$\left\| \int_0^t L(s)dA_j(s)ue(f) \right\| \leq \int_0^t |f_j(s)|\, \|L(s)ue(f)\|ds$$

$$(4) \qquad \left. \begin{aligned} &\left\| \int_0^t L(s)dA_j^+(s)ue(f) \right\| \\ &\text{and} \\ &\left\| \int_0^t L(s)d\Lambda_j^k(s)ue(f) \right\| \end{aligned} \right\} \leq C(f,t)\left[\int_0^t \|L(s)ue(f)\|^2 ds \right]^{1/2}.$$

In the above we have written $ue(f)$ instead of $u \otimes e(f)$ and the above estimates show that the integral w.r.t. $A_j(t)$-process is defined whenever the R.H.S. of the estimate in (4) is finite and that can happen in circumstances more general than the square integrability requirement.

The simplest q.s.d.e. that one can solve is the following

$$(5) \qquad X(t) = X_0 + \int_0^t X(s)[F_k^j d\Lambda_j^k(s) + E_j dA_j^+(s) + G_j dA_j(s) + M ds],$$

where $X_0, F_k^j, E_j, G_j (1 \leq j, k \leq \dim \mathcal{K})$ and M are constant bounded operators in $\mathcal{H}_0$, and we have used the summation convention. In the case when $\dim \mathcal{K} = \infty$, one needs a further condition, viz. $\sum_j \|F_k^j u\|^2,\ \sum_j \|E_j u\|^2 \le C_k \|u\|^2\ \forall\, u \in \mathcal{H}_0$ with some family of positive constants C_k. These issues were studied in [4]. In usual quantum mechanics one is interested in the unitarity of the evolution group and similarly here a natural question would be: under what conditions on the coefficient operators is the solution X of (5) unitary and what are its properties. The answer is that the solution U_t of (5) with the initial value $X_0 = I$ is unitary iff its coefficients satisfy:

$$(6)\qquad \left.\begin{array}{rcl} F_k^j &=& W_k^j - \delta_{jk} \text{ where } W \equiv ((W_k^j)) \text{ is unitary on } \mathcal{H}_0 \otimes \mathcal{K}, \\[4pt] G_j &=& -E_k^* W_j^k, \\[4pt] N &=& -\tfrac{1}{2} E_k^* E_k + iH \text{ where } H\text{is bounded selfadjoint operator in } \mathcal{H}_0 \,. \end{array}\right\}$$

It is useful to replace (5) by a closely related one:

$$U(s,t) = I + \int_s^t U(s,\tau)[F_k^j\, d\Lambda_j^k(\tau) + \cdots + M\,d\tau]. \qquad (5')$$

Then one finds that (i) $U(s,t)$ for $s \le t$ is unitary iff the coefficients satisfy (6) as in the case of (5), (ii) $\{U(s,t)\}$ is an evolution, i.e. $U(r,s)U(s,t) = U(r,t),\quad r \le s \le t;\ U(s,s) = I$. (iii) $U(0,t) = U_t$, (iv) $U(s,t)$ is time-homogeneous, i.e. depends only on $(t-s)$, iff all the coefficients except M are zero, and (v) the expectation operator $P(s,t)$ defined as $\langle u, P(s,t)v \rangle \equiv \langle ue(0),\ U(s,t)ve(0)\rangle$ is time-homogeneous, in fact $P(s,t) \equiv P_{t-s} = e^{M(t-s)}$. Thus $\{P_t\}_{t\ge 0}$ forms a norm-continuous semigroup on $\mathcal{H}$ with bounded generator M. Obviously the situation becomes more complicated if the coefficients still formally satisfy (6) but are *unbounded*. We shall discuss this in the next section.

A quantum stochastic process in Fock space $\mathcal{H}$ is a family of maps $\{j_t\}_{t\ge 0} : \mathcal{A} \to \mathcal{B}(\mathcal{H})$, where $\mathcal{A}$ is a unital $*$-subalgebra of $\mathcal{B}(\mathcal{H}_0)$, satisfying :

(i) $\{j_t(x)\}_{t\ge 0}$ is an adapted family $\forall\, x \in \mathcal{A}$,

(ii) j_t is a $*$-homomorphism from $\mathcal{A}$ into $\mathcal{B}(\mathcal{H})$, i.e. $j_t(x^*) = j_t(x)^*,\ j_t(xy) = j_t(x)j_t(y)\ \forall\, x, y \in \mathcal{A},\ t \ge 0$.

It is said to be *conservative* if $j_t(I) = I\ \forall\, t$ and a *quantum stochastic flow* (q.s.f.) if it satisfies furthermore the q.s.d.e.

(7)

$$j_t(x) = x + \int_0^t [j_s(\theta_k^j(x))d\Lambda_j^k(s) + j_s(\theta_j^0(x))dA_j^+(s) + j_s(\theta_0^j(x))dA_j + j_s(\theta_0^0(x))ds]$$

where the *structure maps* $\theta_\beta^\alpha (0 \le \alpha, \beta \le \dim\ \mathcal{K})$ are linear operators on $\mathcal{A}$ satisfying for $x, y \in \mathcal{A}$

$$(8)\qquad \theta_\beta^\alpha(x^*) = \theta_\alpha^\beta(x)^*,\quad \theta_\beta^\alpha(xy) = x\theta_\beta^\alpha(y) + \theta_\beta^\alpha(x)y + \theta_k^\alpha(x)\theta_\beta^k(y).$$

It is shown in [2] and [5] that for dim $\mathcal{K} < \infty$ if θ^α_β's are norm-bounded and satisfy (8), then equation (7) has a unique solution $j_t(x)$ which defines a q.s.f. such that it is contractive : $\|j_t(x)\| \leq \|x\|$ and $\mathbb{R}_+ \times \mathcal{A} \ni (s, x) \to j_s(x)$ is strongly jointly continuous on $\mathcal{H}$ w.r.t. the strong operator topology on $\mathcal{A} \subseteq \mathcal{B}(\mathcal{H}_0)$. Furthermore, it is conservative iff $\theta^\alpha_\beta(I) = 0$ for all α, β. Most of these results were extended in [4] to the case dim $\mathcal{K} = \infty$ subject to an additional summability condition on θ^α_β 's . As in the preceding paragraph one can pose the question: what happens if θ^α_β 's are *not* norm-bounded on $\mathcal{A}$. Not much is known in this area, but some results can be found in [6, 14].

I shall end this section by giving some applications, viz. the description of classical Markov chains as q.s.f. in Fock space with the degree of freedom equaling the cardinality of the state space. In this context, the following lemma is instructive ([2]).

Lemma 1 (for simplicity dim $\mathcal{K} < \infty$) : Let j_t be a q.s.f. satisfying (7) with bounded structure maps θ^α_β obeying (8). Assume furthermore that $\mathcal{A}$ is abelian. Then $\{j_s(x)|0 \leq s \leq \infty, x \in \mathcal{A}_h$, the self-adjoint part of $A\}$ is a classical process; in fact $[j_s(x), j_t(y)] = 0 \ \forall \ s, t \geq 0$ and $x, y \in \mathcal{A}$.

This lemma allows us to embed classical stochastic processes in the quantum receptacle and this we briefly describe for Markov chain [19]. Let $\mathcal{X}$ be the state space of a countably infinite continuous time Markov porocess and let $p_t(x, y)(x, y \in \mathcal{X})$ be the (stationary) transition probabilities such that $\ell(x, y) \equiv \frac{d}{dt}p_t(x, y)|_{t=0}$ satisfy the Markov conditions:

$$(9) \qquad \ell(x, y) \geq 0 \text{ for } x \neq y \text{ and } \sum_{y \in \mathcal{X}} \ell(x, y) = 0.$$

It is convenient (but not necessary) to put a group structure G on $\mathcal{X}, G$ acting on $\mathcal{X}$ by left translation and let μ be the counting measure on $\mathcal{X}$.

Set $m_x(y) = \sqrt{\ell(y, xy)}$ if $x \neq id$ of G and $= 0$ otherwise, and write for $\phi \in L_\infty(\mathcal{X}, \mu) \equiv \mathcal{A}$:

$$\begin{aligned}
\theta^x_0(\phi)(y) &= \text{Multiplication operator by } m_x(y)[\phi(xy) - \phi(y)], \\
\theta^0_x(\phi)(y) &= \text{Multiplication operator by } \overline{m_x(y)}[\phi(xy) - \phi(y)], \\
\theta^x_{x'}(\phi)(y) &= \text{Multiplication operator by } [\phi(xy) - \phi(y)]\delta_{xx'}, \\
\theta^0_0(\phi)(y) &= \text{Multiplication operator by } \sum_{x \in \mathcal{X}} |m_x(y)|^2[\phi(xy) - \phi(y)].
\end{aligned}$$

Then the q.s.d.e. (7) with the above structure maps on the (abelian) *-algerba $\mathcal{A}$ has a q.s.f. $j_t(\phi)$ as its unique solution if $\sup_{x \in \mathcal{X}} |\ell(x, x)| < \infty$. This is because under this condition, the abovementioned structure maps are norm bounded and we can apply the theory discussed above ([5], [2]). Furthermore the expectation semigroup $T_t(\phi) \equiv \mathbb{E}j_t(\phi) \equiv \langle \cdot e(0), j_t(\phi) \cdot e(0) \rangle$ has the bounded generator θ^0_0 given by $\theta^0_0(\phi)(y) = \sum_{\mathcal{X} \ni x \neq id} \ell(y, xy)[\phi(xy) - \phi(y)] = \sum_{x \neq y} \ell(y, x)\phi(x) - \{\sum_{z \neq y} \ell(y, z)\}\phi(y) = \sum_{x \in \mathcal{X}} \ell(y, x)\phi(x)$, the action is exactly the same as that for the Markov chain.

3. Q.S.D.E. with unbounded operator coefficients

Most of what I shall describe here is part of the thesis of A. Mohari in the Indian Statistical Institute, Delhi, and of publications arising from it [7,8]. For simplicity, we take dim $\mathcal{K} = 1$ and drop the $d\Lambda$ term from q.s.d.e.'s :

$$(10) \qquad V(t) = I + \int_0^t V(s)[EdA^+(s) - E^* dA(s) - \frac{1}{2}E^* E ds].$$

As can be easily seen, the operator coefficients in (10) satisfy formally the unitarity condition (6). The major problem, however, is that Dom $(E) \cap$ Dom (E^*) may be too small, even trivial in which case the equation (10) has hardly any content (see [9] for counter-example). To proceed further, we make

Assumption A : Let E be closed and assume furthermore that there exists a dense subset $\mathcal{D} \subset$ Dom $(E) \cap$ Dom (E^*) and a sequence of bounded operators $\{E_n\}$ in $\mathcal{H}_0$ such that E_n, E_n^* and $E_n^* E_n$ converge strongly on $\mathcal{D}$ to E, E^* and $E^* E$ respectively.

Assumption B : $\mathcal{D}$ is stable under the action of T_t, (the expectation semigroup of $V(t)$ with generator $-\frac{1}{2}E^* E$).
   The basic result is contained in the next theorem.

Theorem 2 : (i) Assume (A) and (B). Then (10) admits a unique adapted contractive solution V.
   (ii) If furthermore, E satisfies the Feller Condition (F) : For some $\lambda > 0$ (and hence for all $\lambda > 0$) the Feller set

$$\beta_\lambda \equiv \{x \in \mathcal{B}(\mathcal{H}_0), 0 \leq x \leq 1 | \langle v, \theta_0^0(x)u \rangle \equiv \langle Ev, xEu \rangle$$

$$-\frac{1}{2}\langle E^* Ev, xu \rangle - \frac{1}{2}\langle x^* v, E^* Eu \rangle = \lambda \langle v, xu \rangle \ \forall \ u, v \in \mathcal{D}\} = \{0\}$$

, then V is unitary.

Sketch of proof : Consider the equation

$$(11) \qquad V_n(t) = I + \int_0^t V_n(s)[E_n dA^+ - E_n^* dA - \frac{1}{2}E_n^* E_n ds],$$

which by the results in section 1 admits a unique adapted unitary solution V_n in $\mathcal{H}$. Let $u \in \mathcal{D}, 0 \leq t_1 < t_2 < T < \infty$, and $f \in C_0(0, \infty) \subseteq h$. Then

$$\|[V_n(t_2) - V_n(t_1)]ue(f)\|^2 \leq C \int_{t_1}^{t_2} \{\|E_n u\|^2 + |f(s)|^2\|E_n^* u\|^2 + \frac{1}{4}\|E_n^* E_n u\|^2\} ds$$

and hence by assumption A we have that for every $\psi \in \mathcal{H}$ $\{\langle \psi, V_n(t)ue(f)\rangle\}$ is a bounded equicontinuous family. Thus we can extract a subsequence converging uniformly on $[0,T]$. Using the separability of $\mathcal{H}$, a diagonal trick, the uniform boundedness of $V_n(t)$ and the totality of vectors of the form $ue(f)$, one can show that $V_n(t)$ converges weakly (by relabelling the subsequence) to an adapted contraction $V(t)$ uniformly on $[0,T]$. From the properties (2) of the basic processes and (11), we have for $u,v \in \mathcal{D}$ and f,g as before,

$$\langle ve(g), V_n(t)ue(f)\rangle = \langle ve(g), ue(f)\rangle$$

$$(12) \qquad + \int_0^t \langle ve(g), V_n(s)[E_n u\bar{g}(s) - E_n^* uf(s) - \frac{1}{2}E_n^* E_n u]e(f)\rangle ds.$$

Choosing an appropriate subsequence, we see that the LHS of (12) converges to $\langle ve(g), V(t)ue(t)\rangle$ whereas by assumption (A), weak convergence of $V_n(s)$ to $V(s)$ uniformly in $[0,T]$ implies that the RHS of (12) converges to $\int_0^t \langle ve(g), V(s)[Eu\bar{g}(s) - E^* uf(s) - \frac{1}{2}E^* Eu]e(f)\rangle ds$. Since $\int_0^t V(s)[EdA^+(s) - E^* dA(s) - \frac{1}{2}E^* Eds]ue(f)$ converges strongly by virtue of the basic estimates (4) and the contractivity of $V(s)$, we have that for all $v \in \mathcal{D}, g \in \mathcal{M}$

$$\langle ve(g), \left\{ V(t) - I - \int_0^t V(s)[EdA^+(s) - E^* dA(s) - \frac{1}{2}E^* Eds] \right\} ue(f)\rangle = 0.$$

This proves that $V(t)$ is a strong solution of (10).

Let $V'(t)$ be another solution of (10) and set $X(t) = V(t) - V'(t)$ so that $\|X(t)\| \leq 2$, $X(0) = 0$ and

$$(13) \qquad X(t) = \int_0^t X(s)[EdA^+(s) - E^* dA(s) - \frac{1}{2}E^* Eds].$$

Fix $f,g \in C_0(0,\infty)$ and define a bounded operator $M(t)$ on $\mathcal{H}_0$ by $\langle u, M(t)v\rangle = \langle ue(f), X(t)ve(g)\rangle$ for $u,v \in \mathcal{D}$. Then by (13) one has

$$(14) \qquad \langle u, M(t)v\rangle = \int_0^t \langle u, M(s)[E\bar{f}(s) - E^* g(s) - \frac{1}{2}E^* E]v\rangle ds.$$

Replacing f and g by αf and βg $(\alpha, \beta \in \mathbb{C})$ respectively, differentiating m times w.r.t. α and n times w.r.t. β and equating coefficients of both sides, we get
$$\langle u, M(t;m,n)v\rangle \equiv \langle uf^{\otimes m}, X(t)vg^{\otimes n}\rangle$$
$$= \int_0^t \{\langle u, M(s;m-1,n)Ev\rangle \bar{f}(s) - \langle u, M(s;m,n-1)E^* v\rangle g(s)$$

$$(15) \qquad\qquad\qquad - \frac{1}{2}\langle u, M(s;m,n)E^* Ev\rangle\}ds.$$

For $m = n = 0$, this leads to $\frac{dM(t;0,0)v}{dt} = -\frac{1}{2}M(t;0,0)E^*Ev \ \forall \ v \in \mathcal{D}$. Thus using (B), $\frac{d}{ds}M(s;0,0)T_{t-s}v = 0$ for all $0 \le s \le T$. Since $M(0;0,0) = 0$ this implies that $M(t;0,0) = 0$ for $0 \le t \le T$. Now by induction let $M(t;k,\ell) = 0$ for $k + \ell \le n$ and consider k, ℓ such that $k + \ell = n + 1$. Then by (15), $\frac{M}{dt}(t;k,\ell)v = -\frac{1}{2}M(t;k,\ell)E^*Ev$ with $M(0;k,\ell) = 0$ and just as above one concludes that $M(t;k,\ell) = 0 \ \forall \ t$ and all k, ℓ which implies $X(t) = 0$ or the uniqueness of the solution of (10).

Let $Y(t) = I - V(t)^*V(t)$ and $Y_\lambda = \int_0^\infty e^{-\lambda t} Y(t)dt, \ \lambda > 0$. Then by the strong continuity of $V(t)$ in $\mathcal{B}(\mathcal{H})$ and since $\|Y(t)\| \le 2$, it is clear that Y_λ is well-defined as a strong Riemann integral and that $\|Y_\lambda\| \le 2/\lambda$ and since $Y(t) \ge 0$, one also has $Y_\lambda \ge 0$. Since $Y(0) = 0$, by the quantum Ito formula (3) one arrives at

$$\langle ve(g), Y(t)ue(f) \rangle = \int_0^t \langle ve(g), \{Y(s)[E\bar{g}(s) - E^*f(s) + G]$$

$$\text{(16)} \qquad + [E^*f(s) - E\bar{g}(s) + G]Y(s) + E^*Y(s)E\}ue(f)\rangle ds,$$

where we have written G for $-\frac{1}{2}E^*E$. As before, going down to the finite particle vectors and considering the diagonal terms only, we have

$$\text{(17)} \qquad \langle vg^{\otimes m}, Y(s)uf^{\otimes m} \rangle = \int_0^t \langle vg^{\otimes m}, \{Y(s)G + GY(s) + E^*Y(s)E\}uf^{\otimes m}\rangle ds.$$

Since $Y(t) \ge 0$, there exists at least one m such that $Y(t)uf^{\otimes m} \ne 0$ for some $f \in M$ and $u \in \mathcal{D}$. Then for such m, f, u one has by integrating by parts and using (17),

$$\langle v, B_\lambda v \rangle \equiv \int_0^\infty e^{-\lambda t} \langle ug^{\otimes m}, Y(t)uf^{\otimes m} \rangle dt = -\frac{1}{\lambda}e^{-\lambda t}\langle vg^{\otimes m}, Y(t)uf^{\otimes m}\rangle|_{t=0}^\infty +$$

$$\lambda^{-1} \int_0^\infty e^{-\lambda t} dt \{ \langle vg^{\otimes m}, Y(t)Guf^{\otimes m} \rangle + \langle Gvg^{\otimes m}, Y(t)uf^{\otimes m} \rangle + \langle Evg^{\otimes m}, Y(t)Euf^{\otimes m} \rangle \}$$

$$\text{(18)} \qquad = \lambda^{-1}\{\langle v, B_\lambda Gu \rangle + \langle Gv, B_\lambda v \rangle + \langle Ev, B_\lambda Eu \rangle\}.$$

Now, if condition (F) is satisfied, i.e. if $\beta_\lambda = \{0\}$ then it follows from (18) that $B_\lambda = 0$ which by the uniqueness of the Laplace transform implies in particular that $\langle uf^{\otimes m}, Y(t)uf^{\otimes m} \rangle = 0$. Since $Y(t) \ge 0$ this means that $Y(t)uf^{\otimes m} = 0$ which is a contradiction. Therefore V is an isometry.

For proving the coisometry of $V(t)$ we use the reflection map [7,9]. On h define a selfadjoint unitary map ρ_T (reflection about $T \ge 0$) by $(\rho_T f)(t) = f(T - t)$ if $t \le T$ and $= f(t)$ if $t > T$, and let R_T be its second quantization to $\Gamma(h)$. Set

$\widetilde{V}_n(s,t) \equiv R_t V_n(t-s)^* R_t$ and then one can compute the q.s.d.e. satisfied by $\widetilde{V}_n$ w.r.t. t as

$$(19) \qquad \widetilde{V}_n(s,t) = I + \int_s^t \widetilde{V}_n(s,\tau)[E_n^* dA(\tau) - E_n dA^+(\tau) - \frac{1}{2}E_n^* E_n d\tau].$$

We can now proceed as in the first part of this proof and see that $\widetilde{V}_n(s,t)$ (or possibly a subsequence of this) converges weakly to an adapted contraction, say $\widetilde{V}(s,t)$. As before one can obtain the q.s.d.e. satisfied by $\widetilde{V}(t) \equiv \widetilde{V}(0,t)$ as :

$$(20) \qquad \widetilde{V}(t) = I + \int_0^t \widetilde{V}(s)[E^* dA(s) - E dA^\dagger(s) - \frac{1}{2}E^* E ds],$$

which is very similar to the equation (10) with E replaced by $-E$. This means that the Feller set $\tilde{\beta}_\lambda$ for the reflected problem is the same as β_λ, the original one. This, by the last paragraph, implies the isometry of $\widetilde{V}(t)$. Finally the definition of $\widetilde{V}$ and the unitarity of R_t shows that this means the coisometry of $V(t)$. ∎

Remark 3 : (i) Classical birth and death processes can be described in the framework of theorem 2, see e.g. [10]. More general results can be derived when the noise is classical [11].

(ii) As can be seen, the Feller set β_λ plays an important role in the analysis. Many examples are known in which β_λ is not trivial and therefore the solutions of (10) even when they exist are not unitary [12]. It is clear that if $\beta_\lambda \neq \{0\}$, then $P_t(I) = EV(t)^* V(t) \neq I$. Then an important question arises: Can one extend the semigroup P_t to a conservative one? Some answers to this can be found in [13].

(iii) The difficulty in satisfying hypothesis (A) in general should be clear. However if E is normal (though unbounded) then (A) can be easily satisfied by taking $E_n = E(E^* E + n)^{-1}$ so that $E_n^* = \overline{(E^* E + n)^{-1} E^*} = E^*(EE^* + n)^{-1}$. However, in this case we can explicitly solve (10) : $V(t) = \int_{\mathbb{C}} P(dz) \otimes W(z\chi_{[0,t]})$, where $E = \int_{\mathbb{C}} z P(dz)$ is the spectral resolution of E and W is the Weyl operator in the Fock space $\Gamma(h)$ ([2]).

4. Applications and Discussion

As is well known, the classical damped harmonic oscillator is described by the equation of motion :

$$(21) \qquad \frac{d^2 q}{dt^2} + 2\alpha \frac{dq}{dt} + \omega^2 q = 0, \quad (0 < \alpha < \omega)$$

and such a (non-conservative) system cannot be described in terms of a Hamiltonian i.e. the above second order equation cannot be recast as a pair of canonical equations of motion.

Nevertheless, we can introduce a pair of "conjugate variables" q and p satisfying a pair of first order differential equations which is equivalent to (21). Set $\delta = \sqrt{\omega^2 - \alpha^2}$ and write

$$(22) \qquad \frac{dq}{dt} = p - \alpha q, \quad \frac{dp}{dt} = -\delta^2 q - \alpha p;$$

and a simple calculation verifies that indeed (22) leads to (21). In fact we would like to introduce 'annihilation' and 'creation' variables.

$$(23) \qquad a = (2\delta)^{-1/2}(p - i\delta q), \ a^+ = (2\delta)^{-1/2}(p + i\delta q)$$

and observe that (22) can be further rewritten in a convenient form

$$(24) \qquad \frac{da}{dt} = (-\alpha - i\delta)a.$$

The equation for a^+ is just the complex conjugate of (24) and does not add any new information. The solution of (24) is simply given as $a(t) = a(0)e^{(-\alpha - i\delta)t}, a^+(t) = a^+(0)^{(-\alpha + i\delta)t}$ or equivalently

$$p(t) = e^{-\alpha t}(p_0 \cos \delta t - \delta q_0 \sin \delta t) \quad \text{and} \quad q(t) = e^{-\alpha t}(q_0 \cos \delta t + \frac{p_0}{\delta} \sin \delta t).$$

From this it follows that even if (q_0, p_0) were a true canonically conjugate pair at $t = 0$, they cannot remain so for any $t > 0$. This is well known and expected. Of course quantization does not bring any change to the above observation and leads to the conclusion that there is *no* unitary time evolution to give rise to the equation of motion (24).

Now we want to change the picture and imagine that the damping in the motion of the quantum harmonic oscillator is due to the presence of some environmental friction which we shall model by quantum Brownian motion as described above. In other words, we consider a q.s.d.e.

$$(25) \qquad U(t) = I + \int_0^t U(s)[\sqrt{2}\alpha(a^* dA(s) - a\, dA^+(s)) + (-\alpha - i\delta)a^* a\, ds].$$

Here a and a^* have the same expression as for a and a^+ in (23), but they are time-independent (unbounded) operators satisfying the CCR : $[a, a^*] = I$ in the initial space $\mathcal{H}_0 = L^2(\mathbb{R})$, the quantum state space for the harmonic oscillator. We also note that if $\alpha = 0$ i.e. if there is no damping that U satisfies $\frac{dU}{dt} = i\delta U a^* a$ whose solution is $U(t) = e^{i\delta a^* a t}$ with $\delta = \omega$ and this is the well known standard quantum harmonic oscillator evolution group.

That equation (25) has a unique unitary solution was proven in [15] using a method different from the one we have described in section 2. Nevertheless we give a sketch of a proof using theorem 2. As has been shown in [12] the map $Q_\lambda(\lambda > 0)$ defined on $\mathcal{B}(\mathcal{H}_0)$ as :

$$(26) \qquad \langle u, Q_\lambda(x)v \rangle = \int_0^\infty e^{-\lambda t} \langle Ee^{-Gt}U, xEe^{-Gt}v \rangle dt$$

is a well-defined completely positive contraction and the Feller set δ_λ of the problem is trivial if $Q_\lambda^n(I)$ converges strongly to 0. In the problem at hand, $E = (2\alpha)^{1/2}a$, $G = (-\alpha - i\delta)a^*a = -\frac{1}{2}E^*E - iH$ (with $H = \delta a^*a$) $= (-\alpha - i\delta)N$ where $N = a^*a$ is the number opeator in $\mathcal{H}_0 = L^2(\mathbb{R})$. It is easy to see that the 'finite particle vectors' form a total set in $\mathcal{H}_0$ and if we choose the (dense) linear span of these to be $\mathcal{D}$, and $E_n = -(2\alpha)^{1/2}na(N+n)^{-1}$ then E_n, E_n^* and $G_n = -\frac{1}{2}E_n^*E_n + i\delta Nn(N+n)^{-1}$ converge strongly and to E, E^* and G respectively. It is also equally easy to prove that $T_t = e^{Gt} = e^{(-\alpha+i\delta)Nt}$ leaves $\mathcal{D}$ invariant, thus verifying assumptions (A) and (B) preceeding theorem 2. In order to prove the unitarity of $U(t)$ it sufficies to show that the Feller set $\delta_\lambda = \{0\}$ for some $\lambda > 0$ and for this we use the observations made in the earlier part of this paragraph.

We can write Q_λ for this problem as:

$$Q_\lambda(x) = 2\alpha \int_0^\infty e^{-\lambda t} e^{-\bar{\gamma}tN} a^* x a e^{-\gamma tN} dt$$

with $\gamma = \alpha - i\delta$ and $\lambda > 0$. Then

$$Q_\lambda(I) \;=\; 2\alpha N \int_0^\infty e^{-(\lambda + 2\alpha N)t} dt$$

$$(27) \qquad\qquad\qquad =\; 2\alpha N(2\alpha N + \lambda)^{-1} = \frac{N}{N + \lambda'},$$

where we have set $\lambda' = \lambda(2\alpha)^{-1}$. A simple calculation shows that

$$(28) \qquad Q_\lambda^n(I) = \frac{N}{N+\lambda'} \frac{N-1}{N-1+\lambda'} \cdots \frac{N-n+1}{N-n+1+\lambda'}.$$

Thus for any $u \in \mathcal{H}_0$,

$$\|Q_\lambda^n(I)u\|^2 \;=\; \sum_{m=0}^\infty \left(\frac{m}{m+\lambda'}, \cdots, \frac{m-n+1}{m-n+t+\lambda'} \right)^2 \|P_m u\|^2$$

$$\equiv\; \sum_{m=0}^\infty g_n(m)\|P_m u\|^2,$$

where P_m is the projection onto the m-particle subspace. It is clear that $g_n(m) = 0$ if $n \geq m+1$ and $|g_n(m)| \leq 1$ for all n, m. Therefore, by the dominated convergence theorem, it follows that $Q_\lambda^n(I)$ converges strongly to zero and as discussed above we have the unitarity of $U(t)$.

If we set $a(t) = U(t)^*(a \otimes I)U(t)$ and $a(t)^* = U(t)^*(a \otimes I)U(t)$ so that $[a(t), a(t)^*] = I_\mathcal{H}$ for all $t \geq 0$, in contrast to what we have discussed earlier. It is clear that the evolved q and p in the presence of noise retains this kind of extended conjugacy. To get back to the description of the harmonic oscillator, we have to take expectation or average out the noise degrees of freedom. This leads to the expectation semigroup T_t on the algebra generated by a and a^* :

$T_t(x) \equiv \mathbb{E}U(t)^* x U(t)$ with its infinitesmal generator, the Lindbladian, given formally as

$$\mathcal{L}(x) = \frac{\alpha}{2}[2a^* x a - \alpha a^* a x - x a^* a] - i\delta[a^* a, x].$$

Thus $\mathcal{L}(a) = (-\alpha + i\delta)a$ or equivalently $T_t(a) = e^{(-\alpha+i\delta)t}a$ just as we had observed earlier. By introducing the noise degree of freedom we have gained back the unitarity of the evolution $U(t)$ though not the group propoerty as in the case with $\alpha = 0$ (viz. $U(t) = \exp(i\delta t\, a^* a)$).

There have been other attempts at various applications, e.g. scattering theory between a class of Markov cocycles ([16]), measurement theory of observables with continuous spectra in quantum mechanics ([17]) and input - output channels in quantum systems ([18]). We also mention that a class of Hamiltonian theories have been studied in the quantum field theoretic set-up to show that in the limit when the coupling constant tends to zero, one can derive an equation similar to (5) if one chooses the scaled macroscopic (or collective) variables appropriately ([20]).

Bibliography

1. R.L. Hudson, K.R. Parthasarathy, Quantum Ito's formula and Stochastic evolutions, Comm. Math. Phys. **93**, 301-323 (1984).

2. K.R. Parthasarathy (1992) An Introduction to Quantum Stochastic Calculus, Birkhauser.

3. P.A. Meyer, Quantum Probability for Probabilists, LNM series No. 1538, Springer 1993.

4. A. Mohari, K.B. Sinha, Quantum stochastic flows with infinite degrees of freedom and countable state Markov chains, Sankhya A **52**, 43-57 (1990).

5. M.P. Evans, Existence of quantum diffusions, Prob. Th. Rel. Fields **81**, 473-483 (1989).

6. F. Fagnola, K.B. Sinha, Quantum flows with unbounded structure maps and finite degrees of freedom, J. London Math. Soc. (2) **48**, 537-551 (1993).

7. A. Mohari, K.R. Parthasarathy, A quantum probabilistic analogue of Feller's condition for the existence of unitary Markovian cocycles in Fock space, 475-497, Statistics and Probability : A Raghu Raj Bahadur Festschrift, Ed: J.K. Ghosh et al., Wiley Eastern 1993.

8. A. Mohari, K.B. Sinha, Stochastic dilation of minimal quantum dynamical semigroup, Proc. Ind. Acad. Sc. (Math. Sc.) **102 (3)**, 159-173 (1992).

9. J.L. Journé, structure des cocycles Markovian sur lépace de Fock, Prob. Th. Rel. Fields **75**, 291-316 (1987).

10. A. Mohari, Quantum stochastic differential equations with unbounded coefficients and dilation of Feller's minimal solution, Sankhya **A 53**, 255-287 (1991).

11. B.V.R. Bhat, K.B. Sinha, A stochastic differential equation with time-dependent unbounded operator coefficients, J. Func. Anal. **114**, 12-31 (1993).

12. B.V.R. Bhat, K.B. Sinha, Examples of unbounded generators leading to non conservative minimal semigroups, 89-104, Quantum Prob. Rel. Topics, Vol. 9, World Scientific 1994.

13. K.B. Sinha, Quantum Dynamical Semigroups, Operator theory — Advances and Applications Vol. **70**, 161-169, Birkhauser 1994.

14. J.M. Lindsay, K.B. Sinha, Feyman-kac representation of some non-commutative elliptic operators, to appear in J. Func. Anal.

15. F. Fagnola, On quantum stochastic differential equatioons with unbounded coefficients, Prob. th. Rel. Fields **86**, 501-517 (1990).

16. F. Fagnola, K.B. Sinha, scattering theory for unitary cocycles, stochastic processes : A festschrift in honour of G. Kallianpur, ed: S. Cambanis et al, 81-88, Springer 1993.

17. K.B. Sinha, On the collapse postulate of Quantum mechanics, Mathematical Physics towards the 21st century, ed: R.N. Sen et al, 344-350, Ben Gurion Univ., Israel 1994.

18. A. Barchielli, Input and output channels in quantum systems and quantum stochastic differential equations, LNM 1303, 37-51, Springer 1988.

19. K.R. parthasarathy, K.B. Sinha, Markov chains as Evans-Hudson diffusions in Fock space, Sem. Prob. XXIV, LNM 1426, 363-369., Springer 1988/89.

20. L. Accardi, A. Frigerio, Y.G.Lu, The weak coupling limit as a quantum functional central limit, Comm. Math. Phys. **131**, 537-570 (1990).

COUPLING

HERMANN THORISSON,[*] *University of Iceland*

Abstract

This paper discusses coupling ideas with focus on equivalences for exact coupling, shift-coupling and ε-couplings of stochastic processes and the generalizations to random fields and topological transformation groups. Applications in regeneration, Markov theory, Palm theory, ergodic theory, exchangeability and self-similarity are indicated and a set of general coupling references provided.

COUPLING; SHIFT-COUPLING; ε-COUPLINGS; TRANSFORMATION COUPLING; TOTAL VARIATION; TAIL σ-ALGEBRA; INVARIANT σ-ALGEBRA; SMOOTH TAIL σ-ALGEBRA; RANDOM FIELD; TOPOLOGICAL TRANSFORMATION GROUP; REGENERATION; PALM THEORY; EXCHANGEABILITY; SELF-SIMILARITY

AMS 1991 SUBJECT CLASSIFICATION: PRIMARY 60G99; 60G60; 60B99

1. Introduction

Coupling means the joint construction of two or more random elements. The aim is usually to establish some distributional relation between the individual elements, or the reverse: to turn a distributional relation into a pointwise relation. Examples are the turning of stochastic domination into pointwise domination, weak convergence into pointwise convergence, and liminf convergence of densities to a density into pointwise convergence where the random elements actually hit the limit. This deepens our understanding of the distributional relation itself, may enable us to establish previously hard-to-prove facts by simple pointwise arguments, and often leads to unexpected new results.

The paper starts [Section 2] with the above mentioned examples and then moves to stochastic processes [Sections 3 – 5]. After sketching an historical example (the classical coupling) we outline how three kinds of coupling (exact coupling, shift-coupling, and ε-couplings) can be linked to three kinds of asymptotic behavior (plain, time-average, and smooth total variation convergence) and to three σ-algebras (the tail, the invariant, and a smooth tail σ-algebra). In all three cases certain coupling inequalities play an essential role. Applications to stationary processes, Markov processes, regenerative processes and in Palm theory are indicated.

---

[*] Postal address: Science Institute, University of Iceland, Dunahaga 3, 107 Reykjavik, Iceland, thoris@rhi.hi.is.

The view is then [Section 6] extended to random fields with applications in Palm Theory, and finally [Section 7] to random elements under a topological transformation group which opens up many new possibilities for applications: self-similarity, exchangeability, . . .

The paper concludes [Section 8] with comments on further reading and a working hypothesis.

2. Definition and three examples

For each i in an index set $\mathbb{I}$ let Y_i be a random element in some measurable space $(H_i, \mathcal{H}_i)$. A *coupling* of $Y_i, i \in I$, is a family $(\hat{Y}_i : i \in I)$ of random elements defined on a common probability space and satisfying

$$\hat{Y}_i \overset{D}{=} Y_i, \quad i \in I. \qquad (\overset{D}{=} \text{ denotes identity in distribution})$$

Note that only the individual $\hat{Y}_i$ are copies of the individual Y_i while the whole family $(\hat{Y}_i : i \in I)$ is typically not a copy of the family $(Y_i : i \in I)$. In other words, the joint distribution of the $\hat{Y}_i$ need not be the same as that of the Y_i. In fact, the Y_i need not even have a joint distribution.

Thus a coupling has fixed marginal distributions (the distributions the individual Y_i) and the trick is to find a dependence structure (joint distribution) that fits ones purposes.

In this paper the Y_i will all be elements in the same space $(H, \mathcal{H}) = (H_i, \mathcal{H}_i)$ independently of $i \in I$. We also assume that all random elements discussed below are defined on a common probability space $(\Omega, \mathcal{F}, \mathbf{P})$; this is no restriction since given any collection of probability spaces we can always form the product space. In particular, this means that there is always at least one coupling, the *independence coupling* consisting of independent copies of the Y_i.

2.1. *First example: stochastic domination - pointwise domination*

Consider a random variable X with distribution function F. Let F^{-1} be the generalized inverse of F, that is, with R the real numbers

$$F^{-1}(u) = \inf\{x \in R : F(x) \geq u\}, \quad 0 < u < 1.$$

Let U be uniformly distributed on $(0, 1)$. Then the random variable $\hat{X} = F^{-1}(U)$ is a copy of X and thus letting F run over the class of all distribution functions yields a coupling of all differently distributed random variables, the *inverse-uniform coupling*.

Let X and X' be two random variables with distribution functions F and G, respectively. If there is a coupling $(\hat{X}, \hat{X}')$ of X and X' such that $\hat{X} \leq \hat{X}'$ then clearly

$$F(x) \geq G(x), \quad x \in R,$$

that is, X is *stochastically dominated* by X', denoted $X \overset{D}{\leq} X'$. Conversely, if $X \overset{D}{\leq} X'$ then $F^{-1}(u) \leq G^{-1}(u)$ which yields $\hat{X} \leq \hat{X}'$ where $(\hat{X}, \hat{X}')$ is the inverse-uniform coupling of X and X'.

This equivalence result can be extended to random elements Y and Y' in a partially ordered Polish space $(E, \mathcal{E})$. In that case $Y \overset{D}{\leq} Y'$ means that $g(Y) \overset{D}{\leq} g(Y')$ for all real valued increasing measurable functions g defined on E (cf. Lindvall [26], Chapter IV.1):

THEOREM 2.1. *Let Y and Y' be random elements in a partially ordered Polish space. Then*

$$Y \overset{D}{\leq} Y'$$

if and only if there is a coupling $(\hat{Y}, \hat{Y}')$ of Y and Y' such that

$$\hat{Y} \leq \hat{Y}'.$$

2.2. Second example: convergence in distribution - pointwise convergence

Let $X_1, \ldots, X_\infty$ be random variables with distribution functions $F_1, \ldots, F_\infty$. If there exists a coupling $(\hat{X}_1, \ldots, \hat{X}_\infty)$ of $X_1, \ldots, X_\infty$ such that $\hat{X}_n$ tends to $\hat{X}_\infty$ pointwise,

$$\hat{X}_n \to \hat{X}_\infty, \quad n \to \infty, \qquad \text{(short for } \hat{X}_n(\omega) \to \hat{X}_\infty(\omega) \text{ for each outcome } \omega)$$

then it is readily checked that

$$F_n(x) \to F_\infty(x), \quad n \to \infty, \qquad \text{for all } x \text{ at which } F_\infty \text{ is continuous,}$$

that is, X_n tends to X_∞ *in distribution*, denoted $X_n \overset{D}{\to} X_\infty$ as $n \to \infty$. Conversely, if $X_n \overset{D}{\to} X_\infty$ as $n \to \infty$ then it is not hard to obtain

$$F_n^{-1}(u) \to F_\infty^{-1}(u), \quad n \to \infty, \qquad \text{for all } u \text{ at which } F_\infty^{-1} \text{ is continuous.}$$

Taking $(\hat{X}_1, \ldots, \hat{X}_\infty)$ the inverse-uniform coupling of $X_1, \ldots, X_\infty$ [based on a uniform variable U not taking the countably many values u at which F_∞^{-1} is discontinuous] yields $\hat{X}_n \to \hat{X}_\infty$.

This equivalence can be extended to random elements $Y_1, \ldots, Y_\infty$ in a Polish space $(E, \mathcal{E})$. In that case $Y_n \overset{D}{\to} Y_\infty$ as $n \to \infty$ means that $\mathbf{E}[g(Y_n)] \to \mathbf{E}[g(Y_\infty)]$ as $n \to \infty$ for bounded continuous functions g (this result is due to Skorohod, for proof see eg. Billingsley [6]):

THEOREM 2.2. *If $Y_1, \ldots, Y_\infty$ are random elements in a Polish space then*

$$Y_n \overset{D}{\to} Y_\infty, \quad n \to \infty,$$

if and only if there exists a coupling $(\hat{Y}_1, \ldots, \hat{Y}_\infty)$ of $Y_1, \ldots, Y_\infty$ such that

$$\hat{Y}_n \to \hat{Y}_\infty, \quad n \to \infty.$$

2.3. *Third example: convergence of densities - hitting the limit*

Consider continuous random variables $X_1, \ldots, X_\infty$ with densities $f_1, \ldots, f_\infty$ and suppose

$$f_n(x) \to f_\infty(x), \quad n \to \infty, \quad x \in R.$$

Let $\hat{X}_\infty$ have density f_∞ and K be a nonnegative finite integer valued random variable which is independent of $\hat{X}_\infty$ and such that

$$\mathbf{P}(K \le n) = \int \inf_{n \le k < \infty} f_k(x)\, dx \qquad (\text{increases to } \int f_\infty(x)\, dx = 1 \text{ as } n \to \infty).$$

Put $\hat{X}_n = \hat{X}_\infty$ when $n \ge K$ and conditionally on the event $\{K > n\}$ let $\hat{X}_n$ have density $(f_n - \inf_{n \le k < \infty} f_n(x))/\mathbf{P}(K > n)$. This yields a coupling such that the $\hat{X}_n$ not only close in on the limit $\hat{X}_\infty$ as in Theorem 2.2 but actually hit it and stay there.

This result cannot be reversed, $f_n \to f_\infty$ as $n \to \infty$ is too strong. Note however that the finiteness of K relies only on the weaker condition $\inf_{n \le k < \infty} f_n \to f_\infty$ as $n \to \infty$ and this result can be reversed (see Thorisson [39]). In fact this equivalence result holds without any restriction on the space where the random elements take values and the proof is the same as for continuous random variables:

THEOREM 2.3. *Let $Y_1, \ldots, Y_\infty$ be random elements in arbitrary space and $f_1, f_2 \ldots$ the densities of $Y_1, Y_2, \ldots$ with respect to some measure (for instance with respect to $\mu = \mu_1/2^1 + \mu_2/2^2 + \ldots$ where μ_n is the distribution of Y_n). Then*

$$\liminf_{n \to \infty} f_n \text{ is a density of } Y_\infty$$

if and only if there exists a coupling $(\hat{Y}_1, \ldots, \hat{Y}_\infty)$ and a finite random integer K such that

$$\hat{Y}_n = \hat{Y}_\infty, \quad n \ge K. \qquad (\textit{pointwise convergence in the discrete topology})$$

We finally note that $\liminf_{n \to \infty} f_n$ is a density of Y_∞ only if (but *not* necessarily *if*)

$$Y_n \overset{tv}{\to} Y_\infty, \quad n \to \infty,$$

where $\overset{tv}{\to}$ denotes *convergence in total variation norm*, that is, with μ_n the distribution of Y_n:

$$(2.1) \qquad \|\mu_n - \mu_\infty\| := 2 \sup_{A \in \mathcal{H}} |\mu_n(A) - \mu_\infty(A)| \to 0, \quad n \to \infty.$$

3. Exact coupling

In 1938 Doeblin [12] proved the asymptotic stationarity of a regular finite state Markov chain $X = (X_k)_0^\infty$ along the following lines. Let a differently started version X' run independently of X until the two chains meet, at a time T say. From T onward let X and X' run together. Regularity means that there is an m and an

$\varepsilon > 0$ such that $\mathbf{P}_i(X_m = j) \geq \varepsilon$ for all initial states i and all j. This implies that $\mathbf{P}(T > km) \leq (1 - \varepsilon)^k \to 0$ as $k \to \infty$. Thus $\mathbf{P}(T < \infty) = 1$, that is, the chains eventually merge, and we obtain

$$|\mathbf{P}(X_n = j) - \mathbf{P}(X'_n = j)| \leq \mathbf{P}(X_n \neq X'_n) \to 0, \quad n \to \infty.$$

Add to this limit result the observation that $\max_i \mathbf{P}_i(X_n = j)$ is non-increasing and $\min_i \mathbf{P}_i(X_n = j)$ is non-decreasing in n to deduce that $\mathbf{P}(X_n = j)$ has a limit. (Nowadays one usually takes X' stationary which makes the last sentence unnecessary.)

In this section we outline the general theory of this kind of coupling and, in fact, the rest of the paper is concerned with modifications and extensions of that theory.

3.1. *Notation*

For the next three sections let $Z = (Z_s)_{s \in [0,\infty)}$ and $Z' = (Z'_s)_{s \in [0,\infty)}$ be stochastic processes on a Polish state space $(E, \mathcal{E})$ with right-continuous paths. We shall regard Z and Z' as random elements in $(H, \mathcal{H})$ where H is the set of all right-continuous functions $z = (z_s)_{s \in [0,\infty)}$ from $[0, \infty)$ to E and $\mathcal{H}$ is the smallest σ-algebra making all the projection mappings $z \mapsto z_t, t \in [0, \infty), \mathcal{H}/\mathcal{E}$ measurable. Define the shift-maps $\theta_t, t \in [0, \infty)$, from H to H by

$$\theta_t z = (z_{t+s})_{s \in [0,\infty)}.$$

For a measure μ on $\mathcal{H}$ and a sub-σ-algebra $\mathcal{A}$ of $\mathcal{H}$ let $\mu_{\mathcal{A}}$ denote the restriction of μ to $\mathcal{A}$. Let $\| \cdot \|$ denote total variation norm, see (2.1).

3.2. *Inequalities*

Let $(\hat{Z}, \hat{Z}')$ be a coupling of Z and Z'. An event C is a *coupling event* if $\hat{Z} = \hat{Z}'$ on C. Clearly

$$\mathbf{P}(Z \in \cdot) - \mathbf{P}(Z' \in \cdot) = \mathbf{P}(\hat{Z} \in \cdot; C^c) - \mathbf{P}(\hat{Z}' \in \cdot; C^c)$$

which yields the following simple but basic inequality.

COUPLING EVENT INEQUALITY. *If C is a coupling event then*

$$\|\mathbf{P}(Z \in \cdot) - \mathbf{P}(Z' \in \cdot)\| \leq 2\mathbf{P}(C^c).$$

A random time T is called a *coupling time* (or coupling epoch) if

$$\theta_T \hat{Z} = \theta_T \hat{Z}' \qquad \text{on } \{T < \infty\}$$

and the triple $(\hat{Z}, \hat{Z}', T)$ is an *exact coupling* of Z and Z' (because $\hat{Z}$ and $\hat{Z}'$ coincide exactly from T onward). Clearly $\{T \leq t\}$ is a coupling event of the coupling $(\theta_t \hat{Z}, \theta_t \hat{Z}')$ of $\theta_t Z$ and $\theta_t Z'$ and thus we obtain the following inequality.

COUPLING TIME INEQUALITY. *If T is a coupling time then*

$$\|\mathbf{P}(\theta_t Z \in \cdot) - \mathbf{P}(\theta_t Z' \in \cdot)\| \leq 2\mathbf{P}(T > t), \quad t \in [0, \infty).$$

3.3. *Asymptotics*

The coupling time inequality is of basic importance for total variation asymptotics. For instance, if T is a.s. finite then

$$(3.1) \qquad \|\mathbf{P}(\theta_t Z \in \cdot) - \mathbf{P}(\theta_t Z' \in \cdot)\| \to 0, \quad t \to \infty,$$

and if, for an $\alpha \in [0, \infty)$, it holds that $\mathbf{E}[T^\alpha] < \infty$ then (since $t^\alpha \mathbf{P}(T > t) \leq \mathbf{E}[T^\alpha; T > t]$) we obtain the rate result

$$(3.2) \qquad t^\alpha \|\mathbf{P}(\theta_t Z \in \cdot) - \mathbf{P}(\theta_t Z' \in \cdot)\| \to 0, \quad t \to \infty,$$

here t^α can be replaced by any increasing function $\phi(t)$ provided $\mathbf{E}[\phi(T)] < \infty$.

If Z' is stationary, that is,

$$\theta_t Z' \overset{D}{=} Z', \quad t \geq 0,$$

then (3.1) can be rewritten as

$$(3.3) \qquad \theta_t Z \overset{tv}{\to} Z', \quad t \to \infty.$$

3.4. *Application*

Let Z be *regenerative in the wide sense*. This means that there are random times

$$0 \leq S_0 < S_1 < \ldots \to \infty$$

splitting Z into an initial delay of length S_0 and a sequence of cycles of length $X_n = S_n - S_{n-1}$ such that the distribution of $(\theta_{S_n} Z, X_{n+1}, X_{n+2}, \ldots)$ does not depend on n and such that $(\theta_{S_n} Z, X_{n+1}, X_{n+2}, \ldots)$ is independent of $S_0, \ldots, S_n$. This is the type of regeneration occuring in Harris chains. [The process Z is regenerative *in the classical sense* if the cycles are i.i.d. and independent of the initial delay, or equivalently, if the distribution of $(\theta_{S_n} Z, X_{n+1}, X_{n+2}, \ldots)$ does not depend on n and $(\theta_{S_n} Z, X_{n+1}, X_{n+2}, \ldots)$ is independent of $(Z_s : s < S_n)$ and $S_0, \ldots, S_n$.]

Let Z' be a *version* of Z, that is, regenerative in the wide sense with the same cycle-sequence distribution as Z. If $\mathbf{E}[X_1] < \infty$ then ([26], [38]) a stationary version of Z is obtained by taking

$$(3.4) \qquad Z' = \theta_{UX_1''} Z'' \quad \text{and} \quad S_n' = S_{n+1}'' - U X_1''$$

where (Z'', S'') has distribution defined by

$$(3.5) \qquad \mathbf{E}[f(Z'', S'')] = \frac{\mathbf{E}[f(Z, S) X_1 | S_0 = 0]}{\mathbf{E}[X_1]} \quad \text{for bounded measurable } f$$

and U is uniformly distributed on $[0, 1]$ and independent of (Z'', S'').

Proofs of the following coupling claims can be found in Lindval [26] in the classical regenerative case and in Thorisson [35] in the wide sense case.

If X_1 is spread-out (that is, for some n the distribution of $X_1 + \ldots + X_n$ has an absolutely continuous component), then there exists an exact coupling of Z and Z'

with a finite T. Thus the limit result (3.1) holds. If also $\mathbf{E}[X_1] < \infty$ then (3.3) holds with Z' defined by (3.4).

If X_1 is spread-out, $\mathbf{E}[X_1] < \infty$ and for an $\alpha \in [0, \infty)$ it holds that $\mathbf{E}[S_0^\alpha], \mathbf{E}[S_0'^\alpha]$ and $\mathbf{E}[X_1^\alpha]$ are all finite, then so is $\mathbf{E}[T^\alpha]$. Thus the rate result (3.2) holds. More general moment results (and thus rate results) can be found in [26] and [35].

3.5. *The tail σ-algebra - Maximality*

Denote the post-t σ-algebra $\mathcal{T}_t = \theta_t^{-1}\mathcal{H}$ and put

$$\mathcal{T} = \bigcap_{t \in [0,\infty)} \mathcal{T}_t = \text{ the tail-}\sigma\text{-algebra.}$$

For any coupling time T

$$(3.6) \qquad \begin{aligned} \|\mathbf{P}(Z \in \cdot)_{\mathcal{T}} - \mathbf{P}(Z' \in \cdot)_{\mathcal{T}}\| &\leq \|\mathbf{P}(Z \in \cdot)_{\mathcal{T}_t} - \mathbf{P}(Z' \in \cdot)_{\mathcal{T}_t}\| \\ &= \|\mathbf{P}(\theta_t Z \in \cdot) - \mathbf{P}(\theta_t Z' \in \cdot)\| \leq 2\mathbf{P}(T > t). \end{aligned}$$

Sending t to infinity yields a $\mathcal{T}$-coupling inequality: *if there exists an exact coupling with time T then*

$$\|\mathbf{P}(Z \in \cdot)_{\mathcal{T}} - \mathbf{P}(Z' \in \cdot)_{\mathcal{T}}\| \leq 2\mathbf{P}(T = \infty).$$

Moreover, identity can be obtained in this inequality (maximality with respect to $\mathcal{T}$):

THEOREM 3.1. *There exists an exact coupling of Z and Z' with time T such that*

$$\|\mathbf{P}(Z \in \cdot)_{\mathcal{T}} - \mathbf{P}(Z' \in \cdot)_{\mathcal{T}}\| = 2\mathbf{P}(T = \infty).$$

This may be proved by applying the analogous result in discrete time (see Proposition 11 in Aldous and Thorisson [2]) to the discrete time processes $(\theta_n Z)_{n \geq 0}$ and $(\theta_n Z')_{n \geq 0}$. The discrete time result relies in turn on the existence of a *maximal exact coupling* of discrete time processes (on a Polish state space): a coupling such that the coupling time inequality is an identity at all times n. This result dates back to Griffeath [20] and Goldstein [17]. In [34] Sverchkov and Smirnov establish the existence of a maximal exact coupling for continuous time D-valued processes (processes on a Polish state space with right-continuous paths having left-hand limits).

3.6. *A limit result - Equivalences*

Taking T as in Theorem 3.1 and sending t to infinity in (3.6) yields the following limit result.

THEOREM 3.2. *It holds that*

$$\|\mathbf{P}(\theta_t Z \in \cdot) - \mathbf{P}(\theta_t Z' \in \cdot)\| \to \|\mathbf{P}(Z \in \cdot)_{\mathcal{T}} - \mathbf{P}(Z' \in \cdot)_{\mathcal{T}}\|, \quad t \to \infty.$$

We can now tie together exact coupling, total variation asymptotics and $\mathcal{T}$ as follows.

THEOREM 3.3. *The following statements are equivalent:*

(a) *there exists an exact coupling of Z and Z' with a finite time;*

(b) $\|\mathbf{P}(\theta_t Z \in \cdot) - \mathbf{P}(\theta_t Z' \in \cdot)\| \to 0$ *as* $t \to \infty$;

(c) $\mathbf{P}(Z \in \cdot)_{\mathcal{T}} = \mathbf{P}(Z' \in \cdot)_{\mathcal{T}}$.

Proof. By the coupling time inequality, (a) implies (b). By Theorem 3.2, (b) implies (c). By Theorem 3.1, (c) implies (a).

For Markov processes we can add tail-triviality to these equivalences.

THEOREM 3.4. *Let Z and Z' be differently started versions of a time-homogeneous Markov process with initial distributions μ and μ'. Then (c) holds for all μ and μ' if and only if*

(d) $\mathbf{P}(Z \in \cdot)_{\mathcal{T}} = 0$ *or* 1 *for all initial distributions* μ.

Proof. Suppose (d) holds. Then $\mathbf{P}(Z \in \cdot)_{\mathcal{T}}$ and $\mathbf{P}(Z' \in \cdot)_{\mathcal{T}}$ are 0-1-measures and so is any mixture of them. Thus (c) holds. Conversely, suppose (c) holds for all μ and μ' and let A be in $\mathcal{T}$. This, together with the Markov property and time-homogeneity, yields the second identity in

$$\mathbf{P}(Z \in A) = \mathbf{P}(\theta_t Z \in \theta_t A) = \mathbf{P}(\theta_t Z \in \theta_t A | Z_s, 0 \le s \le t)$$
$$= \mathbf{P}(Z \in A | Z_s, 0 \le s \le t) \to 1\{Z \in A\} \text{ a.s.,} \quad t \to \infty.$$

Thus $\mathbf{P}(Z \in A)$ must be either 0 or 1, that is, (d) holds.

4. Shift-coupling

A *shift-coupling* of Z and Z' is a coupling $(\hat{Z}, \hat{Z}')$ and two random times T and T' such that

$$\theta_T \hat{Z} = \theta_{T'} \hat{Z}' \text{ on } \{T < \infty\} \quad \text{and} \quad \{T < \infty\} = \{T' < \infty\}.$$

The times T and T' are the *shift-coupling times*. When $T < \infty$ then $T - T'$ is the *shift*. There is no shift if $T \equiv T'$ and then the shift-coupling is an exact coupling.

It turns out that shift-coupling has a theory that parallels that of exact coupling. The concept was introduced by Berbee [5] and further developed by Greven [19], Aldous and Thorisson [2] and Thorisson [37]. For proofs of the coupling claims in this section, see Thorisson [37].

4.1. *Inequality - Asymptotics*

Rather than shifting Z to a non-random t we now shift to a point picked at random in $[0, t]$.

SHIFT-COUPLING INEQUALITY. *If T and T' are shift-coupling times then*

$$\|\mathbf{P}(\theta_{Ut} Z \in \cdot) - \mathbf{P}(\theta_{Ut} Z' \in \cdot)\| \le 2\mathbf{P}(T \vee T' > Ut), \quad t \in [0, \infty),$$

where U is uniformly distributed on $[0, 1]$ and independent of Z, Z', T and T'.

This shifting-at-random formulation can be rewritten in the following time-average form:

$$\|\frac{1}{t}\int_0^t \mathbf{P}(\theta_s Z \in \cdot)\,ds - \frac{1}{t}\int_0^t \mathbf{P}(\theta_s Z' \in \cdot)\,ds\| \leq 2\mathbf{E}\left[\frac{T \vee T'}{t} \wedge 1\right].$$

In the same way as the coupling time inequality is basic for *plain* total variation asymptotics the shift-coupling inequality is basic for *time-average* (or *Cesaro*) total variation asymptotics. For instance, if T is a.s. finite then

$$(4.1) \qquad \|\mathbf{P}(\theta_{Ut} Z \in \cdot) - \mathbf{P}(\theta_{Ut} Z' \in \cdot)\| \to 0, \quad t \to \infty,$$

and if, for an $\alpha \in [0,1)$, both $\mathbf{E}[T^\alpha]$ and $\mathbf{E}[T'^\alpha]$ are finite then

$$(4.2) \qquad t^\alpha \|\mathbf{P}(\theta_{Ut} Z \in \cdot) - \mathbf{P}(\theta_{Ut} Z' \in \cdot)\| \to 0, \quad t \to \infty,$$

since $\mathbf{E}[U^{-\alpha}] < \infty$ for $\alpha < 1$ and thus

$$\mathbf{E}\left[\left(\frac{T \vee T'}{U}\right)^\alpha\right] = \mathbf{E}[U^{-\alpha}]\mathbf{E}[(T \vee T')^\alpha] \leq \mathbf{E}[U^{-\alpha}]\mathbf{E}[T^\alpha + T'^\alpha] < \infty.$$

From the inequality we cannot obtain rates of order $\alpha = 1$ or higher. What can be deduced is that

$$\sup_{t \in [0,\infty)} t\|\mathbf{P}(\theta_{Ut} Z \in \cdot) - \mathbf{P}(\theta_{Ut} Z' \in \cdot)\| \leq 2\mathbf{E}[T \vee T'] \leq 2\mathbf{E}[T] + 2\mathbf{E}[T'].$$

If Z' is stationary then $\theta_{Ut} Z'$ has the same distribution as Z' and (4.1) can be rewritten as

$$(4.3) \qquad \theta_{Ut} Z \overset{tv}{\to} Z', \quad t \to \infty.$$

4.2. *The invariant σ-algebra - Maximality*

The invariant σ-algebra is defined as follows

$$\mathcal{I} = \{B \in \mathcal{H} : \theta_t^{-1} B = B, t \in [0,\infty)\}.$$

For any shift-coupling times T and T'

$$(4.4) \qquad \begin{aligned} \|\mathbf{P}(Z \in \cdot)_\mathcal{I} - \mathbf{P}(Z' \in \cdot)_\mathcal{I}\| &= \|\mathbf{P}(\theta_{Ut} Z \in \cdot)_\mathcal{I} - \mathbf{P}(\theta_{Ut} Z' \in \cdot)_\mathcal{I}\| \\ &\leq \|\mathbf{P}(\theta_{Ut} Z \in \cdot) - \mathbf{P}(\theta_{Ut} Z' \in \cdot)\| \leq 2\mathbf{P}(T \vee T' > Ut). \end{aligned}$$

Sending $t \to \infty$ yields an $\mathcal{I}$-coupling inequality: *if there is a shift-coupling with times T and T' then*

$$\|\mathbf{P}(Z \in \cdot)_\mathcal{I} - \mathbf{P}(Z' \in \cdot)_\mathcal{I}\| \leq 2\mathbf{P}(T = \infty).$$

Again, identity can be obtained in the inequality (maximality with respect to $\mathcal{I}$):

THEOREM 4.1. *There exists a shift-coupling of Z and Z' with times T and T' such that*

$$\|\mathbf{P}(Z \in \cdot)_\mathcal{I} - \mathbf{P}(Z' \in \cdot)_\mathcal{I}\| = 2\mathbf{P}(T = \infty).$$

There cannot in general exist a shift-coupling such that the shift-coupling inequality is an identity for all t since the right-hand side is nonincreasing while the left-hand side need not be. Thus there is no immediate shift-coupling analog of maximal exact coupling; see [2] for further discussion of this problem.

4.3. *A limit result - Equivalences*

Taking T and T' as in Theorem 4.1 and sending t to infinity in (4.4) yields the following.

THEOREM 4.2. *With U uniform on $[0,1]$ and independent of Z and Z' it holds that*

$$\|\mathbf{P}(\theta_{Ut}Z \in \cdot) - \mathbf{P}(\theta_{Ut}Z' \in \cdot)\| \to \|\mathbf{P}(Z \in \cdot)_{\mathcal{I}} - \mathbf{P}(Z' \in \cdot)_{\mathcal{I}}\|, \quad t \to \infty.$$

COROLLARY. *Suppose Z and Z' are both stationary. Then Z and Z' have the same distribution if and only if $\mathbf{P}(Z \in \cdot)_{\mathcal{I}} = \mathbf{P}(Z' \in \cdot)_{\mathcal{I}}$.*

We can now tie together shift-coupling, time-average total variation asymptotics and $\mathcal{I}$ as follows.

THEOREM 4.3. *The following statements are equivalent:*

  (a) there exists a shift-coupling of Z and Z' with finite times;

  (b) with U uniform on $[0,1]$ and independent of Z and Z' it holds that

$$\|\mathbf{P}(\theta_{Ut}Z \in \cdot) - \mathbf{P}(\theta_{Ut}Z' \in \cdot)\| \to 0, \quad t \to \infty;$$

  (c) $\mathbf{P}(Z \in \cdot)_{\mathcal{I}} = \mathbf{P}(Z' \in \cdot)_{\mathcal{I}}$.

The proof is analogous to that of Theorem 3.3. For Markov processes we can add $\mathcal{I}$-triviality to these equivalences, the proof is analogous to that of Theorem 3.4:

THEOREM 4.4. *Let Z and Z' be differently started versions of a time-homogeneous Markov process with initial distributions μ and μ'. Then (c) holds for all μ and μ' if and only if*

  (d) $\mathbf{P}(Z \in \cdot)_{\mathcal{I}} = 0$ *or 1 for all initial distributions μ.*

4.4. *Application*

Suppose there are random times $0 = S_0 < S_1 < \ldots \to \infty$ splitting Z into a stationary sequence of cycles (see Section 3.4). Let C be the sequence of cycles and $\mathcal{J}$ the invariant σ-algebra associated with the state space of C and assume that

$$M = \mathbf{E}[X_1|C^{-1}\mathcal{J}] < \infty \quad \text{a.s.}$$

In Thorisson [38] it is shown that with (Z'', S'') having distribution defined by

$$(4.5) \qquad \mathbf{E}[f(Z'', S'')] = \mathbf{E}[f(Z, S)X_1 M^{-1}] \quad \text{for bounded measurable } f$$

and U uniform on $[0,1]$ and independent of (Z'', S'') it holds that $\theta_{UX_1''}Z''$ is stationary. Since $C^{-1}\mathcal{J}$ contains $Z^{-1}\mathcal{I}$ it follows from (4.5) that $\mathbf{P}(Z \in \cdot)_{\mathcal{I}} = \mathbf{P}(Z'' \in \cdot)_{\mathcal{I}}$. Now $\mathbf{P}(\theta_{UX_1''}Z'' \in \cdot)_{\mathcal{I}} = \mathbf{P}(Z'' \in \cdot)_{\mathcal{I}}$ and thus there is a shift-coupling of Z and $\theta_{UX_1''}Z''$ with finite times and

$$\theta_{Ut}Z \overset{tv}{\to} \theta_{UX_1''}Z'', \quad t \to \infty.$$

5. ε-coupling

An ε-coupling ($\varepsilon > 0$) of Z and Z' is a shift-coupling $(\hat{Z}, \hat{Z}', T, T')$ such that

$$|T - T'| \le \varepsilon \quad \text{on } \{T < \infty\}.$$

This concept was used by Lindvall [25] to prove Blackwell's renewal theorem. Again it turns out there is a theory paralleling those of exact coupling and shift-coupling. The coupling results in this section are established in Thorisson [37].

5.1. *Inequality - Asymptotics*

Rather than shifting to a point picked at random in $[0, t]$ as in the shift-coupling case the appropriate thing to do here is to shift to a t as in the exact coupling case and then blur t slightly (we can also think of the time origin of the processes as blurred slightly).

ε-COUPLING INEQUALITY. *If T and T' are ε-coupling times then for all $h > 0$*

$$\|\mathbf{P}(\theta_{t+Uh}Z \in \cdot) - \mathbf{P}(\theta_{t+Uh}Z' \in \cdot)\| \le 2\mathbf{P}(T > t) + 2\varepsilon/h, \quad t \in [0, \infty),$$

where U is uniformly distributed on $[0, 1]$ and independent of Z and Z'.

The inequality can be rewritten in the following *smooth* form:

$$\left\| \frac{1}{h} \int_t^{t+h} \mathbf{P}(\theta_s Z \in \cdot)\, ds - \frac{1}{h} \int_t^{t+h} \mathbf{P}(\theta_s Z' \in \cdot)\, ds \right\| \le 2\mathbf{P}(T > t) + 2\varepsilon/h.$$

In this case the inequality is basic for *smooth* total variation asymptotics: if for an $\varepsilon > 0$ there is an ε-coupling with finite times then

$$\forall h > 0 : \limsup_{t \to \infty} \|\mathbf{P}(\theta_{t+Uh}Z \in \cdot) - \mathbf{P}(\theta_{t+Uh}Z' \in \cdot)\| \le 2\varepsilon/h,$$

and if for *each* $\varepsilon > 0$ there is an ε-coupling with finite times then sending ε to 0 yields

$$(5.1) \qquad \forall h > 0 : \|\mathbf{P}(\theta_{t+Uh}Z \in \cdot) - \mathbf{P}(\theta_{t+Uh}Z' \in \cdot)\| \to 0, \quad t \to \infty.$$

If Z' is stationary then $\theta_{t+Uh}Z'$ has the same distribution as Z' and (5.1) can be rewritten as

$$(5.2) \qquad \forall h > 0 : \theta_{t+Uh}Z \overset{tv}{\to} Z', \quad t \to \infty.$$

Rate results are harder to obtain so there is no immediate analog of (3.2) and (4.2).

5.2. *Application*

Let Z be regenerative in the wide sense and Z' be a version of Z (see Section 3.4). Suppose the cycle-length X_1 has a distribution that is not supported by any lattice. Then for each $\varepsilon > 0$ there exists an ε-coupling of Z and Z' with finite times: for a proof in the classical regenerative case see Lindvall [26] and extension to the wide sense case is not too hard. Thus (5.1) holds. If, in addition, X_1 has finite mean then (5.2) holds with Z' the stationary version defined at (3.4).

5.3. *The smooth tail σ-algebra - Maximality*

It turns out that there is a σ-algebra playing the same role for ε-couplings ($\varepsilon > 0$) as the tail for exact coupling and the invariant for shift-coupling. Define a class of tail functions by

$$\mathcal{S}^\circ = \{ f \in \mathcal{T} : f(\theta_t z) \to f(z) \text{ as } t \to 0, z \in H \}$$

and define the *smooth tail σ-algebra* by $\mathcal{S} = \sigma\{\mathcal{S}^\circ\}$. Note that $\mathcal{I} \subseteq \mathcal{S} \subseteq \mathcal{T}$ and that $\mathcal{S}$ contains smoothed tail functions:

$$f \in \mathcal{T} \text{ bounded and } h > 0 \quad \Rightarrow \quad \int_0^h f(\theta_s \, \cdot) \, ds \in \mathcal{S}.$$

The analog of the $\mathcal{T}$- and $\mathcal{I}$-coupling inequalities is the following $\mathcal{S}$-coupling inequality: *if for each $\varepsilon > 0$ there is an ε-coupling of Z and Z' with times T_ε and T'_ε then*

$$\|\mathbf{P}(Z \in \cdot)_{\mathcal{S}} - \mathbf{P}(Z' \in \cdot)_{\mathcal{S}}\| \leq 2 \liminf_{\varepsilon \downarrow 0} \mathbf{P}(T_\varepsilon = \infty).$$

Once more identity can be obtained in the inequality (maximality with respect to $\mathcal{S}$):

THEOREM 5.1. *For each $\varepsilon > 0$ there is an ε-coupling of Z and Z' with times T_ε and T'_ε such that*

$$\|\mathbf{P}(Z \in \cdot)_{\mathcal{S}} - \mathbf{P}(Z' \in \cdot)_{\mathcal{S}}\| = 2 \sup_{\varepsilon > 0} \mathbf{P}(T_\varepsilon = \infty).$$

No comment on obtaining identity in the ε-coupling inequality itself is offered.

5.4. *A limit result - Equivalences*

The following limit result is quite different from Theorems 3.2 and 4.2.

THEOREM 5.2. *With U uniform on $[0,1]$ and independent of Z and Z' it holds that*

$$\|\mathbf{P}(\theta_{Uh} Z \in \cdot)_{\mathcal{T}} - \mathbf{P}(\theta_{Uh} Z' \in \cdot)_{\mathcal{T}}\| \to \|\mathbf{P}(Z \in \cdot)_{\mathcal{S}} - \mathbf{P}(Z' \in \cdot)_{\mathcal{S}}\|, \quad h \downarrow 0.$$

We can now tie together ε-couplings ($\varepsilon > 0$), smooth asymptotics and $\mathcal{S}$ as follows.

THEOREM 5.3. *The following statements are equivalent:*

(a) *for each $\varepsilon > 0$, there exists an ε-coupling of Z and Z' with finite times;*

(b) *with U uniform on $[0,1]$ and independent of Z and Z' it holds that, for $h > 0$,*

$$\|\mathbf{P}(\theta_{t+Uh} Z \in \cdot) - \mathbf{P}(\theta_{t+Uh} Z' \in \cdot)\| \to 0, \quad t \to \infty;$$

(c) $\mathbf{P}(Z \in \cdot)_{\mathcal{S}} = \mathbf{P}(Z' \in \cdot)_{\mathcal{S}}.$

Proof. By the ε-coupling inequality, (a) implies (b), see (5.1). By Theorem 3.3, (b) implies $\mathbf{P}(\theta_{Uh} Z \in \cdot)_{\mathcal{T}} = \mathbf{P}(\theta_{Uh} Z' \in \cdot)_{\mathcal{T}}$ for each $h > 0$ which together with Theorem 5.2 yields (c). By Theorem 5.1, (c) implies (a).

For Markov processes we can add $\mathcal{S}$-triviality to these equivalences, the proof is analogous to the proof of Theorem 3.4:

THEOREM 5.4. *Let Z and Z' be differently started versions of a time-homogeneous Markov process with initial distributions μ and μ'. Then (c) holds for all μ and μ' if and only if*

(d) $\mathbf{P}(Z \in \cdot)_{\mathcal{S}} = 0$ *or* 1 *for all initial distributions* μ.

In all three cases (exact coupling, shift-coupling and ε-coupling) statements on harmonic functions and on mixing can in fact be added in the Markov case, see [42].

6. Random fields

Let us return to shift-coupling and consider briefly the two-sided case, $Z = (Z_s)_{s \in R}$ and $Z' = (Z'_s)_{s \in R}$. It turns out that most of the results in Section 4 still hold (with minor modifications) with the shift-maps defined by $\theta_t z = (z_{t+s})_{s \in R}$. In fact the two-sided case is easier to deal with since while the one-sided shifts do not form a group the two-sided shifts do: if we shift the origin to t we have not lost what happened before time t and can shift back. This allows us to replace the assumption that the state space is Polish and the paths right-continuous by a general joint measurability condition. It also allows us to simplify the definition of shift-coupling: if we take $S = T - T'$ and $C = \{T < \infty\}$ then the definition, $\theta_T \hat{Z} = \theta_{T'} \hat{Z}'$ on $\{T < \infty\}$, becomes

$$(6.1) \qquad\qquad \theta_S \hat{Z} = \hat{Z}' \quad \text{on } C.$$

Thus on C the two processes are really the same, only with different centers.

In fact if we replace the one dimensional time R by $R^d (d > 1)$ then the shift-coupling results still hold. Moreover, the Cesaro total variation convergence over intervals $[0, t]$ can be extended to Borel sets $B_h \in \mathcal{B}(R^d), 0 < h < \infty$, having the so-called *Fölner* property:

$$(6.2) \qquad \begin{aligned} & 0 < \lambda(B_h) < \infty \quad \text{and} \\ & \lim_{h \to \infty} \lambda(B_h \triangle (t + B_h))/\lambda(B_h) = 0, \quad t \in R^d, \end{aligned}$$

where λ denotes the Lebesgue measure and $\triangle$ symmetric difference of sets. An example is

$$B_h = hB, \quad 0 < h < \infty,$$

where $B \in \mathcal{B}(R^d)$ and $0 < \lambda(B) < \infty$.

6.1. *The equivalences*

Let $Z = (Z_s)_{s \in R^d}$ and $Z' = (Z'_s)_{s \in R^d}$ be two random fields with a general state space $(E, \mathcal{E})$ having paths in a shift-invariant subset H of E^{R^d}. Regard Z and Z' as random elements in $(H, \mathcal{H})$ where $\mathcal{H}$ is the σ-algebra on H generated by the projection mappings taking $z = (z_s)_{s \in R^d}$ in H to z_t in $E, t \in R^d$, and assume that the mapping taking (z,t) in $H \times R^d$ to z_t in E is $\mathcal{H} \otimes \mathcal{B}(R^d)/\mathcal{E}$ measurable. Define the shift-maps $\theta_t, t \in R^d$, by

$$\theta_t z = (z_{t+s})_{s \in R^d}, \quad z \in H,$$

and the invariant σ-algebra by

$$\mathcal{I} = \{A \in \mathcal{H} : \theta_t A = A, t \in R^d\}.$$

Call a coupling $(\hat{Z}, \hat{Z}')$ of Z and Z' a *shift-coupling* with *shift* S and *event* C if (6.1) holds where now S is a d dimensional random variable. Call the shift-coupling *successful* if $C = \Omega$. The following result is established in Thorisson [40].

THEOREM 6.1. *The following claims are equivalent:*

(a) *there exists a successful shift-coupling of Z and Z';*

(b) *if (6.2) holds and $U(B_h)$ is uniform on B_h and independent of Z and Z' then*

$$\|\mathbf{P}(\theta_{U(B_h)} Z \in \cdot) - \mathbf{P}(\theta_{U(B_h)} Z' \in \cdot)\| \to 0, \quad h \to \infty.$$

(c) $\mathbf{P}(Z \in \cdot)_{\mathcal{I}} = \mathbf{P}(Z' \in \cdot)_{\mathcal{I}}.$

This is an immediate consequence of Theorems 7.1 and 7.2 below. The other results in Section 4 also hold in some form but rather than deriving/stating them here we do so in the more general context of the next section. We shall only mention that the shift-coupling inequality now is: *for $B \in \mathcal{B}(R^d)$ such that $0 < \lambda(B) < \infty$ and with $U(B)$ uniform on B and independent of Z and Z',*

$$\|\mathbf{P}(\theta_{U(B)} Z \in \cdot) - \mathbf{P}(\theta_{U(B)} Z' \in \cdot)\| \leq \mathbf{E}[\lambda(B \triangle (S+B))/\lambda(B); C] + 2\mathbf{P}(C^c).$$

6.2. *On application in Palm theory*

In the ergodic case the distribution of a stationary point process in d dimensions agrees on invariant sets with the distribution of its so-called Palm version. Moreover, in the non-ergodic case the distribution agrees on invariant sets with the distribution of a certain modified Palm version. Thus the stationary point process and its (modified) Palm version are really the same point process with different centers. See Thorisson [41]. This result should extend to the framwork of the next section.

7. Transformation coupling

We end this presentation by extending the view to an abstract setup where general random elements replace random fields and a topological transformation group replaces the shift-maps $\theta_t, t \in R^d$.

7.1. *Transformation coupling - The invariant σ-algebra*

Let Y and Y' be random elements in a measurable space $(H, \mathcal{H})$. Let G be a locally compact second countable topological group of measurable mappings (transformations) from $(H, \mathcal{H})$ to $(H, \mathcal{H})$. Let $\mathcal{G}$ be the Borel subsets of G. Let the mapping from $H \times G$ to H taking (x, γ) to γx be $\mathcal{H} \otimes \mathcal{G}/\mathcal{H}$ measurable. Define the invariant σ-algebra on $(H, \mathcal{H})$ under G by

$$\mathcal{I} = \{A \in \mathcal{H} : \gamma A = A, \gamma \in G\}.$$

Call $(\hat{Y}, \hat{Y}', \Gamma, C)$ a *transformation coupling* of Y and Y' if $(\hat{Y}, \hat{Y}')$ is a coupling of Y and Y', Γ is a random transformation in $(G, \mathcal{G})$, C is an event, and

$$\Gamma\hat{Y} = \hat{Y}' \quad \text{on } C.$$

Call $(\hat{Y}, \hat{Y}', \Gamma)$ a *successful transformation coupling* of Y and Y' if $\Gamma\hat{Y} = \hat{Y}'$. The following result is established in Thorisson [40].

THEOREM 7.1. *The following statements are equivalent:*

 (a) *there exists a successful transformation coupling of Y and Y';*

 (c) $\mathbf{P}(Y \in \cdot)_{\mathcal{I}} = \mathbf{P}(Y' \in \cdot)_{\mathcal{I}}.$

7.2. *Where is (b)?*

In order to add the limit part to these equivalences we must assume that there exists a Fölner family of sets, that is, $B_h \in \mathcal{G}, 0 < h < \infty$, satisfying

$$(7.1) \qquad 0 < \lambda(B_h) < \infty \quad \text{and} \quad \lim_{h \to \infty} \lambda(B_h \triangle B_h \gamma)/\lambda(B_h) = 0, \quad \gamma \in G,$$

where λ is the Haar measure on $(G, \mathcal{G})$. This condition is equivalent to G being amenable, see [18].

THEOREM 7.2. *If (7.1) holds then the equivalent claims (a) and (c) are also equivalent to*

 (b) *with $U(B_h)$ distributed according to $\lambda(\cdot|B_h)$ and independent of Y and Y' it holds that*

$$(7.2) \qquad \|\mathbf{P}(U(B_h)Y \in \cdot) - \mathbf{P}(U(B_h)Y' \in \cdot)\| \to 0, \quad h \to \infty.$$

If Y' is distributionally invariant under G

$$\gamma Y' \overset{D}{=} Y', \quad \gamma \in G,$$

then (7.2) can be written as

$$U(B_h)Y \overset{tv}{\to} Y', \quad t \to \infty.$$

7.3. *The inequality - Proof of Theorem 7.2*

Note that for $A \in \mathcal{H}$ and $B \in \mathcal{G}$

$$(7.3) \qquad \int_B 1\{\gamma\hat{Y} \in A\}\, d\lambda(\gamma) - \int_B 1\{\gamma\Gamma\hat{Y} \in A\}\, d\lambda(\gamma) \le \lambda(B\triangle(B\Gamma))/2.$$

Dividing by $\lambda(B)$ and taking expectations and supremum in A and multiplying by 2 yields the following result.

TRANSFORMATION COUPLING INEQUALITY. *If $(\hat{Y}, \hat{Y}', \Gamma, C)$ is a transformation coupling then, for $B \in \mathcal{G}$ satisfying $0 < \lambda(B) < \infty$ and with $U(B)$ independent of Y and Y' and distributed according to $\lambda(\cdot|B)$,*

$$\begin{aligned}
(7.4) \qquad &\|\mathbf{P}(U(B)Y \in \cdot) - \mathbf{P}(U(B)Y' \in \cdot)\| \\
&\qquad \leq \mathbf{E}[\lambda(B\triangle(B\Gamma))/\lambda(B); C] + 2\mathbf{P}(C^c).
\end{aligned}$$

In particular, if $(\hat{Y}, \hat{Y}', \Gamma)$ is a successful transformation coupling then

$$(7.5) \qquad \|\mathbf{P}(U(B)Y \in \cdot) - \mathbf{P}(U(B)Y' \in \cdot)\| \leq \mathbf{E}[\lambda(B\triangle(B\Gamma))/\lambda(B)].$$

We can now derive Theorem 7.2 from Theorem 7.1. In (7.5) put $B = B_h$ and send h to infinity to obtain that (a) implies (b). Conversely, (b) implies (c) since

$$\begin{aligned}
(7.6) \qquad &\|\mathbf{P}(Y \in \cdot)_{\mathcal{I}} - \mathbf{P}(Y' \in \cdot)_{\mathcal{I}}\| \\
&\qquad = \|\mathbf{P}(U(B_h)Y \in \cdot)_{\mathcal{I}} - \mathbf{P}(U(B_h)Y' \in \cdot)_{\mathcal{I}}\| \\
&\qquad \leq \|\mathbf{P}(U(B_h)Y \in \cdot) - \mathbf{P}(U(B_h)Y' \in \cdot)\|.
\end{aligned}$$

7.4. *Maximality*

Since $\|\mathbf{P}(Y \in \cdot)_{\mathcal{I}} - \mathbf{P}(Y' \in \cdot)_{\mathcal{I}}\| = \|\mathbf{P}(\hat{Y} \in \cdot; C^c)_{\mathcal{I}} - \mathbf{P}(\hat{Y}' \in \cdot; C^c)_{\mathcal{I}}\|$ we have

$$\|\mathbf{P}(Y \in \cdot)_{\mathcal{I}} - \mathbf{P}(Y' \in \cdot)_{\mathcal{I}}\| \leq 2\mathbf{P}(C^c)$$

and once more identity can be obtained in this inequality.

THEOREM 7.3. *There exists a transformation coupling with event C such that*

$$\|\mathbf{P}(Y \in \cdot)_{\mathcal{I}} - \mathbf{P}(Y' \in \cdot)_{\mathcal{I}}\| = 2\mathbf{P}(C^c).$$

Proof. By the Lemma in [36] there is a component μ of $\mathbf{P}(Y \in \cdot)$ and μ' a component of $\mathbf{P}(Y' \in \cdot)$ such that

$$\begin{aligned}
(7.7) \qquad \mu_{\mathcal{I}} &= \mu'_{\mathcal{I}} \\
&= \text{greatest common component of } \mathbf{P}(Y \in \cdot)_{\mathcal{I}} \text{ and } \mathbf{P}(Y' \in \cdot)_{\mathcal{I}}.
\end{aligned}$$

Due to $\mu_{\mathcal{I}} = \mu'_{\mathcal{I}}$ and Theorem 7.1 there are random elements $\hat{V}$ and $\hat{V}'$ and a random transformation Γ such that $\mathbf{P}(\hat{V} \in \cdot) = \mu/\|\mu\|, \mathbf{P}(\hat{V}' \in \cdot) = \mu'/\|\mu\|$ and $\Gamma\hat{V} = \hat{V}'$. Let C be an event such that C is independent of $\hat{V}$ and $\hat{V}'$ and $\mathbf{P}(C) = \|\mu\|$. Put $(\hat{Y}, \hat{Y}') = (\hat{V}, \hat{V}')$ on C and let

$$\mathbf{P}(\hat{Y} \in \cdot, \hat{Y}' \in \cdot; C^c) = (\mathbf{P}(Y \in \cdot) - \mu)(\mathbf{P}(Y' \in \cdot) - \mu')/(1 - \|\mu\|).$$

Then $(\hat{Y}, \hat{Y}', \Gamma, C)$ is a transformation coupling of Y and Y' and the second identity in (7.7) yields the first step in

$$\|\mathbf{P}(Y \in \cdot)_{\mathcal{I}} - \mathbf{P}(Y' \in \cdot)_{\mathcal{I}}\| = \|\mathbf{P}(Y \in \cdot)_{\mathcal{I}} - \mu_{\mathcal{I}}\| + \|\mathbf{P}(Y' \in \cdot)_{\mathcal{I}} - \mu_{\mathcal{I}}\| = 2\mathbf{P}(C^c).$$

7.5. *A limit result*

From (7.6) and (7.4) applied to the coupling in Theorem 7.3 with $B = B_h$ we obtain the following.

THEOREM 7.4. *If* (7.1) *holds then, with* $U(B_h)$ *independent of* Y *and* Y' *and distributed according to* $\lambda(\cdot|B_h)$,

$$\|\mathbf{P}(U(B_h)Y \in \cdot) - \mathbf{P}(U(B_h)Y' \in \cdot)\| \to \|\mathbf{P}(Y \in \cdot)_{\mathcal{I}} - \mathbf{P}(Y' \in \cdot)_{\mathcal{I}}\|, h \to \infty.$$

If the distributions of Y and Y' are invariant under G then the left-hand side does not depend on h and equals $\|\mathbf{P}(Y \in \cdot) - \mathbf{P}(Y' \in \cdot)\|$, and thus we have the following result.

COROLLARY. *If the distributions of* Y *and* Y' *are invariant under* G *and* $\mathbf{P}(Y \in \cdot)_{\mathcal{I}}$ $= \mathbf{P}(Y' \in \cdot)_{\mathcal{I}}$ *then* Y *and* Y' *are identically distributed.*

7.6. *Application in ergodic theory*

Theorem 7.1 has an immediate application in ergodic theory. Suppose Y' is distributionally invariant under G and satisfies, for bounded $\mathcal{H}/\mathcal{B}$ measurable functions f and with B_h as in (7.1),

$$\lambda(B_h)^{-1} \int_{B_h} f(\gamma Y') \, d\lambda(\gamma) \to \mathbf{E}[f(Y')|Y'^{-1}\mathcal{I}] \quad \text{a.s.,} \quad h \to \infty.$$

Then, due to (7.3) and the fact that (c) implies (a), the following holds: for all Y agreeing with Y' in distribution on $\mathcal{I}$ it holds that

$$\lambda(B_h)^{-1} \int_{B_h} f(\gamma Y) \, d\lambda(\gamma) \quad \text{tends a.s. to a limit as } h \to \infty$$

and the limit has the same distribution as $\mathbf{E}[f(Y')|Y'^{-1}\mathcal{I}]$.

7.7. *Application to Brownian motion - Self-similarity*

Let $W = (W_s)_{s \in [0,\infty)}$ be a standard Brownian motion, that is, W is a one-sided continuous time real valued stochastic process with continuous paths and independent increments and W_t is normal with $\mathbf{E}[W_t] = 0$ and $\mathrm{Var}[W_t] = t$ for all $t \in [0, \infty)$. Then W is self-similar:

$$\gamma_r W \stackrel{D}{=} W, \quad 0 < r < \infty,$$

where γ_r is the rescaling defined for a path $z = (z_s)_{s \in [0,\infty)}$ by

$$\gamma_r z = (r^{1/2} z_{s/r})_{s \in [0,\infty)}.$$

Let Z be another one-sided continuous time real valued process with continuous paths.

According to Theorem 7.1, Z and W have the same distribution on measurable sets invariant under the rescalings $\gamma_r, 0 < r < \infty$, *if and only if* there exists a strictly positive finite random variable R such that

$$\gamma_R Z \stackrel{D}{=} W.$$

Further, according to Theorem 7.2 and since W is self-similar, the above equivalent claims hold *if and only if*

$$\gamma_{e^{U(B_h)}} Z \overset{tv}{\to} W, \quad h \to \infty,$$

where B_h are Borel sets such that (6.2) holds and $U(B_h)$ is uniform on B_h and independent of Z.

7.8. *Application in exchangeability*

Let Z and Z' be one-sided discrete time stochastic processes on a general state space. For a path $z = (z_k)_0^\infty$ and with p a finite permutation of $\{0, 1, \dots\}$ define the exchange π by

$$\pi z = (z_{p(k)})_0^\infty.$$

The exchangeable σ-algebra consists of measurable sets invariant under such finite exchanges. A stochastic process is called exchangeable if its distribution is invariant under finite exchanges.

According to Theorem 7.1, Z and Z' have the same distribution on the exchangeable σ-algebra *if and only* if there exists a finite random exchange Π such that

$$\Pi Z \overset{D}{=} Z'.$$

Further, if Z' is exchangeable then, according to Theorem 7.2, the above equivalent claims hold *if and only if*

$$U_n Z \overset{tv}{\to} Z', \quad h \to \infty,$$

where U_n is the random exchange associated with a uniformly distributed random permutation of $\{0, \dots, n\}$ which is independent of Z.

This result can be extended to one-sided continuous time real valued stochastic processes with right-continuous paths having left-hand limits: replace finite permutations by splitting a finite interval into finitely many subintervals (closed to the left and open to the right) and permuting them. Similar extension applies to random fields.

8. Further reading

For a unified presentation of coupling, see Lindvall's book [26]; see also his list of references. For applications in interacting particle systems, see Liggett's book [24]. For coupling and Poisson approximation, see the book by Barbour, Holst and Janson [4]. For card shuffling couplings, see Aldous and Diaconis [1] and Pemantle [28]. For coupling of recursive sequences, see Borovkov and Foss [7, 8] and Baccelli and Brémaud [3]. For the impossible coupling, - the coupling aspect of a (the) problem in quantum physics, - see Gill [15]. For coupling in branching, see Jagers [22]. For decoupling, see de la Peña [11] in this volume. For copulae, see Schweizer and Sklar [32], Scarsini [31] and Cuesta and Matrán [10]. For a semi-group version of transformation coupling, see Georgii [14]. The TES process approach to modeling and forecasting yields a self-coupling of the histogram of the empirical time series, see Melamed [27]. For simulation applications of coupling, see Propp and Wilson [29], Glynn and Wong [16] and Roberts and Rosenthal [30]. For application of coupling to the estimation of the spectral gap, see Chen [9] in this volume. For applications of coupling in various fields, see the collection of articles [23]. The method is finding its way into textbooks, see Grimmett and Stirzaker [21], Durrett [13], and Sigman [33]. Finally, the author of this paper is working on a book with coupling as the main theme [42].

WORKING HYPOTHESIS. *Each meaningful distributional relation should have a coupling counterpart.*

References

[1] ALDOUS, D. AND DIACONIS, P. (1986). Shuffling cards and stopping times. *Amer. Math. Monthly*, **93**, 333-348.

[2] ALDOUS, D. AND THORISSON, H. (1993). Shift-coupling. *Stoch. Proc. Appl.* **44**, 1–14.

[3] BACCELLI, F. AND BRÉMAUD, P. (1994). *Elements of Queueing Theory.* Springer, Berlin.

[4] BARBOUR, A., HOLST, L. AND JANSON, S. (1992). *Poisson Approximation.* Oxford University Press.

[5] BERBEE, H. C. P. (1979). Random walks with stationary increments and renewal theory. *Math. Centre Tract* **112**. Center for Mathematics and Computer Science, Amsterdam.

[6] BILLINGSLEY, P. (1971). *Weak Convergence of Measures: Applications in Probability.* SIAM, Philadelphia.

[7] BOROVKOV, A. A. AND FOSS, S. G. (1992). Stochastically recursive sequences and their generalizations. *Siberian Advances in Mathematics* **2**, No. 1, 16–81.

[8] BOROVKOV, A.A. AND FOSS, S. G. (1994). Two ergodicity criteria for stochastically recursive sequences. *Acta Applicandae Mathematicae* **34**, No. 1&2, 125–134.

[9] CHEN, MU-FA (1997). Trilogy of couplings - new variational formula of spectral gap. *In this Volume.*

[10] CUESTA, J.A. AND MATRÁN, C. (1994). Stochastic convergence through Skorohod representation theorems and Wasserstein distances. *First International Conference on Stochastic Geometry, Convex Bodies and Empirical Measures. Supplemento ai Rendiconti del Circolo Matematico di Palermo*, Serie II, **35**, 89–113.

[11] DE LA PEÑA, V. H. (1997). Decoupling inequalities: a second generation of martingale inequalities. *In this Volume.*

[12] DOEBLIN, W. (1938). Exposé de la theorie des chaînes simple constantes de Markov à un nombre fini d'états. *Rev. Math. Union Interbalkan.* **2**, 77–105.

[13] DURRETT (1991). *Probability: Theory and Examples.* Wadsworth and Broks/Cole.

[14] GEORGII, H.-O. (1997). Orbit coupling. *Ann. Inst. H. Poincaré, Prob. et Statist.*, **33**, 253–268.

[15] GILL, R. (1997). The impossible coupling: on hidden variables models in quantum mechanics. *Proceedings of the 51st Session of the International Statistical Institute.*

[16] GLYNN, P. W. AND WONG, E. W. (1996). Efficient simulation via coupling. *Probability in the Engineering and Informational Sciences*, **10**, 165-186.

[17] GOLDSTEIN, S. (1979). Maximal coupling. *Z. Wahrscheinlichkeitsth.* **46**, 193–204.

[18] GREENLEAF, F. P. (1969). *Invariant Means on Topological Groups.* Van Nostrand, New York.

[19] GREVEN, A. (1987). Coupling of Markov chains and randomized stopping times. Part I and II. *Probab. Th. Rel. Fields* **75**, 195–212 and 431–458.

[20] GRIFFEATH, D. (1978). Coupling methods for Markov processes. *Studies in Probability and Ergodic Theory. Adv. Math. Supplementary Studies* **2**.

[21] GRIMMETT AND STIRZAKER (1992). *Probability and Random Processes.* 2nd ed. Oxford University Press.

[22] JAGERS, P. (1997). Coupling and population dependence in branching processes. *Ann. Appl. Probab.*

[23] KALASHNIKOV, V. AND THORISSON, H. (editors) (1994). Applications of Coupling and Regeneration. Special issue of *Acta Applicandae Mathematicae* **34**, No. 1&2.

[24] LIGGETT, T. M. (1985). *Interacting Particle Systems.* Springer, New York.

[25] LINDVALL, T. (1977). A probabilistic proof of Blackwell's renewal theorem. *Ann. Probab.* **5**, 57–70.

[26] LINDVALL, T. (1992). *Lectures on the Coupling Method.* Wiley, New York.

[27] MELAMED, B. (1993). An Overview of TES Processes and Modeling Methodology. In *Performance Evaluation of Computer and Communications Systems*, (L. Donatiello and R. Nelson, Eds.), 359–393, Springer-Verlag, Lecture Notes in Computer Science.

[28] PEMANTLE, R. (1989). Randomization time for the overhand shuffle. *Journ. Theo. Probab.* **2**, No. 1.

[29] PROPP, J.G. AND WILSON, D.B. Exact sampling with coupled Markov chains and applications to statistical mechanics. *Random Structures and Algorithms* **9**, 223–252.

[30] ROBERTS, G. O. AND ROSENTHAL, J. S. (1997). Shift-coupling and convergence rates of ergodic averages. *Commun. in Statist. - Stochastic Models* **13**, 147–165.

[31] SCARSINI, M. (1989). Copulae of probability measures on product spaces. *Journal of Multivariate Analysis* **31**, 201–219.

[32] SCHWEIZER, B. AND SKLAR, A. (1983). *Probabilistic Metric Spaces.* Elsevier, New York.

[33] SIGMAN, K. (1994). *Stationary Marked Point Processes: An Intuitive Approach.* Chapman and Hall.

[34] SVERCHKOV, M. YU. AND SMIRNOV, S. N. (1990). Maximal coupling of D-valued processes. *Soviet Math. Dokl.* **41**, 352–354.

[35] THORISSON, H. (1983). The coupling of regenerative processes. *Adv. Appl. Probab.* **15**, 531–561.

[36] THORISSON, H. (1986). On maximal and distributional coupling. *Ann. Probab.* **14**, 873–876.

[37] THORISSON, H. (1994). Shift-coupling in continuous time. *Prob. Theo. Rel. Fields.* **99**, 477–483.

[38] THORISSON, H. (1995). On time- and cycle-stationarity. *Stoch. Proc. Appl.* **55**, 183–209.

[39] THORISSON, H. (1995). Coupling methods in probability theory. *Scand. J. Stat.* **22**, 159–182.

[40] THORISSON, H. (1996). Transforming random elements and shifting random fields. *Ann. Probab.* **24**, 2057–2064.

[41] THORISSON, H. Point-stationarity in d dimensions and Palm theory. *Bernoulli* (to appear).

[42] THORISSON, H. (1998). *Coupling, Stationarity and Regeneration*. Springer.

SOME RECENT DEVELOPMENTS FOR QUEUEING NETWORKS

R. J. WILLIAMS,* *University of California, San Diego*

1. Introduction

Early investigations in queueing theory provided detailed analysis of the behavior of a single queue and of networks that in a sense could be decomposed into a product of single queues. Whilst insights from these early investigations are still used, more recent investigations have focussed on understanding how network components interact.

In particular, queueing network models are of current relevance for analyzing congestion and delay in computer systems, communication networks and complex manufacturing systems (see e.g., [1, 54, 66]). Many of these systems have stations that can process more than one class of customer or job (so-called multiclass networks) and/or have complex feedback structures. For example, in a computer communication network one may have voice, video and data being transmitted through each node in the network, and in the manufacture of semiconductor wafers, a job may return to the same machine several times for different stages of processing. Generally such systems cannot be analyzed in closed form and frequently are heavily loaded. One method for analyzing the performance of such systems is to approximate them by more tractable objects.

A certain class of diffusion processes, known as reflecting Brownian motions, have been shown to approximate normalized versions of the queue length or workload processes in single class queueing networks and some multiclass queueing networks under conditions of heavy traffic, i.e., when the networks are heavily loaded (see [65] for an overview). There is now a substantial theory for these diffusion processes which are generally more tractable than the original queueing networks, although some open problems remain (see [64] for a recent survey).

One of the outstanding challenges in contemporary research on approximate models for queueing networks is to understand which multiclass networks can be approximated by reflecting Brownian motions in heavy traffic and to prove limit theorems justifying such approximations. In the last few years there have been some surprises both with regard to conditions for the stability of multiclass queueing networks and to the behavior of such networks in heavy traffic. The aim of this paper is to describe some of the recent developments in this area.

The paper is organized as follows. In section 2 a heavy traffic approximation for a single class tandem queue is described. The purpose of this is to illustrate in a

* Postal address: Department of Mathematics, University of California, San Diego, 9500 Gilman Drive, La Jolla CA 92093-0112, USA.

Research supported in part by NSF Grant GER 9023335.

concrete setting (a) how stability is quantified, and (b) the form of the heavy traffic approximation. In section 3, a brief synopsis of heavy traffic limit theory for single class and some multiclass queueing networks is given. One may extrapolate from this, as was done in [22, 24, 25, 31], to conjecture a general form for the heavy traffic limit of a wide variety of multiclass networks. In section 4, an example similar to one given by Dai and Wang [16] illustrates that care is required with such an extrapolation, in the sense that not all multiclass queueing networks have a heavy traffic approximation of the form conjectured in [22, 24, 25, 31]. A variety of explanations might be proposed for the failure of the approximation in this case. Two possible explanations that might be proposed are (a) the example is not stable, i.e., one has the wrong notion of heavy traffic, or (b) the example may satisfy a different kind of limit theorem than the "conventional" one proposed in [22, 24, 25, 31]. Indeed, contemporaneous work on stability of multiclass networks (see e.g., [42, 44, 53, 3, 4, 55]), illustrates that the problem of determining conditions for the stability of multiclass networks with feedback is more complex than previously supposed. Furthermore, work of Harrison and Williams [32] (see also section 4 of [65]), shows that not all multiclass networks with feedback have conventional heavy traffic behavior. The paper concludes with a final section on open problems.

2. A Single Class Tandem Queue

Consider the tandem queueing network pictured in Fig. 1 under the following assumptions. Customers (or jobs) arrive at station 1 from outside the system according to a renewal process where the i.i.d. interarrival times are assumed to have positive finite mean $1/\alpha$ and finite variance σ_a^2 (α is thus the (long run average) external arrival rate). There is a single server at each of the two stations and the service times at station i are assumed to be i.i.d. with positive finite mean m_i and variance σ_i^2, for $i = 1, 2$. The sequences of interarrival and service times are assumed to be mutually independent. Customers are served on a first-in-first-out (FIFO) basis at each station. After receiving service at station 1, a customer goes next to station 2. Upon completing service there, with probability $p \in (0,1)$, the customer is routed back to join the end of the queue at station 1, and with probability $1-p$ the customer exits the system. Such a tandem queue might be used to model a simple processing facility where a completed job requires total rework with probability p.

Congestion is measured through the behavior of the two-dimensional queue length process $Q(\,\cdot\,) = (Q_1(\,\cdot\,), Q_2(\,\cdot\,))$, where for $i = 1, 2$, $Q_i(t)$ is the number of customers at station i (waiting and being served) at time t. With the general distributional assumptions described above, this system cannot be analyzed exactly. However, when it is stable, but heavily loaded, one can approximate a normalized version of the queue length process by a reflecting Brownian motion living in the positive two-dimensional quadrant.

To describe this approximation, the first question that arises is what does "stable and heavily loaded" mean? Here it will be taken to mean that the system is stable and near the boundary between stability and instability, where "stable" means that the mean queue lengths are bounded for all time. Stability can be quantified in terms of the station level traffic intensity parameters ρ_i, $i = 1, 2$, defined as follows. Let λ

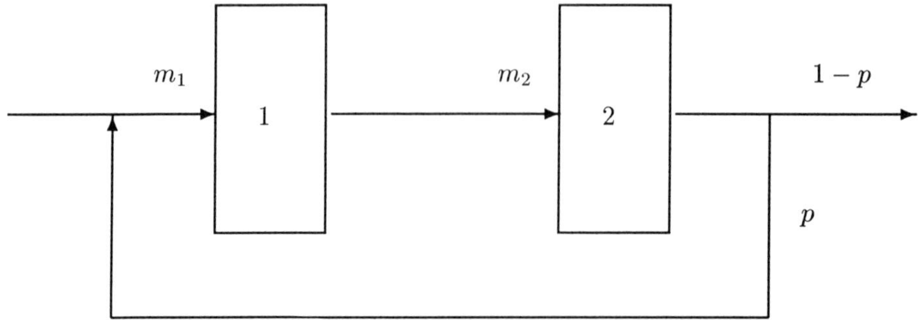

Figure 1. A Single Class FIFO Tandem Queue with Feedback

be the unique solution of the traffic equation

(1) $$\lambda = \alpha + p\lambda.$$

The traffic intensity parameters are defined by

(2) $$\rho_i = \lambda m_i, \quad i = 1, 2.$$

The system is stable if and only if

(3) $$\rho_i < 1, \quad \text{for } i = 1, 2,$$

(see e.g., [47]). Assuming this holds, one can interpret λ as the long run average rate at which customers visit stations 1 and 2 (it is the same for each station because of the tandem structure of the network). Then the flow balance equation (1) is a natural consistency condition. Also, the traffic intensity parameter ρ_i can be interpreted as the long run average rate at which work (measured in units of required service time) arrives at station i, $i = 1, 2$.

A heavy traffic limit theorem for this system may be described as follows. Consider a sequence of systems indexed by n (with associated parameters and processes having a superscript of (n)), all with the same common structure as described above except that the exogenous arrival rate $\alpha^{(n)}$ for the nth system tends to a value α in such a way that the traffic intensity vector $\rho^{(n)} = (\rho_1^{(n)}, \rho_2^{(n)})$ for the nth system tends to the vector $(1, 1)$ in the following manner:

(4) $$\sqrt{n}(\rho_i^{(n)} - 1) \to c_i \quad \text{as } n \to \infty,$$

where c_i is a finite negative constant for $i = 1, 2$. (The negativity of the c_i guarantees that the heavy traffic limit will be positive recurrent [29].) Normalize the two-dimensional queue length process $Q^{(n)}$ for the nth system using a central limit theorem (or diffusion) type of scaling:

(5) $$\hat{Q}^{(n)}(\cdot) = \frac{Q^{(n)}(n \cdot)}{\sqrt{n}}.$$

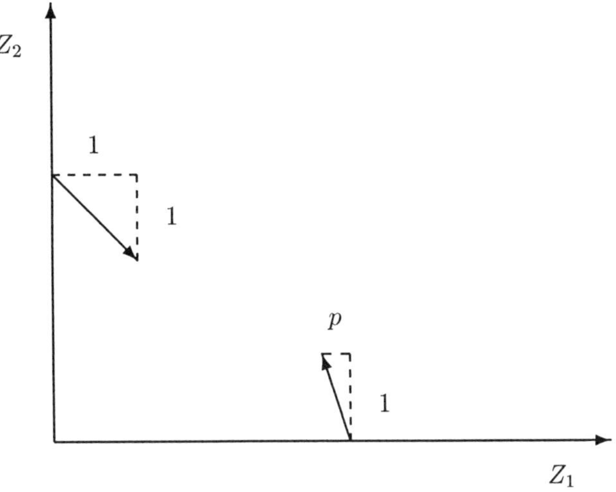

Figure 2. State Space and Directions of Reflection for Z

The process $\hat{Q}^{(n)}$ takes values in the space of two-dimensional r.c.l.l. (right continuous with finite left limits) paths defined on $[0, \infty)$. When this space is endowed with the usual Skorokhod $\mathbf{J_1}$-topology [56, 20], $\hat{Q}^{(n)}$ converges in distribution as $n \to \infty$ to a reflecting Brownian motion Z that lives in the positive quadrant and which has a semimartingale decomposition of the form

$$(6) \qquad Z = X + RY,$$

where X is a two-dimensional Brownian motion with constant drift c and non-degenerate covariance matrix

$$(7) \qquad \Gamma = \begin{bmatrix} \alpha^3 \sigma_a^2 + \mu_1^3 \sigma_1^2 + p\mu_2(1 - p + p\mu_2^2 \sigma_2^2) & -(\mu_1^3 \sigma_1^2 + p\mu_2^3 \sigma_2^2) \\ -(\mu_1^3 \sigma_1^2 + p\mu_2^3 \sigma_2^2) & \mu_1^3 \sigma_1^2 + \mu_2^3 \sigma_2^2 \end{bmatrix},$$

for $\mu_i = 1/m_i$, $i = 1, 2$,

$$(8) \qquad R = \begin{bmatrix} 1 & -p \\ -1 & 1 \end{bmatrix}$$

is called the reflection matrix, and Y is a two-dimensional continuous non-decreasing process that starts from the origin and is such that Y_i can increase only when Z_i is zero (in fact, $m_i Y_i$ is the limit in distribution of the normalized cumulative idletime process for station i, where the normalization is the same central limit theorem type of scaling (5) used for the queue length processes).

Informally, the behavior of the reflecting Brownian motion Z may be described as follows. In the interior of the quadrant, Z behaves like the Brownian motion X. The erratic movements of this Brownian motion are the limit of the up and down movements of the queue length processes corresponding to arrivals and departures, respectively. The process Z is confined to the quadrant by "pushing" at the boundary in the fixed directions shown in Fig. 2. These directions of control are given by the columns of the matrix R and for historical reasons stemming from the one-dimensional case are called directions of "reflection", though one should not think of constructing the process by any kind of mirror reflection. These directions may be loosely interpreted as follows. Consider the case where $Z_1 = 0$. This corresponds to the first queue being empty. Imagine that when the queueing network is in this situation, the server at queue 1 continues working even though there are no customers to serve. The server thereby generates "potential" services. To prevent Q_1 from becoming negative (due to the completion of such a service), each potential service performed by server 1 needs to be corrected by a unit step in the positive Q_1 direction to keep Q_1 zero. This lost potential service also has an effect on the second queue in the sense that there is "lost potential flow" to the second queue. Accordingly, for each corrective unit step in the positive Q_1 direction, there is a corresponding downward unit step in the Q_2 direction. In the heavy traffic limit, this behavior translates to instantaneous pushing at the boundary $Z_1 = 0$ in the direction $(1, -1)$ indicated in Fig. 2. In an analogous manner, on the boundary $Z_2 = 0$ one has pushing in the direction $(-p, 1)$. The term $-p$ comes from the lost potential flow from queue 2 back to queue 1.

The limit result cited above is justified by the heavy traffic theorem of Reiman [50]. Besides varying the arrival rate as $n \to \infty$, one can also allow suitable variations in the initial queue lengths, service rates, arrival and service time variances, and the routing probability p. The reader is refered to [50] for such refinements. Existence and uniqueness of the limiting diffusion process Z follows from a path-by-path construction due to Harrison and Reiman [27].

This tandem queue is a single class network in the sense that at each station there is just one class of customer, i.e., the customers are indistinguishable from one another. Although the result of Reiman [50] was cited to justify the approximation in this two-station case, his result applies to a general d-station ($d \geq 1$) single class FIFO network. Similarly, the existence and uniqueness theorem of Harrison and Reiman [27] applies to the associated reflecting Brownian motion which lives in the positive d-dimensional orthant. This has a form that is the d-dimensional analogue of (6) and in particular the reflection matrix $R = I - P'$, where P is the transition matrix for a transient Markov chain on d states (corresponding to the Markovian routing matrix for the queueing network).

A brief synopsis of the extant heavy traffic limit theory for single and some multi-class networks is given in the next section. For more details the reader is refered to the paper [65] and references therein.

3. Heavy Traffic Limit Theorems: A Brief Synopsis

To facilitate comparison of different results, in the sequel, queueing networks will be assumed to have the following common features. There is a single server at each station, the arrival process for each customer class is a renewal process for which the interarrival times have finite means and variances, the service times for each class are given by a sequence of i.i.d. random variables having finite means and variances, the routing is Markovian, and the arrival processes, service times and customer routing are mutually independent. For single class networks there is just one customer class per station, whereas for multiclass networks there may be several different classes of customers served at a single station and the mapping from classes to stations is many-to-one.

Heavy traffic limit theory has been concerned largely with open networks in which customers arrive from outside the system, receive a finite number of services at various stations and then exit the system. Furthermore the theory is most well developed for single class networks with FIFO (first-in-first-out) service discipline. Let us consider open networks first.

The heavy traffic limit theorem of Iglehart and Whitt [35, 36], for a single class FIFO station, is the prototype for conventional heavy traffic limit theorems. Following their work, Harrison [21] proved a heavy traffic limit theorem for two stations in tandem (without feedback) and first identified a sample path representation for the two-dimensional limit process, which is a reflecting Brownian motion living in the positive quadrant. His limit theorem was generalized by Reiman [50] to single class FIFO networks with Markovian routing which may have feedback. These networks are sometimes referred to as generalized Jackson networks.

In multiclass networks, different classes of customers, perhaps having different service distributions or different routing requirements, may be served at a station. For such networks, it is natural to consider other service disciplines besides FIFO. The extant heavy traffic limit theorems for such networks have largely focussed on those with (static) priority service across classes and FIFO service within a priority class. In multiclass networks, the dimension of the queue length process is equal to the number of classes served in the system (recall that the mapping from classes to stations is many-to-one). Another process of interest is the (immediate) workload process W whose dimension is equal to the number of stations and is such that $W_i(t)$ represents the amount of work (measured in units of service time) embodied in the customers at station i at time t.

Whitt [60] proved a heavy traffic limit theorem for a single multiclass station with two priority classes (high and low). A notable feature here is that the two-dimensional queue length process, normalized with a central limit theorem type of scaling, converges in distribution to a process in which the component corresponding to the high priority class is identically zero and that corresponding to the low priority class is a one-dimensional reflecting Brownian motion. Johnson [37] combined the features of the Reiman [50] and Whitt [60] results to prove a heavy traffic limit theorem for multiclass networks having two types of customers, those of high priority and those of low priority, where a customer retains the same priority designation during its entire sojourn through the network, i.e., once a high priority customer, always a high priority customer etc. A network of this kind is said to have *separated priorities*. Peterson [49] proved a heavy traffic limit theorem for multiclass networks with priority service (two levels, not separated), but he restricted to the case of feedforward routing. In this, the d-dimensional workload process $W^{(n)}$, when normalized with the a central limit theorem type of scaling:

$$\hat{W}^{(n)}(\,\cdot\,) = \frac{W^{(n)}(n\,\cdot\,)}{\sqrt{n}},\tag{9}$$

converges in distribution to a reflecting Brownian motion Z that lives in the positive d-dimensional orthant and has a semimartingale decomposition of the form (6) where X, Y are now d-dimensional and R is a $d \times d$ matrix of the form $(I + G)^{-1}$ where G is non-negative and upper triangular. The class level queue length processes, with the same normalization as for the workload processes, also converge in distribution. As in the single station case [60], for each station, the limit for the high priority queue length processes is identically zero and for the low priority queue length processes is proportional to the limit of the normalized workload processes for the station. Although heavy traffic limit theorems have been proved for a single multiclass *station with feedback* and certain service disciplines such as round-robin [51] and FIFO [13], there is currently no general heavy traffic limit theorem for *open multiclass networks with feedback.*

Closed queueing networks, in which a fixed number of customers or jobs circulate perpetually in the system, are natural models for some manufacturing systems. Analogues of the open network heavy traffic limit theorems of Reiman [50] and Johnson [37] have been proved by Chen and Mandelbaum [8] for closed networks. In particular, they considered closed networks that are single class or have separated priorities. For a closed network, the heavy traffic parameter n has a natural interpretation as the fixed number of customers in the system. Then a natural central limit theorem type of scaling for the queue length process $Q^{(n)}$ is given by

$$\hat{Q}^{(n)}(\,\cdot\,) = \frac{Q^{(n)}(n^2\,\cdot\,)}{n}.\tag{10}$$

Under suitable conditions, the (low priority) d-dimensional queue length process normalized as in (10) converges in distribution as $n \to \infty$ to a reflecting Brownian motion that lives in the d-dimensional simplex (see Harrison, Williams and Chen [33] for some analysis of this limit process), and the normalized high priority queue length process vanishes in the heavy traffic limit. Despite these positive results, as in the open network case, there is currently no general heavy traffic limit theorem for closed multiclass queueing networks.

In contrast to the lack of a general heavy traffic limit theorem for multiclass networks with feedback, there is a rigorous existence and uniqueness theory for semimartingale reflecting Brownian motions (SRBMs). These diffusion processes have a semimartingale form as in (6) where the reflection matrix R need only satisfy a natural feasibility condition that it be completely-$\mathcal{S}$, i.e., for each principal submatrix $\tilde{R}$ of R there is a positive vector $\tilde{y}$ such that $\tilde{R}\tilde{y} > 0$. Reiman and Williams [52] established the necessity of this condition and Taylor and Williams [57] established its sufficiency for existence and uniqueness in law of a SRBM, provided one adds to the definition the mild condition that X minus its drift process is a martingale relative to the filtration generated by X, Y, Z. For an extension of these results to convex polyhedral state spaces, which is relevant to closed network approximations, see Dai and Williams [18].

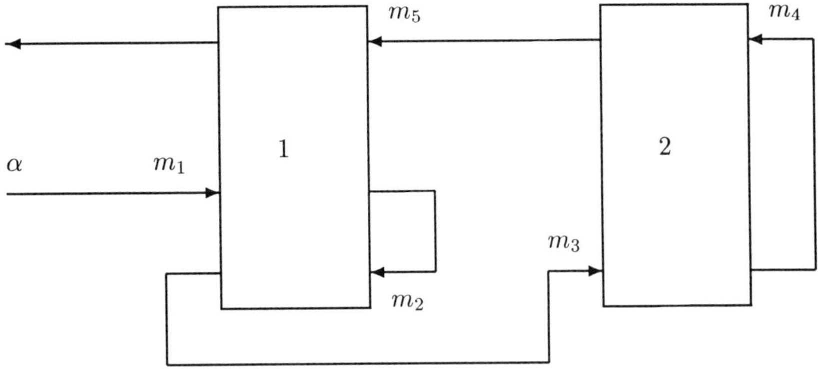

Figure 3. A Multiclass Open Network with FIFO Service

Extrapolating from the extant limit theorems, Harrison [22] and Harrison and Nguyen [24, 25] conjectured a reflecting Brownian motion approximation (or an approximate "Brownian model") for open multiclass queueing networks with FIFO service in heavy traffic. An analogous approximation was proposed in the Appendix to Harrison and Williams [31], where static priority and processor sharing service disciplines are included in addition to FIFO. Despite the appeal of these approximations, apart from the situations covered by [35, 36, 21, 50, 60, 37, 49, 51, 13], no heavy traffic limit theorem has been proved to justify them. However, it was still a surprise when Dai and Wang [16] produced an example which showed that not all open multiclass queueing networks can have such an approximation. A variant of this example and possible explanations for it are described in the next section.

4. Dai-Wang-Type Example and Stability of Open Networks

The following variant of the Dai-Wang example appears in the paper of Dai and Nguyen [14].

Consider the two-station network pictured in Fig. 3. Arrivals to this network are assumed to be given by a Poisson process with arrival rate $\alpha \in (0, 1)$. There is a single server at each of the two stations. Customers are routed through the network in a deterministic manner making visits to the two stations in the following order $1, 1, 2, 2, 1$. A customer awaiting or undergoing its kth service is called a class k customer, $k = 1, 2, 3, 4, 5$. Thus, classes $1, 2, 5$ are served at station 1 and classes $3, 4$ are served at station 2. Customers at a station, regardless of class, are served on a first-in-first-out basis. Thus, for example, after receiving service as a class 3 customer at station 2, a customer changes to class 4 and goes to the end of the line at station 2. The service times are assumed to be independent and exponentially distributed with mean m_k for class k, $k = 1, 2, 3, 4, 5$, where

$$(11) \qquad m = \left(\frac{1}{10}, \frac{1}{10}, \frac{23}{27}, \frac{4}{27}, \frac{4}{5} \right).$$

Interarrival and service times are mutually independent.

The traffic intensity parameters for the two stations are given by

$$(12) \qquad \rho_1 = \alpha(m_1 + m_2 + m_5) \quad \text{and} \quad \rho_2 = \alpha(m_3 + m_4).$$

Traditionally ρ_i has been interpreted as the long run average rate at which work arrives at station i (implicit here is an assumption that such long run average behavior exists). Extrapolating from existing theory, it natural to define heavy traffic for this example as occuring when these traffic intensity parameters are close to one. Indeed, the approximation proposed in [22, 24, 25, 31] would say that with $\alpha = 1 - \frac{1}{\sqrt{n}}$ (and so $\rho_i = 1 - \frac{1}{\sqrt{n}}$, $i = 1, 2$), for n sufficiently large one can approximate the two-dimensional workload process (normalized with the central limit theorem scaling as in (9)), by a reflecting Brownian motion that lives in the positive quadrant and has the form (6), where the reflection matrix is given by

$$(13) \qquad R = \begin{bmatrix} \frac{-310}{27} & 16 \\ 20 & -27 \end{bmatrix}.$$

However, this reflection matrix corresponds to directions of reflection that point *out* of the quadrant. There is no semimartingale reflecting Brownian motion living in the positive quadrant with such directions of reflection. Consequently, the conjectured limit theorem cannot hold in this case. Indeed, investigation by Dai and Nguyen [14] of this example showed that the normalized workload processes cannot converge in distribution to a continuous limit.

Around the time that the Dai–Wang counterexample was produced, there was a separate growing interest in the stability of open multiclass networks. For a single class, d-station, open queueing network satisfying the conditions described at the beginning of section 3, the condition for stability is that

$$(14) \qquad \rho_i < 1 \quad \text{for } i = 1, \dots, d,$$

(see e.g., Meyn and Down [47]). Here the traffic intensity parameters ρ_i are defined by

$$(15) \qquad \rho_i = \lambda_i m_i,$$

where $\lambda = (\lambda_1, \dots, \lambda_d)'$ is the solution of the vector traffic equation

$$(16) \qquad \lambda = \alpha + P'\lambda,$$

$\alpha = (\alpha_1, \dots, \alpha_d)'$ is the vector of exogenous arrival rates (one component for each station), P is the $d \times d$ matrix of Markovian routing probabilities (so that P_{ij} denotes the probability that a customer completing service at station i goes next to station j), and m_i is the mean service time per customer at station i. It is natural to conjecture that (14) is also the condition for stability of *multiclass* networks, provided ρ_i is now defined by

$$(17) \qquad \rho_i = \sum_{k \in \mathcal{C}_i} \lambda_k m_k,$$

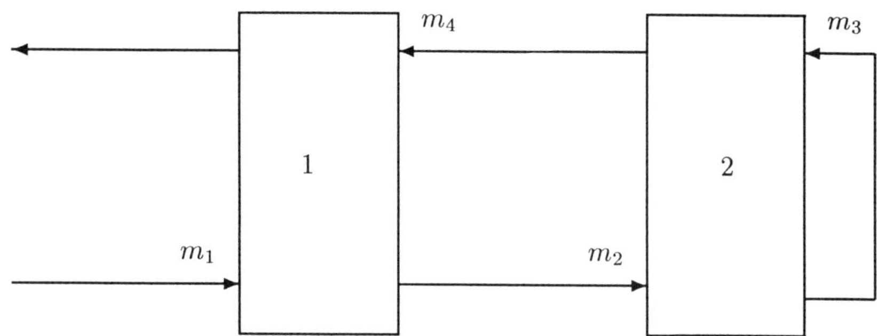

Figure 4. A Multiclass Open Network with Priority Service

where $\mathcal{C}_i$ denotes the constituency of station i, i.e., it consists of those customer classes k that are served at station i, m_k denotes the mean service time for class k customers $k = 1, \ldots, K$, and $\lambda = (\lambda_1, \ldots, \lambda_K)'$ is the solution of the traffic equation (16) now considered to be written at the class level, so that α is the vector of class level exogenous arrival rates and P is the class level matrix of Markovian routing probabilities. Despite some cases [37, 49] where this conjecture is true, Kumar and coworkers [42, 44] have given simple two-station deterministic examples with priority service which show that it is false in general for multiclass networks with feedback, i.e., they have given examples of networks in which $\rho_i < 1$ for each i, but the networks are unstable. Rybko and Stolyar [53] gave the first stochastic counterexample. This is a two-station network with priority service which has similar structure to the Lu-Kumar [44] example.

Another counterexample can be given by considering the network pictured in Fig. 4. This is a hybrid of the Lu-Kumar and Rybko-Stolyar examples. Here the arrivals are given by a Poisson process of rate one. Customers are routed through the network in a deterministic manner, making visits to the two stations in the order $1, 2, 2, 1$. Customers that are awaiting or undergoing their k^{th} service will be called class k customers, $k = 1, 2, 3, 4$. All services are independent and class k service times are exponentially distributed with mean $m_k > 0$, $k = 1, 2, 3, 4$. Each server follows a preemptive resume priority discipline where classes 2 and 4 have priority over classes 3 and 1, respectively. The traffic intensities for the two stations are given by

$$(18) \qquad \rho_1 = m_1 + m_4 \quad \text{and} \quad \rho_2 = m_2 + m_3.$$

Dai and Weiss [17] have shown that this network is stable if

$$(19) \qquad \rho_i < 1 \quad \text{for } i = 1, 2, \quad \text{and} \quad m_2 + m_4 < 1.$$

Furthermore, Dai and VandeVate [15] have recently shown that if any of the inequalities in (19) is violated with strict inequality, then the network is unstable. Thus, for example, if $m_1 = m_3 = 1/4$, $m_2 = m_4 = 2/3$, then the network is unstable, but $\rho_i < 1$ for $i = 1, 2$.

Whilst it might be argued that one can see (with hindsight) that the priorities in the Lu-Kumar/Rybko-Stolyar type examples are "bad" for stability, it was a further surprise when Bramson [3, 4] gave a two-station stochastic counterexample having FIFO service discipline. (Also, Seidman [55] gave a deterministic FIFO counterexample.)

The appearance of these counterexamples was followed by a burst of activity concerned with the stability of open multiclass networks. Frequently Lyapunov functions have been used as a tool for establishing sufficient conditions for stability. Early work used mathematical programming to determine such Lyapunov functions for the queueing networks (see for example, Kumar and Meyn [41] and Bertsimas et al. [2]). A significant advance was made when Dai [10] showed that the stability of associated fluid limits (obtained as limits of a Markovian state descriptor for the queueing network under a law of large numbers type of scaling), was sufficient for stability of the original queueing network. (An analogue of this result for semimartingale reflecting Brownian motions was proved a little earlier by Dupuis and Williams [19].) Thus to determine sufficient conditions for stability, one can seek Lyapunov functions for the often simpler fluid limits rather than for the original queueing networks. This idea has been exploited by a number of authors, especially in combination with piecewise linear Lyapunov functions, to prove sufficient conditions for the stability of open multiclass networks (see Dai [11] for a survey). Networks with FIFO service disciplines can be especially difficult to analyze because of the need to keep track of the order in which customers arrive at each queue. Since the writing of [11], by establishing the stability of associated fluid limits, Bramson [5] has proved the stability of open *Kelly-type* networks with FIFO service discipline, provided $\rho_i < 1$ for each i. Here *Kelly-type* means that the service rate is the same for all customers at a given station, i.e., m_k is the same for all $k \in C_i$. Bramson [6] has also shown stability for networks with a processor sharing service discipline and $\rho_i < 1$ for all i. Here *processor sharing* means that service at each station is equally divided amongst all customers present at the station. In a very recent paper, Chen and Zhang [9] claim to prove stability of open multiclass networks with FIFO service when $\rho_i < 1$ for each i, assuming a spectral radius condition on one of the data matrices. At this time, it is not known if there is a large natural class of multiclass networks satisfying this condition.

In light of this work on stability, one might be tempted to conjecture that the Dai-Wang type examples fail to have Brownian approximations because they are not stable, i.e., one has the wrong notion of heavy traffic. However, a deterministic network example of Whitt [61], which exhibits large oscillations of the queue length processes, suggests that one might also entertain other possible explanations, such as the wrong scaling, the wrong topology on path space, or the wrong limit process. In general, one might consider one or more of the above as explanations for why a multiclass network does not follow a "conventional" heavy traffic limit theorem of the form proposed in [22, 24, 25, 31].

Indeed, in a recent work, Harrison and Williams [32] proved a heavy traffic limit theorem for a queueing network which is a closed network analogue of the Lu-Kumar/Rybko-Stolyar hybrid shown in Fig. 4 and which incorporates all of the possible exceptional features mentioned above. The network is obtained simply by turning off the exogenous arrival process and closing the loop on the left of Fig. 4 so that customers return to class 1 after completing service in class 4. Unconventional features of this limit theorem are as follows: (a) the notion of heavy traffic is different from (or at least a refinement of) that used by Chen and Mandelbaum for closed networks; indeed, the network parameters are chosen so that (19) holds with equality in place of inequality (the motivation here is that since the closed network with a fixed number of customers is automatically stable, it is natural to consider parameters that correspond to the boundary between stability and instability in the open network), (b) the normalization of the queue length processes is different from the usual central limit theorem type of scaling, and (c) the limit of the normalized queue length processes is not obtained from a reflecting Brownian motion, although it is related to Brownian motion. Moreover, to fully describe the limit process, non-trivial limits of all of the normalized queue length processes are needed and for the convergence in distribution, the topology on path space is weaker for some components than the usual Skorokhod $\mathbf{J_1}$-topology. For further details of this example, the reader is refered to the survey paper [65] or the full paper [32].

5. Open Problems

Some of the open problem areas for heavily traffic analysis of queueing networks are described below. For details the reader is refered to the papers cited.

(i) *Stability.* As mentioned in section 4, one of the very active areas of current research on queueing networks is concerned with determining sufficient conditions for the stability of open multiclass networks. Whilst some progress has been made (see e.g., [5, 6, 15, 17, 41]) on identifying service disciplines and network structures for which the conventional condition "$\rho_i < 1$ for all i" is sufficient for stability, we are still a considerable distance from a general classification.

(ii) *Heavy Traffic Approximation of Queueing Networks.* As described in section 3, there is no general heavy traffic limit theorem for multiclass queueing networks, the extant limit theorems being restricted to single class networks (or ones with separated priorities) or to multiclass priority networks with a feedforward structure. It is a compelling problem to identify a "good" class of multiclass networks for which conventional heavy traffic limit theorems can be proved. Conversely, it would be helpful to know the size of the collection of networks that do not have conventional heavy traffic behavior, for example, is it a small set in some measure theoretic sense? In the same vein, it would be interesting to identify the spectrum of possible unconventional heavy traffic behavior. For instance, the example of Harrison and Williams [32] illustrates several ways in which a heavy traffic limit theorem can be unconventional, but are there others?

There are many possible variations on the queueing network models described here. For instance, heavy traffic approximations for state- and/or time-dependent networks with Markovian assumptions have been established by some authors (see e.g., [40, 45, 46, 48]). Also, heavy traffic approximations have been proposed (but no limit theorem has been proved) for queueing networks that incorporate server breakdown and repair [26]. Such models are especially relevant for manufacturing applications.

(iii) *Semimartingale Reflecting Brownian Motions.* Although there is a solid existence and uniqueness theory for semimartingale reflecting Brownian motions (SRBMs), several problems associated with the analysis of these processes remain. A necessary and sufficient condition for positive recurrence is known in the two-dimensional case

[34, 62], but there is no general recurrence classification in dimensions three or more, although some sufficient conditions for positive recurrence are known [29, 30, 63]. The result of Dupuis and Williams [19] implies that one can obtain sufficient conditions for positive recurrence by study of a simpler deterministic dynamical system (see a recent paper of Chen [7] for an illustration of how this can be applied to simplify the proofs of previously known sufficient conditions for positive recurrence). When an SRBM is positive recurrent, its stationary distribution is characterized as the solution of a certain integral relation (see e.g., [64]). Whilst there is a numerical method [12] for analyzing this relation, improvements of this method would facilitate analysis of larger networks. Furthermore, the numerical method would be enhanced by knowledge of the tail behavior of the stationary distribution. A final problem is that of justifying the interchange of limits inherent in using the stationary distribution of a SRBM as an approximation to the equilibrium distribution of the original queueing network (see [38] for a discussion of this in the case of closed networks).

(iv) *Optimization.* The discussion in this paper has been directed to the problem of performance analysis for heavily loaded networks with a fixed structure. However, in some network applications one may have control over some of the system parameters, e.g., service discipline or routing, and want to choose them so as to optimize a measure of performance. Such control problems frequently cannot be analyzed exactly. One possible solution is to optimize an approximate model. This kind of approach, using approximate Brownian models, has been pursued by a few authors (see e.g., [28, 39, 43, 58, 59]), but this is an area with potential for much further development.

References

[1] Bertsekas, D., and Gallagher, R. (1992). *Data Networks.* Prentice-Hall, Englewood Cliffs, N.J.

[2] Bertsimas, D., Paschalidis, I. Ch., and Tsitsiklis, J. N. (1994). Optimization of multiclass queueing networks: Polyhedral and nonlinear characterizations of achievable performance. *Annals of Applied Probability,* **4,** 43–75.

[3] Bramson, M. (1994). Instability of FIFO queueing networks. *Annals of Applied Probability,* **4,** 414–431.

[4] Bramson, M. (1994). Instability of FIFO queueing networks with quick service times. *Annals of Applied Probability,* **4,** 693–718.

[5] Bramson, M. (1995). Convergence to equilibria for fluid models of FIFO queueing networks. *Queueing Systems: Theory and Applications,* to appear.

[6] Bramson, M. (1995). Convergence to equilibria for fluid models of processor sharing queueing networks. Preprint.

[7] Chen, H. (1995). A sufficient condition for the positive recurrence of a semimartingale reflecting Brownian motion in an orthant. Preprint.

[8] Chen, H., and Mandelbaum, A. (1991). Stochastic discrete flow networks: diffusion approximations and bottlenecks. *Annals of Probability,* **4,** 1463–1519.

[9] Chen, H., and Zhang, H. (1995). Stability of multiclass queueing networks under FIFO service discipline. Preprint.

[10] Dai, J. G. (1995). On positive Harris recurrence of multiclass queueing networks: a unified approach via fluid limit models. *Annals of Applied Probability*, **5**, 49–77.

[11] Dai, J. G. (1995). Stability of open multiclass queueing networks via fluid models. In *Stochastic Networks*, IMA Volumes in Mathematics and Its Applications, F. P. Kelly and R. J. Williams (eds.), **71**, Springer-Verlag, New York, 71–90.

[12] Dai, J. G., and Harrison, J. M. (1992). Reflected Brownian motion in an orthant: numerical methods for steady-state analysis. *Annals of Applied Probability*, **2**, 65–86.

[13] Dai, J. G., and Kurtz, T. G. (1995). A multiclass station with Markovian feedback in heavy traffic. *Mathematics of Operations Research*, **20**, 721–742.

[14] Dai, J. G., and Nguyen, V. (1994). On the convergence of multiclass queueing networks in heavy traffic. *Annals of Applied Probability*, **4**, 26–42.

[15] Dai, J. G., and VandeVate, J. The stability of two-station queueing networks. In preparation.

[16] Dai, J. G., and Wang, Y. (1993). Nonexistence of Brownian models of certain multiclass queueing networks. *Queueing Systems: Theory and Applications*, **13**, 41–46.

[17] Dai, J. G., and Weiss, G. (1996) Stability and instability of fluid models for certain re-entrant lines. *Mathematics of Operations Research*, to appear.

[18] Dai, J. G., and Williams, R. J. (1995). Existence and uniqueness of semimartingale reflecting Brownian motions in convex polyhedrons. *Theory of Probability and Its Applications*, **40**, 3–53 (in Russian), to appear in the SIAM translation journal of the same name.

[19] Dupuis, P., and Williams, R. J. (1994). Lyapunov functions for semimartingale reflecting Brownian motions. *Annals of Probability*, **22**, 680–702.

[20] Ethier, S. N., and Kurtz, T. G. (1986). *Markov Processes: Characterization and Convergence*. Wiley, New York.

[21] Harrison, J. M. (1978). The diffusion approximation for tandem queues in heavy traffic. *Adv. Appl. Prob.*, **10**, 886–905.

[22] Harrison, J. M. (1988). Brownian models of queueing networks with heterogeneous customer populations. In *Stochastic Differential Systems, Stochastic Control Theory and Applications*, IMA Volumes in Mathematics and Its Applications, W. Fleming and P.-L. Lions (eds.), Springer-Verlag, New York, 147–186.

[23] Harrison, J. M. (1995). Balanced fluid models of multiclass queueing networks: a heavy traffic conjecture. In *Stochastic Networks*, IMA Volumes in Mathematics and Its Applications, F. P. Kelly and R. J. Williams (eds.), **71**, Springer-Verlag, New York, 1–20.

[24] Harrison, J. M., and Nguyen, V. (1990). The QNET method for two-moment analysis of open queueing networks. *Queueing Systems: Theory and Applications*, **6**, 1–32.

[25] Harrison, J. M., and Nguyen, V. (1995). Brownian models of multiclass queueing networks: current status and open problems. *Queueing Systems: Theory and Applications*, **13**, 5–40.

[26] Harrison, J. M., and Pich, M. T. (1993). Two-moment analysis of open queueing networks with general workstation capabilities. *Operations Research*, to appear.

[27] Harrison, J. M., and Reiman, M. I. (1981). Reflected Brownian motion on an orthant. *Annals of Probability*, **9**, 302–308.

[28] Harrison, J. M., and Wein, L. M. (1989). Scheduling networks of queues: heavy traffic analysis of a simple open network. *Queueing Systems: Theory and Applications*, **5**, 265–280.

[29] Harrison, J. M., and Williams, R. J. (1987). Brownian models of open queueing networks with homogeneous customer populations. *Stochastics*, **22**, 77–115.

[30] Harrison, J. M., and Williams, R. J. (1987). Multidimensional reflected Brownian motions having exponential stationary distributions. *Annals of Probability*, **15**, 115–137.

[31] Harrison, J. M., and Williams, R. J. (1992). Brownian models of feedforward queueing networks: quasireversibility and product form solutions. *Annals of Applied Probability*, **2**, 263–293.

[32] Harrison, J. W., and Williams, R. J. (1995). A multiclass closed queueing network with unconventional heavy traffic behavior. *Annals of Applied Probability*, to appear.

[33] Harrison, J. M., Williams, R. J., and Chen, H. (1990). Brownian models of closed queueing networks with homogeneous customer populations. *Stochastics and Stochastics Reports*, **29**, 37–74.

[34] Hobson, D. G., and Rogers, L. C. G. (1993). Recurrence and transience of reflecting Brownian motion in the quadrant. *Math. Proc. Cambridge Philosophical Society*, **113**, 387–399.

[35] Iglehart, D. L., and Whitt, W. (1970). Multiple channel queues in heavy traffic I. *Adv. Appl. Prob.*, **2**, 150–177.

[36] Iglehart, D. L., and Whitt, W. (1970). Multiple channel queues in heavy traffic II. *Adv. Appl. Prob.*, **2**, 355–364.

[37] Johnson, D. P. (1983). *Diffusion Approximations for Optimal Filtering of Jump Processes and for Queueing Networks*. Ph.D. dissertation, Department of Mathematics, University of Wisconsin, Madison, WI.

[38] Kaspi, H., and Mandelbaum, A. (1992). Regenerative closed queueing networks. *Stochastics and Stochastics Reports*, **39**, 239–258.

[39] Kelly, F. P., and Laws, C. N. (1993). Dynamic routing in open queueing networks: Brownian models, cut constraints and resource pooling. *Queueing Systems: Theory and Applications*, **13**, 47–86.

[40] Krichagina, E. V., Liptser, R. S., and Puhalsky, A. A. (1988). Diffusion approximation for the system with arrival process depending on queue and arbitrary service distribution. *Theory of Probability and Its Applications*, **33**, 124–135.

[41] Kumar, P. R., and Meyn, S. P. (1995). Stability of queueing networks and scheduling policies. *IEEE Transactions on Automatic Control*, **40**, 251–260.

[42] Kumar, P. R., and Seidman, T. I. (1990). Dynamic instabilities and stabilization methods in distributed real-time scheduling of manufacturing systems. *IEEE Transactions on Automatic Control*, **35**, 289–298.

[43] Kushner, H. J. (1995). A control problem for a new type of public transportation system, via heavy traffic analysis. In *Stochastic Networks*, IMA Volumes in Mathematics and Its Applications, F. P. Kelly and R. J. Williams (eds.), **71**, Springer-Verlag, New York, 139–167.

[44] Lu, S. H., and Kumar, P. R. (1991). Distributed scheduling based on due dates and buffer priorities. *IEEE Transactions on Automatic Control*, **36**, 1406–1416.

[45] Mandelbaum, A., and Massey, W. A. (1995). Strong approximations for time-dependent queues. *Mathematics of Operations Research*, **20**, 33–64.

[46] Massey, W. A. (1981). *Nonstationary Queueing Networks*. Ph.D. dissertation, Department of Mathematics, Stanford University, Stanford, CA.

[47] Meyn, S. P., and Down, D. (1994). Stability of generalized Jackson networks. *Annals of Applied Probability*, **4**, 124–148.

[48] Pats, G. (1995). *State Dependent Queueing Networks: Approximations and Applications*. Ph.D. dissertation, Department of Industrial Engineering and Management, Technion, Haifa, Israel.

[49] Peterson, W. P. (1991). Diffusion approximations for networks of queues with multiple customer types. *Mathematics of Operations Research*, **9**, 90–118.

[50] Reiman, M. I. (1984). Open queueing networks in heavy traffic. *Mathematics of Operations Research*, **9**, 441–458.

[51] Reiman, M. I. (1988). A multiclass feedback queue in heavy traffic. *Mathematics of Operations Research*, **20**, 179–207.

[52] Reiman, M. I., and Williams, R. J. (1988–89). A boundary property of semi-martingale reflecting Brownian motions. *Probability Theory and Related Fields*, **77**, 87–97, and **80, 633**.

[53] Rybko, A. N., and Stolyar, A. L. (1991). Ergodicity of stochastic processes describing the operation of an open queueing network. *Problemy Peredachi Informatsil*, **28**, 2–26.

[54] Sauer, C. H., and Chandy, K. M. (1981). *Computer Systems Performance Modeling.* Prentice-Hall, Englewood Cliffs, N.J.

[55] Seidman, T. I. (1994). 'First come, first served' can be unstable! *IEEE Transactions on Automatic Control*, **39**, 2166–2171.

[56] Skorokhod, A. V. (1956). Limit Theorems for Stochastic Processes. *Theory of Probability and Its Applications*, **1**, 261–290.

[57] Taylor, L. M., and Williams, R. J. (1993). Existence and uniqueness of semimartingale reflecting Brownian motions in an orthant. *Probability Theory and Related Fields*, **96**, 283–317.

[58] Wein, L. M. (1990). Scheduling networks of queues: heavy traffic analysis of a two-station network with controllable inputs. *Operations Research*, **38**, 1065–1078.

[59] Wein, L. M. (1992). Scheduling networks of queues: heavy traffic analysis of a multistation network with controllable inputs. *Operations Research*, **40** (suppl.), S312–S334.

[60] Whitt, W. (1971). Weak convergence theorems for priority queues: preemptive resume discipline. *J. Applied Probability*, **8**, 74–94.

[61] Whitt, W. (1993). Large fluctuations in a deterministic multiclass network of queues. *Management Science*, **39, 1020–1028**.

[62] Williams, R. J. (1985). Recurrence classification and invariant measure for reflected Brownian motion in a wedge. *Annals of Probability*, **13, 758–778**.

[63] Williams, R. J. (1987). Reflected Brownian motion with skew symmetric data in a polyhedral domain. *Probability Theory and Related Fields*, **75, 459–485**.

[64] Williams, R. J. (1995). Semimartingale reflecting Brownian motions in an orthant. In *Stochastic Networks*, IMA Volumes in Mathematics and Its Applications, F. P. Kelly and R. J. Williams (eds.), **71**, Springer-Verlag, New York, 125–137.

[65] Williams, R. J. (1996). On the approximation of queueing networks in heavy traffic. In *Stochastic Networks: Theory and Applications*, F. P. Kelly, S. Zachary and I. Ziedins (eds.), Oxford University Press, Oxford.

[66] Yao, D. D. (ed.) (1994). *Stochastic Modeling and Analysis of Manufacturing Systems.* Springer-Verlag, New York.

Lecture Notes in Statistics

For information about Volumes 1 to 53
please contact Springer-Verlag

Vol. 54: K.R. Shah, B.K. Sinha, Theory of Optimal Designs. viii, 171 pages, 1989.

Vol. 55: L. McDonald, B. Manly, J. Lockwood, J. Logan (Editors), Estimation and Analysis of Insect Populations. Proceedings, 1988. xiv, 492 pages, 1989.

Vol. 56: J.K. Lindsey, The Analysis of Categorical Data Using GLIM. v, 168 pages, 1989.

Vol. 57: A. Decarli, B.J. Francis, R. Gilchrist, G.U.H. Seeber (Editors), Statistical Modelling. Proceedings, 1989. ix, 343 pages, 1989.

Vol. 58: O.E. Barndorff-Nielsen, P. Bl¾sild, P.S. Eriksen, Decomposition and Invariance of Measures, and Statistical Transformation Models. v, 147 pages, 1989.

Vol. 59: S. Gupta, R. Mukerjee, A Calculus for Factorial Arrangements. vi, 126 pages, 1989.

Vol. 60: L. Gyorfi, W. Härdle, P. Sarda, Ph. Vieu, Nonparametric Curve Estimation from Time Series. viii, 153 pages, 1989.

Vol. 61: J. Breckling, The Analysis of Directional Time Series: Applications to Wind Speed and Direction. viii, 238 pages, 1989.

Vol. 62: J.C. Akkerboom, Testing Problems with Linear or Angular Inequality Constraints. xii, 291 pages, 1990.

Vol. 63: J. Pfanzagl, Estimation in Semiparametric Models: Some Recent Developments. iii, 112 pages, 1990.

Vol. 64: S. Gabler, Minimax Solutions in Sampling from Finite Populations. v, 132 pages, 1990.

Vol. 65: A. Janssen, D.M. Mason, Non-Standard Rank Tests. vi, 252 pages, 1990.

Vol 66: T. Wright, Exact Confidence Bounds when Sampling from Small Finite Universes. xvi, 431 pages, 1991.

Vol. 67: M.A. Tanner, Tools for Statistical Inference: Observed Data and Data Augmentation Methods. vi, 110 pages, 1991.

Vol. 68: M. Taniguchi, Higher Order Asymptotic Theory for Time Series Analysis. viii, 160 pages, 1991.

Vol. 69: N.J.D. Nagelkerke, Maximum Likelihood Estimation of Functional Relationships. V, 110 pages, 1992.
Vol. 70: K. Iida, Studies on the Optimal Search Plan. viii, 130 pages, 1992.

Vol. 71: E.M.R.A. Engel, A Road to Randomness in Physical Systems. ix, 155 pages, 1992.

Vol. 72: J.K. Lindsey, The Analysis of Stochastic Processes using GLIM. vi, 294 pages, 1992.

Vol. 73: B.C. Arnold, E. Castillo, J.-M. Sarabia, Conditionally Specified Distributions. xiii, 151 pages, 1992.

Vol. 74: P. Barone, A. Frigessi, M. Piccioni, Stochastic Models, Statistical Methods, and Algorithms in Image Analysis. vi, 258 pages, 1992.

Vol. 75: P.K. Goel, N.S. Iyengar (Eds.), Bayesian Analysis in Statistics and Econometrics. xi, 410 pages, 1992.

Vol. 76: L. Bondesson, Generalized Gamma Convolutions and Related Classes of Distributions and Densities. viii, 173 pages, 1992.

Vol. 77: E. Mammen, When Does Bootstrap Work? Asymptotic Results and Simulations. vi, 196 pages, 1992.

Vol. 78: L. Fahrmeir, B. Francis, R. Gilchrist, G. Tutz (Eds.), Advances in GLIM and Statistical Modelling: Proceedings of the GLIM92 Conference and the 7th International Workshop on Statistical Modelling, Munich, 13-17 July 1992. ix, 225 pages, 1992.

Vol. 79: N. Schmitz, Optimal Sequentially Planned Decision Procedures. xii, 209 pages, 1992.

Vol. 80: M. Fligner, J. Verducci (Eds.), Probability Models and Statistical Analyses for Ranking Data. xxii, 306 pages, 1992.

Vol. 81: P. Spirtes, C. Glymour, R. Scheines, Causation, Prediction, and Search. xxiii, 526 pages, 1993.

Vol. 82: A. Korostelev and A. Tsybakov, Minimax Theory of Image Reconstruction. xii, 268 pages, 1993.

Vol. 83: C. Gatsonis, J. Hodges, R. Kass, N. Singpurwalla (Editors), Case Studies in Bayesian Statistics. xii, 437 pages, 1993.

Vol. 84: S. Yamada, Pivotal Measures in Statistical Experiments and Sufficiency. vii, 129 pages, 1994.

Vol. 85: P. Doukhan, Mixing: Properties and Examples. xi, 142 pages, 1994.

Vol. 86: W. Vach, Logistic Regression with Missing Values in the Covariates. xi, 139 pages, 1994.

Vol. 87: J. Müller, Lectures on Random Voronoi Tessellations.vii, 134 pages, 1994.

Vol. 88: J. E. Kolassa, Series Approximation Methods in Statistics. Second Edition, ix, 183 pages, 1997.

Vol. 89: P. Cheeseman, R.W. Oldford (Editors), Selecting Models From Data: AI and Statistics IV. xii, 487 pages, 1994.

Vol. 90: A. Csenki, Dependability for Systems with a Partitioned State Space: Markov and Semi-Markov Theory and Computational Implementation. x, 241 pages, 1994.

Vol. 91: J.D. Malley, Statistical Applications of Jordan Algebras. viii, 101 pages, 1994.

Vol. 92: M. Eerola, Probabilistic Causality in Longitudinal Studies. vii, 133 pages, 1994.

Vol. 93: Bernard Van Cutsem (Editor), Classification and Dissimilarity Analysis. xiv, 238 pages, 1994.

Vol. 94: Jane F. Gentleman and G.A. Whitmore (Editors), Case Studies in Data Analysis. viii, 262 pages, 1994.

Vol. 95: Shelemyahu Zacks, Stochastic Visibility in Random Fields. x, 175 pages, 1994.

Vol. 96: Ibrahim Rahimov, Random Sums and Branching Stochastic Processes. viii, 195 pages, 1995.

Vol. 97: R. Szekli, Stochastic Ordering and Dependence in Applied Probability. viii, 194 pages, 1995.
Vol. 98: Philippe Barbe and Patrice Bertail, The Weighted Bootstrap. viii, 230 pages, 1995.

Vol. 99: C.C. Heyde (Editor), Branching Processes: Proceedings of the First World Congress. viii, 185 pages, 1995.

Vol. 100: Wlodzimierz Bryc, The Normal Distribution: Characterizations with Applications. viii, 139 pages, 1995.

Vol. 101: H.H. Andersen, M.Højbjerre, D. Sørensen, P.S.Eriksen, Linear and Graphical Models: for the Multivariate Complex Normal Distribution. x, 184 pages, 1995.

Vol. 102: A.M. Mathai, Serge B. Provost, Takesi Hayakawa, Bilinear Forms and Zonal Polynomials. x, 378 pages, 1995.

Vol. 103: Anestis Antoniadis and Georges Oppenheim (Editors), Wavelets and Statistics. vi, 411 pages, 1995.

Vol. 104: Gilg U.H. Seeber, Brian J. Francis, Reinhold Hatzinger, Gabriele Steckel-Berger (Editors), Statistical Modelling: 10th International Workshop, Innsbruck, July 10-14th 1995. x, 327 pages, 1995.

Vol. 105: Constantine Gatsonis, James S. Hodges, Robert E. Kass, Nozer D. Singpurwalla(Editors), Case Studies in Bayesian Statistics, Volume II. x, 354 pages, 1995.

Vol. 106: Harald Niederreiter, Peter Jau-Shyong Shiue (Editors), Monte Carlo and Quasi-Monte Carlo Methods in Scientific Computing. xiv, 372 pages, 1995.

Vol. 107: Masafumi Akahira, Kei Takeuchi, Non-Regular Statistical Estimation. vii, 183 pages, 1995.

Vol. 108: Wesley L. Schaible (Editor), Indirect Estimators in U.S. Federal Programs. viii, 195 pages, 1995.

Vol. 109: Helmut Rieder (Editor), Robust Statistics, Data Analysis, and Computer Intensive Methods. xiv, 427 pages, 1996.
Vol. 110: D. Bosq, Nonparametric Statistics for Stochastic Processes. xii, 169 pages, 1996.

Vol. 111: Leon Willenborg, Ton de Waal, Statistical Disclosure Control in Practice. xiv, 152 pages, 1996.

Vol. 112: Doug Fischer, Hans-J. Lenz (Editors), Learning from Data. xii, 450 pages, 1996.

Vol. 113: Rainer Schwabe, Optimum Designs for Multi-Factor Models. viii, 124 pages, 1996.

Vol. 114: C.C. Heyde, Yu. V. Prohorov, R. Pyke, and S. T. Rachev (Editors), Athens Conference on Applied Probability and Time Series Analysis Volume I: Applied Probability In Honor of J.M. Gani. viii, 424 pages, 1996.

Vol. 115: P.M. Robinson, M. Rosenblatt (Editors), Athens Conference on Applied Probability and Time Series Analysis Volume II: Time Series Analysis In Memory of E.J. Hannan. viii, 448 pages, 1996.

Vol. 116: Genshiro Kitagawa and Will Gersch, Smoothness Priors Analysis of Time Series. x, 261 pages, 1996.

Vol. 117: Paul Glasserman, Karl Sigman, David D. Yao (Editors), Stochastic Networks. xii, 298, 1996.

Vol. 118: Radford M. Neal, Bayesian Learning for Neural Networks. xv, 183, 1996.

Vol. 119: Masanao Aoki, Arthur M. Havenner, Applications of Computer Aided Time Series Modeling. ix, 329 pages, 1997.

Vol. 120: Maia Berkane, Latent Variable Modeling and Applications to Causality. vi, 288 pages, 1997.

Vol. 121: Constantine Gatsonis, James S. Hodges, Robert E. Kass, Robert McCulloch, Peter Rossi, Nozer D. Singpurwalla (Editors), Case Studies in Bayesian Statistics, Volume III. xvi, 487 pages, 1997.

Vol. 122: Timothy G. Gregoire, David R. Brillinger, Peter J. Diggle, Estelle Russek-Cohen, William G. Warren, Russell D. Wolfinger (Editors), Modeling Longitudinal and Spatially Correlated Data. x, 402 pages, 1997.

Vol. 123: D. Y. Lin and T. R. Fleming (Editors), Proceedings of the First Seattle Symposium in Biostatistics: Survival Analysis. xiii, 308 pages, 1997.

Vol. 124: Christine H. Müller, Robust Planning and Analysis of Experiments. x, 234 pages, 1997.

Vol. 125: Valerii V. Fedorov and Peter Hackl, Model-oriented Design of Experiments. viii, 117 pages, 1997.

Vol. 126: Geert Verbeke and Geert Molenberghs, Linear Mixed Models in Practice: A SAS-Oriented Approach. xiii, 306 pages, 1997.

Vol. 127: Harald Niederreiter, Peter Hellekalek, Gerhard Larcher, and Peter Zinterhof (Editors), Monte Carlo and Quasi-Monte Carlo Methods 1996, xii, 448 pp., 1997.

Vol. 128: L. Accardi and C.C. Heyde (Editors), Probability Towards 2000, x, 356 pp., 1998.